Phage Display in Biotechnology and Drug Discovery

Drug Discovery Series

Series Editor
Daniel Levy

Phage Display in Biotechnology and Drug Discovery

Second Edition

EDITED BY

Sachdev S. Sidhu
Clarence Ronald Geyer

CRC Press is an imprint of the
Taylor & Francis Group, an **informa** business

CRC Press
Taylor & Francis Group
6000 Broken Sound Parkway NW, Suite 300
Boca Raton, FL 33487-2742

First issued in paperback 2017

© 2015 by Taylor & Francis Group, LLC
CRC Press is an imprint of Taylor & Francis Group, an Informa business

No claim to original U.S. Government works

ISBN-13: 978-1-4398-3649-1 (hbk)
ISBN-13: 978-1-138-89467-9 (pbk)

Visit the Taylor & Francis Web site at
http://www.taylorandfrancis.com

and the CRC Press Web site at
http://www.crcpress.com

Contents

Foreword

Science has always progressed by coupling insightful observations leading to testable hypotheses with innovative technologies that facilitate our ability to observe and test them. In the field of protein science, the technologies for protein display and in vitro selection have had an enormous impact on our ability to probe and manipulate protein functional properties.

The development of site-directed mutagenesis, which allowed one to systematically probe a gene sequence in the late 1970s, gave birth to the field of protein engineering in the early 1980s. Throughout the 1980s, most scientists in the protein engineering field would generate and purify one mutant protein at a time and characterize its functional properties. Some investigators had developed selections and screens that allowed one to test many variants simultaneously, but these tended to be highly specific for certain proteins (notably DNA binding proteins) and focused primarily on studying protein stability. Moreover, the selections were generally done in the context of a living cell, which limited the range of assays that could be performed. While replica plating screens were available to test variant proteins out of the cell, these tended to be quite labor intensive, thus limiting the number of variants that could be screened.

In 1985, George P. Smith published a paper showing that small peptides derived from EcoRI could be inserted into the gene III attachment protein in filamentous bacterial phage, which could then be captured using antibodies to the small peptide. This observation incubated several years, and then, in the late 1980s and early 1990s, other groups showed it was possible to display whole proteins on gene III that were folded and capable of binding their cognate ligands. Moreover, it was shown that by appropriate manipulation of the copy number on the phage, it was possible to select a range of binding affinities, from weak at a high copy number to strong at a low copy number. These selections could all be done in vitro and under a variety of selection conditions, limited only by binding to a support-bound ligand.

Throughout the 1990s up to today, huge improvements have been made to the display technology allowing massive increases in the library number (now routinely $>10^{10}$ variants per selection); recursive mutagenesis cycles allowing one to mutate as one selects; new display formats including other phage species, bacteria, yeast, and ribosomes; and automation to further simplify the process. As with any technology, there are limitations. For example, not all proteins can be readily displayed on phage, and expression effects can bias the outcome of the selection. Nonetheless, phage display has had a huge impact on probing, improving, and designing new functional properties into proteins and peptides including binding affinity, selectivity, catalysis, and chemical and thermal stability, among others. This book edited by Sachdev S. Sidhu provides an excellent review of the state-of-the-art in phage display technology now and in the near future.

James A. Wells
President and Chief Scientific Officer
Sunesis Pharmaceuticals
South San Francisco, California

Preface

Recent years have witnessed the sequencing of numerous genomes, including the all-important human genome itself. While genomic information offers considerable promise for drug discovery efforts, it must be remembered that we live in a protein world. The vast majority of biological processes are driven by proteins, and the full benefits of DNA databases will only be realized by the translation of genomic information into knowledge of protein function. Ultimately, drug discovery depends on the manipulation and modification of proteins, and thus, the genomic panacea comes with significant challenges for life scientists in the field of therapeutic biotechnology. Indeed, it has become clear that success in the modern era of biology will go to those who apply to protein analysis the high-throughput principles that made whole-genome sequencing a reality.

In this context, phage display is an established combinatorial technology that is likely to play an even greater role in the future of drug discovery. The power of the technology resides in its simplicity. Rapid molecular biology methods can be used to create vast libraries of proteins displayed on bacteriophage that also encapsulate the encoding DNA. Billions of different proteins can be screened en masse and individual protein sequences can be decoded rapidly from the cognate gene. In essence, the technology enables the engineering of proteins with simple molecular biology techniques that would otherwise only be applicable to DNA. In addition, the technology is very much suited to the methods currently used for high-throughput screening and thus can be readily adapted to the analysis of multiple targets and pathways.

This book comprises 19 chapters that provide a comprehensive view of the impact and promise of phage display in drug discovery and biotechnology. The chapters detail the theories, principles, and methods current in the field and demonstrate applications for peptide phage display, protein phage display, and the development of novel antibodies. The book as a whole is intended to give the reader an overview of the amazing breadth of the impact that phage display technology has had on the study of proteins in general and the development of protein therapeutics in particular. I hope that this work will serve as a comprehensive reference for researchers in the phage field and, perhaps more importantly, will serve to inspire newcomers to adapt the technology to their own needs in the ever-expanding world of therapeutic biology.

Sachdev S. Sidhu

Contributors

Sara K. Avrantinis
Department of Chemistry
University of California
Irvine, California

Jody D. Berry
Departments of Immunology and
 Medical Microbiology
University of Manitoba
Winnipeg, Manitoba, Canada

Julian Bertschinger
Institute of Pharmaceutical Sciences
Swiss Federal Institute of Technology
Zurich, Switzerland

Robert O. Carlson
Karyopharm Therapeutics
Natick, Massachusetts

Jacob Corn
Department of Protein Engineering
Genentech Inc.
South San Francisco, California

Reto Crameri
Swiss Institute of Allergy and Asthma
 Research (SIAF)
Davos, Switzerland

Mark S. Dennis
Department of Protein Engineering
Genentech, Inc.
South San Francisco, California

Kurt Deshayes
Department of Protein Engineering
Genentech Inc.
South San Francisco, California

Claire L. Dobson
MedImmune Ltd.
Cambridge, United Kingdom

Frederic A. Fellouse
Donnelly Centre
University of Toronto
Toronto, Ontario, Canada

Sabine Flückiger
BioVisioN Schweiz AG
Davos, Switzerland

Clarence Ronald Geyer
Department of Pathology
University of Saskatchewan
Saskatoon, Saskatchewan, Canada

Paul T. Hamilton
Department of Plant and Microbial
 Biology
North Carolina State University
Raleigh, North Carolina

Zhaozhong Han
Alexion Pharmaceuticals, Inc.
Cheshire, Connecticut

Traver Hart
Donnelly Centre
University of Toronto
Toronto, Ontario, Canada

Celia P. Hart-Shorrock
MedImmune Ltd.
Cambridge, United Kingdom

Christian Heinis
Institute of Pharmaceutical Sciences
Swiss Federal Institute of Technology
Zurich, Switzerland

Robin Hyde-DeRuyscher
Biogen Idec
Research Triangle Park, North Carolina

Ece Karatan
Department of Biology
Appalachian State University
Boone, North Carolina

Brian K. Kay
Department of Biological Sciences
University of Illinois at Chicago
Chicago, Illinois

Malgorzata E. Kokoszka
Department of Biological Sciences
University of Illinois at Chicago
Chicago, Illinois

Zoltan Konthur
Max Planck Institute of Molecular
  Genetics
Berlin, Germany

Jianghai Liu
Department of Pathology
University of Saskatchewan
Saskatoon, Saskatchewan, Canada

Henry B. Lowman
Department of Antibody Engineering
Genentech, Inc.
South San Francisco, California

Lee Makowski
Combinatorial Biology Unit
Biosciences Division
Argonne National Laboratory
Argonne, Illinois

Suneeta Mandava
Combinatorial Biology Unit
Biosciences Division
Argonne National Laboratory
Argonne, Illinois

Jonathan S. Marvin
Department of Antibody Engineering
Genentech, Inc.
South San Francisco, California

Megan McLaughlin
Donnelly Centre
University of Toronto
Toronto, Ontario, Canada

Ralph R. Minter
MedImmune Ltd.
Cambridge, United Kingdom

Jason Moffat
Donnelly Centre
and
Department of Molecular Genetics
University of Toronto
Toronto, Ontario, Canada

Dario Neri
Institute of Pharmaceutical Sciences
Swiss Federal Institute of Technology
Zurich, Switzerland

Shuichi Ohkubo
Tsukuba Research Center
Taiho Pharmaceutical Co., Ltd.
Tsukuba, Japan

Gabor Pal
Department of Biochemistry
Eötvös Loránd University
Budapest, Hungary

Valery A. Petrenko
Department of Pathobiology
College of Veterinary Medicine
Auburn University
Auburn, Alabama

Mikhail Popkov
Department of Molecular Biology
The Scripps Research Institute
La Jolla, California

Claudio Rhyner
Swiss Institute of Allergy and Asthma
  Research (SIAF)
Davos, Switzerland

Diane J. Rodi
Combinatorial Biology Unit
Biosciences Division
Argonne National Laboratory
Argonne, Illinois

Jamie K. Scott
Department of Molecular Biology and
  Biochemistry
and
Faculty of Health Sciences
Simon Fraser University
Burnaby, British Columbia, Canada

Sachdev S. Sidhu
Donnelly Centre
and
Department of Molecular Genetics
University of Toronto
Toronto, Ontario, Canada

George P. Smith
Division of Biological Sciences
University of Missouri
Columbia, Missouri

Jelena Tomic
Donnelly Centre
University of Toronto
Toronto, Ontario, Canada

Mihriban Tuna
Department of Biochemistry
School of Life Sciences
University of Sussex
Falmer, United Kingdom

Nienke E. van Houten
Faculty of Health Sciences
Simon Fraser University
Burnaby, British Columbia, Canada

Michael Weichel
Swiss Institute of Allergy and Asthma
  Research (SIAF)
Davos, Switzerland

Gregory A. Weiss
Department of Chemistry
University of California
Irvine, California

Derek N. Woolfson
Department of Biochemistry
School of Life Sciences
University of Sussex
Falmer, United Kingdom

CHAPTER 1

Filamentous Bacteriophage Structure and Biology

Diane J. Rodi, Suneeta Mandava, and Lee Makowski

CONTENTS

1.1 INTRODUCTION

Phage display technology provides a remarkably versatile tool for exploring the interactions between proteins, peptides, and small-molecule ligands. As such, it has become widely adapted for use in epitope mapping, identification of protein–peptide and protein–protein interactions, protein–small molecule interactions, humanization of antibodies, identification of tissue-targeting peptides, and many other applications as outlined throughout this book. However, it must be kept in mind that phage display is a combinatorial *biology* approach, not a combinatorial *chemistry* approach. The great strength of phage display over combinatorial methods that are strictly chemical is that the isolation of a single interacting protein or peptide attached to a phage particle is sufficient to allow the complete characterization of the isolate: the interacting virus can be grown up in bulk and the sequence of the displayed protein or peptide inferred from the DNA sequence carried within the viral particle. The other side of this coin is that phage display technology utilizes living systems, and is therefore constrained in its potential diversity by the molecular requirements of those systems.

The biological limitations that impact phage display technology are defined not simply by viral structure, but by the well-balanced phage–host system as a whole. The display of a protein or peptide on the surface of a bacteriophage particle involves insertion of the corresponding DNA into the gene of a structural protein and the expression of the foreign sequence as a fusion with the structural protein in such a way that it is exposed, at least in part, on the surface of the phage particle. This process perturbs the phage–host system and may result in anything from a negligibly small alteration in phage growth rate to a complete halt of phage production. Disruption of any step along the way between DNA cloning and production of virus, including protein synthesis, protein translocation, viral morphogenesis, viral stability, host cell binding, or subsequent steps in the infection process, can remove a particular display construct from the final phage population. Additionally, in the context of library screening methodology, it is also important to note that different inserts placed at the same site may have very different effects on the rate of viral production, resulting in biases that can seriously impact the diversity of a phage-displayed library and, consequently, the results of affinity selection experiments. Some members of the libraries are present at much lower levels than others whereas others are absent. These biases must be well characterized in order to make optimal use of libraries in affinity selections or other experiments designed to take advantage of the unique properties of display libraries. Therefore, in order to understand the effect of biology on phage display, the way the phage interacts with its host must be considered in detail.

In this chapter, we outline the steps of phage–host interaction and discuss how those interactions may impact the diversity of phage-displayed libraries. Understanding these limitations in more detail should provide a starting point for engineering methods to minimize their effect on the use of phage display technology within the broad range of applications reviewed in this volume. Except for DNA replication, each step appears to have a detectable effect on the expression of some members of some libraries. Some effects appear significant whereas others are barely detectable. At the end, we briefly review identified bottlenecks in the viral life cycle and suggest simple strategies that can be implemented for minimizing the perturbations.

1.2 TAXONOMY AND GENETICS

The filamentous bacteriophages are a family of ssDNA-containing viruses (genus *Inovirus*) that infect a wide variety of gram-negative bacteria, including *Escherichia coli*, *Xanthomonas*, *Thermus*, *Pseudomonas*, *Salmonella*, and *Vibrio*. The best characterized of the filamentous phage are the Ff class of viruses, so named because of their method of host cell entry via the tip of the F conjugative pilus on the surface of male *E. coli* cells. The Ff viruses include M13, fd, and f1, all of which possess a 98% identity at the DNA sequence level. Ff virus particles are long, slender, and flexible rods with a diameter of about 65 Å. The wild-type Ff phages are between 0.8 and 0.9 μm long, giving the virus the proportions of a 4-foot-long pencil. Various engineered strains have somewhat longer genomes with the length of the particle increased proportionate to the length of the encapsulated DNA. Although there is considerable heterogeneity within the family, some similarities of sequence and genome organization are discernable among all group members. An electron micrograph shown in Figure 1.1 gives a rough idea of the proportions of the phage particles. The single-stranded, circular genome occupies the axis of the particle, stretched out for almost the entire length of the virion. Virus lengths are dependent upon both the size of the enclosed genome as stated above and on the physical distribution of the DNA within the capsid (axial distance per base), the latter of which has been demonstrated to be major coat protein charge dependent [1]. Little substructure is visible except at the end involved in host cell attachment. Each cross-section of the virion has an "up" strand and a "down" strand present, but these are not base paired because there is no complementary relationship between the sequences of the two strands except within the hairpin, which acts as the packaging signal that nucleates the initiation of viral assembly.

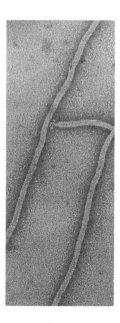

Figure 1.1   Electron micrograph of bacteriophage M13. This micrograph of M13 phage particles visually demonstrates the rationale for their designation as "filamentous" bacteriophage. The amino terminus of at least four copies of the gene *III* protein are visible at the end of the phage particle; the two subtilisin-cleavable N1 and N2 domains are seen as knobby structures at the proximal end of the virus. (Micrograph courtesy of Irene Davidovich.)

Ff viruses are not lytic but rather are parasitic. Productive infections result in viral release via extrusion or secretion across the inner and outer bacterial membranes in the absence of host cell death with the infected cells continuing to grow and divide (albeit at a significantly reduced rate). M13 produces anywhere from 200 to 2000 progeny phage per cell per doubling time [2,3]. This phage production represents a serious metabolic load for the infected *E. coli* with phage proteins making up 1%–5% of total protein synthesis and resulting in a reduction in cell growth of 30%–50% from uninfected cells. The nonlytic nature of Ff infection, along with the simultaneous presence of both single- and double-stranded forms of viral DNA, little size constraint on inserted DNA, and an exceptionally high viral titer capacity (typically 10^{11}–10^{12} particles per ml), has made the filamentous phage, primarily M13, a workhorse for molecular biology for the last 20 years.

1.3 VIRAL GENE PRODUCTS

The Ff phage genome encodes a total of 11 proteins (see Figure 1.2 for genome organization). There are five structural proteins, all of which are inserted into the inner host cell membrane prior to assembly (see Figure 1.3 for overall structural

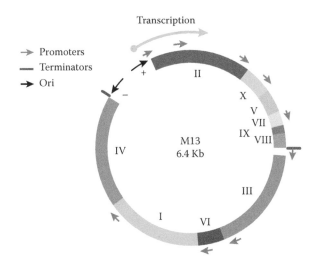

Figure 1.2 Ff phage genome. The location of each viral gene is indicated by number with the direction of transcription shown by arrow. The origin of replication lies within the intergenic region between the genes for pIV and pII. The packaging signal (PS) lies between the (−) strand origin of replication and gene *IV*.

organization of Ff phage). pVIII and pIII are synthesized with signal sequences that are removed subsequent to membrane insertion whereas pVI, pVII, and pIX are absent signal sequences. Three nonstructural phage proteins, pI, pXI, and pIV are required for phage morphogenesis but are not incorporated into the phage structure. pX and pXI are the result of in-frame internal translation initiation events in genes *II* and *I*, respectively, and are identical with the C-terminal portions of pII and pI in amino acid sequence, membrane localization, and topology [4]. In addition to the coding regions, there is an intergenic region, which contains the signals for the initiation of synthesis of both the plus (+) or viral-contained DNA strand, and the minus (−) strand; the initiation of capsid assembly signal (or packaging signal, PS), which lies between the (−) origin and the end of the pIV gene and the signal for termination of RNA synthesis. Parts of the intergenic region have been shown to be dispensable (reviewed in Ref. [3]), but all of the coding region products are necessary for the synthesis of the infectious progeny phage.

1.3.1 Replication Proteins (pII and pX)

pII is a 410 amino acid protein (MW = 46,137), which is required for all phage-specific DNA synthesis other than the formation of the complementary strand of the infecting ssDNA by host enzymes. pII has both endonuclease and topoisomerase activities required during the DNA replication phase of infection. pX is a 111 residue protein (MW = 12,672), which is encoded entirely within gene *II*, initiating at codon 300, an AUG that is in phase with the initiating AUG of gene *II*. Although pX has the same amino acid sequence as the carboxyl-terminal end of pII, it has been

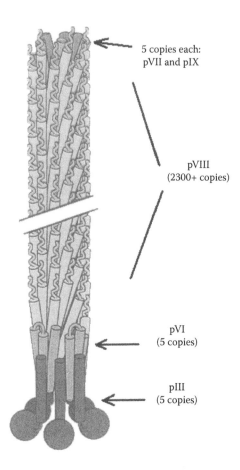

5 copies each:
pVII and pIX

pVIII
(2300+ copies)

pVI
(5 copies)

pIII
(5 copies)

Figure 1.3 Schematic diagram of the Ff bacteriophage. This diagram depicts the structural organization of M13 as a representative of the Ff viral family. At the top of the diagram lies the distal end of the particle at which viral assembly initiates. At the bottom of the figure is the proximal or infectious end of the virus with five copies of the pIII anchored to the particle by five copies of the pVI.

shown to possess unique functions within the viral life cycle, such as inhibition of pII function [5].

1.3.2 Single-Stranded DNA Binding Protein (pV)

Gene *V* codes for an 87 amino acid protein (MW = 9682) that exists as a stable dimer even at a concentration as low as 1 nM [6]. The crystal structure of the protein has been solved to 1.8 Å resolution using multiwavelength anomalous dispersion on a selenomethionine-containing protein and is shown in Figure 1.4 [7]. Each monomer is largely β-structure, with 58 out of 87 residues arranged in a five-stranded antiparallel β-sheet; two antiparallel β-ladder loops protrude from this sheet. The remainder of the molecule is arranged into 3_{10} helices (residues 7–11 and 65–67),

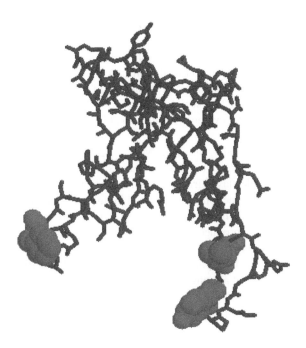

Figure 1.4 Crystal structure of the gene *V* ssDNA binding protein. The crystal structure of pV has been solved to 1.8 Å resolution using multiwavelength anomalous dispersion on a selenomethionine derivative and is shown here in a backbone format [7] (PDB accession code 2GN5). The protein normally exists as a dimer and wraps around the single-stranded form of the viral DNA within the host cell cytoplasm. Residues Tyr26, Leu28, and Phe73 have been shown to be critical for DNA binding [7] (residues shown in spacefill format).

β-bends (residues 21–24, 50–53, and 71–74), and one five-residue loop (residues 38–42). Nuclear magnetic resonance (NMR) analysis of the gene *V* protein [8] suggests that the DNA binding loop (residues 16–28; see Figure 1.4) is flexible in solution. This protein serves the dual functions of sequestering the intracellular ssDNA viral genomes [3] and modulating the translation of gene *II* mRNA [9].

1.3.3 Major Structural Protein (pVIII)

Gene *VIII* codes for the major coat protein of the virus. The major coat proteins of all filamentous phages are short, ranging from 44 to 55 amino acids, with most being encoded with a signal sequence (*Pseudomonas aeruginosa*-infecting phage Pf3 is an example of a pVIII absent a signal sequence). In the Ff group of phage, the major coat protein is 50 amino acids long (MW = 5235) with a 23 amino acid–long signal sequence. Approximately 2800 copies of pVIII are required to coat one full-length wild-type Ff virion. The concentration of pVIII in the inner cell membrane is very high—at least 5×10^5 molecules of pVIII are exported as virions per infected cell per doubling, making it one of the most abundant proteins in the infected cell [10].

1.3.4 Minor Structural Proteins (pIII, pVI, pVII, and pIX)

Each end of the filamentous phage particle is distinctive, both in function and in protein composition. The distal end of the phage contains approximately five copies each of the two small hydrophobic proteins, pVII (33 a.a., MW = 3599) and pIX (32 a.a., MW = 3650). In the absence of either pVII or pIX, almost no phage particles are formed [11], and genetic evidence suggests that both are involved in initiation of assembly [12].

Genes *III* (coding for pIII, 406 a.a., MW = 42,522) and *VI* (coding for pVI, 112 a.a., MW = 12,342) encode two minor coat proteins, which sit at the end of the viral particle that extrudes last during assembly and is also responsible for host cell binding. pIII serves dual functions; it is required for infectivity, and it is necessary for termination of viral assembly. In the absence of pIII, noninfectious multiple-length "polyphage" particles are produced, which contain several unit-length genomes. The crystal structures of the two amino-terminal domains (D1 or N1 and D2 or N2) of pIII have been solved for both phages fd [13] and M13 [14] (see Figure 1.5). These structures demonstrate that although the individual domains are the same, they differ by a rigid body rotation of N2 with respect to N1 around a hinge located at the end of the short antiparallel β-sheet connecting N1 and N2.

Figure 1.5 Crystal structure of the two amino-terminal domains of the gene *III* protein. The crystal structure of N1 and N2 at the amino terminus of pIII has been solved to 1.46 Å resolution using multiwavelength anomalous dispersion on a seleno-methionine derivative and is shown here in cartoon format with the N1 domain highlighted in dark gray [14] (PDB accession code 1G3P). These two domains can be seen as knobs at the proximal end of M13 as seen in the micrograph in Figure 1.1.

1.3.5 Morphogenetic Proteins (pI, pIV, and pXI)

Proteins pI (348 a.a., MW = 39,502), pIV (405 a.a., MW = 43,476), and pXI (108 a.a., MW = 12,424) are required for viral assembly but are not present in the final intact virus particle (for a review, see Ref. [15]). pI spans the inner membrane and functions in multiple ways during phage assembly (see Section 1.5), including interacting with both host cell factors and phage proteins required for morphogenesis, and in helping in the formation of phage-specific adhesion zones. pIV is an integral outer membrane protein whose carboxyl-terminal half mediates formation of a multimer that appears to be composed of between 10 and 12 pIV monomers [16,17]. The morphogenetic proteins pI and pIV, located in different membranes, appear to interact to form an exit structure through which the assembling viral particle extrudes from the host cell [18]. pXI is the result of an internal translational initiation within gene *I*, is more abundant in infected cells than pI, and is believed to be an essential part of a "preassembly complex" consisting of pI, pIV, and pXI [19].

1.4 STRUCTURE OF THE VIRION

1.4.1 Overall Structural Organization

Two symmetry classes of filamentous phage have been identified by x-ray diffraction: Class I, which includes fd, M13, f1, and Ike, and Class II, which includes Pf1 and Xf [20]. Figure 1.1 is an electron micrograph of M13, and Figure 1.3 is a schematic of the organization and relative positions of its protein components. The coat proteins from the two classes have lengths that vary from 44 to 53 amino acids and sequences with similar overall character (each has an amino-terminal end rich in acidic residues, a central hydrophobic region, and a carboxy terminus rich in basic residues) but little conserved sequence. A large fraction of the viral mass (around 85%) is made up of many copies of pVIII, which forms a 15–20 Å thick flexible cylinder about the single-stranded viral genome. There are approximately 0.435 copies of pVIII per nucleotide [21], and the length of the virion is about 3.3 Å per pVIII (or 1.435 Å per nucleotide) plus approximately 175 Å of minor coat proteins [22]. The amino terminus of pVIII is exposed to the surface with the first four to five residues forming a flexible arm that appears to extend away from the virus. The distal or top end of the particle is assembled first and contains approximately three to four copies each of pVII and pIX as determined by labeling studies [23], which may form a hydrophobic plug at the top of the virus [24]. The proximal end (bottom in Figure 1.3) contains approximately five copies each of pVI, which is believed to mediate attachment of approximately five copies of pIII to the body of the virus.

1.4.2 pVIII Structure

In the intact M13 viral particle, pVIII molecules are arranged with a fivefold rotational axis and a twofold screw axis, with a pitch of about 32 Å [25]. Most of the pVIII structure appears to be comprised of a gently curving α-helix extending from Pro6 to near the carboxyl terminus [26–29]. The axis of the pVIII α-helix is tilted about 20° relative to the virion axis and wraps around the virion axis in a right-handed helical sense as can be seen in Figure 1.3. The amino acid sequence of pVIII can be broken into four parts: a mobile surface segment (Ala1–Asp4 or Asp5); an amphipathic α-helix extending from Pro6 to about Tyr24; a highly hydrophobic helix extending from Ala25 to Ala35; and an amphipathic helix forming the inside wall of the coat, extending from Thr36 to Ser50 [27]. This last helical region contains four positively charged side chains, which interact with the phosphate backbone of the DNA, and the DNA bases face inward [28,30].

The amino-terminal half of pVIII is the region in which almost all fusions to pVIII have been constructed. The amino terminus lies at the surface with only the first three residues accessible to digestion by proteases [31], and the remainder of the phage surface is composed largely of the amphipathic helix that extends from Pro6 to about Tyr24. Random peptides inserted at or near the amino terminus of pVIII have shed some light on the packing requirements and structure of the phage particle surface. Fiber diffraction analysis of an insertion mutant with a penta-peptide (GQASG) inserted between residues 4 and 5 indicated that the insert lies in an extended conformation within a shallow groove between two adjacent α-helices on the viral surface [32]. Given the resemblance of this arrangement to the presentation of peptides by major histocompatibility antigens [33], the authors postulate that the three-dimensional conformation of small pVIII inserts may contribute to the high immunogenicity observed for peptides inserted into the gene *VIII* product of M13. The length of the shallow surface groove (17 to 20 Å) corresponds to the upper length limit observed when foreign peptides are fused to all 2800 copies of pVIII (that is, six to eight residues; see Refs. [34–36]). Alternative explanations for the size limitation include reduced signal peptidase cleavage rates for some mutant procoat proteins [35,37], possible interference with pVIII/VII interactions during assembly initiation [38] and/or the physical impossibility of extruding a virion of enlarged diameter through the 7-nm exit pore formed by the pI/pXI/pIV complex [24,39].

The carboxyl terminus of pVIII is at low radius, roughly 20 Å from the surface of the virion in close proximity to the viral DNA. Recent work by Fuh et al. [40] has shown that viable phage particles can be constructed that display, at very low levels, foreign peptides fused to the carboxyl end of pVIII. This display requires a minimum of 8–10 residues as a linker to render the carboxy-terminal fusion product accessible to the surface. These data appear to be consistent with two possible scenarios: either the linker extends through the protein coat and disrupts the local packing of pVIII molecules around the DNA core or, alternatively, the inserts are only tolerated near the proximal end of the virion where the carboxyl terminus of pVIII may be closer to the virus surface. This latter hypothesis is consistent with the fact [41] that display with the carboxy-terminal pVIII fusion was reduced about tenfold relative to amino-terminal fusions on pIII, indicating recombinant pVIII levels of roughly 0.1–0.2 per viral particle (given the roughly 1–2 fusion pIII molecules regularly achieved per particle).

Filamentous phages are notoriously resistant to numerous physicochemical assaults, including prolonged incubation at high temperatures, in nonionic detergents, in high salt, and at low pH. However, Class I viral particles have been shown to be sensitive to inactivation by small organics, such as chloroform, with accompanying structural collapse to rod-shaped particles [42]. Sequence analysis of isolated chloroform-resistant and growth-enhanced mutants of pVIII [36,43,44] have revealed numerous point mutations, which map throughout a single slice of the viral particle with a particular hot spot for mutation at the Pro6/Val31 location at the surface. Modeling studies, based upon fiber diffraction analysis [27,28], indicate the presence of a substantial depression or hole in a large hydrophobic surface patch at this location with Val31 lying exposed at the bottom [43]. The V31L clone isolated by Oh et al. [43] and a growth-enhanced recombinant pVIII, isolated by Iannolo et al. [44], that contained a tripeptide (PFP) insertion, may succeed by "plugging" this hydrophobic hole, which extends to the phage surface, resulting in an overall more stable structure. Studies on the stereochemistry of this site suggest that a chloroform molecule can indeed fit into the Val31 hole [43]. These data are consistent with that of Weiss et al. [45], who isolated numerous point mutants of pVIII with a P6F conversion, which exhibited greater than tenfold enhancement in the efficiency of display of human growth hormone on the surface of M13, possibly the result of stabilized phage particles.

1.4.3 Distal End Structure

The distal end of the phage particle has approximately five copies each of the two small proteins pVII and pIX. This end contains the packaging sequence (PS) and is the part of the phage assembled first. The gene *IX* protein has been shown to exist as 67% α-helix in lipid bilayers with its positively charged carboxyl terminus residing on the outside of the membrane and thus available for binding to the negatively charged DNA hairpin loop of the PS [46,47]. Using sera raised against pVII and pIX, Endemann and Model [38] have shown that pIX is accessible in the intact phage particle whereas at least some epitopes of pVII are not. Immunoprecipitation experiments on detergent-disrupted virus suggest an interaction between pVII and pVIII in the virus particle. Gao et al. [48] have shown that the amino termini of pVII and pIX can be used to display the antibody heavy chain variable domain (V_H) and light chain variable domain (V_L), respectively, and that this heterodimeric presentation affords a viable antibody variable fragment (F_v) with fully functional binding and catalytic activities. Single-chain F_v (scF$_v$) libraries have been displayed as fusions with the amino terminus of pIX [49].

1.4.4 Proximal End

The proximal or infectious end consists of approximately five copies each of pIII and pVI. Antibodies directed against pVI do not interact with intact phage, suggesting that pVI is somewhat buried within the viral particle [38]. A partial model for this end of the virus was postulated by assuming that the amino-terminal half of pVI is structurally homologous to two pVIII molecules [24] as shown in Figure 1.3. At this end of the virus, termination of assembly leaves two layers of pVIII exposed to solvent as there are no continuing pVIII rings to cover them. This potentially unstable situation could be ameliorated by a fold of pVI, much of which forms an amphipathic surface layer, up and around these last two pVIII rings, holding the virion together and helping to maintain viral stability. The remaining parts of pVI may have a portion of pIII folded around them, sequestering them away from solvent and antibody access. Contrast-enhanced electron micrographs of negatively stained microphage particles (E. Bullitt and L. Makowski, unpublished results) exhibit a slightly enlarged diameter near the proximal end, consistent with this model. The C-terminus of pVI has been successfully used as a vehicle for display [50], suggesting that it is accessible on the virion surface.

The gene *III* protein is the largest and most structurally complex component protein of filamentous phage. It consists of three distinct domains, separated by two glycine-rich linkers (residues 68–87 and 218–253), which appear to make portions of pIII somewhat flexible. In electron micrographs of negatively stained phage, the two amino-terminal domains (N1 [or D1] and N2 [or D2]) can be seen as knobby structures at the end of the virion (see Figure 1.1) and can be removed, along with infectivity, with subtilisin digestion [51]. The carboxy-terminal 150 residues of pIII, domain D3 or CT, are proposed to interact with pVIII to form the proximal end of the viral particle as they remain associated after disruption of the virus with detergents [38,52]. The crystal structures of both N1 and N2 from M13 and fd have been determined [13,53,54] (see Figure 1.5) with a comparison showing N1 movement relative to N2 with implications for infection mechanisms (see later discussion).

The amino terminus of the mature pIII was the first location used for the display of foreign proteins and peptides on M13 [55,56]. It is still the most commonly used position. Contrary to common belief, however, polypeptides fused to the carboxy terminus of the M13 gene-3 minor coat protein may be functionally displayed on the phage surface [41]. In a phagemid display system, carboxy-terminal fusion through optimized linker sequences resulted in display levels comparable to those achieved with conventional amino-terminal fusions. The details of the structure of pIII as visualized in the crystallographic analysis of the two N-terminal domains suggest an innocuous structural environment remote from the host cell binding site and generally favorable to insertions that would have little impact on the function of pIII (see Section 1.7 discussion).

1.4.5 DNA Structure within the Viral Particle

The structure of the viral DNA within the phage particle is not well characterized. Solid-state NMR studies of the filamentous phage Pf1 [57] demonstrated a uniform conformation for the phosphate backbone, whereas in the Ff phage, the phosphates take on a larger number of different orientations. Fiber diffraction studies of the intact virion [27,32] suggest a lack of sufficient space within the capsid shell for a B-DNA duplex, suggesting that the DNA in the virion has a somewhat extended conformation. Mutations resulting in a change in the net charge of the C-terminal, basic region of pVIII [1] behaved in a manner indicating a direct but nonspecific electrostatic interaction between the DNA and the coat protein. This was dramatically demonstrated by a series of mutations in which Lys48 was converted to an uncharged amino acid. These mutants were 35% longer than wild type; the total number of basic residues in the viral interior was unchanged (the number of proteins increased in order to compensate for the decreased number of charges per protein); and the DNA structure was stretched by 35% to adapt to the lowered charge density on the interior surface of the protein coat.

Packaging signals (or PSs) earmark viral nucleic acids for encapsidation and have been identified for numerous RNA- and DNA-containing viruses. The PS for filamentous phage was originally identified within the intergenic region of f1 (shown in Figure 1.2) by its ability to enable heterologous ssDNA molecules to form phage-like particles [58]. Bauer and Smith [59] demonstrated that the PS is located at one end of the pV-sequestered/viral ssDNA complex, suggesting that the PS lies exposed in this complex form, ready to initiate encapsulation. Hence, the DNA within filamentous phage particles is oriented within the virion such that the PS is always at the distal end [60]. A hairpin structure has been shown to exist within the virion as it can be cross-linked with psoralen [61]. The PS in its single-stranded form can be drawn as an imperfect hairpin of 32 bp with a small bulge. Not only can the hairpin below the bulge be deleted with no loss of function, but so also can a portion of the upper hairpin as long as the lower is present. In addition, loop sequences at the tip of the hairpin can be altered and short insertions added with no loss of function. Studies published in 1989 [12] on PS(−) mutants showed that although very few phage particles were produced in the absence of a PS, suppressor mutations allowing higher packaging efficiency arose at high frequency. These suppressor mutations mapped to the coding regions of pI, pVII, and pIX, implicating their interaction with the PS during assembly initiation. Given that the mutant proteins function better with PS(+) strains, it is assumed that the mutants have gained the ability to use another small hairpin structure as a "cryptic" PS. The PS of phages Ike and f1, which exhibit little sequence similarity, are functionally interchangeable [12]. A construct with a perfect self-complementary segment of DNA functioned poorly for packaging. Taken together, these studies point toward the conclusion that, although a duplex is essential for the function of the PS in filamentous phage, additional features are also involved.

1.5 FILAMENTOUS BACTERIOPHAGE LIFE CYCLE

1.5.1 Replication of Viral DNA

Filamentous bacteriophages contain a single-stranded DNA genome that replicates in three stages. In stage I, once the phage (+) strand ssDNA (SS) is translocated into the cytoplasm subsequent to infection, bacterial host enzymes synthesize the complementary (−) strand, producing a double-stranded covalently closed supercoiled DNA product called the parental or replicative form (RF) DNA (SS → RF). Stage II occurs when the RF DNA replicates to form a pool of approximately 100 RF DNA molecules per cell (RF → RF). In the final stage, the RF DNA molecules act as a template for the synthesis of progeny single-stranded DNA phage genomes (RF → SS).

Transcription off the (−) strand of the RF DNA occurs in a clockwise direction as shown in Figure 1.2 with gene products produced in the order in which they are required for phage production (that is, pII, pX, pV for replication, etc.). Multiple transcripts are produced with six mRNAs encoding pVIII and four encoding pV, allowing for the production of large amounts of these proteins to coat the emerging virus and intracellular ssDNA genomes, respectively (for a more detailed discussion of transcription, see Ref. [62]). The gene *II* protein is an endonuclease–topoisomerase that introduces a specific nick in the (+) strand of the RF DNA. The resulting 3′ end serves as a primer for synthesis of new (+) strands via a "rolling circle" mode of replication carried out by host cell enzymes and terminated and circularized by pII. The resulting (+) strands can either be used as templates for the synthesis of more RF DNA or be coated by pV dimers and sequestered in the cytoplasm ready to be assembled into progeny phage. Levels of pV thus determine the ratio of RF to (+) strand DNA synthesis; in the presence of adequate pV, (+) strand is coated for virus production, precluding its conversion to the double-stranded form. Protein X is also believed to be involved in regulation of RF/(+) strand synthesis, as it acts as an inhibitor of pII function [5].

The pV dimer binds to DNA in a highly cooperative manner without marked sequence specificity at physiological pH [63–66]. Protein V exhibits two distinct modes of DNA binding, dictated by salt concentration, and leading to either a cooperatively saturated complex through nonspecific binding or an unsaturated complex through specific binding. These two interactions correspond to the two biological functions of pV: sequestering (+) strand and regulating viral mRNA translation [67]. The nonspecific binding of pV to ssDNA produces a superhelical protein–DNA complex. This complex is a flexible structure, approximately 8800 Å long, and approximately 80 Å in diameter [64,68–70] in which the pV dimers appear to form a tightly wound left-handed helix with a pitch of about 90 Å [69]. The radius of gyration of the DNA in the complex is much lower than that of the protein, indicating that the DNA is closer to the axis of the structure. Each turn of the helix contains roughly eight pV dimers, with two antiparallel ssDNA strands occupying the interior of the superhelix [70]. A model for the pV–ssDNA complex, consistent with these observations and with the 1.8 Å resolution crystallographic structure of the pV dimer, has been constructed [7]. This model places the antiparallel DNA single strands within the grooves formed by the β-strands that loop out from the body of the pV dimer, as illustrated in Figure 1.4.

1.5.2 Synthesis of Viral Proteins

The replication proteins (pII, pV, and pX) are synthesized by host cell machinery and reside within the cytoplasm. All other viral proteins are synthesized and then inserted into either the inner (pI, pIII, pVI, pVII, pVIII, pIX, and pXI) or outer (pIV) host cell membrane. Three minor capsid proteins (pVI, pVII, and pIX) are synthesized without signal peptides. The mechanisms of their membrane insertion are unknown, but all three appear to be inserted into the inner membrane with their carboxyl termini on the cytoplasmic side and their amino termini on the periplasmic side. Sequence analysis predicts a single membrane-spanning region in both pVII and pIX. pVI appears to have three membrane-spanning regions and a central, positively charged region protruding into the cytoplasm although this latter orientation is in controversy.

The gene *III* protein is synthesized with an 18 amino acid signal peptide, which is removed after membrane insertion. Membrane translocation is Sec-dependent (see Ref. [71] for a recent review) and results in a single transmembrane domain located five residues from the carboxyl terminus with the carboxy-terminal residues lying within the cytoplasm [72]. Most of its mass (including domains N1, N2, and most of CT) resides in the periplasm. The membrane-anchoring domain consists of the carboxy-terminal part of D3 and is also involved in attaching pIII to the viral particle (see Figure 1.3).

Procoat (pVIII prior to cleavage by signal peptidase) appears to form dimers within the membrane by packing along the hydrophobic face of its amphipathic helix and extending through the membrane-spanning region [73]. During morphogenesis, this L-shaped conformation transforms into the elongated, largely α-helical structure it takes on in the intact phage particle (see Figure 1.6). The major coat protein pVIII was believed until recently to spontaneously insert into the *E. coli* inner membrane by a Sec- and signal recognition particle-independent mechanism. Recent work, however, has shown that inner membrane translocation of pVIII is dependent upon YidC, a homologue of the mitochondrial membrane transporter Oxa 1 p [74] even though pVIII can partition into the membrane in the absence of YidC [75]. Once associated with the inner membrane, procoat appears to have a U-shaped configuration with two transmembrane helices extending from positions −15 to −2 and from 21 to 39 [76], bracketing an amphipathic helix that extends along the periplasmic surface of the inner membrane. Tyr21 and Tyr24 appear to act as anchors to position the second transmembrane helix within the membrane [77]. Signal peptidase cleavage leaves

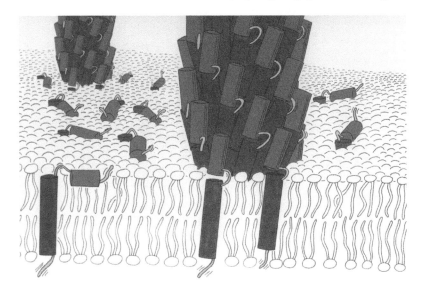

Figure 1.6 Conformational shift of the gene *VIII* protein during viral morphogenesis. This figure is a schematic diagram of the incorporation of coat proteins into a growing Pf1 virion. (From Nambudripad R et al., *Science*, 252:1305–1308, 1991. Reprinted with permission of AAAS.)

the mature pVIII in an L-shaped configuration [78] with five flexible amino-terminal residues, residues 8–16 forming an amphipathic helix extending along the membrane surface, and residues 25–45 making up a transmembrane α-helix that extends into the cytoplasm and includes a portion of the basic carboxy-terminal region of pVIII involved in protein–DNA interactions [79] (see Figure 1.6). The extent of the carboxy-terminal transmembrane helix is probably not altered by signal peptidase cleavage of the amino-terminal signal sequence. The different extents of this helix reported by the two studies quoted from Refs. [78,79] may reflect different conditions under which the measurements were made.

Proteins I, IV, and XI are all required for phage assembly but are not present in the intact virus particle. Proteins I and XI are synthesized without signal peptides and are inserted into the inner membrane in a SecA-dependent manner [80]. They each span the membrane once and are oriented with their carboxy-terminal 75 residues lying within the periplasm [81]. Protein I has a 250 amino acid amino-terminal cytoplasmic domain, all but 10 amino acids of which are missing in pXI (see Figure 1.7). Plasmid-driven

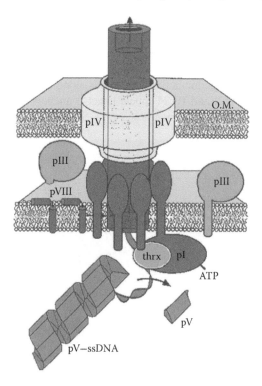

Figure 1.7   Viral morphogenesis through the pIV multimer. This figure depicts the roles of various viral- and host cell-encoded proteins during viral assembly. These complexes are believed to be located at specific adhesion zones, at which the inner and outer bacterial membranes are in closer contact than in nearby areas. (Modified from Makowski L, Russel M, Structure and assembly of filamentous bacteriophages. In: Chiu W, Burnnet RM, Garceia R, eds. *Structural Biology of Viruses*. Oxford University Press, 352–380, 1992.)

production of excess pI results in a loss of membrane potential and cessation of host cell growth, suggesting that it can form some type of channels through the cytoplasmic membrane [82–84]. Excess pXI does not appear to have such an effect. Proteins I and XI appear to interact in the preassembly complex; pXI being the more abundant member. The 250 amino acid cytoplasmic domain of pI contains an ATP binding site and may be actively involved in assembly and extrusion of the phage particle.

The gene *IV* protein is synthesized with a 21 amino acid signal peptide and is translocated into the periplasm via the Sec system [16,80]. The amino-terminal half of pIV forms a protease-resistant domain exposed in the periplasm [80,85], whereas the carboxy-terminal domain mediates outer membrane integration and formation of a multimer composed of between 10 and 12 pIV monomers that form large cylindrical clusters with an internal diameter of 80 Å, and probably comprise a gated channel for phage translocation through the outer membrane [16,17,86]. This channel is sufficient to accommodate the passage of the phage particle, which has an outer diameter of roughly 65 Å. A growing family of bacterial proteins have been identified that share significant sequence homology with the carboxy-terminal half of pIV, the "secretins," which are common to the type II bacterial secretion systems and which similarly form multimeric channels [87]. Protein I has been shown to associate with pIV and this interaction has been implicated in the formation of phage-specific adhesion zones, regions in which the inner and outer membranes of *E. coli* are in close contact (adhesion zones are more common in phage-infected cells and have been identified as sites of phage morphogenesis) [11,18]. Recent work has demonstrated that phage production blocks oligosaccharide transport via pIV gated channels, providing the first definitive evidence for viral exit through a large aqueous channel [88].

1.5.3 Viral Morphogenesis

1.5.3.1 Preassembly Complex and the Initiation of Assembly

Assembly of phage particles begins when the PS interacts with the assembly complex formed by pI, pIV, and pXI and host cell thioredoxin at localized adhesion zones [11,18] (see Figure 1.7). This complex spans the inner and outer membranes and the periplasm and provides both a platform for phage particle assembly and a pathway for phage particle extrusion without lysis of the host cell. Protein IV forms the outer membrane portion of the complex, pI and pXI form the inner membrane portion and the amino-terminal region of pI, and host cell thioredoxin form the cytoplasmic portion of the complex.

Overexpression of pIV has no effect on host cell viability, indicating that the channel formed by pIV is closed in the absence of other phage proteins [39]. In the presence of pI and pXI and amber mutants of pVII or pIX, assembly sites are formed, followed closely by cessation of bacterial growth, suggesting the loss of transmembrane potential [11]. Although pIV from f1 and Ike are not functionally interchangeable, when both pIV and pI are exchanged between the two phage types, some heterologous phage particles are formed, suggesting that pI and pIV interact to promote assembly [18]. This interaction is confirmed by the existence of paired compensatory mutations in the respective periplasmic domains of pI and pIV [18]. Gly355Ala/Ser mutants of pIV that are normally conserved in all known homologues render the host cell sensitive to a large number of antibiotics and detergents that are normally excluded by the outer membrane [89].

Efforts to model the phage ends provide some clues to the process involved in the initiation of assembly. Modeling of the distal end of the phage suggests that the PS may be surrounded by pVIII molecules in the intact virion, rather than by pVII and/or pIX [24]. Some pVIII molecules are associated with pVII in the membranes of infected cells [38], suggesting that it may be a pVII/pVIII complex that associates with the PS during initiation of assembly, followed by docking with the assembly complex.

1.5.3.2 Elongation

An intact assembly apparatus consists of pIV as the outer membrane component and pI/pXI as the inner membrane component, an intact phage "tip" consisting of about five copies each of pVII and pIX plugging the pIV channel, the PS end of the phage genome, and possibly one ring of pVIII molecules. Following formation of the assembly complex, elongation of the phage particle can begin. Assembly requires the presence of pIII, pVI, pVII, pVIII, and pIX in the inner membrane as well as the host protein thioredoxin [15]. Although thioredoxin in its reduced state is a potent reductant of protein disulfide bonds, it is not the redox activity of the protein that is required for assembly as mutations of its two active site cysteine residues do not prevent its participation in phage assembly [90]. One clue to its role in phage assembly is that during infection by the lytic phage T7, thioredoxin complexes with T7 DNA polymerase, conferring processivity on the polymerase, allowing it to polymerize thousands of nucleotides without dropping off the DNA template [91]. It is conceivable that complex formation of thioredoxin with pI enhances processivity in the filamentous phage assembly process also by stabilizing pI binding to DNA [92].

During elongation, pV is stripped from the DNA and replaced with rings of pVIII, and simultaneously, the DNA/pVIII is exported from the host cell. Viral elongation is an ATP-driven process dependent on the presence of an intact Walker nucleotide binding motif ([AG]-XXXX-GK[ST]) near the amino-terminal end of pI [62]. Protein V is removed from the ssDNA in order for pVIII to be added, but there is no evidence of interaction between pI and pV, as gene V proteins from various phages are interchangeable [93]. It has been suggested that ATP hydrolysis by pI can act as a source of energy for pV displacement from the ssDNA during elongation [92]. The 10 residues of pI and pXI that face the cytoplasmic surface of the inner membrane possess an amphiphilic character similar to the 10 carboxy-terminal residues of pVIII that interact with the DNA within the intact virion, suggesting that these residues may interact with ssDNA that has just been stripped of pV, facilitating conformational shifting toward interaction with pVIII [62]. Because the helical symmetry of the DNA in the pV complex is different from that in the viral complex, the two structures must rotate relative to one another during the elongation process.

Studies of pVIII structure both in the intact virion and within detergent micelles by x-ray diffraction, solid-state NMR, neutron diffraction, and Raman spectroscopy have shown that a significant conformational shift must occur within pVIII for the protein to be incorporated into the growing phage particle [27,78,79,94–99]. In addition, each pVIII molecule must now interact with the ssDNA genome and with many other copies of pVIII and, thus, must exchange hydrophobic interactions that anchor it within the membrane for hydrophobic interactions that stabilize it within the phage particle (see Figure 1.6). The shift from the two perpendicular α-helical segments of the membrane-bound form to the single long α-helix of the viral form moves the two ends of the protein approximately 16 Å further apart. A 16 Å axial translocation is precisely what is required to position the viral particle to accept further additions of pVIII rings at its proximal end. The mobility of the residues in the hinge region of the membrane-bound form is likely to enable a smooth transformation from the membrane-bound form to the phage conformation [79,100].

Three separate research groups have performed mutational studies that suggest that certain small residues (such as Ala7, Ala10, Ala18, Leu14, Gly34, and Gly38) are not easily mutated in pVIII or can only be substituted with other small residues [45,79,101]. A model for elongation of phage particles based upon both NMR and x-ray diffraction studies of both position and mobility has been proposed that takes into account the roles of these highly conserved residues [79].

1.5.3.3 Termination

Termination of assembly occurs when the entire DNA molecule has been packaged by pVIII molecules. It seems to be a relatively inefficient process as double-length particles that contain two unit-length genomes occur frequently in a wild-type phage preparation (approximately 5% of the total phage). The addition of a second genome does not represent a second initiation event as ssDNA molecules lacking a PS can be coencapsidated with a PS(+) genome at about the same frequency (5%) as those that contain a PS [12]. Assembly termination involves the addition of pVI and pIII to the tip of the particle. Both pIII and pVI are anchored in the cytoplasmic membrane and cannot be coimmunoprecipitated from nonionic detergent extracted membranes whereas once they are within intact phage particles they coimmuno-precipitate following the same treatment [38,102,103], inferring a strong interaction once within the virus. Although pVI(–) and pIII(–) phages each form noninfectious "polyphage" and are both lacking in pIII molecules, the pVI(–) particles are less stable, tending to unravel at the tip [11]. These observations imply that pVI is needed to both stabilize the proximal end of the virus and attach pIII to the proximal end of the virus particle. This is confirmed by the fact that anti-pVI antibodies do not interact with intact phage [38].

Protein III molecules are anchored within the cytoplasmic membrane via residues 379–401 at their carboxyl termini with most of their residues located within the periplasm (1–378). The amino-terminal domain (residues 1–218) mediates infection (see Section 1.5.4) whereas the carboxy-terminal domain (D3 or CT; residues 253–406) is involved in releasing the phage from the membrane and in capping the virion. A stable but noninfectious phage can be formed with a truncated pIII (residues 197–406), which comprises a portion of the N2 domain, the second glycine-rich spacer and the whole CT domain [104,105]. Recent studies with a series of carboxy-terminal fragments of increasing size has led to the delineation of portions of pIII that are capable of mediating pVI incorporation into the assembling phage, release of sarkosyl-labile viral particles into the culture supernatant, and release of sarkosyl-stable virions, respectively, thus dividing the D3 or CT domain into a C2 subdomain sufficient for host cell release and a C1 subdomain required for virion stability [52]. Coimmunoprecipitation of a certain amount of pVIII from the membranes of infected cells by pVI antiserum suggests that pVI is added to the particle as a pVI/pVIII complex [38]. It has been suggested that once the pVI/pVIII complex is added to the tip of the elongated phage, a hydrophobic patch may be exposed on the pVI surface, which exhibits high affinity for the stretch of residues that anchor pIII within the membrane [106]. Thus, pVI/pIII interactions could replace lipid/pIII interactions, thereby releasing pIII from the inner membrane.

Alternatively, it has been noted that termination of phage assembly results in a change in orientation of pIII relative to pVIII. Prior to termination, pIII is anchored within the cytoplasmic membrane by a carboxy-terminal membrane anchor, placing the vast majority of the protein within the periplasm [72]. Protein VIII is similarly oriented with its carboxyl terminus in the cytoplasm and amino terminus within the periplasm [107,108]. Both prior to and after incorporation into the virion, the two proteins interact, but after incorporation, the amino terminus of pIII points away from pVIII. This change in relative orientation between pIII and pVIII has been used by Rakonjac et al. [52] to postulate a mechanism in which pIII plays the predominant role for termination of phage assembly. Given that the segment of pIII most likely to interact with pVIII is the membrane anchor, the rest of the pIII molecule is free during assembly to bend or flip in the opposite direction from pVIII during incorporation into the growing particle (see Figure 1.8). This flipping motion on the part of the C2 subdomain could then disrupt phage–membrane hydrophobic interactions by allowing C2 to cover up a hydrophobic portion of the pIII carboxyl terminus, and thus resulting in release of the assembled phage from the host cell. This model is corroborated by the fact that a very short carboxy-terminal fragment of pIII that lacks the C2 subdomain just amino-terminal to the membrane anchor can be incorporated into the membrane-associated viral particle but cannot detach the virus from host cells [52]. Additionally, carboxy-terminal fusions of greater than seven amino acid residues have been found to prevent termination of phage assembly, suggesting that although the carboxyl terminus may not be tightly packed, it may be within an enclosed environment (J. Rakonjac and P. Model, unpublished results).

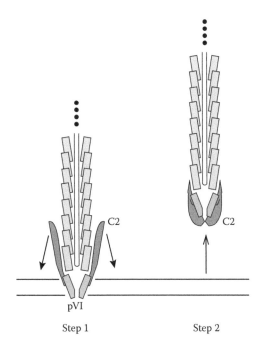

Figure 1.8   Model for termination of viral assembly. This diagram depicts a model proposed by Rakonjac et al. for termination of phage assembly. A conformational shift within pIII is theorized as the primary driving force for phage release based upon the phenotype of a series of pIII deletion mutants. (Modified from Rakonjac J et al., *J Mol Biol*, 289:1253–1265, 1999.)

1.5.4 Infection Process

Filamentous phage infection is a two-step process: (i) recognition—during which the virus binds to its primary bacterial cell surface receptor, the distal tip of the F-pilus, followed by (ii) translocation, which involves pilus retraction, capsid protein integration into the bacterial cell membrane, uncoating of viral DNA, and its concomitant translocation into the host cell cytoplasm.

Only one or a few F-pili are present on the surface of a bacterial cell. Both pilus structural proteins and the proteins required for pili assembly are plasmid encoded with the genes laying within the *tra* operon, or transfer region, of the conjugative plasmid. Different filamentous phages have specificities for different pilus types. The Ff (f1, fd, and M13) bind F-type pili whereas Ike and I2–2 infect *E. coli* strains that produce N- or P-type pili. RNA phages, such as R17 and Qβ, use the side of the F-pilus for attachment to the host cell [109]. The F-pilus consists of a helical array of pilin subunits of 8-nm diameter with a 2-nm lumen [110]. The pilin subunit is about 65%–70% α-helical as determined by circular dichroism [111]. Assembly of the intact F-pilus requires the activity of 11 *tra* gene products [110]. This assembly process is temperature-sensitive; thus, F-specific phage fails to infect or form plaques on bacteria grown at temperatures below around 32°C.

Infection of a male *E. coli* cell is initiated by binding of the N2 domain of pIII to the tip of the F-pilus [112]. This step is followed by pilus retraction into the host cell with pilin subunits believed to depolymerize into the host cell inner membrane [113,114]. It is not known whether phage binding initiates pilus retraction or if phages are brought to the membrane as a consequence of normal cycles of pilus retraction and repolymerization [115]. A single L70S substitution at the last amino acid of F-pilin knocks out both pilus assembly and Ff infection [116]. Three mutations (V48A, F50A, and F60V) have been identified that do not affect pilus assembly but reduce sensitivity to M13KO7 (an M13 derivative commonly used as helper virus for rescue of recombinant pIII phagemids) down to 1%–3% of wild-type levels, suggesting a defect in pilus function, possibly in retraction [110].

The coreceptor for infection by Ff phage, the TolQRA complex, was first identified by the isolation of *tolA* and *tolB* mutants that are insensitive or "tolerant" to the lethal action of a class of proteins termed "colicins," which have antibiotic activity and are produced by some *E. coli* strains to kill noncolicin producing bacteria [117]. The *tolQ* and *tolR* genes were subsequently identified and shown to be not only essential to susceptibility to certain classes of colicins, but for filamentous phage infection as well [118–120]. The Tol system appears to be involved in other types of macromolecular import, given its structural and sequence homologies to the TonB–ExbB–ExbD complex [121,122]. Bacterial cells possessing functional pIII either from a filamentous phage infection [123] or encoded by a plasmid [10] show increased tolerance to the effect of E-type colicins, suggesting that pIII and E colicins have identical or closely adjacent binding sites on the Tol complex. In addition, overexpression of pIII, its amino-terminal fragment [10,124] or proteins possessing the TolA D3 domain [125,126] induces outer membrane leakiness in *E. coli*, leading to loss of periplasmic proteins.

TolA, TolQ, and TolR form a complex within the inner membrane of the bacterial host cell as shown schematically in Figure 1.9. The N-terminus of TolA anchors the protein to the inner membrane whereas the C-terminus, D3, is associated with the outer membrane, connected by an extended helical region, D2 [127]. Analysis of the crystal structures of the tightly interacting N1 and N2 domains of pIII [54] (Figure 1.5) and the complex between N1 of pIII and the carboxy-terminal domain D3 of its coreceptor TolA [14] (see Figure 1.10) demonstrates that during the infection process, the interaction between the N1 and N2 domains of pIII is replaced by the interaction of the D3 domain of TolA with N1. Despite a lack of topological similarity between pIII–N2 and TolA–D3 (compare Figures 1.5 and 1.10), both domains interact with the same region of the pIII–N1 domain and bury comparable accessible surface areas (1768 Å^2 for pIII–N1/TolA–D3 versus 2154 Å^2 for pIII–N1/pIII–N2) [14]. Fourteen of the 21 residues of the pIII–N1 domain that are involved in the interactions with TolA–D3 are also involved in interactions with pIII–N2. Infection of F$^-$ bacteria with wild-type phage can be achieved at low levels (four to five orders of magnitude lower than the rate for F+ hosts) by treatment of the bacteria with 50 mM Ca^{2+}, presumably by altering the outer membrane enough to mediate exposure of TolA–D3 for interaction with the N1 domain of pIII.

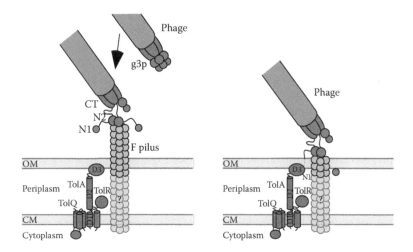

Figure 1.9 The Ff bacteriophage infection process. This figure shows a schematic view of phage infection demonstrating how pilus-mediated separation of N1 from N2 at the amino terminus of pIII frees up N1 for interaction with the D3 domain of the coreceptor TolA, thus mediating viral entry into the host cell. (Reprinted from *Structure*, 7, Lubkowski J, Hennecke F, Plückthun A, Wlodawer A. Filamentous phage infection: Crystal structure of g3p in complex with its coreceptor, the C-terminal domain of TolA, 711–722, Copyright 1999, with permission from Elsevier.)

Figure 1.10 Structure of the N1 domain of pIII cocrystallized with the D3 domain of TolA. The cocrystal structure of the D3 domain of the coreceptor TolA with the amino-terminal domain of pIII has been solved to 1.85 Å resolution [14] (PDB accession code 1TOL). Note that the amino-terminal residue of pIII (arrow) lies far away from the interaction surface between the two proteins.

The mechanism by which capsid proteins are integrated into the host cell cytoplasmic membrane and viral DNA is uncoated and translocated into the bacterial cytoplasm is largely unknown. Gene *VIII* proteins that have become associated with the inner membrane can later be reutilized in the assembly of progeny phage particles as their insertion into the membrane occurs in a manner that gives them the same topology as newly synthesized pVIII molecules [128–130]. Penetration of the DNA has been demonstrated to require host cell TolQRA proteins [126]. It has been suggested that the CT domain of pIII may be involved in the formation of an entrance pore for DNA translocation [131], perhaps as a mirror image of the mechanism by which pIII mediates termination of assembly (see Figure 1.8) [52], with a conformational change of the CT domain of pIII uncapping the virion, exposing hydrophobic surfaces of pIII, pVI, and pVIII, followed by membrane integration. This would be analogous to the mechanism of entry of eukaryotic viruses into host cells via the unmasking of hydrophobic fusogenic peptides [132]. In support of this hypothesis, it has been shown that the insertion of a β-lactamase domain between the amino and carboxyl termini of pIII (thus disrupting their proper distance) decreases infectivity by two orders of magnitude [133].

1.6 PHAGE LIBRARY DIVERSITY

1.6.1 Efficiency as a Biological Strategy for Survival

Considering the effect of insertion mutations on M13 requires an examination of the relationship between M13 and its *E. coli* host. M13 is not a lytic phage. Rather, it parasitizes the host, being carried from generation to generation and producing anywhere from 200 to 2000 progeny phage per cell per doubling time [2,3]. As discussed previously, this phage production represents a serious metabolic load for the infected *E. coli*, corresponding to 1%–5% of total protein synthesis and resulting in a reduction in cell growth to perhaps 30%–50% of uninfected cells. Given this negative effect on host growth, any host mutation that resulted in resistance to phage would seem to be highly favorable, and the resistant host should quickly outgrow the infected host. These mutations are not observed to occur, suggesting that the phage has evolved strategies for preventing its host from developing resistance. What form do these strategies take? Mutations that knock out expression of any of the viral genes (with the exception of pII) result in killing of the host cell [2,134]. This effect appears to be due to the accumulation of phage-encoded proteins III and I in the cytoplasmic membrane that results in the degradation of host cell membranes. These observations suggest that any mutation in either the host cell genome or phage-encoded proteins that results in the halting or *even the slowing* of phage morphogenesis may lead to the buildup of pI and pIII in the host cell and, subsequently, to the death of the host. Similarly, any mutation in a viral protein that blocks or slows phage assembly in such a way as to allow the accumulation of pI or pIII may also lead to host cell death. If this is the case, inserts that slow phage production may be rapidly censored from a phage-displayed library if they result in the buildup of pI and pIII in the host cell membrane. M13 appears to have evolved to prevent the development of resistance in its natural host: any mutation that significantly slows its production represents a fatal mutation. A detailed examination of the diversity and censorship patterns of phage display libraries [135,136] reinforces this hypothesis.

1.6.2 Phage Population Diversity

A number of groups have investigated the biochemical diversity of phage display constructs using various methods, including restriction digestion pattern analysis of small numbers of group members and colony hybridization with primers [137–142]. Statistical methods have been developed and used to quantitate and annotate the sequence diversity of combinatorial peptide libraries on the basis of small numbers (100 or 200) of sequences. Application of these methods in the analysis of commercially available M13 pIII-based phage display libraries [136] demonstrated that these libraries behave statistically as though they correspond to populations containing roughly 4.0% of the random dodecapeptides and 7.9% of the random constrained heptapeptides that are theoretically possible within the phage populations. Analysis of amino acid occurrence patterns in these libraries shows no demonstrable influence on sequence censorship by *E. coli* tRNA isoacceptor profiles or either overall codon or Class II codon usage patterns, suggesting no metabolic constraints on recombinant pIII synthesis. This is in contrast to a clear effect of metabolic limitations to the diversity of pVIII libraries [135]. The pIII libraries exhibit an overall depression in the occurrence of cysteine, arginine, and glycine residues and an overabundance of proline, threonine, and histidine residues, and position-dependent amino acid sequence bias that is clustered at three positions within the inserted peptides of the dodecapeptide library, +1, +3, and +12 downstream from the signal peptidase cleavage site. These sequence limitations can primarily be attributed to two steps during viral assembly: signal peptidase cleavage and incorporation of the recombinant protein into the growing virion from the bacterial inner membrane.

1.7 BIOLOGICAL BOTTLENECKS: SOURCES OF LIBRARY CENSORSHIP

1.7.1 Protein Synthesis

Inserts into the minor structural proteins of M13 are not likely to significantly disrupt or slow their synthesis, and this has been confirmed for combinatorial peptide libraries displayed on pIII [136]. However, the sheer numbers of pVIII molecules that must be synthesized for viral production put the major structural protein of M13 in a separate category. The protein synthesis apparatus of the infected host cell produces from 5×10^5 to 5×10^6 copies of pVIII per doubling time. It is well established that rare codons are used by the *E. coli* bacterium to regulate the rate of synthesis of regulatory proteins [143] and that mRNAs dependent on rare codons can inhibit protein synthesis by robbing the cell of minor tRNA isoacceptors [144,145]. Wild-type pVIII does not include any of the 10 rarest codons in *E. coli*, presumably because their presence would greatly retard the production of pVIII and limit virion production.

Rodi and Makowski [135] analyzed the relative occurrence of codons in a pVIII pentapeptide library constructed with a 32-codon code. The insert was of the form $(NNK)_5$, where N refers to an equimolar mixture of all four bases and K an equimolar mixture of G and T. Seven amino acids have multiple codons in a library of this form. For four of them, a statistically significant correlation between frequency of codon use in the library and abundance of their respective tRNA was observed. These results indicate that the presence of rare codons in an insert into pVIII significantly decrease the likelihood that the insert will be successfully expressed on the surface of the phage. Proteins VIII and V are the only phage proteins synthesized in very large numbers as required for their functions. All other phage proteins are expressed at relatively low levels. These facts are reflected in the lack of rare codons in these proteins. Neither pVIII nor pV have any copies of the rarest *E. coli* codons whereas there are multiple copies of these rare codons in pI and pIII. The presence of these rare codons undoubtedly contributes to the low expression levels of these proteins in host cells. Consequently, the presence of rare codons in an inserted sequence is likely to have a significant effect on the levels at which a protein is expressed.

1.7.2 Protein Insertion in the Inner Membrane

It has been demonstrated multiple times that excess positive charges at the amino-terminal end of a membrane protein reduce their transport across the membrane [146,147] due to the electrical component of the proton motive force (pmf) [148]. This predicts censorship of pIII or pVIII libraries, limiting the positive charges included in inserts at the amino terminus. In addition, it has been demonstrated that an arginine-specific restriction exists that may be due to the interaction of the translocating protein with the SecY protein. Rodi et al. [136] observed a net decrease in charge in two combinatorial peptide libraries displayed at the amino terminus of pIII and noted that this effect was due solely to a decrease in arginine residues as no under-abundance of lysine residues was observed. Peters et al. [149] constructed multiple polyarginine mutants of pIII and found a variable reduction in export of phage particles into the media. No polylysine mutants were constructed in this study. Furthermore, their attempts to rescue export by the addition of negatively charged residues (that is, glutamic and aspartic acids) were to no avail. Export rescue of the arginine-rich mutants was achieved, however, by infection of *prlA* mutants. The *prlA* phenotype has been shown to be the result of mutations within the SecY protein, the largest subunit of the SecYEG translocase complex. Specifically, *prlA* mutants possess a loosened association among the subunits, which facilitates ATP-dependent coinsertion of a portion of SecA (the ATPase subunit loosely associated with SecYEG) with the preprotein [150]. Furthermore, *prlA4* relieved translocation blockage caused by certain folded structures. It has been demonstrated that in order to be translocated, secretory proteins need to be at least partially unfolded [151,152]. These observations suggest that mutants that rescue export of the arginine-rich mutants also provide for export of bulkier, folded structures not translocated by wild-type strains.

1.7.3 Protein Processing

After insertion into the inner host cell membrane, the signal peptides of pIII and pVIII must be cleaved by signal peptidase in order for the proteins to be available for incorporation into the assembling virus particle. There is evidence for sequence preferences in the first two or three positions downstream from the cleavage site [153], a favored position for incorporation of foreign peptides or proteins. Rodi et al. [136] observed a statistically unexpected peak of proline residues at the +3 and +12 positions in a pIII dodecapeptide library and the +3 and +4 positions in a con-strained heptapeptide library. In addition, it has been reported that proline at the +1 position acts as an inhibitor of the signal peptidase enzyme, which cleaves the leader sequence from the mature protein subsequent to membrane insertion [146,154–156]. The mild censorship of proline at +1 observed in the dodecapeptide pool may be explained on the basis of this inhibition.

Furthermore, except for the first position, there is a significant overabundance of proline in combinatorial peptides displayed at the amino terminus of the mature pIII [136]. There is also a dramatic overabundance of proline over most of the length of the peptides in the pIII-displayed libraries [136]. This correlates with both the weak preference for peptides with a high propensity for β-turns in these populations and a preference of signal peptidase for β-turn conformations. Whether the observed overabundance of proline in these libraries is due to the preferred three-dimensional motif for signal peptidase substrates or to a later step in morphogenesis cannot be determined from this data alone. Malik et al. [35] found significantly reduced pro-cessing of a peptide insert, which had a high propensity for α-helix formation. All these data point to the conformational nature of the incorporated peptide being an influence on the efficiency of signal cleavage and consequent inclusion within the display library population.

1.7.4 Display in the Periplasm

E. coli has a highly effective *dsb* system for formation of disulfide bonds in the periplasm. Consequently, inserts at the amino terminus of any of the structural pro-teins may be crosslinked prior to assembly if they contain a cysteine residue. Work done by Haigh and Webster [73] has shown that, prior to incorporation into the grow-ing virion, the close proximity of pVIII molecules within the inner membrane results in a high degree of crosslinking between single cysteines in different pVIII mole-cules, thus precluding their participation in viral morphogenesis. A similar effect has been observed in pIII libraries. The almost complete absence of odd numbers of cys-teine residues in an amino-terminal pIII display library [157] resulted in significant censorship of that library. This crosslinking may be intramolecular, involving one of the other cysteines in the pIII (and potentially resulting in the misfolding of the pIII) or it may be intermolecular, resulting in dimerization, which would preclude incorporation of the pIII into the growing virus particle.

1.7.5 Viral Morphogenesis

Inserts that interfere with the protein–protein interactions that occur during viral morphogenesis have the potential to decrease or eliminate viral production. In spite of a great deal of evidence suggesting that inserts can disrupt viral assembly, the mechanism of this disruption is not well characterized for any specific case. This may involve the interaction of viral proteins as they move actively or passively through the pI–pXI–thioredoxin complex associated with the inner membrane or through the pIV pore in the outer membrane. Peptides or proteins displayed at the amino terminus of pIII or pVIII are exposed to the periplasm prior to assembly. From that position, it would seem relatively unlikely for them to disrupt protein–protein interactions involving pI or pXI. They could, however, pose a problem for phage during extrusion through the outer membrane pIV pore. Display on pIII may not be limited by the pIV pore because pIII appears to be "dragged" through the outer membrane at the tail end of the virus particle. Display on pVIII, however, could be severely limited by the pore.

Early work indicated that the display of peptides on pVIII (when present on every copy of pVIII) was limited to around six amino acids [158,159]. Iannolo et al. [34] reported that the majority of six residue inserts, 40% of eight residue inserts, 20% of 10 residue inserts, and only 1% of 16 residue inserts, were tolerated within pVIII. This seems consistent with the demonstration that a peptide inserted near the amino terminus of mature pVIII occupies a shallow groove on the surface of the phage particle and that this groove is capable of accommodating about eight residues [32]. The presumptive conclusion is that peptides that cannot be accommodated by the shallow groove will not allow passage of the particle through the outer membrane pore.

Other data, however, suggest that this may not be the correct interpretation. In hybrid display systems, it is possible to display large proteins at the amino terminus of pVIII as long as both mutant and native pVIII are present. If the extrusion of the virus through the outer membrane is so restrictive as to prevent the display of relatively short peptides on every pVIII, how can the extrusion of phage displaying a limited number of large proteins on pVIII be possible? Given how little we know about the process, we cannot exclude the possibility that the mechanics of viral extrusion will allow for export of a few large displayed proteins but exclude export of virions displaying many short peptides, somewhat akin to moving large irregular pieces of furniture through a small doorway.

Equally compatible with these data is the theory that the outer membrane pore is not so tight as to exclude virions displaying peptides of about 10 or more amino acids in length but that the exclusion of longer peptides may be due to limitations imposed by other phases of the phage life cycle. It is possible that the metabolic demand upon *E. coli* becomes limiting for long inserts if present within every copy of pVIII (see the previous discussion). It has also been pointed out that pVIII is involved in special interactions at the two ends of the virus particle [24,160]—those needed to attach pVI, pVII, and pIX to the particle—and that these interactions may have special requirements on either length or nature of the insert that cannot be readily met. Finally, the interaction of pVIII with pI during assembly [18] may not involve every pVIII. An insert that disrupts this interaction may represent a fatal mutation if every pVIII harbors it, but not in a hybrid phage with native pVIII present to perform that function.

1.7.6 Infection Process

For those viral particles that escape host cell one (that is, the cell that acquired the electroporated DNA) via successful virion assembly, the last step of growth is reinfection of F+ bacterial cells. This process is initiated by the attachment of the N2 domain of pIII to the tip of the F-pilus. Following pilus retraction, the N1 domain of pIII (to which the random peptide is attached) interacts with the third domain of the bacterial surface receptor TolA (TolA–D3). The cocrystallization of these two purified domains, as shown in Figure 1.5, delineated the interaction surfaces required for infection by M13 [53]. Given the position of the amino terminus of pIII, only a fully extended dodecapeptide would be likely to impart significant interference of the infection process. The position of both the TolA binding site and the intermolecular interface with pIII–N2 lay opposite the amino terminus, explaining why fusions of peptides and proteins to the amino terminus of pIII do not preclude phage infection [53]. A possible rationale for the large overabundance of proline at +12 is that a proline residue at that position (arrow in Figure 1.10) might direct the inserted peptide away from the binding site between TolA–D3 and N1 of pIII, minimizing loss of infectivity.

1.8 QUANTITATIVE DIVERSITY ESTIMATION

The biology of the phage–host system as reviewed above acts as a censor that inevitably limits the diversity of a phage-displayed library. Because the utility of a library is proportional to the diversity of the library, it is important to have quantitative measures of diversity to assess library quality.

As a surrogate for a true measure of diversity, the number of independent clones in a library and the inferred number of copies of each peptide/protein are sometimes quoted as a measure of library complexity (for example, Ref. [142]). Scott and Smith [56] calculated the probability of peptides being present in a library of 2.3×10^6 clones assuming Poisson statistics and equal probability of occurrence for all possible clones. Cwirla et al. [137] recognized that the apparent diversity of a peptide library will be limited by the fact that each of the 20 amino acids is coded for by different numbers of distinct codons. They further observed that roughly most amino acids occurred at most positions in their hexapeptide library, leading them to conclude that viral morphogenesis did not impose severe constraints on the diversity of their library. The effect of viral morphogenesis is, in fact, observable and significant but not severe [136]. DeGraaf et al. [138] carried out a more extensive analysis of diversity in a phage-displayed library of random decapeptides. They analyzed the sequence of 52 clones selected at random from a population of 2×10^6 individual clones and demonstrated that the frequency of amino acid occurrence in this library had a rough correlation with that expected from the number of distinct codons corresponding to each amino acid. They further identified the presence of 250 of the 400 dipeptides theoretically possible in the 52 decapeptides selected and analyzed. Because only 468 dipeptides are included in this limited population of 52 decapeptides, this observation is not significantly different from that expected from random sampling. Although each of these analyses was motivated by the need to measure the diversity, or complexity, of a peptide library, each falls short of a true quantitative measure of library diversity.

There are two possible approaches one can take to define the diversity of a population with multiple copies of individual members: "technical diversity" or completeness, corresponding to the percentage of possible members of a population that exist at any copy number within a population, and "functional diversity," which additionally takes into account the copy numbers of each distinct member of the population. In the latter scenario, if the copy numbers of the members present in the population are dramatically different, the diversity is intrinsically lower. Real phage-displayed peptide populations invariably contain unequal numbers of different peptides and any useful measure of sequence diversity must take this into account. Experiments that utilize sequence information from limited numbers of population members to estimate peptide population diversity cannot provide accurate estimates of completeness because very rare members of the population will inevitably go unsampled.

Limited sequence information, however, is capable of estimating the functional diversity of a peptide library. Makowski and Soares [161] introduced an analytical expression for the "functional" diversity of a population of peptides, not only demonstrating that this expression is consistent with intuitive expectations for the properties of population diversity, but providing a means to calculate the diversity on the basis of a limited number of peptide sequences. Whereas the diversity of a population containing equal numbers of half of all possible peptides (and no copies of the remaining half) is intuitively equal to 50% (or 0.5), the diversity of populations containing unequal numbers of different peptides is less easy to calculate. Their measure of library diversity is directly linked to the probability of selecting the same library member twice during random selection from the population. As a result, the "functional" diversity is equal for all populations in which this probability is the same.

For a population in which there is a theoretical maximum of N possible members, the diversity, d, is defined as

$$d = 1 / \left(N \sum_k p_k^2 \right)$$

where the sum is over all possible members, k, and p_k is the probability of the kth member being selected in any random selection from the population—a direct measure of the relative abundance of the peptides. Note that for any population, $\sum_k p_k = 1$.

If all possible members are present in equal numbers (equal probability of being chosen), then $p_k = 1/N$ for all k. The sum in the equation is then equal to $N(1/N^2) = (1/N)$, and $d = 1.0$ as expected. If half $(N/2)$ of the members are present in equal numbers and the other half are missing, then for half the population, $p_k = 2/N$, and for the other half, $p_k = 0$. The sum in Equation 1.1 is then $(N/2) (2/N)^2 = 2/N$, and it follows that $d = 0.5$ as one would intuitively expect.

For combinatorial peptide libraries, this diversity can be readily estimated from the frequency of occurrence of each amino acid at each position in the library [161]. Furthermore, by combining a quantitative estimation of the diversity of peptide libraries with peptide sequence pattern analysis, it is possible to measure the impact of various steps in the viral life cycle on library quality [136]. According to the diversity definition of Makowski and Soares [161], the sequence diversity of a population of peptides in which each amino acid residue is present at each position 1/20 or 5% of the time is estimated to be 1.0 (100%). However, the relative levels of each of the 20 amino acids at each position within the random peptides is not random but is determined by the numbers of corresponding codons within a 32 (or 61) codon system, not just a simple factor of 1/20. As some of the amino acids are coded by two or three codons, this gives an enhanced abundance of certain residues, such as leucine (six codons) and glycine (four codons). Let us use a genetically random dodecapeptide insert at the amino terminus of pIII as a test case for diversity quantitation and censorship source assignment. An in silico computationally constructed dodecapeptide library based upon a 32-codon code behaves statistically as though only 11.8% of all possible peptides are present. These values indicate that due to the redundant nature of the genetic code, the diversity of a random dodecapeptide library drops from 100% to 11.8%, a decrease in diversity greater than that caused by other host–virus factors [136].

The second largest quantitative effect on diversity of combinatorial peptide libraries on M13 is the almost complete absence of odd numbers of cysteine residues [141,157]. McConnell et al. [141] and Lowman and Wells [162] postulated that unpaired cysteine residues within an amino-terminal extension on pIII could form disulfide bonds with one of the four native cysteine residues in the N1 domain of pIII and partially or completely interfere with the infection process. Within 100 randomly selected peptide sequences from a dodecamer library, although one would expect to see 37.5 cysteine residues as dictated by codon frequency, only 11 residues were observed [136]. Work done by Haigh and Webster [73] has shown that pVIII molecules are close enough together within the inner membrane prior to incorporation into the growing virion that single cysteine residues have a tendency to crosslink between different pVIII molecules, precluding their participation in viral morphogenesis. A similar scenario may be possible with unpaired cysteine residues within pIII molecules in a pentavalent fusion display library, in which intermolecular disulfide bridges could interfere with either proper morphogenesis and/or infection. It is also possible that single cysteine residues may occasionally be displayed in a structural context that allows them to avoid the crosslinking activity of the *E. coli* periplasmic *dsb* system. Although cysteine is one of only 20 amino acids, 63% of all possible 12-mer peptides would be expected to have an odd number of cysteine residues. Censorship of *all* odd-numbered cysteine-containing peptides would reduce by 63% the 11.8% diversity remaining after accounting for codon bias, dropping the 11.8% further down to 4.4% of the total number possible in a random dodecapeptide pIII library.

Finally, an antiarginine bias within the first half of the random peptide sequence contributes a relatively minor additional bias (see the previous discussion). If we approximate that bias to maximal levels (that is, *no* arginine residues at +1), we lose another 0.2% in diversity and are down to 4.3% total. An assumption of 100% loss of glycine at +2 removes another 0.2%, arriving at a value of 4.2% diversity, well within the estimated 4.0% ± 1.6% number calculated in Rodi et al. [136].

The obligate steps of membrane insertion and signal peptidase cleavage result in patterns of censorship that are reflected in the statistical properties of the libraries. Quantitatively, however, it is viral morphogenesis that is the predominant biological source of sequence bias within the random dodecapeptide library. Due to the presence of odd numbers of cysteine residues in the dodecapeptide library, as many as 63% of pIII molecules are either trapped at the inner membrane unable to participate in viral assembly due to crosslinking via disulfide bonds, or are unable to undergo successful infection due to improperly folded pIII–N1 domains.

1.9 IMPROVED LIBRARY CONSTRUCTION

The quantitative analysis described in Section 1.8 indicated that, for combinatorial peptide libraries displayed on pIII, the nature of the genetic code has the greatest impact on library diversity, followed by the censorship of odd numbers of cysteines and an amino-terminal censorship of arginines. Comparison of the amino acid composition of in silico constructed human proteome-derived and codon-dictated libraries alongside actual in vivo assembled random peptide libraries demonstrates that the sequence biases seen in real-life phage libraries do not bias them either toward or away from the composition or statistical properties of real proteins as compared to computationally constructed random libraries [136]. These biases simply increase the chance of some motifs and decrease the chance of other motifs being observed. The predominant reasons for this are the methods of construction (that is, representation is strictly codon based) and the biological bottlenecks imposed primarily by the viral morphogenesis process, which necessitates translocation and assembly of the virus through two membranes and the periplasmic space of *E. coli*.

A more diverse population could be obtained by using specifically constructed trinucleotide cassettes for the synthesis of random inserts. By incorporating the trinucleotides corresponding to the most common tRNA iso-acceptors for each amino acid in the ratio of amino acid occurrences as measured within the human genome, even given the assembly-associated sequence censorship, a library of random dodecapeptides could be constructed with a functional diversity approaching 33%, an improvement of threefold over present capabilities. Insertion of a sequence that is a good substrate for signal peptidase between the signal peptidase cleavage site and the peptide library may further increase the functional diversity of the library toward a value approaching 37%. Propagation within a *prlA* host with its loosened translocase complex would also contribute to a more complex population of displayed peptides on M13 although a quantitation of this effect is difficult to estimate.

Consideration of the phage–host biology provides significant guidance for the design of improved phage-displayed libraries. Given the impact and widespread application of the existing libraries, incorporation of improvements based on an understanding of the biology of the phage–host system should substantially improve the success rates for experiments involving the use of these libraries as outlined in the remainder of this volume.

REFERENCES

1. Symmons MF, Welsh LC, Nave C, Marvin DA, Perham RN. Matching electrostatic charge between DNA and coat protein in filamentous bacteriophage. Fibre diffraction of charge-deletion mutants. *J Mol Biol* 1995; 245:86–91.
2. Marvin DA, Hohn BA. Filamentous bacterial viruses. *Bacteriol Rev* 1969; 33:172–209.
3. Model P, Russel M. Filamentous bacteriophage. In: Calendar R, ed. *The Bacteriophages*, Vol. 2. New York: Plenum Publishing, 1998:375–456.
4. Haigh NG, Webster RE. The pI and pXI assembly proteins serve separate and essential roles in filamentous phage assembly. *J Mol Biol* 1999; 293:1017–1027.
5. Fulford W, Model P. Gene *X* of bacteriophage f1 is required for phage DNA synthesis-mutagenesis of in-frame overlapping genes. *J Mol Biol* 1984; 178:137–153.
6. Terwilliger TC. Gene *V* protein dimerization and cooperativity of binding of poly (dA). *Biochemistry* 1996; 35:16652–16664.
7. Skinner MM, Zhang H, Leschnitzer DH, Guan Y, Bellamy H, Sweet RM, Gray CW et al. Structure of the gene *V* protein of bacteriophage f1 determined by multiwavelength x-ray diffraction on the selenomethionyl protein. *Proc Natl Acad Sci U S A* 1994; 91:2071–2075.
8. Prompers JJ, Folmer RHA, Nilges M, Folkers PJM, Konings RNH, Hilbers CW. Refined solution structure of the Tyr41 → His mutant of the M13 gene *V* protein: A comparison with the crystal structure. *Eur J Biochem* 1995; 232:506–514.
9. Michel B, Zinder ND. Translational repression in bacteriophage f1: Characterization of the gene *V* protein target on the gene II mRNA. *Proc Natl Acad Sci U S A* 1989; 86:4002–4006.
10. Boeke JD, Model P, Zinder ND. Effects of bacteriophage f1 gene *III* protein on the host cell membrane. *Mol Gen Genet* 1982; 186:185–192.
11. Lopez J, Webster RE. Assembly site of bacteriophage f1 corresponds to adhesion zones between the inner and outer membranes of the host cell. *J Bacteriol* 1985; 163:1270–1274.
12. Russel M, Model P. Genetic analysis of the filamentous bacteriophage packaging signal and of the proteins that interact with it. *J Virol* 1989; 63:3284–3295.
13. Holliger P, Riechmann L, Williams RL. Crystal structure of the two N-terminal domains of g3p from filamentous phage fd at 1.9 Å: Evidence for conformational lability. *J Mol Biol* 1999; 288:649–657.
14. Lubkowski J, Hennecke F, Plückthun A, Wlodawer A. Filamentous phage infection: Crystal structure of g3p in complex with its coreceptor, the C-terminal domain of TolA. *Structure* 1999; 7:711–722.
15. Russel M. Filamentous phage assembly. *Mol Microbiol* 1991; 5:1607–1613.

16. Russel M, Kazmierczak B. Analysis of the structure and subcellular location of filamentous phage pIV. *J Bacteriol* 1993; 175:3998–4007.
17. Kazmierczak BI, Mielke DM, Russel M, Model P. pIV, a filamentous phage protein that mediates phage export across the bacterial cell envelope, forms a multimer. *J Mol Biol* 1994; 238:187–198.
18. Russel M. Protein–protein interactions during filamentous phage assembly. *J Mol Biol* 1993; 231:689–697.
19. Feng J-N, Model P, Russel M. A trans-envelope protein complex needed for filamentous phage assembly and export. *Mol Microbiol* 1999; 34:745–755.
20. Marvin DA. Model-building studies of *Inovirus*: Genetic variations on a geometric theme. *Int J Biol Macromol* 1990; 12:125–138.
21. Berkowitz SA, Day LA. Mass, length, composition and structure of the filamentous bacterial virus fd. *J Mol Biol* 1976; 102:531–547.
22. Specthrie L, Bullitt E, Horiuchi K, Model P, Russel M, Makowski L. Construction of a microphage variant of filamentous bacteriophage. *J Mol Biol* 1992; 228:720–724.
23. Simons GF, Konings RN, Schoenmakers JG. Genes *VI, VII* and *IX* of phage M13 code for minor capsid proteins of the virion. *Proc Natl Acad Sci U S A* 1981; 78:4194–4198.
24. Makowski L. Terminating a macromolecular helix: Structural model for the minor proteins of bacteriophage M13. *J Mol Biol* 1992; 228:885–892.
25. Makowski L, Caspar DLD. The symmetries of filamentous phage particles. *J Mol Biol* 1981; 145:611–617.
26. Opella SJ, Stewart PL, Valentine KG. Protein structure by solid-state NMR spectroscopy. *Q Rev Biophys* 1987; 19:7–49.
27. Glucksman MJ, Bhattacharjee S, Makowski L. Three-dimensional structure of a cloning vector: X-ray diffraction studies of filamentous bacteriophage M13 at 7 Å resolution. *J Mol Biol* 1992; 226:455–470.
28. Marvin DA, Hale RD, Nave C, Citterich MH. Molecular models and structural comparisons of native and mutant class I filamentous bacteriophages Ff (fd, f1, M13), If1 and Ike. *J Mol Biol* 1994; 235:260–286.
29. Marvin DA. Filamentous phage structure, infection and assembly. *Curr Opin Struct Biol* 1998; 8:150–158.
30. Greenwood J, Hunter GJ, Perham RN. Regulation of bacteriophage length by modification of electrostatic interactions between coat protein and DNA. *J Mol Biol* 1991; 217:223–227.
31. Terry TD, Malik P, Perham RN. Accessibility of peptides displayed on filamentous virions: Susceptibility to proteases. *Biol Chem* 1997; 378:523–530.
32. Kishchenko G, Batliwala H, Makowski L. Structure of a foreign peptide displayed on the surface of bacteriophage M13. *J Mol Biol* 1994; 241:208–213.
33. Brown JH, Jardetzky TS, Gorga JC, Stern LJ, Urban RG, Strominger JL, Wiley DC. Three-dimensional structure of the human class II histocompatibility antigen HLA-DR1. *Nature (London)* 1993; 364:33–39.
34. Iannolo G, Minenkova O, Petruzzelli R, Cesareni G. Modifying filamentous phage capsid: Limits in the size of the major capsid protein. *J Mol Biol* 1995; 248:835–844.
35. Malik P, Terry TD, Gowda LR, Langara A, Petukhov SA, Symmons MF, Welsh LC et al. Role of capsid structure and membrane protein processing in determining the size and copy number of peptides displayed on the major coat protein of filamentous bacteriophage. *J Mol Biol* 1996; 260:9–21.
36. Petrenko VA, Smith GP, Gong X, Quinn T. A library of organic landscapes on filamentous bacteriophage. *Protein Eng* 1996; 9:797–801.

37. Malik P, Terry TD, Bellintani F, Perham RN. Factors limiting display of foreign peptides on the major coat protein of filamentous bacteriophage capsids and a potential role for leader peptidase. *FEBS Lett* 1998; 436:263–266.

38. Endemann H, Model P. Location of filamentous phage minor coat proteins in phage and in infected bacteria. *J Mol Biol* 1995; 250:496–506.

39. Marciano DK, Russel M, Simon SM. An aqueous channel for filamentous phage export. *Science* 1999; 284:1516–1519.

40. Fuh G, Pisabarro MT, Li Y, Quan C, Lasky LA, Sidhu SS. Analysis of PDZ domain-ligand interactions using carboxyl-terminal phage display. *J Biol Chem* 2000; 275:21486–21491.

41. Fuh G, Sidhu SS. Efficient phage display of polypeptides fused to the carboxy-terminus of the M13 gene-3 minor coat protein. *FEBS Lett* 2000; 480:231–234.

42. Roberts LM, Dunker AK. Structural changes accompanying chloroform-induced contraction of the filamentous phage fd. *Biochemistry* 1993; 32:10479–10488.

43. Oh JS, Davies DR, Lawson JD, Arnold GE, Dunker AK. Isolation of chloroform-resistant mutants of filamentous phage: Localization in models of phage structure. *J Mol Biol* 1999; 287:449–457.

44. Iannolo G, Minenkova O, Gonfloni S, Castagnoli L, Cesareni G. Construction, exploitation and evolution of a new peptide library displayed at high density by fusion to the major coat protein of filamentous phage. *Biol Chem* 1997; 378:517–521.

45. Weiss GA, Wells JA, Sidhu SS. Mutational analysis of the major coat protein of M13 identifies residues that control protein display. *Protein Sci* 2000; 9:647–654.

46. Houbiers MC, Spruijt RB, Demel RA, Hemminga MA, Wolfs CJAM. Spontaneous insertion of gene 9 minor coat protein of bacteriophage M13 in model membranes. *Biochim Biophys Acta* 2001; 1511:309–316.

47. Houbiers MC, Wolfs CJAM, Spruijt RB, Bollen YJM, Hemminga MA, Goormaghtigh E. Conformation and orientation of the gene 9 minor coat protein of bacteriophage M13 in phospholipids bilayers. *Biochim Biophys Acta* 2001; 1511:224–235.

48. Gao C, Mao S, Low C-HL, Wirsching P, Lerner RA, Janda KD. Making artificial antibodies: A format for phage display of combinatorial heterodynamic arrays. *Proc Natl Acad Sci U S A* 1999; 96:6025–6030.

49. Gao C, Mao S, Ditzel HJ, Farnaes L, Wirsching P, Lerner RA, Janda KD. A cell-penetrating peptide from a novel pVII-pIX phage-displayed random peptide library. *Bioorg Med Chem* 2002; 10:4057–4065.

50. Fransen M, Van Veldhoven PP, Subramani S. Identification of peroxisomal proteins by using M13 phage protein VI phage display: Molecular evidence that mammalian peroxisomes contain a 2,4-dienoyl-CoA reductase. *Biochem J* 1999; 340:561–568.

51. Gray CW, Brown RS, Marvin DA. Adsorption complex of filamentous fd virus. *J Mol Biol* 1981; 146:621–627.

52. Rakonjac J, Feng J-N, Model P. Filamentous phage are released from the bacterial membrane by a two-step mechanism involving a short C-terminal fragment of pIII. *J Mol Biol* 1999; 289:1253–1265.

53. Holliger P, Riechmann L. A conserved infection pathway for filamentous bacteriophages is suggested by the structure of the membrane penetration domain of the minor coat protein g3p from phage fd. *Structure* 1997; 5:265–275.

54. Lubkowski J, Hennecke F, Plückthun A, Wlodawer A. The structural basis of phage display elucidated by the crystal structure of the N-terminal domains of g3p. *Nat Struct Biol* 1998; 5:140–147.

55. Smith GP. Filamentous fusion phage: Novel expression vectors that display cloned antigens on the virion surface. *Science* 1985; 228:1315–1317.

56. Scott JK, Smith GP. Searching for peptide ligands with an epitope library. *Science* 1990; 249:386–390.

57. Cross TA, Tsang P, Opella SJ. Comparison of protein and deoxyribonucleic acid backbone structures in fd and Pf1 bacteriophages. *Biochemistry* 1983; 22:721–726.

58. Dotto GP, Enea V, Zinder ND. Functional analysis of the bacteriophage f1 intergenic region. *Virology* 1981; 114:463–473.

59. Bauer M, Smith GP. Filamentous phage morphogenetic signal sequence and orientation of DNA in the virion and gene *V* protein complex. *Virology* 1988; 167:166–175.

60. Webster RE, Grant RA, Hamilton LAW. Orientation of the DNA in the filamentous bacteriophage f1. *J Mol Biol* 1981; 152:357–374.

61. Shen C, Ikoku A, Hearst JE. Specific DNA orientation in the filamentous bacteriophage fd as probed by psoralen crosslinking and electron microscopy. *J Mol Biol* 1979; 127:163–175.

62. Webster RE. Filamentous phage biology. In: Barbas CF III, Burton DR, Scott JK, Silverman GJ, eds. *Phage Display: A Laboratory Manual.* New York: Cold Spring Harbor Laboratory Press, 2001:1.1–1.37.

63. Fulford W, Model P. Bacteriophage f1 DNA replication genes: II. The roles of gene V protein and gene II protein in complementary strand synthesis. *J Mol Biol* 1988; 203:39–48.

64. Gray CW. Three-dimensional structure of complexes of single-stranded DNA-binding proteins with DNA. IKe and fd gene 5 proteins form left-handed helices with single-stranded DNA. *J Mol Biol* 1989; 208:57–64.

65. Model P, McGill C, Mazur B, Fulford WD. The replication of bacteriophage f1: Gene *V* protein regulates the synthesis of gene *II* protein. *Cell* 1982; 29:329–335.

66. Zaman GJR, Kaan AM, Schoenmakers JGG, Konings RNH. Gene *V* protein-mediated translational regulation of the synthesis of gene *II* protein of the filamentous bacteriophage M13: A dispensable function of the filamentous-phage genome. *J Bacteriol* 1992; 174:595–600.

67. Wen J-D, Gray CW, Gray DM. SELEX selection of high-affinity oligonucleotides for bacteriophage Ff gene 5 protein. *Biochemistry* 2001; 40:9300–9310.

68. Torbet J, Gray DM, Gray CW, Marvin DA, Siegrist H. Structure of the fd DNA-gene 5 protein complex in solution—A neutron small-angle scattering study. *J Mol Biol* 1981; 146:305–320.

69. Gray CW, Kneale GG, Leonard KR, Siegrist H, Marvin DA. A nucleoprotein complex in bacteria infected with pf1 filamentous phage—Identification and electron-microscopic analysis. *Virology* 1982; 116:40–52.

70. Gray DM, Gray CW, Carlson RD. Neutron-scattering data on reconstituted complexes of fd deoxyribonucleic acid and gene *V* protein show that the deoxyribonucleic acid is near the center. *Biochemistry* 1982; 21:2702–2713.

71. Manting EH, Driessen AJM. *Escherichia coli* translocase: The unraveling of a molecular machine. *Mol Microbiol* 2000; 37:226–238.

72. Davis NG, Boeke JD, Model P. Fine structure of a membrane anchor domain. *J Mol Biol* 1985; 181:111–121.

73. Haigh NG, Webster RE. The major coat protein of filamentous bacteriophage f1 specifically pairs in the bacterial cytoplasmic membrane. *J Mol Biol* 1998; 279:19–29.

74. Samuelson JC, Chen M, Jiang F, Möller I, Wiedmann M, Kuhn A, Phillips GJ et al. YidC mediates membrane protein insertion in bacteria. *Nature* 2000; 406:637–641.

75. Nelson JC, Jiang F, Yi L, Chen M, de Gier J-W, Kuhn A, Dalbey RE. Function of YidC for the insertion of M13 procoat protein in *Escherichia coli. J Biol Chem* 2001; 276:34847–34852.

76. Eisenhawer M, Cattarinussi S, Kuhn A, Vogel H. Fluorescence resonance energy transfer shows a close helix–helix distance in the transmembrane M13 procoat protein. *Biochemistry* 2001; 40:12321–12328.

77. Yuen CTK, Davidson AR, Deber CM. Role of aromatic residues at the lipid–water interface in micelle-bound bacteriophage M13 major coat protein. *Biochemistry* 2000; 39:16155–16162.

78. McDonnell PA, Shon K, Kim Y, Opella SJ. fd coat protein structure in membrane environments. *J Mol Biol* 1993; 233:447–463.

79. Papavoine CHM, Christiaans BEC, Folmer RHA, Konings RNH, Hilbers CW. Solution structure of the M13 major coat protein in detergent micelles: A basis for a model of phage assembly involving specific residues. *J Mol Biol* 1998; 282:401–419.

80. Raposa MP, Webster RE. The filamentous phage assembly proteins require the bacterial SecA protein for correct localization to the membrane. *J Bacteriol* 1993; 175:1856–1859.

81. Guy-Caffey JK, Rapoza MP, Jolley KA, Webster RE. Membrane localization and topology of a viral assembly protein. *J Bacteriol* 1992; 174:2460–2465.

82. Horabin JL, Webster RE. An amino acid sequence which directs membrane insertion causes loss of membrane potential. *J Biol Chem* 1992; 263:11575–11583.

83. Guy-Caffey JK, Webster RE. The membrane domain of a bacteriophage assembly protein: Membrane insertion and growth inhibition. *J Biol Chem* 1993; 268:5496–5503.

84. Raposa MP, Webster RE. The products of gene *I* and the overlapping in-frame gene *XI* are required for filamentous phage assembly. *J Mol Biol* 1995; 248:627–638.

85. Brissette JL, Russel M. Secretion and membrane integration of a filamentous phage-encoded morphogenetic protein. *J Mol Biol* 1990; 211:565–580.

86. Linderoth NA, Simon MN, Russel M. The filamentous phage pIV multimer visualized by scanning transmission electron microscopy. *Science* 1997; 278:1635–1638.

87. Russel M. Macromolecular assembly and secretion across the bacterial envelope: Type II protein secretion systems. *J Mol Biol* 1998; 279:485–499.

88. Marciano DK, Russel M, Simon SM. Assembling filamentous phage occlude pIV channels. *Proc Natl Acad Sci U S A* 2001; 98:9359–9364.

89. Russel M. Mutants at conserved positions in gene *IV*, a gene required for assembly and secretion of filamentous phages. *Mol Microbiol* 1994; 14:357–369.

90. Russel M, Model P. The role of thioredoxin in filamentous phage assembly. Construction, isolation, and characterization of mutant thioredoxins. *J Biol Chem* 1986; 261:14997–15005.

91. Tabor S, Huber HE, Richardson CC. *Escherichia coli* thioredoxin confers processivity on the DNA polymerase activity of the gene 5 protein of bacteriophage T7. *J Biol Chem* 1987; 262:16212–16233.

92. Russel M. Moving through the membrane with filamentous phages. *Trends Microbiol* 1995; 3:223–228.

93. Russel M. Interchangeability of related proteins and autonomy of function. The morphogenetic proteins of filamentous phage f1 and IKe cannot replace one another. *J Mol Biol* 1992; 227:453–462.

94. Cross TA, Opella SJ. Protein structure by solid state nuclear magnetic resonance. Residues 40 to 45 of bacteriophage fd coat protein. *J Mol Biol* 1985; 182:367–381.

95. Stark W, Glucksman MJ, Makowski L. Conformation of the coat protein of filamentous bacteriophage Pf1 determined by neutron diffraction from magnetically oriented gels of specifically deuterated virions. *J Mol Biol* 1988; 199:171–182.

96. Overman S, Thomas G Jr. Raman spectroscopy of the filamentous virus Ff (fd, f1, M13): Structural interpretation for coat protein aromatics. *Biochemistry* 1995; 34:5440–5451.

97. Overman SA, Tsuboi M, Thomas GJ Jr. Subunit orientation in the filamentous virus Ff (fd, f1, M13). *J Mol Biol* 1996; 259:331–336.

98. Henry GD, Sykes BD. Assignments of the amide ^{1}H and ^{15}N NMR resonances in detergent-solubilized M13 coat protein: A model for the coat protein dimer. *Biochemistry* 1992; 31:5284–5297.

99. van de Ven FJM, van Os JWM, Aelen JMA, Wymenga SS, Remerowski ML, Konings RNH, Hilbers CW. Assignment of ^{1}H, ^{15}N, and backbone ^{13}C resonances in detergent-solubilized M13 coat protein via multinuclear multidimensional NMR: A model for the coat protein monomer. *Biochemistry* 1993; 32:8322–8328.

100. Nambudripad R, Stark W, Opella S, Makowski L. Membrane-mediated assembly of filamentous bacteriophage Pf1 coat protein. *Science* 1991; 252:1305–1308.

101. Williams KA, Glibowicka M, Li Z, Li H, Khan AM, Chen YMY, Wang J et al. Packing of coat protein amphipathic and transmembrane helices in filamentous bacteriophage M13: Role of small residues in protein oligomerization. *J Mol Biol* 1995; 252:6–14.

102. Gailus V, Rasched I. The adsorption protein of bacteriophage fd and its neighbor minor coat protein build a structural entity. *Eur J Biochem* 1994; 222:927–931.

103. Rakonjac J, Model P. The roles of pIII in filamentous phage assembly. *J Mol Biol* 1998; 282:25–41.

104. Nelson FK, Friedman SM, Smith GP. Filamentous phage DNA cloning vectors—A non-infective mutant with a nonpolar deletion in gene *III*. *Virology* 1981; 108:338–350.

105. Crissman JW, Smith GP. Gene-*III* protein of filamentous phages: Evidence for a carboxyl-terminal domain with a role in morphogenesis. *Virology* 1984; 132:445–455.

106. Makowski L, Russel M. Structure and assembly of filamentous bacteriophages. In: Chiu W, Burnnet RM, Garceia R, eds. *Structural Biology of Viruses*. UK: Oxford University Press, 1992:352–380.

107. Wickner W. Asymmetric orientation of a phage coat protein in cytoplasmic membrane of *Escherichia coli*. *Proc Natl Acad Sci U S A* 1975; 72:4749–4753.

108. Ohkawa I, Webster RE. The orientation of the major coat protein of bacteriophage f1 in the cytoplasmic membrane of *Escherichia coli*. *J Biol Chem* 1981; 256:9951–9958.

109. Paranchych W. Attachment, ejection and penetration stages of the RNA phage infectious process. In: Zinder ND, ed. *RNA Phages*. Cold Spring Harbor, NY: Cold Spring Harbor Laboratory Press, 1975:85–111.

110. Manchak J, Anthony KG, Frost LS. Mutational analysis of F-pilin reveals domains for pilus assembly, phage infection and DNA transfer. *Mol Microbiol* 2002; 43:195–205.

111. Paranchych W, Frost LS. The physiology and biochemistry of pili. *Adv Microb Physiol* 1988; 29:53–114.

112. Stengele I, Bross P, Garces X, Giray J, Rasched I. Dissection of functional domains in phage fd adsorption protein. Discrimination between attachment and penetration sites. *J Mol Biol* 1990; 212:143–149.

113. Jacobson A. Role of F-pili in the penetration of bacteriophage f1. *J Virol* 1972; 10:835–843.

114. Novotny CP, Fives-Taylor P. Retraction of F-pili. *J Bacteriol* 1974; 117:1306–1311.

115. Firth N, Ippen-Ihler K, Skurray R. Structure and function of the F factor and mechanism of conjugation. In: Neidhardt FC, ed. *Escherichia coli and Salmonella Cellular and Molecular Biology*. Washington, DC: American Society of Microbiology, 1996: 2377–2401.

116. Rondot S, Anthony KG, Dübel S, Ida N, Wiemann S, Beyreuther K, Frost LS et al. Epitopes fused to F-pilin are incorporated into functional recombinant pili. *J Mol Biol* 1998; 279:589–603.

117. Nagle de Zwaig R, Luria SE. Genetics and physiology of colicin-tolerant mutants of *Escherichia coli*. *J Bacteriol* 1967; 94:1112–1123.

118. Sun T, Webster RE. *fii*, a bacterial locus required for filamentous phage infection and its relation to colicin-tolerant *tolA* and *tolB*. *J Bacteriol* 1986; 165:107–111.

119. Sun T, Webster RE. Nucleotide sequence of a gene cluster involved in the entry of the E colicins and the single-stranded DNA of infecting filamentous phage into *Escherichia coli*. *J Bacteriol* 1987; 169:2667–2674.

120. Click EM, Webster RE. Filamentous phage infection: Required interactions with the TolA protein. *J Bacteriol* 1997; 179:6464–6471.

121. Webster RE. The *tol* gene products and the import of macromolecules into *Escherichia coli*. *Mol Microbiol* 1991; 5:1005–1011.

122. Braun V, Hermann C. Evolutionary relationship of uptake systems for biopolymers in *Escherichia coli*: Cross-complementation between the TonB-ExbB-ExbD and the TolA-TolQ-TolR proteins. *Mol Microbiol* 1993; 8:261–268.

123. Zinder ND. Resistance to colicins E3 and K induced by infection with bacteriophage f1. *Proc Natl Acad Sci U S A* 1973; 70:3160–3164.

124. Rampf B, Bross P, Vocke T, Rasched I. Release of periplasmic proteins induced in *E coli* by expression of an N-terminal proximal segment of the phage fd gene 3 protein. *FEBS Lett* 1991; 280:27–31.

125. Levengood-Freyermuth SK, Click EM, Webster RE. Role of the carboxyl-terminal domain of TolA in protein import and integrity of the outer membrane. *J Bacteriol* 1993; 175:222–228.

126. Click EM, Webster RE. The TolQRA proteins are required for membrane insertion of the major capsid protein of the filamentous phage f1 during infection. *J Bacteriol* 1998; 180:1723–1728.

127. Levengood SK, Beyer WF Jr, Webster RE. TolA: A membrane protein involved in colicin uptake contains an extended helical region. *Proc Natl Acad Sci U S A* 1991; 88:5939–5943.

128. Trenkner E, Bonhoeffer F, Gierer A. The fate of the protein component of bacteriophage fd during infection. *Biochem Biophys Res Commun* 1967; 28:932–939.

129. Smilowitz H. Bacteriophage f1 infection: Fate of the parental major coat protein. *J Virol* 1974; 13:94–99.

130. Armstrong J, Hewitt JA, Perham RN. Chemical modification of the coat protein in bacteriophage fd and orientation of the virion during assembly and disassembly. *EMBO J* 1983; 2:1641–1646.

131. Glaser-Wuttke G, Keppner J, Rasched I. Pore forming properties of the adsorption protein of filamentous phage fd. *Biochim Biophys Acta* 1989; 985:239–247.

132. White JM. Membrane fusion. *Science* 1992; 258:917–924.

133. Krebber C, Spada S, Desplancq D, Krebber A, Liming G, Plückthun A. Selectively-infective phage (SIP): A mechanistic dissection of a novel in vivo selection for protein-ligand interactions. *J Mol Biol* 1997; 268:607–618.

134. Pratt D, Tzagoloff H, Erdahl WS. Conditional lethal mutants of the small filamentous coliphage M13. I. Isolation, complementation, cell killing, time of cistron action. *Virology* 1966; 30:397–410.

135. Rodi DJ, Makowski L. Transfer RNA isoacceptor availability contributes to sequence censorship in a library of phage displayed peptides. *Proceedings of the 22nd Tanaguchi International Symposium*, Seika, Japan, November 18–21, 1996.

136. Rodi DJ, Soares A, Makowski L. Quantitative assessment of peptide sequence diversity in M13 combinatorial peptide phage display libraries. *J Mol Biol* 2002; 322(5):1039–1052.

137. Cwirla SE, Peters EA, Barrett RW, Dower WJ. Peptides on phage: A vast library of peptides for identifying ligands. *Proc Natl Acad Sci U S A* 1990; 87:6378–6382.

138. DeGraaf ME, Miceli RM, Mott JE, Fischer HD. Biochemical diversity in a phage display library of random decapeptides. *Gene* 1993; 128:13–17.

139. Marks JD, Hoogenboom HR, Bonnert TP, McCafferty J, Griffiths AD, Winter G. By-passing immunization: Human antibodies from V-gene libraries displayed on phage. *J Mol Biol* 1991; 222:581–597.

140. Christian RB, Zuckermann RN, Kerr JM, Wang L, Malcolm BA. Simplified methods for construction, assessment and rapid screening of peptide libraries in bacteriophage. *J Mol Biol* 1992; 227:711–718.

141. McConnell SJ, Uveges AJ, Fowlkes DM, Spinella DG. Construction and screening of M13 phage libraries displaying long random peptides. *Mol Divers* 1995; 1:165–176.

142. Noren KA, Noren CJ. Construction of high-complexity combinatorial phage display peptide libraries. *Methods* 2001; 23:169–178.

143. Ikemura T. Correlation between the abundance of *Escherichia coli* transfer RNAs and the occurrence of the respective codons in its protein genes: A proposal for a synonymous codon choice that is optimal for the *E coli* translational system. *J Mol Biol* 1981; 151:389–409.

144. Konigsberg W, Godson GN. Evidence for use of rare codons in the dnaG gene and other regulatory genes of *Escherichia coli*. *Proc Natl Acad Sci U S A* 1983; 80:687–691.

145. Zahn K. Overexpression of an mRNA dependent on rare codons inhibits protein synthesis and cell growth. *J Bacteriol* 1996; 178:2926–2933.

146. Yamane K, Mizushima S. Introduction of basic amino acid residues after the signal peptide inhibits protein translocation across the cytoplasmic membrane of *Escherichia coli*. *J Biol Chem* 1988; 263:19690–19696.

147. Andersson H, von Heijne G. A 30-residue-long "export initiation domain" adjacent to the signal sequence is critical for protein translocation across the inner membrane of *Escherichia coli*. *Proc Natl Acad Sci U S A* 1991; 88:9751–9754.

148. Schuenemann TA, Delgado-Nixon VM, Dalbey RE. Direct evidence that the proton motive force inhibits membrane translocation of positively charged residues within membrane proteins. *J Biol Chem* 1999; 274:6855–6864.

149. Peters EA, Schatz PJ, Johnson SS, Dower WJ. Membrane insertion defects caused by positive charges in the early mature region of protein pIII of filamentous phage fd can be corrected by prlA suppressors. *J Bacteriol* 1994; 176:4296–4305.

150. Duong F, Wickner W. The PrlA and PrlG phenotypes are caused by a loosened association among the translocase SecYEG subunits. *EMBO J* 1999; 18:3263–3270.

151. Randall LL, Hardy SJS. Correlation of competence for export with lack of tertiary structure of the mature species: A study in vivo of maltose-binding protein in *E Coli*. *Cell* 1986; 46:921–928.

152. Arkowitz RA, Joly JC, Wickner W. Translocation can drive the unfolding of a preprotein domain. *EMBO J* 1993; 12:243–253.

153. Von Heijne G. A new method for predicting signal sequence cleavage sites. *Nucleic Acids Res* 1986; 14:4683–4690.

154. Pluckthün A, Knowles JR. The consequences of stepwise deletions from the signal-processing site of beta-lactamase. *J Biol Chem* 1987; 262:3951–3957.

155. Barkocy-Gallagher GA, Cannon JG, Bassford PJ Jr. Beta-turn formation in the processing region is important for efficient maturation of *Escherichia coli* maltose-binding protein by signal peptidase I in vivo. *J Biol Chem* 1994; 269:13609–13613.

156. Nilsson I, von Heijne G. A signal peptide with a proline next to the cleavage site inhibits leader peptidase when present in a sec-independent protein. *FEBS Lett* 1992; 299:243–246.

157. Kay BK, Adey NB, He Y-S, Manfredi JP, Mataragnon AH, Fowlkes DM. An M13 library displaying 38-amino acid peptides as a source of novel sequences with affinity to selected targets. *Gene* 1993; 128:59–65.

158. Ilyichev AA, Minenkova OO, Tatkov SI, Karpyshev NN, Eroshkin AM, Petrenko VA, Sandakhchiev LS. Production of the M13 phage viable variant with a foreign peptide inserted into the coat basic-protein. *Dokl Akad Nauk USSR* 1989; 307:481–483.

159. Greenwood J, Willis AE, Perham RN. Multiple display of foreign peptides on a filamentous bacteriophage: Peptides from *Plasmodium falciparum* circumsporozoite protein as antigens. *J Mol Biol* 1991; 220:821–827.

160. Makowski L. Structural constraints on the display of foreign peptides on filamentous bacteriophages. *Gene* 1993; 128:5–11.

161. Makowski L, Soares A. Estimating the diversity of peptide populations from limited sequence data. *Bioinformatics* 2003; 19:483–489.

162. Lowman HB, Wells JA. Affinity maturation of human growth hormone by monovalent phage display. *J Mol Biol* 1993; 234:564–578.

CHAPTER 2

Vectors and Modes of Display

Valery A. Petrenko and George P. Smith

CONTENTS

2.1 INTRODUCTION

Construction of molecular chimeras from different sources has been routine in molecular biology since gene splicing began in the middle of the 1970s. For a detailed discussion of the genetics and biochemistry of molecular cloning systems, refer to the comprehensive survey of vectors edited by Rodriguez and Denhardt [1]. In 1985, recombinant DNA techniques were used to fashion a new type of chimera that underlies today's phage display technology [2]. To create one of these chimeras, a foreign coding sequence is spliced in-frame into a phage coat protein gene, so that the "guest" peptide encoded by that sequence is fused to a coat protein and thereby displayed on the exposed surface of the virion. A phage display library is an ensemble of up to about 10 billion such phage clones, each harboring a different foreign coding sequence, and therefore displaying a different guest peptide on the virion surface. The foreign coding sequence can derive from a natural source, or it can be deliberately designed and synthesized chemically. For instance, phage libraries displaying billions of random peptides can be readily constructed by splicing degenerate synthetic oligonucleotides into the coat protein gene.

Surface exposure of guest peptides underlies affinity selection, a defining aspect of phage display technology. A target binding molecule, which we will call generically the "selector," is immobilized on a solid support of some sort (e.g., on a magnetic bead or on the polystyrene surface of an enzyme-linked-immunosorbent serologic assay [ELISA] well) and exposed to a phage display library. Phage particles whose displayed peptides bind the selector are captured on the support and can remain there while all other phages are washed away. The captured phage—generally a minuscule fraction of the initial phage population—can then be eluted from the support without destroying phage infectivity and propagated or cloned by infecting fresh bacterial host cells. A single round of affinity selection is able to enrich for selector-binding clones by many orders of magnitude; a few rounds suffice to survey a library with billions or even trillions of initial clones for exceedingly rare guest peptides with particularly high affinity for the selector. After several rounds of affinity selection, individual phage clones are propagated and their ability to bind the selector confirmed.

Affinity selection represents a sort of in vitro evolution: the phage library is analogous to a natural population of "organisms"; affinity for the selector is an artificial analogue to the "fitness" that governs an individual's survival in the next generation. By introducing ongoing mutation into the "evolving" phage population, the analogy with natural selection can be made even closer.

General principles and numerous applications of phage display technology are covered in this book and summarized in a recent review [2]. Here, we will focus specifically on the development and logic of phage display vectors.

2.2 MOST DISPLAY VECTORS ARE BASED ON FILAMENTOUS PHAGES

Although useful display systems based on bacteriophages T4, T7, and λ have been introduced [3–5], the technology is most fully developed in the Ff class of filamentous phages, which includes three wild-type strains: f1, first isolated in New York City [6]; M13, isolated in Munich, Germany [7]; and fd, isolated in Heidelberg, Germany [8].

These phages are flexible, threadlike particles approximately 1 mm long and 6 nm in diameter. The bulk of their protective tubular capsid (outer shell) consists of 2700 identical subunits of the 50-residue major coat protein pVIII arranged in a helical array with a fivefold rotational axis and a coincident twofold screw axis with a pitch of 3.2 nm; the major coat protein constitutes 87% of total virion mass [9]. Each pVIII subunit is largely α-helical and rod shaped; its axis lies at a shallow angle to the long axis of the virion. About half of its 50 amino acids are exposed to the solvent, the other half being buried in the capsid (for a review of phage structure, see Ref. [10]). At the leading tip of the particle—the end that emerges first from the cell during phage assembly—the outer tube is capped with five copies each of minor coat proteins pVII and pIX (encoded by genes *VII* and *IX*); five copies each of minor coat proteins pIII and pVI (encoded by genes *III* and *VI*) cap the trailing end; it is assumed, but not proven, that the minor proteins form rings that match the fivefold rotational symmetry of the pVIII array. The capsid encloses a single-stranded DNA (ssDNA)—the viral or plus strand, whose length is 6407–6408 nucleotides in wild-type strains but is not constrained by the geometry of the helical capsid. Longer or shorter plus strands—including recombinant genomes with foreign DNA inserts—can be accommodated in a capsid whose length matches the length of the enclosed DNA by including proportionally fewer or more pVIII subunits.

The wild-type genome is very compact, consisting of 11 genes: the five mentioned coat protein genes and six genes for proteins involved in viral replication and assembly (Figure 2.1). An intergenic region of 508 bases encompasses a packaging (morphogenesis) signal sequence PS and the origins of replication for the minus and plus strands (see Figure 2.2 for more details).

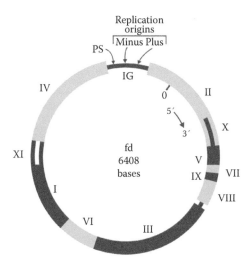

Figure 2.1 The genome of the fd bacteriophage. Genes are marked I–XI. Viral single-stranded DNA has 6408 nucleotides [11], which are numbered clockwise from the unique *Hind*II site located in gene II (represented by 0). The arrow indicates the polarity of the single-stranded viral DNA. IG, intergenic region (see diagram in Figure 2.2 for more details); PS, packaging signal.

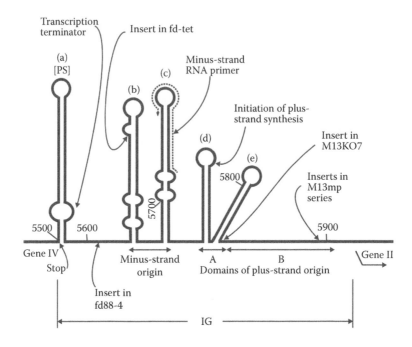

Figure 2.2 Schematic diagram of the intergenic region (IG) of the Ff genome. Position numbers are as in Figure 2.1, according to Ref. [11]. The bracketed letters indicate hypothetical hairpin loops in single-stranded DNA as postulated by van Wezenbeek et al. [12] (reviewed in Ref. [13]). The first large hairpin (a) located immediately distal to gene IV at positions 5500–5577 contains a rho-dependent transcription termination signal (discussed in Ref. [12]) and the packaging signal PS. Hairpins (b) and (c) are considered the minus-strand origin; RNA polymerase recognizes this site and synthesizes the short RNA (shown as a waved dotted line) that primes synthesis of the minus strand [14]. Hairpins (d) and (e) constitute domain A of the plus-strand origin. It contains the site at which pII (product of phage gene *II*) nicks the plus strand of double-stranded RF DNA (between positions 5780 and 5781 of f1 and M13) to initiate rolling-circle extension of the plus strand. Domain B of the plus-strand origin increases the initiation of synthesis but is not absolutely required.

The phage infects strains of *Escherichia coli* that harbor the conjugative F episome, which encodes a threadlike appendage called the F pilus. As summarized in detail in Chapter 1, the pilus mediates the infection process, which culminates in penetration of the plus strand into the cell. A complementary ssDNA strand (the minus strand) is then synthesized by host polymerases to form the double-stranded replicative form (RF), as illustrated in Figure 2.3 [15]. Minus-strand synthesis is initiated with high efficiency by host RNA polymerase at a special minus-strand origin in the intergenic region of the ssDNA (shown schematically in Figure 2.2) but can occur at low efficiency in the absence of the minus-strand origin. Rolling-circle replication of the double-stranded RF produces progeny plus strands, a process that requires the phage replication protein pII acting at a plus-strand origin that also lies

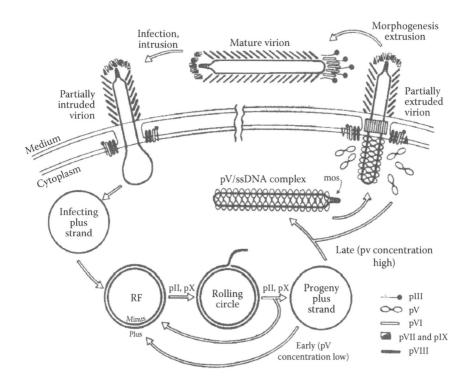

Figure 2.3 Infection cycle of filamentous phage Ff (described in Section 2.1). (Adapted from Smith GP, Filamentous phage as cloning vectors. In: Rodriquez RL, ed. *Vectors: A Survey of Molecular Cloning Vectors and Their Uses.* Boston, MA: Butterworth, 61–83, 1987.)

in the intergenic region (Figure 2.2). Early in infection, progeny plus strands serve as template for minus-strand synthesis; balanced plus- and minus-strand synthesis thus results in the accumulation of double-stranded RF molecules to a copy number of 100 or more. Late in infection and in chronically infected cells, however, the phage ssDNA-binding protein pV sequesters nearly all progeny plus strands into a filament-shaped complex. These ssDNAs are extruded through the cell envelope, concomitantly shedding pV and acquiring the virion coat proteins from the inner membrane to emerge as completed virions. Extrusion of progeny virions does not kill the cell; chronically infected cells continue to divide, albeit at a slower rate than uninfected cells; it is the slowing of cell division, not cell lysis, that explains plaque formation by these phages.

All five coat proteins are incorporated into the nascent virion from the inner membrane of the cell (for review, see Ref. [10]). Both the major coat protein pVIII and the minor coat protein pIII are synthesized with N-terminal signal peptides, which are cleaved from the mature polypeptides (50 amino acids for pVIII, 406 for pIII) as they are inserted in the inner membrane. Short hydrophobic membrane-anchoring domains span the inner membrane, separating the bulk of the protein from a short cytoplasmic C-terminal segment. The other three coat proteins (pVI, pVII, and pIX) are also membrane proteins, though they are not synthesized with signal peptides.

As shown schematically in Figure 2.4, the mature pIII protein consists of three distinct domains, D1, D2, and D3, linked by glycine-rich tetrapeptide and pentapeptide repeats L1 and L2; in addition, the C-terminal membrane-anchoring hydrophobic segment is present. There are significant interactions between D1, D2, and D3 that probably lead to a compact, stable organization of the ring of five pIII subunits at the tip of the particle [16].

Phage infection is initiated by interaction of the middle domain D2 with the tip of the pilus, as illustrated in Figure 2.5. The pilus is then hypothesized to retract, drawing the bound phage to the cellular envelope and allowing the N-terminal domain D1 to interact with periplasmic protein TolA, the coreceptor for phage infection. It is thought that the D1 domain is displaced from the D1/D2 complex during phage binding to the pilus and thus becomes available for interaction with the coreceptor [17–19] (for review, see Refs. [10] and [20]). Interdomain linker L1, separating domains D1 and D2, also participates in phage infection, probably by interacting with the pilus and assisting the D1 domain to bind to the TolA coreceptor [21].

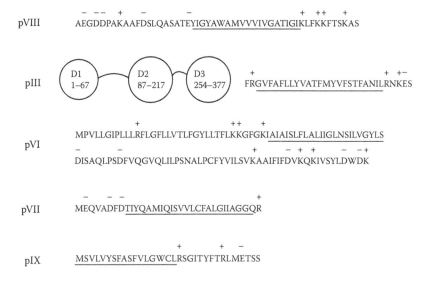

Figure 2.4 Mature forms of fd coat proteins. The N-terminus is to the left. The hydrophobic domains that span the bacterial inner membrane during phage assembly are underlined. The charged amino acids are marked by + or –. The mature pVIII, pIII, pVI, pVII, and pIX proteins have 50, 406, 112, 33, and 32 amino acids, respectively. The first 377 residues of pIII consist of three domains separated by glycine-rich tandem repeat linkers of GGGS and EGGGS; numbers within the circles indicate the residues assigned to each domain. (Reproduced from Marvin DA, *Curr Opin Struct Biol*, 8:150–158, 1998. With permission of Current Biology Ltd.)

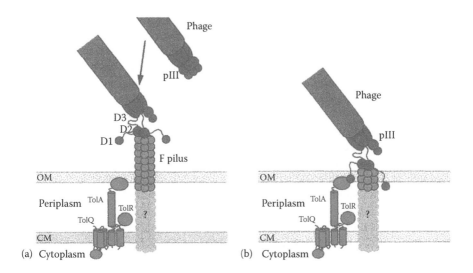

Figure 2.5   Initiation of phage infection. D1–D3 domains correspond to D1–D3 domains in the Marvin model (Figure 2.4). (a) Phage binds to the tip of an F pilus. (b) Domain D1 of pIII binds to the TolA receptor. CM, cytoplasmic (inner) membrane; OM, outer membrane. (Reprinted from *Structure*, 7, Lubkowski J, Hennecke F, Pluckthun A, Wlodawer A, Filamentous phage infection: Crystal structure of g3p in complex with its coreceptor, the C-terminal domain of TolA, 711–722, Copyright 1999, with permission from Elsevier.)

2.3 GENERAL CLONING VECTORS BASED ON FILAMENTOUS PHAGES

As DNA sequencing by the chain termination method and oligonucleotide-directed mutagenesis came to the fore in the 1970s, researchers familiar with filamentous phages soon came to realize that they have a special virtue as DNA cloning vectors: they deliver one of the two strands of vector DNA to the researcher in a very convenient form—the virion. The virions are particularly easy to purify in high yield, and the ssDNA extracted from them can serve as a template for chain termination sequencing and oligonucleotide-directed mutagenesis without interference by the complementary strand. The intergenic region readily tolerates foreign DNA inserts without interfering with viral functions; the tubular capsid, unlike spherical capsids, imposes no geometric constraint on the overall length of the viral DNA it encloses.

In 1977, Messing [22] introduced a family of filamentous phage vectors, the M13mp series, that rapidly came to dominate this field. These vectors carry an insert of *lac* DNA that spans the promoter–operator region and the first 146 codons of the *lacZ* gene, which encode an N-terminal peptide of β-galactosidase called the α-peptide. The α-peptide by itself does not exhibit enzyme activity, but it restores enzymatic activity to a defective β-galactosidase that is missing the N-terminal amino acids as a result of a deletion in the *lacZ* gene (deletion Δ*M15*). This phenomenon is known as α-complementation. Vectors in the M13mp series contain unique restriction sites within the α-peptide coding region (different vectors in the series carry different combinations of sites). When foreign inserts are spliced into these restriction sites, they generally disrupt the α-peptide and, thus, α-complementation. Presence or absence of α-complementation serves as a convenient cloning indicator. When M13mp vectors without a foreign insert are plated on medium containing the chromogenic β-galactosidase substrate X-gal, using a bacterial host whose resident *lacZ* gene carries the Δ*M15* deletion, the plaques are blue because of α-complementation, whereas the lawn of uninfected cells is colorless. In contrast, clones whose α-peptide is disrupted by a foreign insert form colorless plaques. The same research group installed this cloning indicator system in a family of high-copy-number plasmid vectors, the pUC series, which are the basis of many plasmid vectors in use today.

When the plus-strand origin of filamentous phages came to be functionally analyzed in detail, a puzzle emerged: it turns out that the *lac* insert in the M13mp series disrupts domain B of that origin (Figure 2.2), yet these vectors replicate. The paradox was resolved with the discovery that the M13mp phage harbors a compensating gene *II* mutation, which bypasses the requirement for domain B [23]. Elsewhere in the intergenic region, inserts are tolerated with no apparent effect on phage function; for example, as shown in Figure 2.2, vector fd88-4 has a large insert between the packaging signal PS and the minus-strand origin.

Many advantages of filamentous phage vectors can be combined with the convenience of very high copy number in a special kind of plasmid vector called a "phagemid" (reviewed in Refs. [24] and [25]). Phagemids are the basis of many phage display systems. These plasmids have an ampicillin (or other antibiotic) resistance gene and two origins of replication: a plasmid replication origin—usually derived from the pUC plasmid series—that allows them to replicate to an extremely high copy number in an *E. coli* host and a filamentous phage replication origin, which is inactive until the phagemid-bearing cell is infected with a filamentous phage acting as a "helper" (Figure 2.6). In a helper-infected cell harboring both phagemid and helper phage genomes, phage-encoded proteins act at the phagemid's phage origin of replication as well as at the helper's, leading to production of single-stranded phagemid DNA that can be efficiently packaged into virion particles. Therefore, two types of infective virions are secreted: particles carrying single-stranded helper phage DNA and particles carrying single-stranded phagemid DNA. Phagemids secreted as virions in these circumstances are said to have been "rescued" by the helper phage. When a helper virion infects a cell, the cell acquires the phage DNA and produces progeny helper virions—usually in sufficient yield and with sufficient infectivity to give visible plaques on agar medium. In contrast, when a phagemid virion infects a cell, the cell acquires the antibiotic resistance conferred by the phagemid (usually ampicillin resistance) but does not produce virions—until, of course, the phagemid is rescued by superinfection with additional helper phages, as illustrated in Figure 2.6.

Commonly used as a helper phage, M13K07 is a modified M13 phage with the mutant gene *II* of M13mp1, a phage origin of replication, an additional origin of replication from plasmid p15A, and a kanamycin resistance gene [26] (Figure 2.7). The kanamycin resistance gene allows cells harboring both a phagemid and a helper to be specifically selected by culturing them in medium containing both kanamycin and

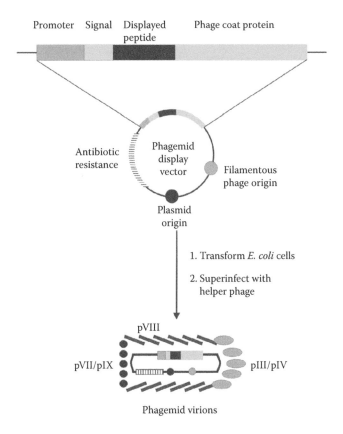

Figure 2.6   Phagemid display vector. A "typical" phagemid display vector contains *origins of replication* for double- and single-stranded DNA synthesis (plasmid and fil-amentous phage origins), an *antibiotic resistance gene* providing selection of transformed bacteria, and a *fusion gene* under the control of a regulated *promoter*. If the fusion gene derives from phage gene *VIII* or *III*, a *signal* sequence fused to the coat protein directs secretion of the coat protein and is subsequently cleaved by signal peptidase, leaving the coat protein spanning the inner membrane. A phagemid vector is converted into an infective phage by superinfection of phagemid-bearing cells with a helper phage, as described in the text.

ampicillin. Utilizing the p15A origin, M13K07 is able to replicate independently of any phage proteins. This allows it to maintain adequate levels of proteins needed for phage production and to overcome the effects of competition with the phage origin of replication on the phagemid. In M13K07, as in the M13mp phage, domain B of the phage origin is disrupted, creating a defective origin that may be less active than the wild-type origin on the phagemid. This, plus the high copy number of phage-mids, leads to preferential packaging of phagemid DNA into viral particles. When M13K07 is propagated by itself, however, the M13mp1 pII, having the compensating mutation mentioned earlier in this section, functions well enough at the defective phage origin to maintain a good yield of helper virions.

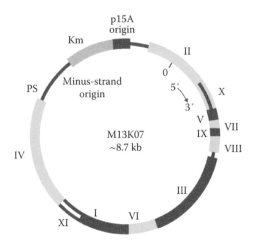

Figure 2.7 Structure of helper phage M13K07. M13K07 is an M13 derivative, which carries the mutation Met40Ile in gene *II*, the origin of replication from p15A and the kanamycin resistance gene (Km) from Tn903 both inserted at the *Ava*I site within the M13 origin of replication (position 5825, see Figure 2.2). I–XI: Genes of the M13 phage. Viral single-stranded DNA has about 8700 nucleotides [26], which are numbered clockwise from the unique *Hin*dII site located in gene II (represented by 0). IG, intergenic region (see diagram in Figure 2.2 for more details); PS, packaging signal. The arrow shows the polarity of the single-stranded viral DNA.

2.4 CLASSIFICATION OF FILAMENTOUS PHAGE DISPLAY SYSTEMS

Before cataloging filamentous phage display systems in the remainder of this review, it will be useful to introduce a nomenclature that succinctly captures some of their most essential features [27]. In type 3, 8, 6, 7, and 9 systems (generically, type *n* systems), the guest peptide is fused to every copy of the pIII, pVIII, pVI, pVII, or pIX coat protein, respectively; there is a single phage vector genome that includes the recombinant coat protein gene (*III*, *VIII*, *VI*, *VII*, or *IX*). So far, the only type *n* systems to be actually developed are types 3 and 8. Figure 2.8 illustrates types of display systems exploiting recombinant gene *VIII*.

A type 88 vector differs from a type 8 vector in that the phage vector genome harbors two genes *VIII*. One gene encodes the wild-type pVIII subunit; the other has the cloning sites and encodes the recombinant pVIII with the fused guest peptide. Type 88 virions are therefore mosaics, their capsids being composed of a mixture of recombinant and wild-type pVIII subunits (along with all the other phage coat proteins). Type 33, 66, 77, and 99 systems (generically, type *nn*) are defined analogously.

Type 8 + 8 systems differ from type 88 systems in that the two genes *VIII* are on separate genomes (Figure 2.8): the recombinant version is on a phagemid, while the wild-type version is on a helper phage (phagemid/helper systems were described in Section 2.3). Both the helper and rescued phagemid virions, like the type 88 virions described in the previous paragraph, have mosaic capsids composed of a mixture of recombinant and wild-type pVIII molecules. Type 3 + 3, 6 + 6, 7 + 7, and 9 + 9

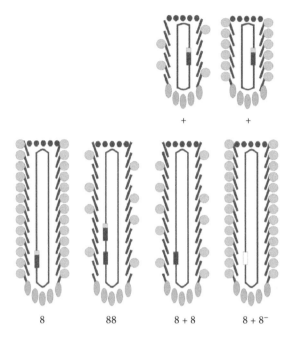

Figure 2.8 Classification of phage display vectors. Eight of the 20 theoretically possible types are discussed in the text.

are like type 8 + 8 systems, except that the phagemid carries an insert-bearing recombinant gene *III*, *VI*, *VII*, or *IX*, respectively. We will call this class of systems generically type *n* + *n*.

Type 3 + 3 phagemids must meet a special design requirement, stemming from the fact that cells making pIII molecules cannot be superinfected by filamentous phages—including helpers. For this reason, it is imperative that expression of the phagemid-borne recombinant gene *III* be controllable. Gene *III* expression is shut off as phagemid-bearing cells are being prepared for helper infection and then is turned on again so the cell can secrete mosaic virions with both types of pIII subunit.

In 8 + 8⁻ (or 3 + 3⁻) type, the helper phage has no gene *VIII* (or gene *III*), as illustrated in Figure 2.8. As a consequence, all pVIII (pIII) subunits are recombinant, exactly as in type 8 (type 3) vectors (discussed in Section 2.9).

2.5 PHAGE f1—THE FIRST PHAGE DISPLAY VECTOR

In his first report demonstrating phage display, Smith [28] engineered a type 3 "fusion phage" (as he called it then) from phage f1 using the unique *Bam*HI site in gene *III* as the cloning site. A foreign DNA fragment encoding a 57-amino acid segment of *Eco*RI endonuclease was ligated into the *Bam*HI site, and *E. coli* cells were transformed with the resulting recombinant DNA. The transfected cells released progeny phage particles that displayed the 57-residue guest peptide on their surface as a new segment within the pIII protein.

Wild-type f1 turned out to be of little practical use as a type 3 display vector, however. The fusion phage formed very small or even invisible plaques on the host strains, were 25 times less infective than the wild-type phage, and produced a deletion mutant phage lacking the foreign insert under propagation. A plausible reason for these defects became clear when the D1 and D2 domains of pIII were later defined (Section 2.1, Figure 2.4). By chance, the guest peptide had been introduced into a loop in domain D2 that is bridged by the Cys188–Cys201 disulfide, 20 amino acids away from the domain's C-terminus. It is not known if this loop is directly involved in binding the F pilus, but the fact that the antibody to the guest peptide blocks infection by the fusion phage hints that either it or neighboring amino acids may participate in the infection process. Thus, when a guest peptide is placed into the loop, it probably hampers the interaction of pIII with the pilus and makes the infection process less efficient. The inserted peptide may well influence assembly and propagation of the phage as well. Subsequent filamentous display vectors fuse the foreign peptide to either the N- or C-terminus of the coat protein and, as a rule, do not exhibit the severe defects just described.

2.6 LOW-DNA-COPY-NUMBER DISPLAY VECTORS BASED ON fd-tet

The first practical type 3 display vectors [29] were derived from a phage called fd-tet [30] by introducing unique restriction sites into gene *III*. This allowed foreign peptides to be fused to the amino terminus of the mature pIII protein, where they seemed to have much less effect on viral function than when they disrupted domain D2. Vectors derived from fd-tet were used to construct the first random peptide and antibody libraries [31–34] and many other applications (for review, see Ref. [2]). Some of the fd-tet-derived vectors are included in Table 2.1.

The DNA copy number—the number of double-stranded phage RF DNA molecules per cell—is very low in the fd-tet family. This is because a 2.8 kbp tetracycline resistance determinant disrupts the minus-strand origin (in hairpin [b] in Figure 2.2), forcing synthesis of this strand to occur through the inefficient alternative pathway. Plaques are so small that it is impractical to quantify infection in terms of plaque-forming units (pfus). However, because infection transduces the infected cell to tetracycline resistance, infectious units can be effectively quantified as transducing units (TUs) by spreading infected cells on tetracycline-containing nutrient agar. These phages can be propagated independently of infection, even in an uninfectable F⁻ host, by culturing the phage-bearing cells in medium containing the antibiotic.

In some respects, the replication defect of the fd-tet family is an advantage for phage display because it largely averts a complication called "cell killing." When phage assembly is fully or partially blocked, intracellular phage DNA and gene products accumulate to toxic levels, and the host cell is killed without releasing progeny phages [35]. Cell killing is largely averted in fd-tet because of its low RF copy number; even severe morphogenetic defects are readily tolerated [36].

Table 2.1 Type 3 Vectors

Vector	Antibiotic Resistance	Cloning Sites	References
	Low DNA Copy Number Vectors		
fAFF1	Tetracycline	*Bst*XI	[32]
fd-*Sfi*/*Not*I	Tetracycline	*Sfi*I, *Nco*I, *Xho*I, *Not*I	[37]
fd-tet-DOG1	Tetracycline	*Apa*LI, *Pst*I, *Sac*I, *Xho*I, *Not*I	[38]
fd-tetGIIID	Tetracycline	*Apa*LI, *Bbs*I, *Bgl*II, *Not*I	[39]
fdTET/Sfi/Not	Tetracycline	*Sfi*I, *Nco*I, *Xho*I, *Not*I	[37]
fUSE1	Tetracycline	*Pvu*II	[29]
fUSE2	Tetracycline	*Bgl*II	[29]
fUSE5	Tetracycline	*Sfi*I	[31]
	High DNA Copy Number Vectors		
Wild-type f1	None	*Bam*HI	[28,40]
CYT-VI	None	*Xho*I, *Xba*I	[41]
fd-CAT1		*Pst*I, *Xho*I	[33]
m655	Tetracycline	*Xho*I, *Xba*I	[42]
m666	None	*Xho*I, *Xba*I	[42]
M13KBst	Kanamycin	*Bst*XI	[43]
M13KBstX	Kanamycin		[44]
M13KE	None	*Kpn*I/*Acc*65I, *Eag*I	[45]
M13LP67	Ampicillin	*Eag*I, *Kpn*I	[46]
M13-PL6	Kanamycin	*Kpn*I, *Bst*XI	[47]
M13stufferbb	Ampicillin	*Sac*I, *Kpn*I	[48]
MAEX	Ampicillin	*Eag*I, *Xba*I, *Nar*I	[49]
MKTN	Kanamycin	*Acc*III, *Cel*II	[49]

Display vectors based on this phage therefore accommodate recombinant coat proteins that impair phage assembly or that are directly toxic in their own right.

This advantage of fd-tet-based vectors is offset by significant limitations. Their infectivity—the ratio of successfully infected cells to the input of physical particles—is only about 0.05 TU/virion, compared to about 0.5 pfu/virion for wild-type phage. The 10-fold reduction in infectivity is a result of the defect in minus-strand synthesis, which is required for conversion of the incoming viral ssDNA to double-stranded RF DNA, and thus for expression of the tetracycline resistance gene. In the interval—typically about 45 min—between infection and challenge with tetracycline, the incoming ssDNA has been successfully converted to RF in only a small proportion of the cells. The replication defect in fd-tet also results in a fourfold decrease in particle yield in comparison with the wild-type yield of approximately 2×10^{12} virions/mL. The overall yield of infectious units is therefore reduced by a factor of approximately 40-fold. There are also indications that fd-tet-derived vectors are genetically unstable when propagated in the absence of tetracycline, resulting in clones lacking the tetracycline resistance marker [50].

A number of type 3 vectors that do not have this replication defect have been developed in subsequent years.

2.7 DIVERSITY OF TYPE 3 VECTORS

Type 3 phage vectors differ chiefly with regard to the DNA copy number (low-copy-number vectors derived from fd-tet or high-copy-number vectors based on the wild-type phage or M13mp), antibiotic resistance marker, and cloning sites, as summarized in Table 2.1 and previously reviewed [2]. Including an antibiotic resistance gene allows isolation of a phage whose infectivity is too poor to permit formation of visible plaques—either because of a defect in replication (see the description of fd-tet in Section 2.6) or because the fused foreign peptide partially impairs pIII function. Such phage clones can be revealed as colonies of infected (or transfected) bacteria on agar medium containing the antibiotic. However, for display of relatively small peptides (12-mers, for example) in M13mp-derived type 3 vectors, this precaution is not necessary; even small plaques can be easily identified by their blue appearance on X-gal-containing media [45]. Unique cloning sites allow directional in-frame splicing of foreign peptides somewhere between the signal peptide and the mature sequence of the coat protein.

Guest peptides displayed on all five pIII subunits are constrained to lie very close to each other in a ring at one tip of the virion, but their attachment to the virion surface is probably quite flexible. For these reasons, it is likely that such displayed peptides can form multivalent interactions with an immobilized selector during affinity selection—both multivalent selectors like IgG antibody molecules and monovalent selectors arrayed on the solid support at high surface density. Multivalent binding leads to an "avidity effect": a vast increase in overall affinity resulting from summation of two or more monovalent interactions that are individually weak. The avidity effect is an advantage in some applications, but in others, it may undermine selection for peptide ligands with high monovalent affinity for a target receptor.

2.8 TYPE 8 VECTORS: FIRST LESSONS

Foreign peptides were displayed on pVIII soon after pIII display was introduced [51]. The first type 8 constructs were motivated by the need for polyvalent components in antiviral vaccines and immunodiagnostics; the guest peptide was thus an epitope recognized by an antibody. These constructs displayed the guest peptide on every pVIII subunit [52,53], increasing the virion's total mass by 10%. Yet, remarkably, such particles could retain their ability to infect *E. coli* and form phage progeny. Such particles were eventually given the name "landscape" phage to emphasize the dramatic change in surface architecture caused by arraying thousands of copies of the guest peptide in a dense, repeating pattern around the tubular capsid [54–58].

Not surprisingly, phages in which every pVIII subunit bears a guest peptide are defective to some degree. Some recombinant precoat proteins cannot be processed normally at the inner membrane of the *E. coli* cells [59,60], and it is likely that other types of defect are operative as well. In a few cases, this problem can be countered by including an antibiotic resistance gene in the vector so that clones can be isolated without the need for plaque formation; this was the design of the first type 8 display vector, M13B, which included a β-lactamase gene [51,52]. Still, most guest peptides—even small ones—are not tolerated when displayed on every pVIII subunit in a high-copy-number vector, and it was not until the introduction of vectors based on the replication-defective phage fd-tet (Section 2.5) that large landscape libraries, with 10^9 or more clones, could be successfully constructed [54].

Foreign peptides in landscape phages can subtend as much as 25–30% of the virion surface, dramatically changing the particle's surface architecture and properties. Depending upon the particular foreign peptide sequence, the landscape phage can bind organic ligands, proteins, antibodies, or cell receptors [54–58,61,62]; interact with proteases [63]; induce specific immune responses in animals [52,64–66]; resist stress factors such as chloroform or high temperature [54]; or migrate differently in an electrophoretic gel (Petrenko, unpublished). The avidity effect can be even more pronounced in this display system than in the pIII display systems discussed in Section 2.7.

Table 2.2 summarizes type 8 display vectors. The most advanced of them—f8-5 and f8-6—allow for insertion of a foreign peptide into any exposed site in pVIII using unique restriction sites *Pst*I, *Bam*HI, *Nhe*I, and *Mlu*I [57]. Vector f8-6 features two TAG amber stop codons between the cloning sites, which prevent contamination of

Table 2.2 Type 8 Vectors

Name	Parent Phage	DNA Copy Number	Antibiotic Resistance	Applications	References
f8-1	fd-tet	Low	Tetracycline	Billion-clone 8-mer peptide library	[54]
f8-5	fd-tet	Low	Tetracycline	Hundred million-clone α-helical peptide library	[57]
fdAMPLAY8	fd	High	Ampicillin	Cloning of peptides	[60]
fdH	fd	High	None	Cloning of 4- and 6-mer peptides	[53]
fdISPLAY	fd	High	None	Cloning of peptides	[67]
M13B	M13mp10	High	Ampicillin	Cloning of 5-mer peptides	[51,52]
PM48	f1	High	None	Ten million-clone 8-mer peptide library; small 9-mer library	[55,68]
PM54	fd-tet	Low	Tetracycline	Small 6–16-mer peptide libraries	[68]
PM52	fd-tet	Low	Tetracycline	Small 6–16-mer peptide libraries	[68]

(Continued)

Table 2.2 (Continued)   Type 8 Vectors

Sequences of the cloning sites

```
        ... -1   1   2   3   4   5   6   7   8   9  10  11  12  13  14  15  16  17  18  19  20  21  22  23  24  25  26
        ...  A   A   E   G   E   D   P   A   K   A   A   F   D   S   L   Q   A   S   A   T   E   Y   I   G   Y   A   W
fd-tet  ...GCTGCTGAGGGTGACGATCCCGCAAAAGCGGCCTTTGACTCCCTGCAAGCCTCAGCGACCGAATATATCGGTTATGCGTG
```

```
f8-5   ...GCTGCAGAGGGTGAGGATCCCGCAAAAGCGGCCTTTGACTCCCTGCAAGCTAGCGCGACCGAATATATCGGTTACGCGTG...
            PstI        BamHI                                Nhe1                        MluI
```

```
f8-6   ...GCTGCAGAGTAGTAGGATCCCGCAAAAGCGGCCTTTGACTCCCTGCAAGCTAGCGCGACCGAATATATCGGTTACGCGTG...
            PstI        BamHI                                Nhe1                        MluI
```

```
f8-1, PM52
       ...GCTGCAGAGGGTGAGGATCCCGCAAAAGCGG
            PstI        BamHI
```

```
M13B   ...GCTGCTGAGGGTGAGGATCCCGCAAAAGCGG...
                        BamHI
```

```
            V*  N*
fdH    ...GCTGCTGAGGTTAACGATCCCGCAAAAGCGG...
                     HpaI
```

```
fdISPLAY, fdAMPLAY8
       ...GCCGCGGAGGGTGACGATCCCGCCAAGGCGG...
            SacII                   StyI
```

```
PM45, PM54
   ... -1   1   2   3   4       5   6   7   8   9  10  11
   ...  A   A   E   G   E   F*  N*  D   P   A   K   A   A   F
       ...GCTGCTGAGGGTGAATTCTAGGATCCCGCAAAAGCGGCCTTT
               EcoRI       BamHI
```

Note: Amber TAG stop codons in the vectors PM48 and f8-6 are underlined; the signal peptidase cleaves the precoat proteins between A-1 and A1.

recombinant phages with the wild-type vector phages. The vector is propagated in an amber-suppressing *E. coli* strain, while the progeny recombinant phage clones are grown in a nonsuppressor strain, blocking production of noninsert-bearing vector phages.

Even in a replication-defective vector, there is a stringent limit to the size and composition of peptides that can be tolerated on every pVIII subunit [51,53,68,69]. This is not because of any blanket restriction on incorporation of large recombinant pVIII subunits into the virion, however. Using type 88 and 8 + 8 vector systems, guest peptides spanning hundreds of amino acids can be displayed on chimeric particles that include wild-type as well as recombinant pVIII subunits (exemplified in our review [2]), as will be discussed in Sections 2.8 and 2.9. Rather, there seems to be some particular stage of virion assembly that cannot accommodate large guest peptides. Initiation is an obvious candidate for this hypothetical special stage: perhaps large recombinant subunits cannot form functionally active complexes with the minor coat proteins pVII/pIX during initiation of phage assembly [70]. The "ban" against inserting longer guest peptides into every pVIII subunit can sometimes be overcome by randomizing pVIII amino acids not involved in the phage assembly [54,57,71] (see Section 2.11).

2.9 MOSAIC DISPLAY IN TYPE *nn* SYSTEMS

Mosaic display using type 33 and 88 systems overcomes two potential disadvantages of type 3 and type 8 systems. First, much larger, more complex guest peptides can be successfully displayed on pIII or pVIII when wild-type versions of the coat proteins are also available. This is especially so of pVIII display: type 8 vectors cannot accommodate guest peptides longer than about 10 amino acids, whereas type 88 vectors can display foreign peptides containing hundreds of amino acids. Second, the avidity effect is lessened because fewer copies of the guest peptides are displayed on the virion surface.

The phage genes are tightly packed into two transcriptional domains, as illustrated in Figure 2.1. There are accordingly only two noncoding segments of the genome available to accommodate extra genes. The first noncoding area is the intergenic region, which lies between genes *IV* and *II* and which contains the morphogenesis (packaging) signal PS and the plus and minus origins of DNA replication (Figure 2.2). A large tetracycline resistance determinant interrupts the minus-strand origin in fd-tet (Section 2.5), while a large *lac* insert disrupts the plus-strand origin in M13mp vectors (Section 2.6). Large inserts can also be accommodated between the packaging signal PS and the minus origin of replication, apparently without compromising any phage functions. The second noncoding segment lies between genes *VIII* and *III* and has also been used to accommodate extra genes [72,73].

The type 88 vectors of the MB series [74] contain a synthetic recombinant gene *VIII* inserted into the intergenic region of M13mp18 (replacing the *lacZ'* gene and multiple cloning site). The recombinant gene comprises a strong *lac* (or *tac*) promoter, a symmetrical *lac* operator, the signal peptide coding sequence from *phoA*, cloning sites for the insertion of a guest peptide coding sequence, a synthetic coding sequence for the mature form of pVIII, and a strong transcriptional terminator. To minimize recombination between the recombinant and wild-type *VIII* genes, the sequence of the recombinant gene is designed to be very different from the wild-type *VIII* gene, while specifying the same amino acid sequence. MB virions are mosaics with both wild-type and guest peptide-bearing recombinant pVIII subunits, the proportion of which can be regulated by inducing the recombinant gene *VIII*. In contrast to type 8 vectors, which can accommodate only relatively short peptides, functional proteins up to approximately 20 kDa have been displayed on the MB phages as fusions to the recombinant pVIII protein [75].

Similarly, the JC-M13-88 display vector [76] was constructed by modifying M13mp18. The α-peptide encoding region of the *lacZ'* gene was replaced with the ompA leader and a synthetic gene *VIII* cassette (from vector f88-4; see the next paragraph), which are expressed from a *tac* promoter as fusion with foreign peptides. Under induction of the *tac* promoter with Isopropyl β-D-1-thiogalactopyranoside (IPTG), the phage gains an average of 1–4 guest peptide-bearing pVIII subunits per virion, 50% of phage particles being without guest peptides. Other examples of M13-derived type 88 vectors are listed in Table 2.3 (see also more comprehensive catalog of vectors in our review [2]).

Table 2.3 Type *nn* Vectors

Vector	Type	Antibiotic Resistance	Cloning Sites	References
		Low DNA Copy Number Vectors		
f88-4	88	Tetracycline	*Hind*III, *Pst*I	[77]
fth1	88	Tetracycline	*Sfi*I	[73]
		High DNA Copy Number Vectors		
fd88-4	88	None	*Hind*III, *Pst*I	[78]
fdAMPLAY88	88	Ampicillin	*Sac*II, *Sty*I	[79]
JC-M13-88	88	None	*Xba*I, *Hind*III	[76]
MB	88	None	a	[74]
M13IXL604	88	None	*Nco*I, *Xba*I, *Xho*I, *Spe*I	[80,81]
M13702, M13MK100	33	None	b	[82]
pM13Tsn::III	33	None	a	[83]
pM13Tsn::VIII	88	None	a	[83]
	33			[84]

a These vectors were used for display of unique products and have no universal cloning sites.
b Data are not available.

The type 88 vector f88-4 is derived from the low-copy-number phage fd-tet [77]. In addition to the wild-type gene *VIII*, it harbors an artificial gene *VIII* in the minus origin, next to the tetracycline resistance gene (Figure 2.9). This gene has an IPTG-regulated *tac* promoter and unique *Pst*I and *Hind*III sites for directional cloning of foreign DNA and expression as a fusion between the signal peptide and the mature pVIII. The vector has been extensively applied for preparation of numerous random and natural peptide libraries [77,85–87].

Guest peptides are likely to be very stable in low-copy-number type 88 vectors. To investigate this supposition, an f88-4 clone displaying the FLAG epitope (DYKDDDDKL [88]) was subjected to six rounds of propagation without selection; then about 50 individual subclones were tested for retention of the FLAG epitope. In two independent repetitions of the experiment, all subclones retained the epitope. In contrast, when the same experiment was carried out with the high-copy-number vector fd88-4, which carries the recombinant *VIII* expression cassette between the packaging signal PS and the minus origin of replication (Figure 2.2), only 8% and 41% of the subclones retained the epitope in the two independent repetitions. The observed rate of epitope loss in the high-copy-number vector is undoubtedly low enough to be overcome by selection—even very weak selection. But the rate of loss of large guest peptides may well be so high in high-copy-number type 88 vectors as to mandate use of a low copy number alternative like f88-4.

Similarly, type 33 vectors [82–84] bear two genes *III*: one full-length and one truncated (amino acids 198–408). The former expresses a functional pIII that supports infectivity of the chimeric phage, while the second gene produces a fusion protein that can be built into the capsid during phage assembly. The type 33 vectors

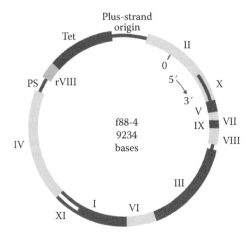

Figure 2.9   Vector f88-4. The type 88 vector is derived by splicing a 379 bp semisynthetic major coat protein gene into filamentous phage vector fd-tet. The resulting phage has two major coat protein genes: the wild-type gene *VIII* and the artificial, recombinant *VIII*. The artificial gene is driven by a *tac* promoter and is therefore inducible by IPTG; under fully induced conditions (1 mM IPTG), roughly 300 of the 3900 coat protein subunits stem from the recombinant gene *VIII*, the remainder stemming from the wild-type gene *VIII*. The recombinant gene *VIII* has unique *Hind*III and *Pst*I cloning sites separated by a 21 bp "stuffer." Replacement of the stuffer with an appropriate synthetic or natural DNA insert can result in display of up to 300 copies of a peptide encoded by the insert on the virion surface, fused to the recombinant gene *VIII*-derived coat protein subunits. Genes are marked I–XI. Viral single-stranded DNA has 9234 nucleotides, which are numbered clockwise from the unique *Hind*II site located in gene II (represented by 0). The arrow shows the polarity of the single-stranded viral DNA. IG, intergenic region (see diagram in Figure 2.2 for more details); PS, packaging signal.

may be genetically unstable, unlike type 88 vectors that use different nucleotide sequences to encode the recombinant and wild-type pVIII polypeptides. There are no reports of type 66, 77, and 99 vectors.

2.10 MOSAIC DISPLAY IN PHAGEMID SYSTEMS

A "typical" phagemid display vector shown in Figure 2.6 contains both phage and plasmid origins of replication, an *antibiotic resistance gene* providing selection of transformed bacteria, and a *fusion gene* under the control of a regulated *promoter*. The *signal* peptide fused to the coat protein directs secretion of the coat protein and is subsequently cleaved by signal peptidase, leaving the coat protein spanning the inner membrane. From this position, the protein is then incorporated into assembled virions. Some specialized phagemid display vectors are supplied with additional features—for targeting eukaryotic cells and gene transfer, among other things. The phagemid pEGFP-lacp8, for example, contains green fluorescence protein and *neo* markers that can be used for selection of transduced eukaryotic cells. It also bears the SV40 origin of replication, which allows amplification of plasmid DNA in mammalian cells expressing the SV40 large T antigen [58].

A major advantage of phagemid vectors is that they can be propagated as a plasmid, when the recombinant coat gene is silent and there is no selection pressure to remove inserted DNA. This contrasts to fusion phages, where faster proliferating deletion mutants could quickly dominate. Furthermore, when combining various phagemid and helper partners, this system allows for control of display multiplicity from monovalent to multivalent and can achieve thousands of fusion proteins per virion [58]. Monovalent display is used for the generation of protein phage display libraries and selection of the most prominent binders using affinity selection [89,90] (reviewed in Chapter 4). Polyvalent display, in contrast, allows for selection of a broad spectrum of binding candidates. An advanced selection strategy combines the advantages of both monovalent and polyvalent display systems, first using polyvalent display for selection of primary lead candidates, followed by monovalent display of the leads to identify the most avid clones [36,91,92] (reviewed in Chapter 4).

Monovalent display exploits pIII [37] or its C-terminal domain (amino acids 198–406 or 230–406 [93]) for fusion of displayed peptides, and provides controlled expression of the pIII fusion, which can be as low as one copy per 100 phage particles [90]. Since N-terminal domains of pIII expressed in *E. coli* cells hinder infection of the cells with phages [94], deletion of these domains allows for superinfection of the cells with the helper phage. The multiplicity of display can be controlled through several different means: (1) using inducible promoters [95]; (2) switching between pVIII (8 + 8) and pIII (3 + 3) display systems [69]; (3) mutating the recombinant coat protein [96]; or (4) using "hyper" helper phages with gene *III* or *VIII* deleted. The latter type of display may be called 3 + 3$^-$ or 8 + 8$^-$, emphasizing that the second gene *III* or *VIII* coming from a helper phage is missing (illustrated in Figure 2.8) [58,97–99]. If the helper phage is lacking gene *VIII*, the phagemid DNA is coated in the homogenous landscape format with the recombinant pVIII [58], exactly like type 8 vectors. The defective helper phages can be produced in cells that harbor the plasmids containing gene *III* or *VIII* under control of the *tac* or phage shock protein (*psp*) promoters [58,100]. The salient feature of the *psp* promoter is that it is activated in the presence of bacteriophage protein IV and is therefore silent until the moment of infection.

Minor coat proteins pVII and pIX have been used in phagemid format for display of heavy and light antibody chains [101]. Fusion proteins were expressed from the phagemid as procoats with *ompA* and *pelB* leaders, unlike the native pVII and pIX that are synthesized without leaders and require a processing step. Since these two proteins appear to interact with one another in the phage capsid, they may be ideal for the display of dimeric proteins such as antibodies or integrins. Later, aspects of this technology were invoked and extended to construct a large, human singlechain Fv (scFv) antibody library displayed on pIX [102]. The type 6 + 6 format that allows C-terminal display is described separately in Section 2.10. Some examples of phagemid display systems are presented in Table 2.4.

Table 2.4   Phagemid Display Systems

Name	Recombinant Phage Gene	Antibiotic Resistance	Cloning Sites	References
PC89	*VIII*	None	*Eco*RI, *Bam*HI	[69]
pCANTAB 5E	III	Ampicillin	*Sfi*I, *Not*I	Pharmacia
pCES1	*III*	Ampicillin	*Apa*LI, *Not*I	[103]
pCGMT-1b	*VII, IX*		*Sac*I, *Nco*I	[101]
pComb3	*D3-III*	Ampicillin	*Xho*I, *Spe*I	[104]
pComb3H, pComb3X	*D3-III*	Ampicillin	*Sfi*I	[93]
pComb3-M3	*D3-III*	Ampicillin	Multiple	[105]
pComb8	*VIII*	Ampicillin		[106]
pDONG6	*VI*	Ampicillin	*Sfi*I, *Not*I, *Bam*HI	[107]
pEGFP-lacp8	*VIII*	Kanamycin	*Bpu*AI, *Bam*HI	[58]
pETT7GSTgp8	*VIII*	Ampicillin	*Sac*II, *Stu*I, *Xho*I, *Not*I	[108]
pEXmide 3	*III*	Ampicillin	Multiple	[109]
pGEM-gIII	*D3-III*	Ampicillin	*Sac*I, *Xho*I, *Nhe*I	[110]
pGP-F100	*III*	Ampicillin	*Sfi*I	[111]
pGZ1	*III*	Ampicillin	*Not*I, *Sfi*I	[112]
phGH-M13gIII	*D3-III*	Ampicillin	a	[89]
pHEN1	*III*	Ampicillin	*Sfi*I, *Not*I	[38]
pSEX	*III*	Ampicillin, chloramphenicol	*Bam*HI, *Pst*I	[113]
pSEX40	*III*	Ampicillin	Multiple	[114]

[a] These vectors were used for display of unique products and have no universal cloning sites.

2.11   VECTORS FOR C-TERMINAL DISPLAY

Since the commonly used display systems utilize fusion of foreign peptides to the N-terminus of pIII or pVIII, they are unsuitable for surface expression of full-length cDNA bearing stop codons. However, another coat protein, pVI, can tolerate C-terminal fusion to foreign peptides without abolishing phage viability [107]. Thus, cDNAs can be ligated to the 3′ end of gene *VI* using phagemid vector pDONG6, allowing fusion of foreign coding sequences to gene *VI* in all three possible frames.

C-terminally fused peptides have also been displayed indirectly on pIII via a leucine zipper "fastener" [115,116], as illustrated in Figure 2.10. The phagemid bears two recombinant genes: the coding sequence for the Jun half of the AP-1 zipper fused N-terminally to phage gene *III* and the coding sequence for the Fos half of the zipper fused N-terminally to a cDNA library. Both fusion proteins are preceded by signal peptides and expressed from IPTG-inducible *lac* promoters. When cells harboring the phagemid are superinfected with helper phages and induced with IPTG, the two proteins are secreted into the periplasm without their signal peptides, where the two "half-zippers" join prior to, or concomitantly with, incorporation into the secreted virions.

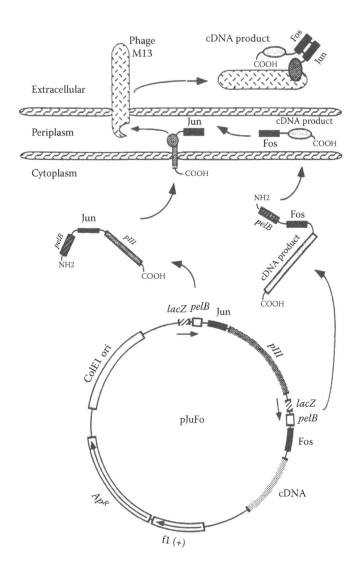

Figure 2.10   pJuFo phagemid—a vector for C-terminal display. The diagram illustrates the proposed pathway for display of cDNA products on the phage surface explained in
the text. (Reprinted from *Gene*, 137, Crameri R, Suter M. Display of biologically
active proteins on the surface of filamentous phages: A cDNA cloning system for
selection of functional gene products linked to the genetic information responsible
for their production, 69–75, Copyright 1993, with permission from Elsevier.)

Recently, a new approach was developed allowing display of foreign proteins on
the C-terminus of an artificial coat protein [71]. Normally, the negatively charged
N-terminus of pVIII is exposed on the surface of the phage while the C-terminus
forms a positively charged lumen. Using a combination of rational design and extensive mutagenesis, Sidhu and colleagues [71,117] developed an artificial pVIII with

a reversed sequence of functionally important amino acids, which exposes its nega-tively charged C-terminus on the surface of the capsid. It is unlikely that the protein itself supports phage assembly, but it can be introduced into a phage capsid with low probability during phage morphogenesis supported by an excess of the normal pVIII [117].

2.12 PHAGE PROTEINS AS CONSTRAINING SCAFFOLDS

It is commonly accepted that the binding affinities of peptides may be higher when they are displayed as segments of folded proteins, called "scaffolds." A scaf-fold displays foreign peptides in a specific constrained conformation, which allows the peptide to bind a receptor without paying an entropy penalty. For example, the natural immunoglobulin scaffold in antibodies brings six antigen-binding peptides together and helps them to collaborate in binding of an antigen. Such scaffolds, along with fused random peptides, can be displayed on the phage surface and used for selection of binding entities. At the same time, a phage itself may serve as a scaffold for the constrained presentation of foreign peptides if these peptides replace exposed segments of the coat proteins without disturbing their general architecture and phage viability. The minor coat protein pIII and the major coat protein pVIII can be con-sidered as primary candidate scaffolds because their structures and functions have been studied in great detail. However, pIII was used only for N-terminal presentation of peptides and, in its truncated form, for display of functional domains of foreign proteins in their intrinsic conformation, as described in Section 2.6. Study of the pVIII protein as a constraining scaffold is more advanced.

The conformation of major coat protein pVIII in a phage capsid has been studied using physical methods and confirmed by mutagenesis (reviewed in Ref. [118]). It consists of four structural domains A–D, as shown below.

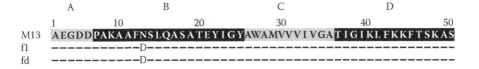

The N-terminal mobile surface domain A (Ala-1 to Asp-5) has a nonhelical, pos-sibly disordered conformation [119]. Domain B is an amphipathic, gradually curving α-helix, extending from Pro-6 to about Tyr-24 [120]. Domain C, a highly hydro-phobic α-helix extending from Ala-25 to Ala-35, is entirely buried in the interior of the protein coat. The remainder of protein D, from Thr-36 to Ser-50, constitutes an amphipathic α-helix forming the inside wall of the protein coat. Four basic residues near the C-terminus interact with the viral DNA. The N-terminus of pVIII is exposed on the outer surface, and its C-terminus lies in the lumen [10]. Neighboring pVIII subunits make numerous intersubunit contacts that impart remarkable physical sta-bility to the structure. The length of the capsid is directly proportional to the length of the enclosed viral DNA and inversely proportional to the number of positively charged residues at the luminal (C-terminal) end of the pVIII polypeptide [121]. According to the α-helical model of segment B, 11 residues (Lys-8, Asp-12, Ser-13, Gln-15, Ala-16, Ser-17, Thr-19, Glu-20, Tyr-21, Gly-23, and Tyr-24) comprise a polar area exposed on the surface of the phage, while the other residues constitute a non-polar region that interacts with the phage body [118,122,123].

In type 8 systems, where only one kind of pVIII is available, only minor pVIII sequence variants can be tolerated without abolishing phage assembly or infectivity, including point mutations [123], deletion of a few N-terminal amino acids [57], or insertion of very short peptides into the N-terminus [51,53–55,63,68,69,124]. Much more drastic changes in pVIII can be tolerated in type 8 + 8 systems, however. In particular, it is possible to incorporate a few copies of recombinant pVIII bearing a large foreign domain with hundreds of amino acids into a capsid that is mainly composed of wild-type pVIII subunits. In fact, some mutations in the recombinant gene *VIII* even increase incorporation efficiency, enabling the development of improved phage display platforms [96,125].

In order to explore pVIII as a scaffold, two domains, A and B, were replaced by foreign peptides using type 8 vectors, described in Section 2.7. Peptides fused to N-terminal domain A can probably adopt many different conformations depending on their sequence, while peptides loaded into domain B conform to the α-helical architecture of the flanking wild-type residues [57]. In this way, it is possible to generate very large libraries of surface-accessible α-helical ligands that can serve as a source of potential drugs, vaccines, and substitute antibodies [57].

2.13 CONCLUSION

Table 2.5 summarizes salient features of the display systems described in this chapter. For type *n* and type *nn* vectors, DNA copy number is an important parameter. High-DNA-copy-number vectors are convenient because of their high infectivity and phage production. However, inserts are much less stable in these vectors, and they may not be suitable for long inserts such as antibody fragments. Insert stability has not been systematically investigated in *n* + *n* phagemid systems, but since phagemids can have very high DNA copy numbers, insert instability is likely to be a concern in these systems too.

In many display systems, there is no sharp limit on the length of the displayed peptide. However, type 8 vectors will only accommodate very short peptides—especially in high-DNA-copy-number systems. In type 3 systems, long inserts tend to impair infectivity. In all display systems, the number of peptides displayed intact on the virion surface tends to be much lower for long peptides than for short ones. The poorer display density of long peptides may be due to less efficient incorporation or to more efficient degradation by the potent cytosolic and periplasmic proteases that eliminate malfolded proteins in *E. coli*.

Some display systems are effectively "monovalent," in that the number of peptides displayed per virion is one or even much less. Low display density can be the

Table 2.5   Summary of Display Systems

Type	DNA Copy Number	Length of Peptides Accepted	Number of Peptides per Virion	Position of Peptide Fusion	Special Considerations
3	High	Unknown	5	N-terminus	Long peptides may be genetically unstable
3	Low	No limit	5	N-terminus	
8	High	Very short	> 2700	N-terminus	No cysteines [54]
8	Low	~10	> 2700	N-terminus	No cysteines [54]; highly constrained α-helical scaffold
88	High	Unknown	~300	N-terminus	Long peptides may be genetically unstable
88	Low	No limit	~300	N-terminus	
33	Low	No limit	< 1	N-terminus	
33	High		< 1	N-terminus	
3 + 3		No limit	< 1	N-terminus	Often used for monovalent display; C-terminal fusion possible through leucine zipper fastener (Section 2.10)
8 + 8		No limit	100–1000	N-terminus	
8 + 8		Unknown	< 1	C-terminus	See Section 2.10
6 + 6		No limit	< 1	N-terminus	
6 + 6		No limit	< 1	C-terminus	See Section 2.10
9 + 9		No limit	< 1	N-terminus	

result of poor incorporation or high degradation (e.g., in C-terminal type 6 + 6 or 8 + 8 systems), or purposely fostered by providing wild-type subunits to compete with peptide-bearing subunits for incorporation into virions (as in type 3 + 3 systems). Monovalent display is an important advantage when it is desired to select for high monovalent affinity. Multivalent display undermines selection for this property, by greatly increasing avidity (effective affinity) to the point where weak and strong ligands cannot be distinguished. In other applications, however, multivalency is a distinct advantage. Such is the case when it is desirable to accumulate a broad spectrum of peptides as potential leads; the identified peptides can be ordered subsequently in accordance with their affinity and selectivity using monovalent display. Polyvalent display vectors are particularly effective for targeting multiple receptors on cell surfaces, especially for triggering receptor-mediated endocytosis [58,61,126,127]. "Landscape" phages using type 8 vectors are an extreme case of multivalency, in which thousands of copies of the displayed peptide are arranged in a regular array on the virion surface. This ultrahigh-density display refashions a substantial fraction of the surface chemical architecture and can lead to "emergent" properties of the virion as a whole that would be hard to foresee from the properties of the individual peptides. Such virions can be looked on as a new, biologically selectable type of "nanofiber."

In preparing very large peptide or antibody phage libraries, phagemid vectors may be preferred because their DNA copy number is much higher than that of phages.

This allows easy purification of large amounts of high-purity double-stranded vector DNA for library construction. On the other hand, the necessity for superinfection with helper phages introduces a significant complication into their use—a complication that is encountered many thousands of times for a given library, not just once at the stage of its construction. Expression of the fusion genes in phagemid systems is also reported to be less efficient than in phage vectors [37].

Phage display is constantly being applied to new experimental and practical problems, motivating the engineering of vectors with new unique characteristics. For example, development of advanced gene-delivery systems would require vectors of small size with enhanced ability for transfection of eukaryotic cells [58,128]. Design of specific probes for detection would require environmentally stable vectors that can self-assemble to generate bioselective layers on biosensors [129]. Other applications of phage display technology will demand even more specialized new vector systems.

REFERENCES

1. Rodriguez RL, Denhardt DT, eds. *Vectors: A Survey of Molecular Cloning Vectors and Their Uses.* Stoneham, MA: Butterworth Publishers, 1987.
2. Smith GP, Petrenko VA. Phage display. *Chem Rev* 1997; 410:97–391.
3. Ansuini H, Cicchini C, Nicosia A, Tripodi M, Cortese R, Luzzago A. Biotin-tagged cDNA expression libraries displayed on lambda phage: A new tool for the selection of natural protein ligands. *Nucleic Acids Res* 2002; 30:e78.
4. Houshmand H, Froman G, Magnusson G. Use of bacteriophage T7 displayed peptides for determination of monoclonal antibody specificity and biosensor analysis of the binding reaction. *Anal Biochem* 1999; 268:363–370.
5. Malys N, Chang DY, Baumann RG, Xie D, Black LW. A bipartite bacteriophage T4 SOC and HOC randomized peptide display library: Detection and analysis of phage T4 terminase (gp17) and late sigma factor (gp55) interaction. *J Mol Biol* 2002; 319:289–304.
6. Loeb T. Isolation of bacteriophage specific for the F+ and Hfr mating types of *Escherichia coli* K12. *Science* 1960; 131:932–933.
7. Hofschneider PH. Untersuchungen uber "kleine" *E. coli* K12 bacteriophagen M12, M13, und M20. *Z Naturforschg* 1963; 18b:203–205.
8. Marvin DA, Hoffman-Berling H. Physical and chemical properties of two new small bacteriophages. *Nature (London)* 1963; 197:517–518.
9. Berkowitz SA, Day LA. Mass, length, composition and structure of the filamentous bacterial virus fd. *J Mol Biol* 1976; 102:531–547.
10. Marvin DA. Filamentous phage structure, infection and assembly. *Curr Opin Struct Biol* 1998; 8:150–158.
11. Beck E, Zink B. Nucleotide sequence and genome organisation of filamentous bacteriophages fl and fd. *Gene* 1981; 16:35–58.
12. van Wezenbeek PM, Hulsebos TJ, Schoenmakers JG. Nucleotide sequence of the filamentous bacteriophage M13 DNA genome: Comparison with phage fd. *Gene* 1980; 11:129–148.
13. Baas PD, Jansz HS. Single-stranded DNA phage origins. *Curr Top Microbiol Immunol* 1988; 136:31–70.

14. Higashitani N, Higashitani A, Horiuchi K. Nucleotide sequence of the primer RNA for DNA replication of filamentous bacteriophages. *J Virol* 1993; 67:2175–2181.

15. Smith GP. Filamentous phage as cloning vectors. In: Rodriquez RL, ed. *Vectors: A Survey of Molecular Cloning Vectors and Their Uses.* Boston, MA: Butterworth, 1987:61–83.

16. Chatellier J, Hartley O, Griffiths AD, Fersht AR, Winter G, Riechmann L. Interdomain interactions within the gene 3 protein of filamentous phage. *FEBS Lett* 1999; 463:371–374.

17. Holliger P, Riechmann L, Williams RL. Crystal structure of the two N-terminal domains of g3p from filamentous phage fd at 1.9 Å: Evidence for conformational lability. *J Mol Biol* 1999; 288:649–657.

18. Lubkowski J, Hennecke F, Pluckthun A, Wlodawer A. Filamentous phage infection: Crystal structure of g3p in complex with its coreceptor, the C-terminal domain of TolA. *Structure* 1999; 7:711–722.

19. Deng LW, Malik P, Perham RN. Interaction of the globular domains of pIII protein of filamentous bacteriophage fd with the F-pilus of *Escherichia coli. Virology* 1999; 253:271–277.

20. Webster R. Filamentous phage biology. In: Barbas CF et al., eds. *Phage Display. A Laboratory Manual.* Cold Spring Harbor, NY: Cold Spring Harbor Laboratory Press, 2001:37, 1.1–1.

21. Nilsson N, Malmborg AC, Borrebaeck CAK. The phage infection process: A functional role for the distal linker region of bacteriophage protein 3. *J Virol* 2000; 74:4229–4235.

22. Messing J. Cloning single-stranded DNA. *Mol Biotechnol* 1996; 5:39–47.

23. Dotto GP, Zinder ND. Reduction of the minimal sequence for initiation of DNA synthesis by qualitative or quantitative changes of an initiator protein. *Nature* 1984; 311:279–280.

24. Mead DA, Kemper B. Chimeric single-stranded DNA phage-plasmid cloning vectors. In: Rodriguez RL, Denhardt DT, eds. *Vectors: A Survey of Molecular Cloning Vectors and Their Uses.* Stoneham, MA: Butterworth Publishers, 1988:85–102.

25. Cesareni G. Phage-plasmid hybrid vectors. In: Rodriguez RL, Denhardt DT, eds. *Vectors: A Survey of Molecular Cloning Vectors and Their Uses.* Stoneham, MA: Butterworth Publishers, 1988:103–111.

26. Vieira J, Messing J. Production of single-stranded plasmid DNA. *Methods Enzymol* 1987; 153:3–11.

27. Smith GP. Preface. Surface display and peptide libraries. *Gene* 1993; 128:1–2.

28. Smith GP. Filamentous fusion phage: Novel expression vectors that display cloned antigens on the virion surface. *Science* 1985; 228:1315–1317.

29. Parmley SF, Smith GP. Antibody-selectable filamentous fd phage vectors: Affinity purification of target genes. *Gene* 1988; 73:305–318.

30. Zacher ANI, Stock CA, Golden JWI, Smith GP. A new filamentous phage cloning vector: fd-tet. *Gene* 1980; 9:127–140.

31. Scott JK, Smith GP. Searching for peptide ligands with an epitope library. *Science* 1990; 249:386–390.

32. Cwirla SE, Peters EA, Barrett RW, Dower WJ. Peptides on phage: A vast library of peptides for identifying ligands. *Proc Natl Acad Sci U S A* 1990; 87:6378–6382.

33. McCafferty J, Griffiths AD, Winter G, Chiswell DJ. Phage antibodies: Filamentous phage displaying antibody variable domains. *Nature* 1990; 348:552–554.

34. Clackson T, Hoogenboom HR, Griffiths AD, Winter G. Making antibody fragments using phage display libraries. *Nature* 1991; 352:624–628.

35. Pratt D, Tzagoloff H, Erdahl WS. Conditional lethal mutants of the small filamentous coliphage M13. I. Isolation, complementation, cell killing, time of cistron action. *Virology* 1966; 30(3):397–410.

36. Smith GP. Filamentous phage assembly: Morphogenetically defective mutants that do not kill the host. *Virology* 1988; 167(1):156–165.

37. O'Connell D, Becerril B, Roy-Burman A, Daws M, Marks JD. Phage versus phagemid libraries for generation of human monoclonal antibodies. *J Mol Biol* 2002; 321:49–56.

38. Hoogenboom HR, Griffiths AD, Johnson KS, Chiswell DJ, Hudson P, Winter G. Multi-subunit proteins on the surface of filamentous phage: Methodologies for displaying antibody (Fab) heavy and light chains. *Nucleic Acids Res* 1991; 19:4133–4137.

39. MacKenzie R, To R. The role of valency in the selection of anti-carbohydrate single-chain Fvs from phage display libraries. *J Immunol Methods* 1998; 220:39–49.

40. de la Cruz VF, Lal AA, McCutchan TF. Immunogenicity and epitope mapping of foreign sequences via genetically engineered filamentous phage. *J Biol Chem* 1988; 263:4318–4322.

41. McConnell SJ, Uveges AJ, Fowlkes DM, Spinella DG. Construction and screening of M13 phage libraries displaying long random peptides. *Mol Divers* 1996; 1:165–176.

42. Fowlkes DM, Adams MD, Fowler VA, Kay BK. Multipurpose vectors for peptide expression on the M13 viral surface. *Biotechniques* 1992; 13:422–428.

43. Burritt JB, Quinn MT, Jutila MA, Bond CW, Jesaitis AJ. Topological mapping of neutrophil cytochrome b epitopes with phage-display libraries. *J Biol Chem* 1995; 270:16974–16980.

44. Burritt JB, Bond CW, Doss KW, Jesaitis AJ. Filamentous phage display of oligopeptide libraries. *Anal Biochem* 1996; 238:1–13.

45. Noren KA, Noren CJ. Construction of high-complexity combinatorial phage display peptide libraries. *Methods* 2001; 23:169–178.

46. Devlin JJ, Panganiban LC, Devlin PE. Random peptide libraries: A source of specific protein binding molecules. *Science* 1990; 249:404–406.

47. O'Neil KT, Hoess RH, Jackson SA, Ramachandran NS, Mousa SA, DeGrado WF. Identification of novel peptide antagonists for GPIIb/IIIa from a conformationally constrained phage peptide library. *Proteins* 1992; 14:509–515.

48. Hammer J, Takacs B, Sinigaglia F. Identification of a motif for HLA-DR1 binding peptides using M13 display libraries. *J Exp Med* 1992; 176:1007–1013.

49. McLafferty MA, Kent RB, Ladner RC, Markland W. M13 bacteriophage displaying disulfide-constrained microproteins [review]. *Gene* 1993; 128:29–36.

50. Enshell-Seijffers D, Smelyanski L, Vardinon N, Yust I, Gershoni JM. Dissection of the humoral immune response toward an immunodominant epitope of HIV: A model for the analysis of antibody diversity in HIV plus individuals. *FASEB J* 2001; 15:2112–2120.

51. Ilyichev AA, Minenkova OO, Tatkov SI, Karpyshev NN, Eroshkin AM, Petrenko VA, Sandakhchiev LS. Construction of M13 viable bacteriophage with the insert of foreign peptides into the major coat protein. *Dokl Biochem (Proc Acad Sci USSR) Engl Transl* 1989; 307:196–198.

52. Minenkova OO, Ilyichev AA, Kishchenko GP, Petrenko VA. Design of specific immunogens using filamentous phage as the carrier. *Gene* 1993; 128:85–88.

53. Greenwood J, Willis AE, Perham RN. Multiple display of foreign peptides on a filamentous bacteriophage. Peptides from Plasmodium falciparum circumsporozoite protein as antigens. *J Mol Biol* 1991; 220:821–827.

54. Petrenko VA, Smith GP, Gong X, Quinn T. A library of organic landscapes on filamentous phage. *Protein Eng* 1996; 9:797–801.

55. Iannolo G, Minenkova O, Gonfloni S, Castagnoli L, Cesareni G. Construction, exploitation and evolution of a new peptide library displayed at high density by fusion to the major coat protein of filamentous phage. *Biol Chem* 1997; 378:517–521.

56. Petrenko VA, Smith GP. Phages from landscape libraries as substitute antibodies. *Protein Eng* 2000; 13:589–592.

57. Petrenko VA, Smith GP, Mazooji MM, Quinn T. Alphahelically constrained phage display library. *Protein Eng* 2002; 15:943–950.

58. Legendre D, Fastrez J. Construction and exploitation in model experiments of functional selection of a landscape library expressed from a phagemid. *Gene* 2002; 290:203–215.

59. Malik P, Terry TD, Gowda LR, Langara A, Petukhov SA, Symmons MF, Welsh LC et al. Role of capsid structure and membrane protein processing in determining the size and copy number of peptides displayed on the major coat protein of filamentous bacteriophage. *J Mol Biol* 1996; 260:9–21.

60. Malik P, Terry TD, Bellintani F, Perham RN. Factors limiting display of foreign peptides on the major coat protein of filamentous bacteriophage capsids and a potential role for leader peptidase. *FEBS Lett* 1998; 436:263–266.

61. Romanov VI, Durand DB, Petrenko VA. Phage display selection of peptides that affect prostate carcinoma cells attachment and invasion. *Prostate* 2001; 47:239–251.

62. Bishop-Hurley SL, Mounter SA, Laskey J, Morris RO, Elder J, Roop P, Rouse C et al. Phage-displayed peptides as developmental agonists for Phytophthora capsici zoospores. *Appl Environ Microbiol* 2002; 68:3315–3320.

63. Terry TD, Malik P, Perham RN. Identification of Hcv core mimotopes—Improved methods for the selection and use of disease-related phage-displayed peptides. *Biol Chem* 1997; 378:495–502.

64. di Marzo Veronese F, Willis AE, Boyer-Thompson C, Appella E, Perham RN. Structural mimicry and enhanced immunogenicity of peptide epitopes displayed on filamentous bacteriophage. The V3 loop of HIV-1 gp120. *J Mol Biol* 1994; 243:167–172.

65. Perham RN, Terry TD, Willis AE, Greenwood J, di Marzo Veronese F, Appella E. Engineering a peptide epitope display system on filamentous bacteriophage. *FEMS Microbiol Rev* 1995; 17:25–31.

66. De Berardinis P, Sartorius R, Fanutti C, Perham RN, Del Pozzo G, Guardiola J. Phage display of peptide epitopes from HIV-1 elicits strong cytolytic responses. *Nat Biotechnol* 2000; 18:873–876.

67. Malik P, Perham RN. New vectors for peptide display on the surface of filamentous bacteriophage. *Gene* 1996; 171:49–51.

68. Iannolo G, Minenkova O, Petruzzelli R, Cesareni G. Modifying filamentous phage capsid: Limits in the size of the major capsid protein. *J Mol Biol* 1995; 248:835–844.

69. Felici F, Castagnoli L, Musacchio A, Jappelli R, Cesareni G. Selection of antibody ligands from a large library of oligopeptides expressed on a multivalent exposition vector. *J Mol Biol* 1991; 222:301–310.

70. Endemann H, Model P. Location of filamentous phage minor coat proteins in phage and in infected cells. *J Mol Biol* 1995; 250:496–506.

71. Sidhu SS. Engineering M13 for phage display. *Biomol Eng* 2001; 18:57–63.

72. Krebber C, Spada S, Desplancq D, Pluckthun A. Co-selection of cognate antibody–antigen pairs by selectively-infective phages. *FEBS Lett* 1995; 377:227–231.

73. Enshell-Seijffers D, Smelyanski L, Gershoni JM. The rational design of a "type 88" genetically stable peptide display vector in the filamentous bacteriophage fd. *Nucleic Acids Res* 2001; 29:E50.

74. Markland W, Roberts BL, Saxena MJ, Guterman SK, Ladner RC. Design, construction and function of a multicopy display vector using fusions to the major coat protein of bacteriophage M13. *Gene* 1991; 109:13–19.

75. Roberts BL, Markland W, Ladner RC. Affinity maturation of proteins displayed on surface of M13 bacteriophage as major coat protein fusions. *Methods Enzymol* 1996; 267:68–82.
76. Chappel JA, He M, Kang AS. Modulation of antibody display on M13 filamentous phage. *J Immunol Methods* 1998; 221:25–34.
77. Zhong G, Smith GP, Berry J, Brunham RC. Conformational mimicry of a chlamydial neutralization epitope on filamentous phage. *J Biol Chem* 1994; 269:24183–24188.
78. Smith GP, Fernandez AM. Effect of DNA copy number on genetic stability of phage-displayed peptides. *Biotechniques* 2004; 36(4):610–614, 616, 618.
79. Malik P, Perham RN. Simultaneous display of different peptides on the surface of fila-mentous bacteriophage. *Nucleic Acids Res* 1997; 25:915–916.
80. Huse WD. Combinatorial antibody expression libraries in filamentous phage. In: Borrebaeck CAK, ed. *Antibody Engineering: A Practical Guide*. New York: W.H. Freeman and Company, 1991:103–120.
81. Huse WD, Stinchcombe TJ, Glaser SM, Starr L, MacLean M, Hellstrom KE, Hellstrom I et al. Application of a filamentous phage pVIII fusion protein system suitable for effi-cient production, screening, and mutagenesis of F(ab) antibody fragments. *J Immunol* 1992; 149:3914–3920.
82. McConnell SJ, Kendall ML, Reilly TM, Hoess RH. Constrained peptide libraries as a tool for finding mimotopes. *Gene* 1994; 151:115–118.
83. Wang CI, Yang Q, Craik CS. Phage display of proteases and macromolecular inhibitors. *Methods Enzymol* 1996; 267:52–68.
84. Corey DR, Shiau AK, Yang Q, Janowski BA, Craik CS. Trypsin display on the surface of bacteriophage. *Gene* 1993; 128:129–134.
85. Kouzmitcheva GA, Petrenko VA, Smith GP. Identifying diagnostic peptides for Lyme disease through epitope discovery. *Clin Diagn Lab Immunol* 2001; 8:150–160.
86. Matthews LJ, Davis R, Smith GP. Immunogenically fit subunit vaccine components via epitope discovery from natural peptide libraries. *J Immunol* 2002; 169:837–846.
87. Bonnycastle LLC, Shen J, Menendez A, Scott JK. Production of peptide libraries. In: Barbas CF III et al., eds. *Phage Display. A Laboratory Manual*. Cold Spring Harbor, NY: Cold Spring Harbor Laboratory Press, 2001:16.11–16.28.
88. Hopp TP, Prickett KS, Price VL, Libby RT, March CJ, Ceretti DP, Urdal DL, Conlon PJ. A short polypeptide marker sequence useful for recombinant protein identification and purification. *Biotechnology* 1988; 6:1204–1210.
89. Bass S, Greene R, Wells JA. Hormone phage: An enrichment method for variant proteins with altered binding properties. *Proteins* 1990; 8:309–314.
90. Lowman HB. Structural and mechanistic determinants of affinity and specificity of ligands discovered or engineered by phage [review]. *Annu Rev Biophys Biomol Struct* 1997; 26:27–45.
91. Wrighton NC, Farrell FX, Chang R, Kashyap AK, Barbone FP, Mulcahy LS, Johnson DL et al. Small peptides as potent mimetics of the protein hormone erythropoietin. *Science* 1996; 273:458–464.
92. Yanofsky SD, Baldwin DN, Butler JH, Holden FR, Jacobs JW, Balasubramanian P, Chinn JP et al. High affinity type I interleukin 1 receptor antagonists discovered by screening recombinant peptide libraries. *Proc Natl Acad Sci U S A* 1996; 93:7381–7386.
93. Barbas CF III, Burdon DR, Scott JK, Silverman GJ. Phage-display vectors. In: Barbas CF III, Buton DR, Scott JK, Silverman GJ, eds. *Phage Display: A Laboratory Manual*. Cold Spring Harbor, NY: Cold Spring Harbor Laboratory Press, 2001:2.1–2.19.
94. Boeke JD, Model P, Zinder ND. Effects of bacteriophage f1 gene III protein on the host cell membrane. *Mol Gen Genet* 1982; 186:185–192.

95. Lutz R, Bujard H. Independent and tight regulation of transcriptional units in *Escherichia coli* via the LacR/O, the TetR/O and AraC/I1-I2 regulatory elements. *Nucleic Acids Res* 1997; 25:1203–1210.

96. Sidhu SS, Weiss GA, Wells JA. High copy display of large proteins on phage for functional selections. *J Mol Biol* 2000; 296:487–495.

97. Griffiths AD, Malmqvist M, Marks JD, Bye JM, Embleton MJ, McCafferty J, Baier M et al. Human anti-self antibodies with high specificity from phage display libraries. *EMBO J* 1993; 12:725–734.

98. Duenas M, Borrebaeck CA. Novel helper phage design: Intergenic region affects the assembly of bacteriophages and the size of antibody libraries. *FEMS Microbiol Lett* 1995; 125:317–321.

99. Rondot S, Koch J, Breitling F, Dubel S. A helper phage to improve single-chain antibody presentation in phage display. *Nat Biotechnol* 2001; 19:75–78.

100. Rakonjac J, Jovanovic G, Model P. Filamentous phage infection-mediated gene expression: Construction and propagation of the gIII deletion mutant helper phage R408d3. *Gene* 1997; 198:99–103.

101. Gao C, Mao S, Lo CH, Wirsching P, Lerner RA, Janda KD. Making artificial antibodies: A format for phage display of combinatorial heterodimeric arrays. *Proc Natl Acad Sci U S A* 1999; 96:6025–6030.

102. Gao C, Mao S, Kaufmann G, Wirsching P, Lerner RA, Janda KD. A method for the generation of combinatorial antibody libraries using pIX phage display. *Proc Natl Acad Sci U S A* 2002; 99:12612–12616.

103. Hoogenboom HR, de Bruine AP, Hufton SE, Hoet RM, Arends JW, Roovers RC. Antibody phage display technology and its applications. *Immunotechnology* 1998; 4:1–20.

104. Barbas CF III, Kang AS, Lerner RA, Benkovic SJ. Assembly of combinatorial antibody libraries on phage surfaces: The gene III site. *Proc Natl Acad Sci U S A* 1991; 88:7978–7982.

105. Tang Y, Jiang N, Parakh C, Hilvert D. Selection of linkers for a catalytic single-chain antibody using phage display technology. *J Biol Chem* 1996; 271:15682–15686.

106. Kang AS, Barbas CF, Janda KD, Benkovic SJ, Lerner RA. Linkage of recognition and replication functions by assembling combinatorial antibody Fab libraries along phage surfaces. *Proc Natl Acad Sci U S A* 1991; 88:4363–4366.

107. Jespers LS, Messens JH, De Keyser A, Eeckhout D, Van den Brande I, Gansemans YG, Lauwereys MJ et al. Surface expression and ligand-based selection of cDNAs fused to filamentous phage gene VI. *Biotechnology* 1995; 13:378–382.

108. Shin YC, Kim YE, Cho T-J. A novel phage display vector for easy monitoring of expressed proteins. *J Biochem Mol Biol* 2000; 33:242–248.

109. Soderlind E, Simonsson AC, Borrebaeck CA. Phage display technology in antibody engineering: Design of phagemid vectors and in vitro maturation systems. *Immunol Rev* 1992; 130:109–124.

110. Ridder R, Schmitz R, Legay F, Gram H. Generation of rabbit monoclonal antibody fragments from a combinatorial phage display library and their production in the yeast Pichia pastoris. *Biotechnology* 1995; 13:255–260.

111. Paschke M, Zahn G, Warsinke A, Hohne W. New series of vectors for phage display and prokaryotic expression of proteins. *Biotechniques* 2001; 30:720–724, 726.

112. Zahn G, Skerra A, Hohne W. Investigation of a tetracycline-regulated phage display system. *Protein Eng* 1999; 12:1031–1034.

113. Breitling F, Dubel S, Seehaus T, Klewinghaus I, Little M. A surface expression vector for antibody screening. *Gene* 1991; 104:147–153.

114. Dubel S, Breitling F, Fuchs P, Braunagel M, Klewinghaus I, Little M. A family of vectors for surface display and production of antibodies. *Gene* 1993; 128:97–101.

115. Crameri R, Suter M. Display of biologically active proteins on the surface of filamentous phages: A cDNA cloning system for selection of functional gene products linked to the genetic information responsible for their production. *Gene* 1993; 137:69–75.

116. Crameri R, Jaussi R, Menz G, Blaser K. Display of expression products of cDNA libraries on phage surfaces. A versatile screening system for selective isolation of genes by specific gene-product/ligand interaction. *Eur J Biochem* 1994; 226:53–58.

117. Weiss GA, Sidhu SS. Design and evolution of artificial M13 coat proteins. *J Mol Biol* 2000; 300:213–219.

118. Marvin DA, Hale RD, Nave C, Helmer-Citterich M. Molecular models and structural comparisons of native and mutant class I filamentous bacteriophages Ff (fd, f1, M13), If1 and IKe. *J Mol Biol* 1994; 235:260–286.

119. Kishchenko G, Batliwala H, Makowski L. Structure of a foreign peptide displayed on the surface of bacteriophage M13. *J Mol Biol* 1994; 241:208–213.

120. Glucksman MJ, Bhattacharjee S, Makowski L. Three-dimensional structure of a cloning vector. X-ray diffraction studies of filamentous bacteriophage M13 at 7 Å resolution. *J Mol Biol* 1992; 226:455–470.

121. Hunter GJ, Rowitch DH, Perham RN. Interactions between DNA and coat protein in the structure and assembly of filamentous bacteriophage fd. *Nature* 1987; 327:252–254.

122. Makowski L. Structural constraints on the display of foreign peptides on filamentous bacteriophages. *Gene* 1993; 128:5–11.

123. Williams KA, Glibowicka M, Li Z, Li H, Khan AR, Chen YM, Wang J et al. Packing of coat protein amphipathic and transmembrane helices in filamentous bacteriophage M13: Role of small residues in protein oligomerization. *J Mol Biol* 1995; 252:6–14.

124. Kishchenko GP, Minenkova OO, Ilyichev AA, Gruzdev AD, Petrenko VA. Study of the structure of phage-M13 virions containing chimeric B-protein molecules. *Mol Biol Engl Trans* 1991; 25:1171–1176.

125. Weiss GA, Wells JA, Sidhu SS. Mutational analysis of the major coat protein of M13 identifies residues that control protein display. *Protein Sci* 2000; 9:647–654.

126. Ivanenkov V, Felici F, Menon AG. Targeted delivery of multivalent phage display vectors into mammalian cells. *Biochim Biophys Acta* 1999; 1448:463–472.

127. Huie MA, Cheung MC, Muench MO, Becerril B, Kan YW, Marks JD. Antibodies to human fetal erythroid cells from a nonimmune phage antibody library. *Proc Natl Acad Sci U S A* 2001; 98:2682–2687.

128. Larocca D, Baird A. Receptor-mediated gene transfer by phage-display vectors: Applications in functional genomics and gene therapy [review]. *Drug Discov Today* 2001; 6:793–801.

129. Petrenko VA, Vodyanoy VJ. Phage display for detection of biological threat agents. *J Microbiol Methods* 2003; 1768:1–10.

CHAPTER 3

Methods for the Construction of Phage-Displayed Libraries

Frederic A. Fellouse and Gabor Pal

CONTENTS

3.1 INTRODUCTION

In this chapter, we present different methods for making libraries of recombinant polypeptides. First, we describe how to create highly defined libraries by using synthetic DNA. Second, we present alternative approaches that can be used to introduce mutations randomly. In these cases, the position and nature of the mutations are not defined by the experimenter. Both types of methods enable the production of libraries that can be used to select polypeptides with desired functions. Finally, we describe techniques that have been developed to further increase the diversity of these libraries by either recombination or shuffling.

3.2 OLIGONUCLEOTIDE-DIRECTED MUTAGENESIS

Oligonucleotide-directed mutagenesis is a highly controlled method for the introduction of mutations into proteins. The technique is usually used when the region to be randomized is confined to a limited area, but protocols have also been developed to allow for the mutagenesis of entire genes. By introducing degeneracy into a synthetic oligonucleotide sequence, the position and degree of randomization can be precisely controlled. This section presents various strategies for the design and synthesis of degenerate oligonucleotides and methods for the introduction of oligonucleotide libraries into phage display vectors.

3.2.1 Strategies for the Design and Synthesis of Degenerate Codons

3.2.1.1 Saturation Mutagenesis

"Saturation mutagenesis," or "hard randomization," refers to the replacement of one or more codons by a codon that encodes for all 20 natural amino acids. One degenerate codon that encodes for all 20 amino acids contains an equimolar mix of all four nucleotides at each position within the codon. The NNN codon (Table 3.1) contains all 64 natural codons, and this high redundancy leads to a highly biased protein diversity, since some amino acids are represented by only one codon, while others are represented by as many as six codons. Furthermore, three of the unique codons are stop codons, which cause premature polypeptide termination and the loss of functional library members. As a less redundant alternative, the third codon position is allowed to vary either as G/C or G/T (NNK or NNS codons, Table 3.1), and this results in 32 unique codons that cover all 20 amino acids and contain only one stop codon.

Saturation mutagenesis was successfully applied to generate the first synthetic antibody library [1], and it has been routinely used for the construction of highly diverse polypeptide libraries. This simple strategy is very successful in applications where a thorough search of a localized region is required, for example, for antibody hypervariable loops that have evolved to display diverse sequences [1] or for naïve peptide libraries [2–5]. The theoretical diversities of libraries increase exponentially with the number of residues that are randomized, while the practical diversities

Table 3.1   Useful Degenerate Codons

Codon	Description	Amino Acids[a]	Stop Codons	No. of Codons[b]
NNN	All 20 amino acids	All 20	TAA, TAG, TGA	64
NNK or NNS	All 20 amino acids	All 20	TAG	32
NNC	15 amino acids	A, C, D, F, G, H, I, L, N, P, R, S, T, V, Y	None	16
NWW	Charged, hydrophobic	D, E, F, H, I, K, L, N, Q, V, Y	TAA	16
RVK	Charged, hydrophilic	A, D, E, G, H, K, N, R, S, T	None	12
DVT	Hydrophilic	A, C, D, G, N, S, T, Y	None	9
NVT	Charged, hydrophilic	C, D, G, H, N, P, R, S, T, Y	None	12
NNT	Mixed	A, D, G, H, I, L, N, P, R, S, T, V	None	16
VVC	Hydrophilic	A, D, G, H, N, P, R, S, T	None	9
NTT	Hydrophobic	F, I, L, V	None	4
RST	Small side chains	A, G, S, T	None	4
TDK	Hydrophobic	C, F, L, W, Y	TAG	6

[a] IUB code nomenclature (N = G/A/T/C, K = G/T, S = G/C, W = A/T, R = A/G, V = G/A/C, D = G/A/T).
[b] The number of unique codons contained in the degenerate codon. Due to the redundancy of the genetic code, the number of unique codons can be higher than the number of unique amino acids encoded by a degenerate codon.

accessible through phage display ($\sim 10^{11}$) are limited by bacterial transformation efficiencies [6]. Consequently, only six or seven positions can be subjected simultaneously to saturation mutagenesis with a reasonable probability of completely representing all the possible amino acid combinations [7] (Table 3.2). As alternatives to saturation mutagenesis, many different strategies have been developed for the construction of libraries with reduced or more controlled chemical diversity.

Table 3.2   Diversities of DNA Encoded Protein Libraries

Randomized Positions	DNA Diversity[a] (32^n)	Protein Diversity (20^n)	Required Library Size[b]
1	32	20	1.5×10^2
2	1.0×10^3	4.0×10^2	4.8×10^3
3	3.3×10^4	8.0×10^3	1.5×10^5
4	1.1×10^6	1.6×10^5	4.9×10^6
5	3.4×10^7	3.2×10^6	1.6×10^8
6	1.1×10^9	6.4×10^7	5.0×10^9
7	3.4×10^{10}	1.3×10^9	1.6×10^{11}
8	1.1×10^{12}	2.6×10^{10}	5.1×10^{12}

[a] Calculations based on the use of an NNK or NNS degenerate codon.
[b] Library size required for a complete representation of all possible amino acid sequences with a 99% confidence using a Poisson distribution.

3.2.1.2 *Mutagenesis with Spiked Oligonucleotides*

It is possible to synthesize mutagenic oligonucleotides by deliberately con-
taminating the wild-type sequence by a defined level of the other nucleotides. The
result is a "spiked" oligonucleotide that encodes predominantly a wild-type protein
sequence but also encodes mutations at a frequency dependent upon the levels of
nonwild-type nucleotide contamination [8,9]. In these synthesis schemes, codons are
altered predominantly by the mutation of only one nucleotide, and consequently,
substitutions requiring more than a single nucleotide change will have a lower probability of occurrence. Libraries produced in this way are highly biased toward the
wild-type protein sequence, and thus, this strategy is sometimes referred to as a
"soft" randomization approach since it produces subtle changes in sequence that are
particularly useful for affinity maturation applications [6].

3.2.1.3 *Mutagenesis with Tailored Oligonucleotides*

Another strategy for reducing the library diversity is to randomize with degenerate codons that encode for only a selected subset of the natural amino acids. In this
approach, the amino acid subset is chosen to favor functional polypeptides and is
selected on the basis of several possible criteria.

By comparing the sequences of homologous proteins from different species, one
can define a list of amino acid residues that are likely to be well tolerated at a particular position. If the three-dimensional structure of the protein is known, rational
design methods can be applied to generate a list of mutations likely to be beneficial.
Similarly, certain amino acids can also be excluded from positions where they are
likely to be deleterious.

Tailored diversity can also be designed based on the results of previous selections.
For example, subsets of positions within a polypeptide can be subjected to saturation mutagenesis in several separate libraries. Statistical analysis of the sequences of
functional clones selected from these libraries can identify positions that are biased
toward certain subsets of amino acids [5]. This information can then be used to
design tailored degenerate codons that encode for the preferred amino acids, and
these codons can be used in second-generation libraries.

Many amino acid subsets can be readily designed by simple inspection of the
genetic code (Figure 3.1) and the use of stoichiometric mixtures of nucleotides in
degenerate codons (Table 3.1). However, degenerate codons for some amino acid
distributions cannot be elucidated in this manner. Thus, nonstoichiometric mixtures
of bases can be useful to obtain particular amino acid combinations, and computer
algorithms have been developed for this purpose. Arkin and Youvan [10] analyzed
the possible amino acid distributions encoded by all the possible degenerate codon
sets that can be obtained with a fractional resolution of 10% (i.e., setting the proportion of each nucleotide type from 0% to 100% in 10% increments). LaBean
and Kauffman [11] developed a semiautomated approach to designing codon sets
that mimic the overall amino acid compositions characteristic of natural globular
proteins.

Second position

		T	C	A	G	
		Phe	Ser	Tyr	Cys	T
	T	Phe	Ser	Tyr	Cys	C
		Leu	Ser	*	*	A
		Leu	Ser	*	Trp	G
		Leu	Pro	His	Arg	T
	C	Leu	Pro	His	Arg	C
First position		Leu	Pro	Gln	Arg	A
		Leu	Pro	Gln	Arg	G
		Ile	Thr	Asn	Ser	T
	A	Ile	Thr	Asn	Ser	C
		Ile	Thr	Lys	Arg	A
		Met	Thr	Lys	Arg	G
		Val	Ala	Asp	Gly	T
	G	Val	Ala	Asp	Gly	C
		Val	Ala	Glu	Gly	A
		Val	Ala	Glu	Gly	G

Third position

Figure 3.1   The genetic code. The 64 genetic codons encode for 20 natural amino acids shown in the three-letter code. Three codons (TAA, TAG, TGA) encode for stop codons shown as asterisks.

3.2.1.4 Mutagenesis with Advanced Oligonucleotide Synthesis Methods

Automated DNA synthesis is traditionally performed on resin beads packed into a reaction column. After any step of the synthesis, the beads can be split and repacked into two or several new columns. Each of these columns can be individually coupled with single or mixed nucleotides and then pooled and repacked into different columns again [12]. In principle, this resin-splitting method can be used to obtain exact predetermined combinations of codons. In contrast, by using standard methods, synthesis of some codon subsets requires the inclusion of codons for additional undesired amino acids. For example, Figure 3.2 illustrates how resin splitting can be used to obtain only tyrosine and tryptophan codons. Unfortunately, the method becomes increasingly complicated as more complex codon sets are required, and a method that can generate any composition of the 20 amino acids requires the use of 10 synthesis columns and a tediously frequent resin-splitting and mixing procedure [13,14].

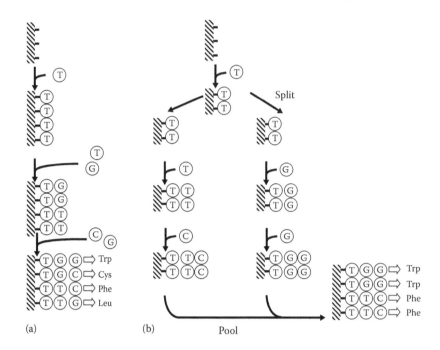

Figure 3.2 Oligonucleotide synthesis (a) with a single column or (b) by the split-pool method. Each growing oligonucleotide is covalently linked to a bead surface. In the example, the goal is to synthesize a degenerate codon that encodes for only Trp and Phe. With single-column synthesis, codons for two additional amino acids (Cys and Leu) are generated, but split-pool synthesis can be used to obtain the binomial distribution. Synthetic DNA is typically synthesized in the 3′–5′ direction, but 5′–3′ synthesis is depicted for clarity.

Since amino acids are coded by nucleotide triplets, the ideal way to synthesize degenerate oligonucleotides would be the use of trinucleotides as building blocks instead of the traditional mononucleotides. This would eliminate the codon biases introduced by the redundant nature of the genetic code and would allow for the introduction of precise compositions of the amino acids at positions to be mutagenized. The applicability of this technique relies on two prerequisites: availability of high-quality trinucleotide precursors in large quantities and high efficiency of coupling with the trinucleotide blocks. Several research groups have reported the successful synthesis and utilization of trinucleotides [5,15–17], and a commercial source has recently become available (Glen Research, Sterling, Virginia). Thus, it seems likely that this method should become more generally applicable in the near future.

3.2.2 Integration of Mutagenic Oligonucleotides into Phage Display Vectors

To produce libraries from synthetic DNA, it is necessary to insert the DNA into a suitable phage display vector and to introduce the resulting recombinant DNA into a bacterial host for phage production. There are several strategies whereby this can be achieved. Perhaps the simplest method involves the annealing of a mutagenic oligonucleotide to a single-stranded DNA (ssDNA) vector and subsequent enzyme-mediated incorporation into a double-stranded DNA (dsDNA) form. Alternatively, dsDNA cassette mutagenesis can be used, and many polymerase chain reaction (PCR) methods have also been developed. The principles behind these various approaches are described in Section 3.5.1.

3.2.2.1 ssDNA Method

The ssDNA method is ideally suited to the construction of M13 phage librar-ies, since the viral DNA is packaged in an ssDNA form and can be easily purified from phage particles [6]. An appropriately designed oligonucleotide is annealed to the ssDNA template and enzymatically extended and ligated to produce a dsDNA heteroduplex (Figure 3.3). For library construction, the mutagenic oligonucleotide contains degenerate DNA in the region to be mutagenized flanked at both ends by approximately 18 bases that are complementary to the regions preceding and fol-lowing the target region. Thus, the complementary regions ensure correct anneal-ing around the target region, and the degenerate DNA incorporates the designed mutations. As oligonucleotides containing up to 100 bases can be synthesized with standard methods, a single oligonucleotide can introduce mutations in up to approxi-mately 20 consecutive codons. Furthermore, multiple oligonucleotides can be simul-taneously annealed to allow for mutations in regions that are far apart in the primary sequence [18,19].

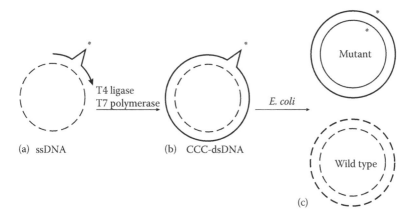

(a) ssDNA

T4 ligase
T7 polymerase

(b) CCC-dsDNA

E. coli

Mutant

Wild type

(c)

Figure 3.3 Oligonucleotide-directed mutagenesis with an ssDNA template. (a) A synthetic oligonucleotide (solid line) is annealed to the ssDNA template (dashed line). The oligonucleotide is designed to encode mutations (shown as asterisks) in the mis-matched variable region, which is flanked by perfectly complementary sequences. (b) Heteroduplex, covalently closed circular dsDNA (CCC-dsDNA) is enzymati-cally synthesized by T7 DNA polymerase and T4 DNA ligase. (c) Heteroduplex dsDNA is introduced into an *Escherichia coli* host where the mismatched region is repaired to either the wild type or mutant sequence.

The original ssDNA mutagenesis method [20,21] suffered from low mutation frequencies because the *Escherichia coli* host preferentially replicated the parental DNA strand rather than the in vitro synthesized mutagenic strand. A major advance in the method was achieved by producing template ssDNA from *E. coli* dut⁻/ung⁻ strains, which contain mutations that result in DNA containing significant amounts of uracil bases in place of thymine bases [22,23]. The use of templates of this type results in heteroduplexes in which the wild-type strand contains uracil, while the mutagenic strand does not. Introduction of these heteroduplexes into *E. coli* dut⁺/ung⁺ results in the preferential inactivation of the uracil-containing strand and, thus, greatly increases the mutation frequency. Using optimized protocols based on this strategy, high mutation frequencies (>80%) and large library diversities (>10^{10}) can be readily achieved [6].

3.2.2.2 Cassette Methods

A mutagenic cassette is a dsDNA fragment in which the region containing mutations is flanked by constant regions that contain restriction enzyme cleavage sites for cloning into phage display vectors. The cassette can be synthesized by different methods, but one of the two strands is always synthesized chemically. The other strand can be synthesized chemically [24] or by enzyme-mediated extension [25–27]. In the case of complete chemical synthesis, both strands contain degenerate positions, and perfectly complementary annealing occurs only in the flanking regions. In the case of enzyme-mediated extension, the two strands are complementary since the synthetic strand is used as a template for DNA polymerase. Most cassette mutagenesis strategies require cloning into enzymatically cleaved dsDNA vectors, but a method that achieves cassette cloning with ssDNA vectors also has been described [28].

Cassette mutagenesis methods can achieve mutation efficiencies close to 100% provided that the enzymatic cleavage reactions are highly efficient. However, there are inherent limitations accompanied with these methods. The most important one is the strict requirement for unique restriction endonuclease cleavage sites flanking the region to be mutated. Furthermore, while cassette methods are well suited for mutations within a short continuous stretch of about 30 codons or less, simultaneous mutagenesis of positions located at greater distances within a gene would require multiple cassettes, each with its own requirements for unique flanking restriction sites. Thus, these methods are best suited for mutations in a localized region, which can be obtained with a single cassette.

3.2.2.3 PCR Methods

Several methods use the PCR as a means for incorporating a mutagenic oligonucleotide into a cassette that can then be cloned into a phage display vector. In the overlap-extension method, two independent PCRs are performed to amplify two DNA fragments with overlapping termini [29,30]. Mutations are inserted in both PCR products at the overlapping regions by using mutagenic primers that also contain a region of complementarity (Figure 3.4). In a second step, the two PCR products are annealed to each other, and the resulting overlap product is extended to produce a

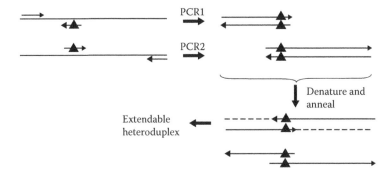

Figure 3.4 Overlap-extension PCR. Two separate PCRs are performed (PCR1 and PCR2). The oligonucleotides encoding the mutations (triangles) are partially complementary. The two PCR products are mixed, denatured, annealed, and extended with DNA polymerase to generate the full-length sequence.

mutagenic cassette that can be cloned into a vector using restriction sites introduced by the flanking, nonmutagenic primers. A significant advantage of this method relative to standard cassette mutagenesis is that the restriction sites need not be proximal to the sites of mutation, since they are contained within the flanking primers.

An alternative "megaprimer" method [31,32] uses three primers and two successive rounds of PCR (Figure 3.5). In the first round, the central mutagenic primer and one of the flanking primers are used to amplify a so-called megaprimer. The megaprimer is then used in a second PCR with the second flanking primer to generate the complete mutagenic cassette with mutations incorporated at the annealing

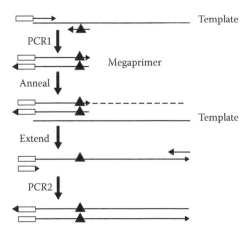

Figure 3.5 The megaprimer PCR method. A PCR (PCR1) is performed with a flanking primer and a central primer encoding for mutations (triangle). The PCR product is used as a megaprimer that anneals to the template and reconstitutes the full-length product. A second PCR (PCR2) preferentially amplifies the mutated DNA by using a primer that anneals to a flanking sequence (white box) added by the flanking primer used in PCR1.

Figure 3.6   Inverse PCR. Primers are designed to anneal to opposite strands with the two annealing sites close to each other. PCR results in amplification of the entire vector, which can be regenerated by ligation. Mutations can be incorporated by mismatches in one of the primers, as shown.

site of the central primer. Again, the flanking primers are designed to incorporate restriction sites for convenient cassette cloning.

In the inverse PCR method [33], the primers are designed to anneal to the opposite template strands with their 5′ ends near to each other (Figure 3.6). One of the primers can be designed to incorporate mutations, while the other can be perfectly complementary. A PCR with primers of this type results in amplification of the entire vector. Circularization of this product by ligation regenerates the vector and incorporates the designed mutations.

3.3 RANDOM MUTAGENESIS

The oligonucleotide-directed methods target distinct areas of a gene and are thus ideal for highly controlled mutagenesis. However, methods have also been developed for random mutagenesis throughout a gene, and these can be useful when it is not clear which region of a protein should be mutated to achieve a particular effect. This section describes methods of this type.

3.3.1 In Vitro Chemical Mutagenesis

Chemical mutagenesis methods were instrumental tools for classical genetics experiments that required high mutation rates [34]. The development of powerful site-directed methods described above have limited the use of these less controlled methods in phage display. Nonetheless, chemical mutagenesis may be a viable alternative in specialized applications.

Many chemicals used for whole organism mutagenesis can also be used in vitro to modify DNA in a manner that results in mutations in vivo. Nitrous acid, hydroxylamine, methoxyamine, sodium bisulfite, and many other chemicals can be used to induce mutations [35–38]. However, one issue with in vitro chemical mutagenesis is how to confine the mutations to the gene of interest so as not to complicate the later selection cycles with mutations in other regions of the vector. For mutagens that act only on ssDNA, a target segment can be generated in a dsDNA vector by the successive action of an endonuclease and an exonuclease [39]. Alternatively, following treatment with a mutagen, the DNA of interest can be excised from the vector and cloned into nonmutagenized vector DNA.

In any case, the method needs to be carefully optimized in order to achieve the desired levels of mutation. Furthermore, because different chemicals produce different types of mutations, mixtures of chemicals might be required to achieve a sufficiently random distribution of mutations.

3.3.2 *E. coli* Mutator Strains

E. coli mutator strains are deficient in DNA editing or proofreading enzymes. The most common mutated genes are *mutD*, *mutS*, or *mutT*. The *mutD* and *mutT* mutations induce transitions and transversions, respectively, while the *mutS* mutation induces both transitions and transversions [40]. Mutator strains may contain only a single mutant gene, but strains that contain multiple mutant genes are also available [41].

The use of mutator strains presents the advantage of being extremely simple; phage need simply be passed through a mutator strain without any chemical or enzymatic manipulation. However, it is important to be aware of the instability of the mutator strain, as loss of the mutator phenotype presents a growth advantage for the bacteria [41]. In order to confine mutations to the gene of interest, a two-plasmid system has been developed [42]; one plasmid expresses an error-prone DNA polymerase I, and the other plasmid bears the gene of interest downstream of an origin of replication specifically recognized by DNA polymerase I.

3.3.3 Error-Prone PCR

The thermostable *Taq* DNA polymerase lacks a $3'–5'$ proofreading function, and thus, its error frequency ($\sim 10^{-4}$ per base) is far higher than that of in vivo DNA synthesis ($\sim 10^{-9}$ per base) [43]. The error frequency can be further enhanced by using suboptimal reaction conditions, and this low fidelity can be exploited to produce fairly random mutations through error-prone PCR.

There are several parameters that can be manipulated to alter the fidelity of *Taq* DNA polymerase. The buffer can be altered by changing the pH, changing the ratios of the four nucleotides, increasing the concentration of Mg^{2+} ions, or adding Mn^{2+} ions [44–46]. In addition, the reaction cycle parameters can be adjusted to facilitate errors; the number of cycles or the extension times can be increased, or the annealing temperature can be lowered [47,48].

Ideally, error-prone PCR would result in perfectly random substitutions evenly distributed along the entire DNA fragment, but in reality, the process is biased. In general, transitions are more frequent than transversions, and mutations tend to cluster in particular regions of the DNA sequence [49]. Another important limitation is that the relatively low mutation rate results in a strong bias toward amino acid substitutions that arise from single nucleotide mutations. Despite these drawbacks, error-prone PCR has been used quite extensively in in vitro evolution experiments [48,50,51].

3.4 COMBINATORIAL INFECTION AND RECOMBINATION

In vivo recombination strategies have been developed to expand the diversities of phage display libraries beyond the limits imposed by *E. coli* transformation efficiencies. The methods are particularly effective for heterodimeric proteins such as antigen-binding fragments (Fabs). However, variations of the methods also allow for applications with single-chain proteins in which two distinct regions of the sequence are mutated.

In the simplest recombination method (Figure 3.7), two libraries are created independently, and each library codes for one chain of a heterodimeric protein. One library is carried by a phagemid packaged into phage particles, while the other is maintained as a plasmid in bacteria. Bacteria carrying the plasmid library are infected with the phagemid library, and site-specific recombination between the phagemid and the plasmid forms a new phagemid that contains both genes of the heterodimeric protein. Thus, a huge library can be produced by combining two moderate-size libraries, and the size of the library is only limited by the number of infected cells.

The concept was first demonstrated [52] by using combinatorial infection and the site-specific Cre-lox recombination system [53] to generate a Fab that contained light and heavy chains encoded by different vectors. The loxP recombination sites persist after the recombination event, and thus, heterodimeric Fab proteins are ideally suited for this method, because the recombination sites can be introduced into a nontranscribed DNA sequence anywhere between the two genes. Mixed pairs of nonhomologous loxP sequences were used, as Cre recombinase does not catalyze recombination between the wild-type and the mutant loxP sites [54]. Each vector contained one wild-type and one mutant sequence to allow recombination between the two vectors while avoiding deletion of DNA segments through cis recombination. The same group subsequently used the strategy to generate a Fab library containing 6.5×10^{10} members [55].

The loxP site is 34 base pairs long, and if it is introduced into a gene encoding for a monomeric protein, its sequence has to be in the same reading frame as that of the gene, and its translation must not interfere with the function of the protein. In an application of this type, the loxP site was used as a linker between the V_L and

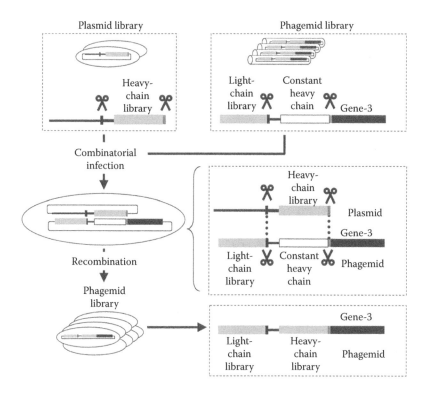

Figure 3.7   Generation of a Fab library by combinatorial infection and in vivo recombination. Bacteria harboring a plasmid carrying a heavy-chain library are infected with a phagemid library of light chains with a constant heavy chain. Subsequent infection with a P1 bacteriophage that provides Cre recombinase results in the exchange of DNA fragments between plasmids and phagemids (represented by scissors). The in vivo recombination produces a phagemid library with 6.5×10^{10} members with combined light-chain and heavy-chain diversity. (From Griffiths AD et al., *EMBO J*, 13:3245–3260, 1994.)

V_H domains of a single-chain variable fragment (scFv) [56]. In another application, the second complementarity determining region (CDR2) was replaced by a loxP site and separate CDR1, and CDR3 libraries were recombined [57]. The number of amino acids inserted by the loxP site has been reduced by placing the site within a self-splicing intron; following recombination, self-splicing of the intron resulted in a remaining insertion of only 15 base pairs [58].

The in vivo recombination system has also been simplified so as to require only a single vector [59]. This strategy was applied to an scFv library; mutant and wild-type loxP sites flanked the region encoding the V_H domain and the coat protein fusion (Figure 3.8). Bacteria expressing Cre recombinase were infected with extremely high concentrations of phage, and this resulted in individual cells being infected by multiple phage particles. Subsequent recombination resulted in V_H domain exchange between different vectors, and a recombined library of 3×10^{11} members was obtained from a primary library of 7×10^7 members.

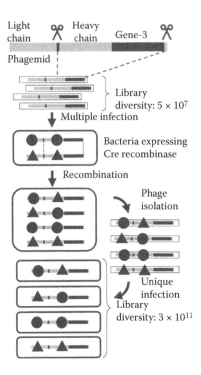

Figure 3.8 Generation of an scFv library by multiple infection and in vivo recombination. An scFv phagemid library with diversity in the light and heavy chains was used to infect *E. coli* at a very high multiplicity of infection. As a result, each bacterial cell harbored multiple phagemids, and in vivo recombination (represented by scissors) mediated by Cre recombinase resulted in the shuffling of light and heavy chains from different phagemids. The resulting recombined library had an estimated diversity of 3×10^{11} members. (From Sblattero D, Bradbury A, *Nat Biotechnol*, 18:75–80, 2000.)

3.5 DNA SHUFFLING

Oligonucleotide-directed and random mutagenesis methods can be used to generate mutations, but they lack an important element of natural evolution; namely, they do not allow for recombination of mutations from different selected sequences. Genetic recombination is a powerful means whereby beneficial traits engendered by individual mutations can be combined to potentially generate even greater beneficial effects. DNA shuffling techniques have been developed to mimic genetic recombination, and these methods have greatly enhanced the power of in vitro evolution strategies.

In the original DNA shuffling method [60,61] (Figure 3.9a), a DNA fragment containing the gene of interest was digested with Dnase I to yield a pool of small, random fragments. These fragments were then reassembled with a self-priming PCR in which the fragments primed each other to regenerate the full-length gene. By using a pool of DNA fragments that contained mutations selected for a particular

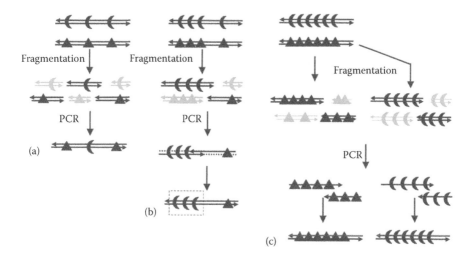

Figure 3.9 Principles and limitations of DNA shuffling. (a) Parental genes are fragmented. The fragments are mixed and reassembled by self-priming PCR, which results in the shuffling of sequences from different parents. (b) Mutations that are close together will recombine only rarely (low-resolution problem) and will tend to remain clustered together (dashed box). (c) If the homology between parental genes is too low, the fragments will reassemble mainly as the original parental genes.

trait, the DNA shuffling procedure resulted in a new pool that contained not only the starting sequences but also products that resulted from the recombination of different sequences within the pool. In this way, mutations from different clones were effectively combined to generate genes containing multiple mutations that could then be tested for further improvements in the selected trait. Additional mutations could also be introduced by the error-prone nature of DNA replication by *Taq* DNA polymerase. DNA shuffling has also been applied to combine sequences of related homologous genes rather than mutants of a single gene, and this process of "DNA family shuffling" has been very effective in generating novel functional proteins that combine sequence elements from closely related natural homologues [62].

Although very straightforward and used with great success, the original DNA shuffling technique suffers from several limitations. First of all, the Dnase I cleavage reaction is not perfectly random, as the enzyme prefers to hydrolyze dsDNA at sites adjacent to pyrimidine nucleotides [63]. The number of crossovers that can be created is also restricted by limitations on the fragment size, which has to provide a large enough overlap to ensure stable annealing between the fragments. Moreover, crossover occurs preferentially at regions of high sequence identity [64,65]. As a result of these limitations, blocks of parental sequences tend to be conserved as the method provides a low crossover resolution (Figure 3.9b). In the case of DNA family shuffling of natural homologues, the sequence identity must exceed a certain threshold (~70%) to allow for productive annealing between homologues [66]. Below the threshold value, very little recombination occurs, and the original parental genes are the major products of the reaction (Figure 3.9c). Several alternatives and improvements to the DNA shuffling method have been developed to address these limitations, as described in Sections 3.5.1 and 3.5.2.

3.5.1 Methods for Improved Crossover Resolution

Several methods have been developed in an effort to increase the frequency of crossovers in DNA shuffling. In the random-priming in vitro recombination strategy, random hexanucleotides are annealed to two parental templates and used to prime DNA synthesis [67]. The short fragments are then reassembled, by self-priming PCR, and it was shown that up to six crossovers were obtained. Alternatively, a staggered extension process (StEP) utilizes primers that anneal to one end of the parental genes and PCR with very short annealing and extension steps (Figure 3.10) [68]. As a result, only short extensions occur at each cycle, and the growing strands can switch templates multiple times as the replication of the full-length genes requires multiple cycles. In a method called "random chimeragenesis on transient

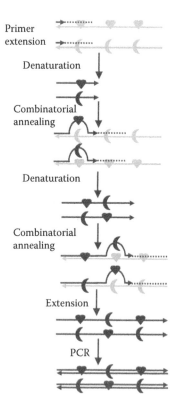

Figure 3.10 The staggered extension process (StEP). A primer is annealed to one end of each parental template and is extended in very short cycles of denaturation and rean-nealing. As a result, the elongating strand pairs with different templates in each extension cycle and the template-switching process results in chimeric genes. (From Zhao H et al., *Nat Biotechnol*, 16:258–261, 1998.)

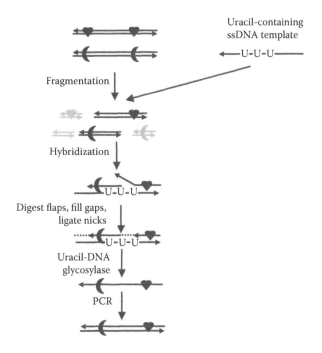

Figure 3.11   Random chimeragenesis on transient templates (RACHITT). Fragments from multiple genes are annealed onto a uracil-containing ssDNA template. Unannealed flaps are removed by nuclease treatment, the gaps are filled, and the nicks are ligated. The ssDNA template is inactivated by treatment with uracil-DNA glycosylase, and the reassembled chimeric genes are amplified by PCR. (From Coco WM et al., *Nat Biotechnol*, 19:354–359, 2001.)

templates" (RACHITT), a uracil-containing parental ssDNA strand is used as a template onto which fragments generated by Dnase digestion are annealed (Figure 3.11) [69]. Unhybridized ends on the annealed fragments are removed by nuclease digestion, gaps are filled in by DNA polymerase, and full-length genes are regenerated by sealing nicks with DNA ligase. The transient template is rendered unamplifiable by treatment with uracil-DNA glycosylase, and PCR can be used to amplify the chimeric genes. It was demonstrated that this method produced much higher crossover frequencies in comparison with conventional DNA shuffling.

3.5.2 Incorporation of Synthetic DNA

The earliest reports of DNA shuffling demonstrated that diversity could be increased by the addition of synthetic oligonucleotides that were incorporated into the reassembled chimeric genes [60,70]. Indeed, several reports have demonstrated that it is feasible to perform DNA shuffling with synthetic oligonucleotides as the only source of DNA [71,72], as exemplified by the degenerate homoduplex recombination (DHR) method [73]. In this method, "top-strand" oligonucleotides are synthesized so that together, they span the entire length of a gene with the exception of small gaps between the oligonucleotides (Figure 3.12). The gaps are bridged by shorter "bottom-strand" oligonucleotides that are designed to anneal to the ends of two adjacent top-strand oligonucleotides. The oligonucleotides are modified during chemical synthesis such that only the top-strand oligonucleotides are competent for enzyme-mediated extension and ligation. Chimeric gene assembly is achieved by annealing all of the oligonucleotides together, followed by extension and ligation to fill in the gaps to produce full-length, recombined genes.

Methods of this type combine the precision of oligonucleotide-directed mutagenesis with the combinatorial power of DNA shuffling, and they provide significant advantages over conventional DNA shuffling. The use of degenerate synthetic DNA enables precise design of sequence variations and greatly increases the levels of diversity that can be introduced. Second, the crossovers are generated in a designed manner through chemical synthesis, and the process can be more precisely controlled to achieve optimal, unbiased crossover resolution.

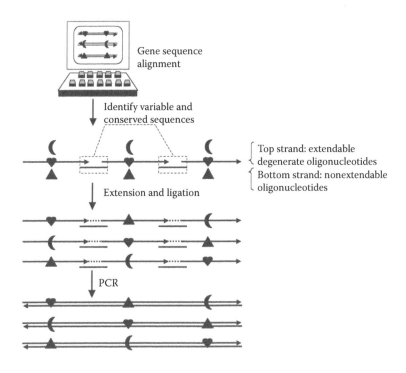

Figure 3.12 Degenerate homoduplex recombination (DHR). Conserved regions are identified by sequence alignment of homologous genes. Synthetic oligonucleotides are synthesized as either extendable top strands or as nonextendable bottom strands. The bottom strand oligonucleotides serve to assemble the top strand oligonucleotides (dashed boxes). Enzyme-mediated extension and ligation generates full-length chimeric genes, which can be amplified by PCR. (From Coco WM et al., *Nat Biotechnol*, 20:1246–1250, 2002.)

REFERENCES

1. Barbas CF III, Bain JD, Hoekstra DM, Lerner RA. Semisynthetic combinatorial antibody libraries: A chemical solution to the diversity problem. *Proc Natl Acad Sci U S A* 1992; 89:4457–4461.
2. Scott JK, Smith GP. Searching for peptide ligands with an epitope library. *Science* 1990; 249:386–390.
3. Devlin JJ, Panganiban LC, Devlin PE. Random peptide libraries: A source of specific protein binding molecules. *Science* 1990; 249:404–406.
4. Cwirla SE, Peters EA, Barrett RW, Dower WJ. Peptides on phage: A vast library of peptides for identifying ligands. *Proc Natl Acad Sci U S A* 1990; 87:6378–6382.
5. Deshayes K, Schaffer ML, Skelton NJ, Nakamura GR, Kadkhodayan S, Sidhu SS. Rapid identification of small binding motifs with high-throughput phage display: Discovery of peptidic antagonists of IGF-1 function. *Chem Biol* 2002; 9:495–505.
6. Sidhu SS, Lowman HB, Cunningham BC, Wells JA. Phage display for selection of novel binding peptides. *Methods Enzymol* 2000; 328:333–363.
7. Clackson T, Wells JA. In vitro selection from protein and peptide libraries. *Trends Biotechnol* 1994; 12:173–184.
8. Hutchison CA III, Nordeen SK, Vogt K, Edgell MH. A complete library of point substitution mutations in the glucocorticoid response element of mouse mammary tumor virus. *Proc Natl Acad Sci U S A* 1986; 83:710–714.
9. Hermes JD, Parekh SM, Blacklow SC, Koster H, Knowles JR. A reliable method for random mutagenesis: The generation of mutant libraries using spiked oligodeoxyribonucleotide primers. *Gene* 1989; 84:143–151.
10. Arkin AP, Youvan DC. Optimizing nucleotide mixtures to encode specific subsets of amino acids for semi-random mutagenesis. *Biotechnology (N Y)* 1992; 10:297–300.
11. LaBean TH, Kauffman SA. Design of synthetic gene libraries encoding random sequence proteins with desired ensemble characteristics. *Protein Sci* 1993; 2:1249–1254.
12. Glaser SM, Yelton DE, Huse WD. Antibody engineering by codon-based mutagenesis in a filamentous phage vector system. *J Immunol* 1992; 149:3903–3913.
13. Huse WP, Yelton DE, Glaser SM. Increased antibody affinity and specificity by codon-based mutagenesis. *Int Rev Immunol* 1993; 10:129–137.
14. Neuner P, Cortese R, Monaci P. Codon-based mutagenesis using dimer-phosphoramidites. *Nucleic Acids Res* 1998; 26:1223–1227.
15. Ono A, Matsuda A, Zhao J, Santi DV. The synthesis of blocked triplet-phosphoramidites and their use in mutagenesis. *Nucleic Acids Res* 1995; 23:4677–4682.
16. Sondek J, Shortle D. A general strategy for random insertion and substitution mutagenesis: Substoichiometric coupling of trinucleotide phosphoramidites. *Proc Natl Acad Sci U S A* 1992; 89:3581–3585.
17. Virnekas B, Ge L, Pluckthun A, Schneider KC, Wellnhofer G, Moroney SE. Trinucleotide phosphoramidites: Ideal reagents for the synthesis of mixed oligonucleotides for random mutagenesis. *Nucleic Acids Res* 1994; 22:5600–5607.
18. Pal G, Kossiakoff AA, Sidhu SS. The functional binding epitope of a high affinity variant of human growth hormone mapped by shotgun alanine-scanning mutagenesis: Insights into the mechanisms responsible for improved affinity. *J Mol Biol* 2003; 332:195–204.
19. Vajdos FF, Adams CW, Breece TN, Presta LG, de Vos AM, Sidhu SS. Comprehensive functional maps of the antigen-binding site of an anti-ErbB2 antibody obtained with shotgun scanning mutagenesis. *J Mol Biol* 2002; 320:415–428.

20. Zoller MJ, Smith M. Oligonucleotide-directed mutagenesis of DNA fragments cloned into M1 3 vectors. *Methods Enzymol* 1983; 100:468–500.
21. Zoller MJ, Smith M. Oligonucleotide-directed mutagenesis using M13-derived vectors: An efficient and general procedure for the production of point mutations in any fragment of DNA. *Nucleic Acids Res* 1982; 10:6487–6500.
22. Kunkel TA. Rapid and efficient site-specific mutagenesis without phenotypic selection. *Proc Natl Acad Sci U S A* 1985; 82:488–492.
23. Kunkel TA, Roberts JD, Zakour RA. Rapid and efficient site-specific mutagenesis without phenotypic selection. *Methods Enzymol* 1987; 154:367–382.
24. Wells JA, Vasser M, Powers DB. Cassette mutagenesis: An efficient method for generation of multiple mutations at defined sites. *Gene* 1985; 34:315–323.
25. Oliphant AR, Nussbaum AL, Struhl K. Cloning of random-sequence oligodeoxynucleotides. *Gene* 1986; 44:177–183.
26. Reidhaar-Olson JF, Bowie JU, Breyer RM, Hu JC, Knight KL, Lim WA, Mossing MC et al. Random mutagenesis of protein sequences using oligonucleotide cassettes. *Methods Enzymol* 1991; 208:564–586.
27. Goldman ER, Youvan DC. An algorithmically optimized combinatorial library screened by digital imaging spectroscopy. *Biotechnology (N Y)* 1992; 10:1557–1561.
28. Bonnycastle LL, Mehroke JS, Rashed M, Gong X, Scott JK. Probing the basis of antibody reactivity with a panel of constrained peptide libraries displayed by filamentous phage. *J Mol Biol* 1996; 258:747–762.
29. Higuchi R, Krummel B, Saiki RK. A general method of in vitro preparation and specific mutagenesis of DNA fragments: Study of protein and DNA interactions. *Nucleic Acids Res* 1988; 16:7351–7367.
30. Ho SN, Hunt HD, Horton RM, Pullen JK, Pease LR. Site-directed mutagenesis by overlap extension using the polymerase chain reaction. *Gene* 1989; 77:51–59.
31. Kammann M, Laufs J, Schell J, Gronenborn B. Rapid insertional mutagenesis of DNA by polymerase chain reaction (PCR). *Nucleic Acids Res* 1989; 17:5404.
32. Ke SH, Madison EL. Rapid and efficient site-directed mutagenesis by single-tube "megaprimer" PCR method. *Nucleic Acids Res* 1997; 25:3371–3372.
33. Hemsley A, Arnheim N, Toney MD, Cortopassi G, Galas DJ. A simple method for site-directed mutagenesis using the polymerase chain reaction. *Nucleic Acids Res* 1989; 17:6545–6551.
34. De Lucia P, Cairns J. Isolation of an E. coli strain with a mutation affecting DNA polymerase. *Nature* 1969; 224:1164–1166.
35. Solnick D. An adenovirus mutant defective in splicing RNA from early region 1A. *Nature* 1981; 291:508–510.
36. Chu CT, Parris DS, Dixon RA, Farber FE, Schaffer PA. Hydroxylamine mutagenesis of HSV DNA and DNA fragments: Introduction of mutations into selected regions of the viral genome. *Virology* 1979; 98:168–181.
37. Busby S, Irani M, Crombrugghe B. Isolation of mutant promoters in the Escherichia coli galactose operon using local mutagenesis on cloned DNA fragments. *J Mol Biol* 1982; 154:197–209.
38. Kadonaga JT, Knowles JR. A simple and efficient method for chemical mutagenesis of DNA. *Nucleic Acids Res* 1985; 13:1733–1745.
39. Shortle D, Botstein D. Directed mutagenesis with sodium bisulfite. *Methods Enzymol* 1983; 100:457–468.
40. Cox EC. Bacterial mutator genes and the control of spontaneous mutation. *Annu Rev Genet* 1976; 10:135–156.

41. Greener A, Callahan M, Jerpseth B. An efficient random mutagenesis technique using an E. coli mutator strain. *Mol Biotechnol* 1997; 7:189–195.

42. Camps M, Naukkarinen J, Johnson BP, Loeb LA. Targeted gene evolution in Escherichia coli using a highly error-prone DNA polymerase I. *Proc Natl Acad Sci U S A* 2003; 100:9727–9732.

43. Zakour RA, Loeb LA. Site-specific mutagenesis by error-directed DNA synthesis. *Nature* 1982; 295:708–710.

44. Vartanian JP, Henry M, Wain-Hobson S. Hypermutagenic PCR involving all four transitions and a sizeable proportion of transversions. *Nucleic Acids Res* 1996; 24: 2627–2631.

45. Zhou YH, Zhang XP, Ebright RH. Random mutagenesis of gene-sized DNA molecules by use of PCR with Taq DNA polymerase. *Nucleic Acids Res* 1991; 19:6052.

46. Fromant M, Blanquet S, Plateau P. Direct random mutagenesis of gene-sized DNA fragments using polymerase chain reaction. *Anal Biochem* 1995; 224:347–353.

47. Leung DW, Cachianes G, Kuang WJ, Goeddel DV, Ferrara N. Vascular endothelial growth factor is a secreted angiogenic mitogen. *Science* 1989; 246:1306–1309.

48. Cadwell RC, Joyce GF. Randomization of genes by PCR mutagenesis. *PCR Methods Appl* 1992; 2:28–33.

49. Botstein D, Shortle D. Strategies and applications of in vitro mutagenesis. *Science* 1985; 229:1193–1201.

50. Moore JC, Arnold FH. Directed evolution of a para-nitrobenzyl esterase for aqueous–organic solvents. *Nat Biotechnol* 1996; 14:458–467.

51. Zhao H, Arnold FH. Optimization of DNA shuffling for high fidelity recombination. *Nucleic Acids Res* 1997; 25:1307–1308.

52. Waterhouse P, Griffiths AD, Johnson KS, Winter G. Combinatorial infection and in vivo recombination: A strategy for making large phage antibody repertoires. *Nucleic Acids Res* 1993; 21:2265–2266.

53. Sternberg N, Hamilton D. Bacteriophage P1 site-specific recombination. I. Recombination between loxP sites. *J Mol Biol* 1981; 150:467–486.

54. Hoess RH, Wierzbicki A, Abremski K. The role of the loxP spacer region in P1 site-specific recombination. *Nucleic Acids Res* 1986; 14:2287–2300.

55. Griffiths AD, Williams SC, Hartley O, Tomlinson IM, Waterhouse P, Crosby WL, Kontermann RE et al. Isolation of high affinity human antibodies directly from large synthetic repertoires. *EMBO J* 1994; 13:3245–3260.

56. Tsurushita N, Fu H, Warren C. Phage display vectors for in vivo recombination of immunoglobulin heavy and light chain genes to make large combinatorial libraries. *Gene* 1996; 172:59–63.

57. Davies J, Riechmann L. An antibody VH domain with a lox-Cre site integrated into its coding region: Bacterial recombination within a single polypeptide chain. *FEBS Lett* 1995; 377:92–96.

58. Fisch I, Kontermann RE, Finnern R, Hartley O, Soler-Gonzalez AS, Griffiths AD, Winter G. A strategy of exon shuffling for making large peptide repertoires displayed on filamentous bacteriophage. *Proc Natl Acad Sci U S A* 1996; 93:7761–7766.

59. Sblattero D, Bradbury A. Exploiting recombination in single bacteria to make large phage antibody libraries. *Nat Biotechnol* 2000; 18:75–80.

60. Stemmer WP. DNA shuffling by random fragmentation and reassembly in vitro recombination for molecular evolution. *Proc Natl Acad Sci U S A* 1994; 91:10747–10751.

61. Stemmer WP. Rapid evolution of a protein in vitro by DNA shuffling. *Nature* 1994; 370:389–391.

62. Crameri A, Raillard SA, Bermudez E, Stemmer WP. DNA shuffling of a family of genes from diverse species accelerates directed evolution. *Nature* 1998; 391:288–291.

63. Murakami H, Hohsaka T, Sisido M. Random insertion and deletion of arbitrary number of bases for codon-based random mutation of DNAs. *Nat Biotechnol* 2002; 20:76–81.

64. Moore GL, Maranas CD, Lute S, Benkovic SJ. Predicting crossover generation in DNA shuffling. *Proc Natl Acad Sci U S A* 2001; 98:3226–3231.

65. Patrick WM, Firth AE, Blackburn JM. User-friendly algorithms for estimating completeness and diversity in randomized protein-encoding libraries. *Protein Eng* 2003; 16:451–457.

66. Joern JM, Meinhold P, Arnold FH. Analysis of shuffled gene libraries. *J Mol Biol* 2002; 316:643–656.

67. Shao Z, Zhao H, Giver L, Arnold FH. Random-priming in vitro recombination: An effective tool for directed evolution. *Nucleic Acids Res* 1998; 26:681–683.

68. Zhao H, Giver L, Shao Z, Affholter JA, Arnold FH. Molecular evolution by staggered extension process (StEP) in vitro recombination. *Nat Biotechnol* 1998; 16:258–261.

69. Coco WM, Levinson WE, Crist MJ, Hektor HJ, Darzins A, Pienkos PT, Squires CH et al. DNA shuffling method for generating highly recombined genes and evolved enzymes. *Nat Biotechnol* 2001; 19:354–359.

70. Crameri A, Stemmer WP. Combinatorial multiple cassette mutagenesis creates all the permutations of mutant and wild-type sequences. *Biotechniques* 1995; 18:194–196.

71. Ness JE, Kim S, Gottman A, Pak R, Krebber A, Borchert TV, Govindarajan S et al. Synthetic shuffling expands functional protein diversity by allowing amino acids to recombine independently. *Nat Biotechnol* 2002; 20:1251–1255.

72. Zha D, Eipper A, Reetz MT. Assembly of designed oligonucleotides as an efficient method for gene recombination: A new tool in directed evolution. *ChemBioChem* 2003; 4:34–39.

73. Coco WM, Encell LP, Levinson WE, Crist MJ, Loomis AK, Licato LL, Arensdorf JJ et al. Growth factor engineering by degenerate homoduplex gene family recombination. *Nat Biotechnol* 2002; 20:1246–1250.

CHAPTER 4

Selection and Screening Strategies

Mark S. Dennis

CONTENTS

4.1 INTRODUCTION

A multitude of selection and screening strategies for phage-displayed libraries have been developed since the technique was first described by Scott and Smith [1]. Each relies on the ability to discriminate desired phage from the library either through selective capture to allow the removal of unbound phage or through selective elution, as in the case of substrate phage to specifically elute desired phage (see Chapter 8). Ultimately, the degree to which target-directed phage can be separated from the rest of the phage library will determine the effectiveness of the panning method. This chapter summarizes many of these strategies and discusses the considerations that must be kept in mind when using them.

4.2 GENERAL CONSIDERATIONS

The selection of phage-displayed ligands to a specific target requires that the target be presented in a native form and at a sufficient concentration to allow enrichment over background binding phage. If the target concentration is too low, the number of selectively bound phage recovered following a round of selection will have no amplification advantage relative to the background binding phage and will thus never take over the library population. Particularly in the initial rounds of selection, the concentration of any given phage in the library is extremely low, and thus, the rate of phage binding to the target is relatively slow. Capture can be improved by exposure of the phage pool to the target for several hours at room temperature or overnight at 4°C.

The anticipated affinity of the interaction between the displayed ligand and the target is also an important consideration (Table 4.1). If the interaction between a ligand and its target is very tight, a target presented at a relatively high concentration or the displayed ligand presented in a polyvalent format on phage will make selection of tighter binding variants more difficult. Inversely, if the interaction is likely to be weak, as with naïve peptide libraries, polyvalent display of the ligand or the presentation of a relatively high concentration of the target is desired in order to boost the chances of finding an interaction.

Another important aspect of sorting is the number of phage to include in the selection round. This will depend upon the diversity and valence of the library. Generally, when sorting, enough phage should be added so that every member of the library is represented 100–1000 times. Monovalently displayed libraries require higher representation depending upon the ligand display level relative to polyvalent phage libraries for which every phage particle is expected to present at least a few copies of the displayed ligand. Bass et al. [2], for example, have estimated that monovalently displayed growth hormone is present on about 10% of the phage; thus, a 100-fold overrepresentation of each library member results in each being present only 10 times.

To reduce nonspecific phage binding, irrelevant proteins, such as bovine serum albumin (BSA), gelatin, casein, ovalbumin, or powdered milk are often used to coat

Table 4.1   Suggested Phage Valence and Target
Concentration for Selections

Valence	Target Concentration (nM)	Affinities Selected
Polyvalent	1–10	0.1–100 µM
Monovalent	100	0.1–10 µM
Monovalent	1–10	0.1–100 nM
Monovalent	0.1	10–100 pM

the surfaces in the sorting reaction and can also be included in the sorting buffer along with nonionic detergents such as Tween 20 or Triton X-100. Occasionally, phage libraries contain members that will bind to these blocking agents, and this results in a high background after a few rounds of selection and amplification. In such cases, a change or rotation of blocking agents and repetition of the previous round of selection generally avoids or reduces this increase in background. Obviously, when choosing a sorting buffer, the stability of the target must be considered as well as any requirements for reducing agent, divalent cations, or other cofactors. Additionally, trace amounts of biotin present in casein or powdered milk can interfere with selections using an avidin capture (see Section 4.4.1). In selection strategies discussed in Section 4.4, the presentation of target in a diverse environment can serve to weed out nonspecific phage.

Depending upon the method employed, the selection stringency is an important consideration. This is particularly true in the first rounds of selection, when the library diversity is the greatest and the representation of each member is the lowest. In order to maintain this diversity through successive rounds, stringency should begin low and gradually increase with successive rounds as diversity is reduced and the population of selectively bound phage increases. Thus, in the early rounds, efforts to maximize the capture of all potentially interesting clones should be made. As enrichment of selective binding clones is observed in later rounds, stringency can be increased. An effective means of increased selection stringency is through reduction of the target concentration, causing only the tightest binding clones to be selected. However, a minimum target concentration—one that provides a higher number of selective versus background phage to be recovered—is required for enrichment to occur.

Besides selecting phage based upon binding affinity, phage-displayed libraries can be directed to discriminate between two or more closely related targets using a technique called competitive selection. Competitive selection involves providing an undesired target in solution while capturing discriminating phage that are bound to the desired immobilized target. This technique has been used successfully to increase the selectivity of a general serine protease inhibitor for the coagulation protease, factor VIIa (FVIIa) [3], and also, to generate receptor-selective variants of atrial natriuretic peptide and vascular endothelial growth factor [4,5]. In many cases, variants exhibiting greater than 1000-fold selectivity have been obtained. To establish initial selection conditions, the binding of phage bearing the wild-type ligand to the desired immobilized target is monitored in the presence of increasing concentrations of an undesired competing target. The concentration of the soluble competing target to add during the selection depends upon its affinity for the initial lead and is determined empirically. The point at which 75–95% of the phage are prevented from binding to the immobilized target provides a good starting concentration of the competing target to add in the first round of selection (Figure 4.1). For successive rounds, an increase in enrichment can be countered with an increase in stringency, generated by increasing the concentration of the competing target. In many cases, subtle sequence changes to the binding interface can dramatically alter ligand-binding specificity.

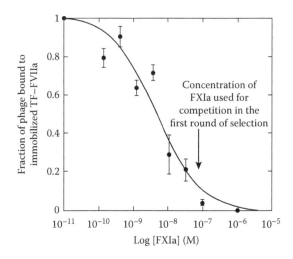

Figure 4.1 Establishing conditions for competitive selection of phage binding to an immobilized target. Phage displaying the wild-type ligand are added to target (tissue factor [TF]–FVIIa)-coated wells containing increasing concentrations of a competitive target (FXIa) in solution. Selection conditions for libraries based on the displayed ligand should begin with the concentration of the competitive target required to reduce phage capture by 75%–95% (indicated by arrow) and is dependent on the affinity of the ligand for the competitive target. As the rounds of selection progress and enrichment is observed, the concentration of the competing target can be increased to further drive the selection of specific binding variants.

4.3 THE SELECTION PROCESS

4.3.1 Washing

As mentioned in Section 4.2, in the early rounds of selection, when library diversity is high and the number of desired clones is low, it is important to maximize the recovery of desired clones. Thus, stringent washing to lower background binding phage in the initial rounds of selection is generally not prudent. Wash stringency can be increased in later rounds, as diversity is reduced through the loss of nonbinding clones and the population of desired clones is increased. This can be achieved through multiple washings or increased washing times or by adding increasing concentrations of a competing ligand to the wash buffer to allow clones with slower dissociation rate constants to be selected.

4.3.2 Elution Methods

The most commonly used method for recovering phage bound to a target is through simple disruption of the binding interaction between the displayed ligand

and the target by altering the pH. Other disruptive techniques are also suitable as long as they are reversible to the point that they do not interfere with the amplification of the phage in *Escherichia coli* (i.e., infection). Although this method generally does not provide complete elution of the bound phage, a sufficient amount of phage is eluted to allow for amplification, and good enrichments are achieved.

The use of a known ligand to selectively displace bound phage can be successful, although very high concentrations of ligand and long elution times are generally required. This is particularly true for tight-binding ligands with slow off rates. Ligand elution is unlikely to work with ligands displayed on protein 8 (p8) as a result of the high display and avidity provided by polyvalent display.

Selective proteolytic elution is another useful method that has been used to identify selective ligands. For the selection of peptides that bound to the erythropoietin receptor, the target was presented as an immunoadhesin [6]. This can frequently result in unintended selection of clones that bind to the Fc region of the immunoadhesin. To avoid selection of these nontarget-directed clones, a thrombin cleavage site was introduced between the Fc and the erythropoietin receptor. Thus, upon the addition of thrombin and release of the immobilized target, phage that bound selectively to the receptor could be selectively eluted [6].

4.3.3 Amplification

Following elution, the eluted pool of phage can be amplified in *E. coli* prior to further rounds of selection. This amplification process can lead to artifacts such as the biased production of better-expressing clones or faster-growing "monster" phage clones that lack any displayed ligand. Generally, however, if specific binding clones are present in the library and are captured in sufficient numbers, these artifacts remain a minor component of the amplified library.

To avoid a loss of clonal diversity, an excess of rapidly growing *E. coli* host must be provided to capture all the phage present in the pool. Following infection with the eluted phage, a helper phage is added for the production of new phage particles.

4.3.4 Monitoring the Selection Process

By monitoring the enrichment ratio (the ratio of clones eluted from a target-coated well to clones eluted from a blank well), the progress of the selection process can be tracked. With naïve libraries, enrichment is usually not observed during the first two rounds of selection; usually, in the third round, the enrichment ratio is greater than 10 and can be as high as 1000 or more in later rounds. For biased libraries, or those based on a preexisting ligand, enrichment may be observed sooner. The number of phage added and recovered during a typical phage selection experiment is shown in Table 4.2.

Table 4.2   Phage (cfu) Present in Pools from a Typical Panning Experiment

Phage Pool	Round 1	Round 2	Round 3	Round 4
Starting phage library diversity	1×10^9	5×10^5	1×10^6	1×10^7
Phage input		1×10^{11}		
Target eluted phage	1×10^5	5×10^5	1×10^6	1×10^7
Nontarget eluted phage	1×10^5	1×10^5	1×10^5	1×10^5
Enrichment	1×	5×	10×	100×
Amplified phage pool		1×10^{14}		

Note: cfu, colony forming units.

4.3.5 Sequence Analysis

The decision of when to sequence clones from selected libraries is based on two general strategies and depends upon the nature of the library and the information desired. For maturing existing phage–target associations, such as ligand–receptor interactions, often, only the sequence of the best variant is important. Generally, four to six rounds of selection may be required to reduce the population to a handful of "winners." Since phage selection is a function of display level (i.e., expression) as well as binding affinity, excessive rounds of selection can lead to artifacts, as mentioned in Section 4.3.3. For example, identical clones or "siblings" present at a high frequency may result not only from enhanced binding properties but also from higher expression and display. Thus, sequencing clones from a selection round that still retains some diverse sequences followed by an assessment of the purified ligands is recommended to avoid missing the best variant. In contrast, when searching for an initial lead, as with naïve libraries, often, a diverse set of potential leads can be useful. Here, examination of sequences following just two to three rounds of selection can yield a surprising array of initial leads. In this case, a large number of potential leads can be rapidly evaluated by assessing specific phage binding en masse before progressing further [7].

While sequence information can be useful and interesting, what ultimately matters is a ligand possessing the properties being sought. Several individual phage clones can be selected and tested for target-specific binding. Although difficult to achieve when using polyvalent phage, monovalent phage can be assessed for specific binding and an estimate of affinity can be determined by titrating phage binding with a soluble target [8].

4.4 SELECTION METHODS

4.4.1 Purified Targets

The most frequently used method of screening against purified protein targets involves direct immobilization on a solid support such as a bead or microtiter plate. This enables the physical separation of bound and unbound phage simply by washing the support. Numerous supports have been used, including modified affinity resins, glass beads, modified magnetic beads, covalently modified plastics, and chelating matrices. Supports should be chosen based upon their low background for nonspecific phage binding and their ability to present the target in a native conformation and at a desirable concentration. Characteristics of the target molecule, such as stability, solubility, and availability of free thiols, free amines, or carbohydrates for covalent matrix attachment, are also important considerations. For example, mutations engineered on the surface of a target protein may provide a means to chemically modify a specific site to facilitate capture and present the target in a specific orientation [9]. Protein complexes can be screened through direct immobilization of the complex or through the specific capture of the target through an immobilized associating partner as was demonstrated in the capture of FVIIa by immobilized TF to generate an immobilized TF–FVIIa complex [10].

Special consideration should be made when panning phage libraries against biotinylated targets captured on avidin-coated surfaces, since this can lead to the unintended selection of biotin mimics [11,12]. To avoid the selection of biotin mimics such as the tripeptide sequence HPQ, excess biotin can be added following capture of the immobilized target to block remaining free biotin sites prior to the addition of the phage library. In addition, rotation between avidin, streptavidin, and neutravidin as the capture medium between successive rounds of selection can prevent selection of clones selective for other regions of these proteins (unpublished data).

Alternatively, phage binding to biotinylated targets can be performed in solution followed by a very brief selective capture of the target on avidin-coated plates or magnetic beads [13]. Magnetic beads can be used to reduce avidity effects resulting from displaying multiple copies of the ligand on phage since independent binding interactions are not tied to a matrix. The brief capture often leads to a dramatically lower background, and the ability to bind the target in solution provides a convenient way to control the target concentration. Importantly, selection and blocking buffers containing trace amounts of biotin (present in casein and powdered milk) should be avoided.

As evident in Table 4.1, selections for ligand affinities in the subnanomolar range are difficult. By utilizing a solution-binding strategy, the target concentration can be reduced to less than 1 nM. This in concert with increased stringency, such as a dramatically extended the wash period, can often lead to the selection of affinities in the picomole range. When target concentrations below 1 nM are used, however, very low specific phage titers are recovered. Thus, this method should only be used in later selection rounds after diversity has been reduced and the populations of binding phage are high. Although phage will be recovered, the values determined for enrichment are not likely to be meaningful, and thus, further characterization of selected clones is required to identify those with desired properties.

4.4.2 Cell-Surface Targets

Several methods have been developed for panning libraries against whole cells. These techniques are useful for protein targets that cannot be easily expressed and purified or presented in an active from. Alternatively, libraries can be directed toward specific cell types to identify particular antigens that differentiate them. For example, by panning against a cancer cell line, a cell-specific marker may be identified. In such an instance, the selected target is only relevant in that it may distinguish a specific cell line from other types of cells. This approach may have particular utility in identifying tumor-specific antigens. Ligands or antibodies directed toward these antigens may in turn be armed with drugs for specific delivery to these cell types.

The ability to pan a particular target expressed on a cell surface can be significantly more challenging than panning against a purified target. Because of the vast heterogeneity presented on the cell surface and the generally low target concentration available, elaborate methods are typically required to reduce background binding and improve enrichment. Methods to overcome these limitations have relied on overexpression of the receptor as a means of biasing the selection, as well as additional steps to ensure selectivity. Direct selection against a specific target on cells has only been successful in a few cases, and involve libraries that are biased toward the target [14,15]. Ligand elution can be used to selectively identify phage that bind to a specific cell target [16].

Subtractive panning is the most common method for which there are a number of variations. Generally, phage libraries are preincubated with similar cells lacking the receptor of interest in order to remove nonspecific binding phage from the library. Rotation between multiple cell lines expressing the receptor from one round to the next, thus ensuring that the background is constantly changing, can facilitate the selection of specific clones [17,18]. For example, De Lorenzo et al. [18] rotated their selection against two different target-expressing cell lines in the presence of the same cells lacking a target to attain clones that bound specifically to ErbB2 from a single chain variable fragment (scFv) phage library. Phage bound to fluorescently labeled target-bearing cells were separated from the mixture by fluorescence-activated cell sorting (FACS).

In comparison to selecting against specific targets expressed on cells, the ability of phage libraries to identify antigens that distinguish between cell types has been much more successful through competitive selections to preabsorb or competitively absorb undesired phage [19–23]. Alternatively, while selecting against a target receptor presented on immobilized cells, the same cell type lacking the targeted receptor can be presented in suspension (Figure 4.2). These methods tend to competitively eliminate nonselective phage and allow for the selection of clones that bind to distinguishing regions on the cell surface.

The success of cell selection strategies can depend upon the method used for separating bound and unbound phage. In addition to washing directly immobilized target cells, centrifugation through a density gradient [24] or nonmiscible organic phase [22] has been used to separate cells and bound phage from unbound phage (Figure 4.3). These methods rapidly remove nonspecific phage, providing lower backgrounds and higher cell-bound phage recoveries than traditional washing of adherent cells.

Following three rounds of this selective process, Giordano et al. [22] found that 67% of clones from a polyvalent peptide library selected against vascular endothelial growth factor-stimulated human endothelial cells displayed enrichments of less than 10-fold, yet when compared for binding to directly immobilized receptors,

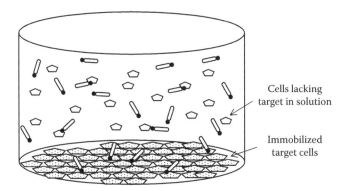

Figure 4.2   Selective binding of phage to target cells by competitive selection. The phage library is mixed with nontarget cells in suspension. The mixture is added to target cells immobilized on plastic. The nontarget cells adsorb phage library members that recognize epitopes present on both cell types. Only unique targets on the immobilized cells recruit phage binding.

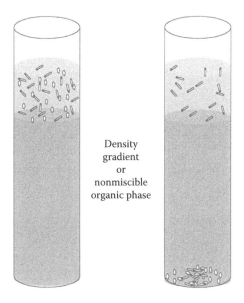

Figure 4.3   Phage separation by selective centrifugation. The phage library is mixed with cells in suspension. Cells, along with bound phage, are separated by centrifugation. Nonprecipitated phage can be used in subtractive panning. Alternatively, phage associated with centrifuged cells can be isolated.

enrichments of up to 1000-fold were observed. This illustrates that while lower backgrounds and higher specific phage recoveries can be obtained in comparison to traditional washing of adherent cells, the ability to identify rare or lower-affinity clones may still be expected to be much more challenging using cells as opposed to standard target immobilization methods, outlined in Section 4.4.1.

To separate phage bound to target cells from competing nontarget cells, a number of methods, including FACS and selective capture (biotinylation), have been used. For example, the separation of phage bound to a mixture of particular cell types, each distinguished by fluorescently labeled antibodies directed to cell-specific markers, can be accomplished using FACS [25]. A selectable marker such as random biotinylation of the cell surface can identify cells bearing the target antigen, allowing cells to be mixed with nontarget cells for selection in solution followed by specific capture of target cells [26].

The pathfinder approach offers a novel and highly selective means of identifying phage that bind to a cell-surface receptor. This approach utilizes a ligand or antibody (the pathfinder) to the receptor conjugated to horseradish peroxidase (HRP), in order to colocalize HRP in the vicinity of the target (Figure 4.4a and b) [27]. HRP is then used to catalyze the conversion of biotin tyramine (BT) to a free radical that will biotinylate the nearest neighboring nucleophile. Target-bound phage are biotinylated and can be preferentially recovered from nonspecific, nonbiotinylated phage using streptavidin-coated magnetic beads [27]. Thus, the method offers the ability to select for phage binding to epitopes neighboring existing ligand-binding sites. Alternatively, ligands identified in this manner can be used as pathfinders themselves to drive selection of new ligands to the original pathfinder site (Figure 4.4c and d) [28].

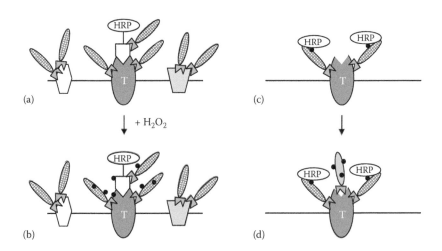

Figure 4.4 The pathfinder approach. (a) Phage are bound to the surface of a cell expressing target antigen (T) in the presence of an HRP-conjugated antibody or natural ligand directed to the target. (b) In the presence of hydrogen peroxide, biotin tyramine (●) is covalently linked to phage binding in the vicinity of the HRP. Biotinylated phage can then be eluted and recovered using streptavidin-coated beads. These phage likely recognize the target antigen or the ligand–target antigen complex. (c) Additionally, the eluted biotinylated phage can be added to fresh target-presenting cells in the presence of streptavidin–HRP. (d) A new aliquot of the phage library is added, and a second biotinylation reaction is performed. All biotinylated phage are eluted from the cells, recovered using streptavidin-coated beads, and screened for the ability to bind to the original pathfinder site.

Frequently, a pathfinder to deliver HRP to the vicinity of the target is unavailable. In such instances, the addition of a flag tag to the target could potentially allow generation of an artificial pathfinder delivery system.

For the targeted delivery of therapeutics to cells, the ability of ligands to be internalized into cells is an important feature. This method relies on the ability of phage bound to cellular receptors to be internalized and illustrates many of the challenges in panning against cell-surface antigens. By stringently stripping phage from the cell surface, causing only internalized phage to be recovered, this method has the added potential advantage of lowering nonspecific background phage [29–31]. Becerril et al. [29] panned against cells that overexpress ErbB2 and found that recovery of internalized phage could provide an additional (~10-fold) enrichment when compared to recovery of phage from the cell surface. In another example, Heitner et al. [30] have used this technique to identify internalizing phage sorted against cells expressing the epidermal growth factor receptor (EGFR). In this example, both adherent (Chinese hamster ovary cells [CHO]/EGFR) and nonadherent (A431) cell lines expressing the target receptor were used in selections incorporating two styles of subtractive panning. For the adherent cell line, untransfected CHO cells were added in solution to selectively remove the nontarget-directed phage. For the nonadherent cell line, the phage library was preselected against fibroblasts prior to exposure to target-bearing cells. In both cases, binding was performed at 4°C, the cells were washed, and internalization was initiated by warming the cells to 37°C. Following an extensive wash, the cells were lysed with triethylamine, and recovered phage were amplified for the next round of panning. Increased phage titers were observed with successive rounds of panning. After three rounds, individual clones were selected and screened for binding to the EGFR. Although target-specific clones were identified, only 10% of the phage recovered from nonadherent cells and 1% of the phage recovered from the adherent cells were target specific. The success of this approach requires that phage binding the receptor of interest are preferentially internalized over other cell-surface binding phage, and factors such as the target receptor expression level and rate of internalization relative to other competing receptors are likely important. Despite this clever combination of subtractive panning and selection, these examples point to the difficulties involved in panning whole cells.

An interesting enhancement of the selection for internalizing phage involves the inclusion of a reporter gene such as green fluorescent protein (GFP) under the control of a mammalian promoter on the phagemid. Upon internalization, transcription and translation of the reporter gene allows selective sorting by FACS and recovery of the internalized phage [32].

4.4.3 Integral Membrane Targets

The extracellular domains of many cell membrane receptors cannot be expressed and purified to allow selection against a purified target. For example, this approach is not feasible for multitransmembrane-spanning receptors such as the G-coupled receptors. These receptors represent a large and important family of proteins with a proven history of being excellent drug targets. As a result, integral membrane proteins have not been pursued with phage-displayed libraries.

One approach for selecting against integral membrane targets is to present them in liposomes. For example, the envelope glycoproteins from HIV-1 were first solubilized using a nonionic detergent and then reconstituted into vesicles by the removal of detergent using gel filtration [33]. The vesicles were then captured for panning in microtiter plates using a lectin known to bind the envelope glycoproteins. Potentially, many other detergent solubilized membrane proteins could be constituted into liposomes and used for phage selections as well.

A method that captures these receptors properly oriented in their native lipid environment as liposome-encased paramagnetic beads has recently been demonstrated

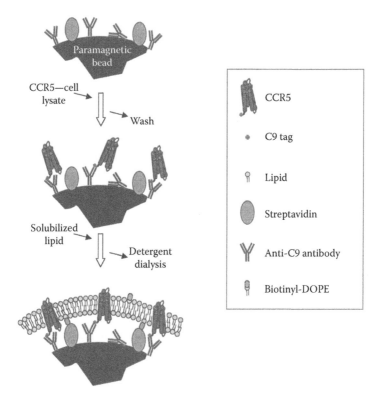

Figure 4.5 Formation of paramagnetic C-C chemokine receptor type 5 (CCR5) proteoliposomes. The surface of nonporous paramagnetic beads was covalently conjugated with streptavidin and an antibody that recognizes the C-terminal C9 tag on CCR5. The conjugated beads were used to capture the C9-tagged CCR5 from the cell lysate. After extensive washing, the beads were mixed with detergent-solubilized lipid containing 0.1%–1% biotinyl 1,2-dioleoyl-*sn*-glycero-3-phosphoethanolamine (DOPE). During the removal of detergent by dialysis, the lipid bilayer membrane self-assembles around the beads and CCR5 is returned to its native environment.

[34]. Paramagnetic beads coated with streptavidin and a flag tag-directed antibody are used to capture flag-tagged receptor from cell lysates. Detergent-solubilized lipid containing 0.1%–1% biotinyl-DOPE is added, and upon dialysis to remove the detergent, a lipid bilayer assembles around the bead and provides a native environment for the receptor (Figure 4.5). Further studies have found, however, that the biotin–streptavidin bridge between the bead and the lipid bilayer is not essential, thus simplifying generation of the paramagnetic proteoliposome [35]. Following exposure of these beads to phage-displayed libraries, a magnetic separator is used to capture receptor-specific phage. Putative receptor-binding phage clones can be analyzed by FACS for binding to the receptor expressed in cell culture, and this can also be used to confirm the integrity of the receptor displayed on paramagnetic liposomes. Aside from providing pure receptor in a native form, the concentration of the receptor in liposomes is considerably increased relative to the levels typically expressed in cells. Additionally, the flag tag provides selective orientation of the receptor and gives this method an advantage over standard liposomes.

A method that combines competitive cell selection and selective sedimentation techniques, mentioned in Section 4.4.2, involved screening of an scFv phage library against thymic tissue fragments in the presence of lymphocytes and spleen cells [23]. Tissue fragments were allowed to sediment along with bound phage, while unbound and lymphocyte- or spleen-cell-associated phage were removed. Isolated phage clones displayed scFvs that selectively bound to the thymic tissue fragments and were used to identify specific thymic stromal markers.

4.4.4 In Vivo Selections

An ingenious way to identify phage ligands capable of homing to a specific cell type or tissue is to actually inject phage libraries into live animals [36]. Nonspecific phage are distributed and absorbed to surfaces throughout the entire animal, while harvesting the tissue of interest can preferentially retrieve selective targeting ligands. This technology does not rely on binding to a known target but, rather, upon the local expression of unidentified targets to achieve selectivity. The phage-derived ligands can, in turn, be used to identify these markers that can provide endothelial targets useful for selective delivery of drugs to specific tissues [37]. It is possible that these same targets provide the homing signals used by tumor cells and leukocytes to target-specific organs and tissues. A map for many of these endothelial markers has recently been described for humans using a phage-displayed peptide library injected into a patient meeting the formal definition of brain-based determination of death [38]. Peptide motifs that provided specific homing to bone marrow, fatty tissue, skeletal muscle, prostate, and skin were identified and suggest potential candidate proteins that may act on these tissues. As tumor targeting technology progresses, the relatively short in vivo half-life of peptides may make tumor-selective targeting peptides useful delivery vehicles for radioimmunotherapy [39].

REFERENCES

1. Scott JK, Smith GP. Searching for peptide ligands with an epitope library. *Science* 1990; 249(4967):389–390.
2. Bass S, Greene R, Wells JA. Hormone phage: An enrichment method for variant proteins with altered binding properties. *Proteins* 1990; 8:309–314.
3. Dennis MS, Lazarus RA. Kunitz domain inhibitors of tissue factor—factor VIIa. II. Potent and specific inhibitors by competitive phage selection. *J Biol Chem* 1994; 269(35):22137–22144.
4. Cunningham BC et al. Production of an atrial natriuretic peptide variant that is specific for type A receptor. *EMBO J* 1994; 13(11):2508–2515.
5. Li B et al. Receptor-selective variants of human vascular endothelial growth factor. Generation and characterization. *J Biol Chem* 2000; 275(38):29823–29828.
6. Wrighton NC et al. Small peptides as potent mimetics of the protein hormone erythropoietin. *Science* 1996; 273(5274):458–463.
7. Krebs B et al. High-throughput generation and engineering of recombinant human antibodies. *J Immunol Methods* 2001; 254:67–84.
8. Deshayes K et al. Rapid identification of small binding motifs with high–throughput phage display: Discovery of peptidic antagonists of IGF-1 function. *Chem Biol* 2002; 9:495–505.
9. Kirkham PM, Neri D, Winter G. Towards the design of an antibody that recognises a given protein epitope. *J Mol Biol* 1992; 285(3):909–915.
10. Dennis MS, Lazarus RA. Kunitz domain inhibitors of tissue factor—factor VIIa. I. Potent inhibitors selected from libraries by phage display. *J Biol Chem* 1994; 269(35):22129–22136.
11. Devlin JJ, Panganiban LC, Devlin PE. Random peptide libraries: A source of specific protein binding molecules. *Science* 1990; 249:404–406.
12. Lam KS et al. A new type of synthetic peptide library for identifying ligand-binding activity. *Nature* 1991; 354:82–86.
13. Hawkins RE, Russell SJ, Winter G. Selection of phage antibodies by binding affinity mimicking affinity maturation. *J Mol Biol* 1992; 226:889–896.
14. Portolano S, McLachlan SM, Rapoport B. High affinity, thyroid-specific human autoantibodies displayed on the surface of filamentous phage use V genes similar to other autoantibodies. *J Immunol* 1993; 151:2839–2851.
15. Szardenings M et al. Phage display selection on whole cells yields a peptide specific for melanocortin receptor 1. *J Biol Chem* 1997; 272(44):27943–27948.
16. Doorbar J, Winter G. Isolation of a peptide antagonist to the thrombin receptor using phage display. *J Mol Biol* 1994; 244:361–369.
17. Goodson RJ et al. High-affinity urokinase receptor antagonists identified with bacteriophage display. *Proc Natl Acad Sci U S A* 1994; 91:7129–7133.
18. De Lorenzo C et al. A new human antitumor immunoreagent specific for ErbB2. *Clin Cancer Res* 2002; 8(6):1710–1719.
19. Marks JD et al. Human antibody fragments specific for human blood group antigens from a phage display library. *Biotechnology (N Y)* 1993; 11(10):1145–1149.
20. Noronha EJ et al. Limited diversity of human scFv fragments isolated by panning a synthetic phage-display scFv library with cultured human melanoma cells. *J Immunol* 1998; 161:2968–2976.

21. Ridgway JBB et al. Identification of a human anti-CD55 single-chain Fv by subtractive panning of a phage library using tumor and nontumor cell lines. *Cancer Res* 1999; 59:2718–2723.

22. Giordano RJ et al. Biopanning and rapid analysis of selective interactive ligands. *Nat Med* 2001; 7(11):1249–1253.

23. Van Ewijk W et al. Subtractive isolation of phage-displayed single-chain antibodies to thymic stromal cells by using intact thymic fragments. *Proc Natl Acad Sci U S A* 1997; 94:3903–3908.

24. Williams BR, Sharon J. Polyclonal anti-colorectal cancer Fab phage display library selected in one round using density gradient centrifugation to separate antigen-bound and free phage. *Immunol Lett* 2002; 81:141–148.

25. de Kruif J et al. Rapid selection of cell subpopulation-specific human monoclonal antibodies from a synthetic phage antibody library. *Proc Natl Acad Sci U S A* 1995; 92:3938–3942.

26. Siegel DL et al. Isolation of cell surface-specific human monoclonal antibodies using phage display and magnetically-activated cell sorting: Applications in immunohematology. *J Immunol Methods* 1997; 206:73–85.

27. Osborn JK et al. Pathfinder selection: In situ isolation of novel antibodies. *Immunotechnology* 1998; 3:293–302.

28. Osborn JK et al. Directed selection of MIP-1a neutralizing CCR5 antibodies from a phage display human antibody library. *Nat Biotechnol* 1998; 16:778–781.

29. Becerril B, Poul M-A, Marks JD. Toward selection of internalizing antibodies from phage libraries. *Biochem Biophys Res Commun* 1999; 255:386–393.

30. Heitner T et al. Selection of cell binding and internalizing epidermal growth factor receptor antibodies from a phage display library. *J Immunol Methods* 2001; 248:17–30.

31. Barry MA, Dower WJ, Johnston SA. Toward cell-targeting gene therapy vectors: Selection of cell-binding peptides from random peptide-presenting phage libraries. *Nat Med* 1996; 2(3):299–305.

32. Poul M-A, Marks JD. Targeted gene delivery to mammalian cells by filamentous bacteriophage. *J Mol Biol* 1999; 288:203–211.

33. Labrijin AF et al. Novel strategy for the selection of human recombinant Fab fragments to membrane proteins from a phage-display library. *J Immunol Methods* 2002; 261:37–48.

34. Mirzabekov T et al. Paramagnetic proteoliposomes containing a pure, native, and oriented seven-transmembrane segment protein, CCR5. *Nat Biotechnol* 2000; 18(6):649–654.

35. Babcock GJ et al. Ligand binding characteristics of CXCR4 incorporated into paramagnetic proteoliposomes. *J Biol Chem* 2001; 276(42):38433–38440.

36. Pasqualini R, Ruoslahti E. Organ targeting in vivo using phage display peptide libraries. *Nature* 1996; 380(6572):364–366.

37. Rajotte D, Ruoslahti E. Membrane dipeptidase is the receptor for a lung-targeting peptide identified by in vivo phage display. *J Biol Chem* 1999; 274(17):11593–11598.

38. Arap W et al. Steps toward mapping the human vasculature by phage display. *Nat Med* 2002; 8(2):121–127.

39. Kennel SJ et al. Labeling and distribution of linear peptides identified using in vivo phage display selection for tumors. *Nucl Med Biol* 2000; 27:815–825.

CHAPTER 5

Leveraging Synthetic Phage-Antibody Libraries for Panning on the Mammalian Cell Surface

Jelena Tomic, Megan McLaughlin, Traver Hart,
Sachdev S. Sidhu, and Jason Moffat

CONTENTS

5.1 INTRODUCTION

The vast majority of monoclonal antibodies available in the market as research reagents, diagnostic tools, or therapeutic antibodies have been produced using hybridoma technology, which has changed little since its invention in the 1970s

(Kohler and Milstein 1975). This statement is particularly true for cell surface and secreted proteins. Although hybridoma technology is used to generate most monoclonal antibodies for biological research, the technology suffers from several fundamental drawbacks that have limited discovery of functional antibodies for cell surface and secreted proteins. For example, the need for animal immunization means that selections occur in an uncontrolled environment, which may be suitable for stable antigens but not for sensitive antigens, including membrane proteins and conformational epitopes. The animal strongly influences the antibody repertoire that results from introducing a foreign antigen, so there is no guarantee that immunodominant epitopes will be those of interest. By extension, the mouse immune repertoire is restricted to eliminate antibodies that would recognize mouse proteins. Therefore, it can be a challenge to raise effective antibodies against conserved epitopes in other species, which is highly problematic since many functionally important epitopes, including protein–protein interaction sites, are highly conserved. Moreover, hybridomas provide antibodies and not the encoding DNA. This makes it difficult to alter or improve an antibody of interest without complicated procedures that convert the antibody into a recombinant form that can be genetically engineered. Last, hybridoma technology requires the use of animals, which is cumbersome and expensive. The limits of hybridoma technology have become rate limiting in the postgenomics era as antibodies grow in number and potential as therapeutic agents. More efficient and sophisticated panning procedures using human antibody frameworks that are displayed as libraries on model display systems should help to relieve this bottleneck.

The accumulation of detailed knowledge of antibody structure and function has enabled phage-displayed antibody libraries to emerge as a powerful in vitro alternative to hybridoma methods (Bradbury and Marks 2004; Hoogenboom 2002). When combined with diverse selection procedures, phage-displayed antibody libraries can give rise to antibodies with desirable characteristics such as high affinity, specificity, solubility, and stability (Hoogenboom et al. 1998; Roovers et al. 1998). Phage display selections or pannings are most commonly done by incubating phage-displayed antibody libraries with immobilized antigen on a solid support (Sidhu and Fellouse 2006); however, this approach sometimes suffers from misrepresentation of antigen due to protein impurity, protein denaturation, and epitope inaccessibility (Butler et al. 1992; Koide et al. 2009). Nevertheless, phage-based selections on immobilized antigen is an attractive approach due to simplicity of the method and the high frequency of success at yielding clones that bind purified protein both in vitro and in vivo (Hoogenboom 2005; Sidhu and Fellouse 2006). In addition to purified antigen, it has also been possible to use mixtures of proteins, cell fractions, tissue sections, whole cells, and even single rare cells to screen for potential antibodies to a specific target molecule (Hoogenboom 2005; Sorensen and Kristensen 2011). However, robust methods that combine state-of-the-art synthetic human antibody libraries displayed on phage and ultradeep sequencing for systematic discovery of affinity reagents that bind to specific features on cell surfaces or secreted proteins in complex cell mixtures have not been described.

In this chapter, we outline a method for performing phage display selections on populations of whole cells in vitro, in order to identify phage-antibody clones that bind to cell surface features. The goal of this chapter is to provide a general summary of the cell-based selection or panning strategy we have developed termed "CellectSeq," and discuss practical considerations when using this approach (Figure 5.1). CellectSeq can be used for targeting multiple, complex, fully functional, cell surface epitopes in their natural environment without fear of epitope loss due to immobilization methods. Although the surface of the cell is a dynamic environment, phage libraries can be used for selections on cells under defined states such as target gene overexpression or physiological perturbations. Practically speaking,

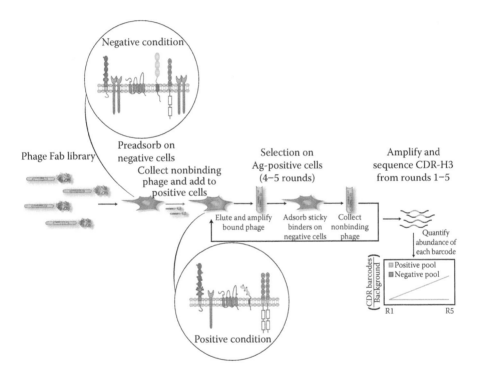

Figure 5.1 A schematic showing how antibody fragments (i.e., Fabs) are obtained with diverse synthetic antibody phage displayed libraries using the CellectSeq approach under negative and positive selection conditions. The selections or pannings begin with a preabsorption step in which the library phage are incubated with "negative" cells. After a brief incubation, the mixture is pelleted to remove the library clones bound to the cells. These clones are likely specific for cell surface epitopes that are not of interest (i.e., nonspecific binding clones). The library phage remaining in the supernatant are incubated with the "positive" cells, and nonbinding phage are then washed away. The phage bound to the positive cells are amplified in an *E. coli* host, and the amplified phage are then purified and used in the next round of panning. Several variations on this theme can be used to extract specific "binders" from synthetic human phage-displayed antibody fragment libraries.

identification of affinity agents for cell surface features and secreted proteins has broad applicability for research reagents, diagnostics, and even, potentially, therapeutics.

5.2 GENERAL AND PRACTICAL CONSIDERATIONS FOR CellectSeq

Current methods for generating recombinant antibodies against cell surface proteins typically rely on the use of recombinant proteins, but this can be a problem as purification of integral membrane proteins is challenging and antibodies generated against recombinant purified proteins may not bind native protein expressed on the cell surface of interest (Chames 2012). Although several methods have been developed for panning libraries against whole cells to obtain antibodies that recognize cell surface features (Edwards et al. 2000; Huie et al. 2001; Sidhu 2005; Yamashita et al. 2010; Zhou et al. 2010), there has been no comprehensive method described premised on the idea of molecular barcoding that combines the use of (1) phage-displayed synthetic antibody libraries, (2) direct selection on engineered mammalian cell lines, subpopulations of mammalian cells or mammalian cells maintained under varying growth conditions, (3) high-throughput DNA sequencing technology, (4) novel antibody recovery approaches, and (5) target validation/discovery methods to identify unique synthetic antibody fragments that bind specific features on the mammalian cell surface.

Just as with selections involving immobilized antigen, in order for phage display selections on cell surfaces to be successful and yield antibodies with high affinity, specificity, and potency, there are a few important things to consider. For example, library size and quality, the selection process per se, and the state of the antigen are all critical factors to consider for success. The library size, along with quality, is a major determinant of the availability of highly functional antibody clones, and the development of synthetic antibody libraries with diversity greater than that of natural immune repertoires (i.e., in excess of 10^{10} unique clones) has enabled generation of specificities not otherwise attainable (Persson et al. 2013). By using synthetic antibody libraries, the inherent limitations of natural immune repertoires are mitigated. The diversity that synthetic antibody libraries possess is derived from man-made sources, rather than from natural repertoires. Consequently, these libraries are constructed from scratch, which permits incorporation of features to stabilize human framework, enhance antibody performance, and minimize risk of immunogenicity for successful therapeutic applications (Persson et al. 2013). The choice of gene segments (i.e., V-gene segments) for the construction of a synthetic antibody library can be guided by factors such as expressibility, folding, and low toxicity, which can ultimately increase the overall performance of the library (Hoogenboom et al. 1998). By designing a synthetic antibody library, one has precise control over the composition of diversity incorporated into antigen-binding sites. This capability offers the key to optimization of any library that follows. One of the fundamental strengths of synthetic libraries is in that they possess the potential to reflect diversity foreign to natural immune repertoires, without inherent biases, and as a result, enable recognition of targets outside the natural scope (Miersch and Sidhu 2012).

Aside from the synthetic antibody library, the most important consideration in obtaining desirable antibodies is the precise selection conditions and panning process. The selection conditions determine which target epitopes are displayed on the cell surface, both in the "negative" and "positive" cell populations (Figure 5.1). During cell-based selections, it is important to first incubate cells from the negative selection pool (i.e., cells that do not express or have reduced expression of the antigen[s] of interest) with the phage-antibody library in order to remove phage-antibody clones that bind to cells in a nonspecific fashion. This preincubation step is crucial in eliminating sticky and nonspecific binders. In order for the preincubation step to be successful at depleting sticky binders, and leave behind the pool of specific binders, antigen presentation on the cell surface in the natural context of plasma membrane must be relatively low in the negative cells compared with the positive cells. Since cell-based selections are performed using synthetic phage-antibody libraries on mammalian cells in vitro, the antibody-expressing phage can recognize nonimmunogenic and functionally conserved epitopes on the positive cells in an unbiased manner.

The state of the target antigen is also critical in the selection process. That is, the antigen must be properly folded and displayed in order to obtain clones that bind functional protein, which is more likely to occur in the native context of the plasma membrane for cell surface proteins. The cell-culture conditions can influence both the conformation of the antigen, which controls epitope accessibility, and the affinity of selected binding clones. Interestingly, manipulation of the cell growth conditions and the in vitro selection environment can promote desired molecular target configurations and binding events, which increase the chance of obtaining clones with desired binding properties such as high specificity, high affinity, conformational specificity, thermostability, and species cross-reactivity. Clever and creative design of the selection conditions, along with library design and quality, can yield a plethora of unique binders following these methods.

5.3 SELECTION PROCESS FOR ENRICHING POOLS OF PHAGE-BASED ANTIBODIES TARGETING CELL SURFACE FEATURES

CellectSeq aims to deplete antibody-based phage libraries of clones that are specific for surface antigens shared between negative and positive cell populations. Because this approach suffers from nonspecific phage binding to mammalian cells, which limits the enrichment that can be achieved per round of panning, at least three (but ideally four to five) rounds of panning and amplification are needed to enrich the quality of binders from each of the synthetic phage-displayed antibody libraries we have tried (Figure 5.1). Each round of selection is followed by recovery of antigen-bound phage and subsequent infection in bacteria. Just as the selection stringency is an important consideration during phage selections involving purified target antigen, there are a few things to consider when performing cell-based selections with phage-based antibody libraries.

5.3.1 Establishing Positive and Negative Cell Populations for Direct Panning

There are two critical rules to follow when performing the CellectSeq procedure. First, positive and negative cell populations must be significantly and measurably different. Second, consistent positive and negative cell populations are absolutely critical during each round of panning. The antibody-based phage library diversity is the greatest during the first rounds of panning, where the representation of each phage clone is low. In order to ensure enrichment of phage-antibody clones specific for features on positive cells, the cells that are used must be grown such that they are as identical and synchronized as possible from round 1 through 4 (or 5). For example, the cell numbers should be kept the same from rounds 1 through 4 (or 5), and the cell confluency should be maintained at 75%–80% so that cell crowding, cell death, and media exhaustion are avoided. It is equally important to wash away any cellular debris before the selection process begins. To maximize the capture of all potentially interesting phage-antibody clones in the early rounds, and to enrich for selective binding clones in the later rounds, the consistent presentation of antigen on the positive and negative cell populations is crucial. It has to be noted that in most cases, antigens are present at very low concentrations/densities on the cell surface, and measures are taken to estimate the abundance of target proteins on the cell surface of positive and negative cells when matched cell lines are engineered to overexpress and underexpress the target protein of interest for a specific selection. Even in instances when antigen concentration is sufficient for antibody binding, antigen inaccessibility through steric hindrance caused by glycosylation or other proteins may prevent the selection of antibodies specific for the target antigen. We note that overexpression of target cell surface proteins has been successful across multiple families of single-pass and integral membrane proteins that have been targeted in our hands.

If the selection conditions are made to be very stringent such that very few phage particles are recovered in the first round (for example, total cell number used is less than desired), then chances are that different subsets of antibodies will be selected every time the selection is repeated. One may consider decreasing the antigen concentration when progressing through rounds of panning as this may result in only the tightest-binding clones to be selected. A minimum target concentration should be maintained at each round in order for the enrichment to occur. Ultimately, the success of the selection strategy depends on the antigen density and concentration as well as target antigen accessibility. Cell-based selections targeting "difficult" antigens, such as receptors that are only functionally retained in the lipid bilayer, may require an additional understanding of the parameters that govern the outcome of the selection process.

5.3.2 Washing Unbound Phage Antibodies

As with phage selections involving immobilized target antigen, the goal is to maximize the recovery of specific antibody-phage clones with high affinities to the target antigen. During cell-based selections or CellectSeq, stringent washing was not found to be necessary; cells were only washed with phosphate-buffered saline three to four times. In cases where it is necessary to reduce background binding phage, multiple washings with increased washing times can be performed. The goal with this step is to retain the relevant binders, while washing away nonadherent or loosely bound antibody-phage particles.

5.3.3 Elution Method and Phage Amplification

Bound phage can be recovered from the surface, reinfected into bacteria, and regrown for further enrichment. The enrichment of specific binding phage from a large excess of nonbinding clones can be achieved by multiple rounds of phage binding to the target, washing to remove nonspecific or loosely bound phage, and elution to retrieve specific binding phage (Figure 5.1). Phage antibodies bound to antigen can be eluted in different ways (Hoogenboom et al. 1998), but the most commonly used solution to elute phage bound to the cell surface target is 0.1 M hydrochloric acid. Eluted phage can be amplified in *Escherichia coli* prior to further rounds of selection using standard phage amplification protocols. As with phage eluted from an immobilized target, the amplification process can lead to the biased production of better-expressing clones or faster-growing phage clones that lack any displayed ligand. However, our experience with cell-based selections is that these artifacts are relatively minor and can be mitigated with information from deep sequencing.

5.3.4 Monitoring the Cell-Based Selection Process

The enrichment ratio is the ratio of binders eluted from a positive selection pool and the binders eluted from a negative selection pool compared from each round of panning. This ratio is used to monitor the progress of the selection process for both the positive and negative selection pools. With synthetic antibody-phage libraries, the enrichment is usually seen in the third, fourth, and/or fifth round of panning (Figure 5.1).

5.4 SEQUENCE ANALYSIS OF ENRICHED POOLS OF PHAGE ANTIBODIES

The use of a single-antibody framework in synthetic antibody libraries facilitates the use of high-throughput DNA sequencing to identify antibodies specific to cell surface proteins, as sequencing runs need only recover complementarity determining region (CDR) H3 and/or L3 information to identify unique antibodies. Paired-end sequencing of synthetic antibody libraries where the framework is fixed captures a portion of the constant region flanking the H3 and L3 variable regions. We perform sequence analysis using custom scripts to extract the variable region sequences, which can vary substantially in length as well as nucleotide composition, from the total read. This is done by finding sequences within the read that match the constant regions 5′ and 3′ of the variable region, within a small error tolerance (e.g., one mismatch in the 30–36 bases upstream of the variable region and one mismatch in the 10–15 bases downstream).

The variable region sequences are extracted and counted. Reads observed with high frequency (that is, high fraction of total reads) in experimental samples are good candidates for specific binders. However, at high sequencing read depth, even a low error rate at the polymerase chain reaction (PCR) amplification or sequencing steps could yield erroneous observations that occur frequently enough to be considered candidate binders. To minimize this problem, we combine closely related sequences into a single representative sequence using a simple error model. First, the most abundant sequence is considered to be a "true" sequence. Then each candidate sequence, in descending order of frequency, is compared to each of the true sequences, and the Hamming distance (number of mismatches) between them is calculated. The observed sequence frequency is compared to the frequency expected if the read were an error given the Hamming distance, the number of reads of the true sequence, and some reasonable estimate of the sequencing error rate (~1% per base). For example, if the observed frequency is roughly the expected error frequency, that read and its associated counts are merged with the true read; otherwise, if the candidate matches no true sequences, it is added to the true list. This process is repeated until the candidate sequence frequency drops below some threshold (e.g., 0.01% of total reads), after which reads are unlikely to be considered candidate hits and further clustering is unnecessary to obtain "enriched" sequences. After clustering, DNA sequences are translated to peptide sequences. Candidate binders can be identified by simple read counts, by testing for enrichment relative to the negative selection pool, or by comparing across multiple samples.

5.5 RECOVERING ANTIBODIES FROM ENRICHED POOLS OF PHAGE ANTIBODIES

Individual phage-based antibodies can be identified based on the CDRs described in Section 5.4 given their high diversification in synthetic phage-based antibody libraries. As a consequence of this diversification, individual antibodies can be recovered from the late rounds of panning using a PCR-based method in which primers anneal specifically to the CDR-H3, or CDR-H3 and CDR-L3 regions in combination. Following amplification of these clones using PCR, unique (and rare) phage clones can be recovered from the selected pools of phage antibodies.

5.6 TARGET IDENTIFICATION/CONFIRMATION/VALIDATION

The outcome of the cell-based selection procedure can yield a mixture of antibody clones with different target-binding properties. This requires each clone to be individually screened. Antigen binding of each phage antibody is usually tested by performing ELISAs first (ELISAs can also be done to test the affinity and epitope binding), followed by flow cytometry (to test the expression level of the target on the cell surface) or immunofluorescence. In order to confirm that the antibody clones bind to the target of interest, an approach of immunoprecipitation (IP) followed by mass spectrometry (MS) can be performed. In the instances that the binders were selected against an unknown antigen, the IP-MS method can also be applied to identify the target. Ultimately, cell-based bioassays are performed to test the function of the antibodies as binding affinity and potency are not always correlated, especially if selected antibodies neutralize, agonize, or antagonize the receptor binding.

5.7 CONCLUSION

Unlike conventional approaches where phage-displayed antibody libraries are screened against immobilized purified protein antigens, CellectSeq enables generation of high-affinity antibodies against transmembrane proteins by direct selection on live cells expressing the antigen of interest. This methodology is coupled with deep sequencing analysis and PCR-based amplification, which allows for comprehensive identification and recovery of specific antigen-binding clones. The potential impact of CellectSeq is significant because the methodology overcomes difficulties associated with targeting integral membrane protein receptors, which are a therapeutically rich class of proteins that are notoriously difficult to purify and target with Abs. The relative ease of constructing and screening antibody libraries comes from many well-established published protocols. Still, there are limitations to antibody phage display (one example is that clones of interest may be missed as a result of poor selection strategies or recovery), but the combination of antibody phage display with high-throughput methods such as next-generation sequencing offers a powerful combination of approaches for building unique affinity reagents to decipher mechanisms of disease. The versatility, robustness, and proven capabilities of these synthetic antibody libraries displayed on phage will be a cornerstone of modern biotechnology and drug development for the foreseeable future. So far, this approach has allowed us to isolate and engineer fully human antibodies of high affinity and specificity to a broad range of antigens, without using animals and hybridomas, but the true potential of this technology is yet to be fully realized.

ACKNOWLEDGMENTS

The CellectSeq methodology has been filed (US patent application no. 61/539,546). The authors would like to acknowledge Patricia Mero, for generating Figure 5.1, and all previous and current members of the Moffat and Sidhu laboratories for helpful discussions. This work was supported by grants to JM and SS from the Ontario Research Fund and the Canadian Institutes of Health Research. JT is supported by a Canadian Institute for Health Research (CIHR) postdoctoral fellowship. JM is a senior fellow at the Canadian Institute for Advanced Research and a Tier II Canada Research Chair in Functional Genomics of Cancer.

REFERENCES

Bradbury, A.R., and Marks, J.D. (2004). Antibodies from phage antibody libraries. *J Immunol Methods 290*, 29–49.

Butler, J.E., Ni, L., Nessler, R., Joshi, K.S., Suter, M., Rosenberg, B., Chang, J. et al. (1992). The physical and functional behavior of capture antibodies adsorbed on polystyrene. *J Immunol Methods 150*, 77–90.

Chames, P. (2012). *Antibody Engineering: Methods and Protocols*, 2nd edn (New York: Humana Press).

Edwards, B.M., Main, S.H., Cantone, K.L., Smith, S.D., Warford, A., and Vaughan, T.J. (2000). Isolation and tissue profiles of a large panel of phage antibodies binding to the human adipocyte cell surface. *J Immunol Methods 245*, 67–78.

Hoogenboom, H.R. (2002). Overview of antibody phage-display technology and its applications. *Methods Mol Biol 178*, 1–37.

Hoogenboom, H.R. (2005). Selecting and screening recombinant antibody libraries. *Nat Biotechnol 23*, 1105–1116.

Hoogenboom, H.R., de Bruine, A.P., Hufton, S.E., Hoet, R.M., Arends, J.W., and Roovers, R.C. (1998). Antibody phage display technology and its applications. *Immunotechnology 4*, 1–20.

Huie, M.A., Cheung, M.C., Muench, M.O., Becerril, B., Kan, Y.W., and Marks, J.D. (2001). Antibodies to human fetal erythroid cells from a nonimmune phage antibody library. *Proc Natl Acad Sci U S A 98*, 2682–2687.

Kohler, G., and Milstein, C. (1975). Continuous cultures of fused cells secreting antibody of predefined specificity. *Nature 256*, 495–497.

Koide, A., Wojcik, J., Gilbreth, R.N., Reichel, A., Piehler, J., and Koide, S. (2009). Accelerating phage-display library selection by reversible and site-specific biotinylation. *Protein Eng Des Sel 22*, 685–690.

Miersch, S., and Sidhu, S.S. (2012). Synthetic antibodies: Concepts, potential and practical considerations. *Methods 57*, 486–498.

Persson, H., Ye, W., Wernimont, A., Adams, J.J., Koide, A., Koide, S., Lam, R. et al. (2013). CDR-H3 diversity is not required for antigen recognition by synthetic antibodies. *J Mol Biol 425*, 803–811.

Roovers, R.C., Henderikx, P., Helfrich, W., van der Linden, E., Reurs, A., de Bruine, A.P., Arends, J.W. et al. (1998). High-affinity recombinant phage antibodies to the pan-carcinoma marker epithelial glycoprotein-2 for tumour targeting. *Br J Cancer 78*, 1407–1416.

Sidhu, S.S. (2005). *Phage Display in Biotechnology and Drug Discovery* (Boca Raton, FL: CRC Press/Taylor & Francis).

Sidhu, S.S., and Fellouse, F.A. (2006). Synthetic therapeutic antibodies. *Nat Chem Biol 2*, 682–688.

Sorensen, M.D., and Kristensen, P. (2011). Selection of antibodies against a single rare cell present in a heterogeneous population using phage display. *Nat Protoc 6*, 509–522.

Yamashita, T., Utoguchi, N., Suzuki, R., Nagano, K., Tsunoda, S., Tsutsumi, Y., and Maruyama, K. (2010). [Development of anti-tumor blood vessel antibodies by phage display method]. *Yakugaku Zasshi 130*, 479–485.

Zhou, Y., Zou, H., Zhang, S., and Marks, J.D. (2010). Internalizing cancer antibodies from phage libraries selected on tumor cells and yeast-displayed tumor antigens. *J Mol Biol 404*, 88–99.

CHAPTER 6

Phage Libraries for Developing Antibody-Targeted Diagnostics and Vaccines

Nienke E. van Houten and Jamie K. Scott

CONTENTS

6.1 INTRODUCTION

Phage library technology is a powerful tool for targeting antibodies (Abs) of known or unknown specificity. In this chapter, we examine techniques for applying this technology to Ab-dependent applications, specifically epitope mapping and the development of diagnostics and vaccines. Abs produced during bacterial and viral infections, allergic responses, or chronic illnesses, such as autoimmune disease and cancer, can provide information about the immune response and sometimes disease etiology. A proportion of serum Abs may be biologically active. For example, some may neutralize a virus or block bacterial infection, whereas others may trigger cell lysis by complement or mediate Ab-dependent cellular cytotoxicity (ADCC). Phage display libraries can be screened with Abs to map epitopes on protein antigens (Agns) and to study ligand interactions. Moreover, ligands for biologically active Abs can serve as candidate leads for therapeutic and vaccine development.

By screening phage libraries with an Ab or mix of Abs, highly specific ligands can be identified. Different types of phage display libraries are available for a diversity of applications. Whole-antigen (Agn) libraries, encoded by full-length cDNA, can be used to identify proteins that bind to an Ab or serum of interest. Agn-fragment libraries (AFLs), produced from fragmented cDNAs, can be used to identify subregions of a protein that bind Abs. Random peptide libraries (RPLs), made from degenerate oligonucleotides, can identify ligands for a wide variety of Abs. In addition to protein-binding Abs, RPLs can be used to identify peptide ligands for Abs that bind nonprotein Agns, such as carbohydrate (CHO) or DNA. In the following, we discuss the molecular basis of Ab–ligand interactions in the context of phage library screening.

Agn is defined as "any molecule that can bind specifically to an Ab" [1]. Agns include proteins, CHOs, DNA, lipids, and a variety of haptens; the latter are small molecules that, on their own, are not immunogenic but can be made immunogenic by being chemically coupled to an immunogenic carrier protein. Organisms and cells can be Agns, including infectious organisms, plant and animal components, and even components of oneself (self-Agns). Abs bind specific regions of an Agn at a site referred to as the antigenic determinant or epitope. Conversely, the Agn-binding site on an Ab is referred to as the Ab combining site or paratope. Typically, there are many unique epitopes on protein Agns, whereas fewer unique epitopes exist on Agns composed of repeating units, such as CHO or DNA. Haptens, being very small and typically buried within an Ab paratope, comprise a single epitope. The term *Agn* is not synonymous with immunogen. An immunogen is an Agn that can elicit specific Ab; however, not all Agns are immunogenic. For example, self-Agns, which are non-immunogenic in healthy individuals, elicit Abs in some autoimmune diseases and cancers. Antiself Abs may also be intentionally produced by immunizing with self-Agn mixed with an adjuvant and/or by immunizing with the self-Agn chemically coupled to an immunogenic carrier protein.

An Ab that is produced by a single B-cell clone and is expressed from a single set of heavy- and light-chain genes is said to be a monoclonal antibody (MAb). Each MAb binds a single epitope on the corresponding immunogen; in addition, some MAbs may have multiple reactivities (i.e., they bind to more than one Agn). Screening libraries with a MAb will result in a restricted set of ligands that bind the whole paratope or a subsite on it. Thus, ligands that bind an Ab *functionally* mimic the Ab-binding epitope on the corresponding cognate Agn. In some cases, these ligands can be used as immunogens to elicit the production of specific Ab in vivo.

The serum Ab response to immunization comprises polyclonal antibodies (PCAbs), whose diversity depends upon the complexity of the immunogen. PCAb responses consist of multiple Abs that target a variety of epitopes on the Agn surface. Typically, oligoclonal Ab responses are elicited by immunization with Agns having a limited number of epitopes (haptens, DNA, lipids, and CHOs); such Ab responses are sometimes genetically restricted to a few V_H–V_L gene combinations. In contrast, whole proteins and organisms (e.g., viruses and bacteria) elicit highly diverse and complex PCAb responses. PCAbs typically isolate multiple sets of related peptides or AFs (i.e., peptides or AFs that share sequence motifs or sequences, respectively) from RPLs and AFLs. Each set of related clones is specific for a single Ab reactivity, such that clones sharing related sequences will compete with each other in binding to an identical subset of PCAbs, whereas clones bearing unrelated sequences will not. It is important to note that there is bias in the type of peptide a PCAb will select from an RPL; peptides that cross-react best with epitopes on the immunogen will emerge as dominant, even though the Abs that select them may not be the most dominant in the serum Ab response [2,3]. Similarly, the AFs selected from AFLs by PCAbs will be those that best mimic epitopes on the folded protein and therefore cross-react with the PCAb; again, these may not correspond to the most immunodominant epitopes.

Protein epitopes bind Abs via specific residues that contribute to binding affinity by making low- and high-energy contacts with the Ab paratope. Critical-binding residues (CBRs) in a protein epitope make high-energy contacts with the Ab paratope. Amino acid replacements can be used to identify CBRs on a protein, but it must be demonstrated that they do not affect the global protein fold [4]. Protein epitopes can be described in a number of ways. Conformational epitopes rely on protein folding for binding activity, and Agn denaturation destroys Ab binding to these sites. Linear epitopes (also called continuous epitopes) consist of CBRs that are close together in a short polypeptide sequence. They are often resistant to denaturation, and most of them have been defined by cross-reactivity with peptides (see Section 6.4). In contrast, discontinuous epitopes result from protein folding that brings distant CBRs close together [5]; these are typically conformational epitopes. Linear epitopes can sometimes have a conformational component, with Ab binding being ablated by Agn denaturation, especially reduction of disulfide bridging. In the Ab response against folded protein Agn, Abs against conformational (mainly discontinuous) epitopes typically dominate, as evidenced by a large decrease in serum Ab binding to denatured protein Agn.

Peptide mimicry is defined by ligand interactions with the Ab paratope. Functional mimics of an epitope are cross-reactive in that they compete with cognate Agn for binding to Abs. CBRs in peptide mimics are defined by amino acid replacements that strongly reduce affinity [6,7]. Peptide CBRs can mediate binding either through direct, high-energy contacts with the paratope or by promoting a conformation of the epitope that will allow these contacts to be made. Structural mimics both cross-react with Agn and make the same contacts with the Ab paratope as the cognate epitope; they may or may not have the same contact residues or structure as the cognate epitope. Structural mimic peptides are more apt to mimic linear epitopes than discontinuous ones. (The word "mimotope" was originally coined to refer to structural peptide mimics of discontinuous epitopes but has become synonymous with functional peptide mimics.) Immunogenic mimics of an epitope are peptides that, when used as immunogens, elicit Abs that bind the cognate epitope. Although a peptide probably has to be a functional mimic of its cognate epitope to be an immunogenic mimic, it is not clear that it also has to be a structural mimic.

In general, peptides isolated from different types of phage display libraries bind Abs by different mechanisms. Peptides derived from AFLs with antiprotein Abs bind by a mechanism similar or identical to the cognate epitope. In many circumstances, these will be linear epitopes or discontinuous epitopes within a small domain, as many epitope determinants for discontinuous epitopes may be missing from an AFL. Ligands from RPL screening can mimic the function and/or the structure of a native protein epitope. Typically, the binding mechanism for functional mimics of discontinuous epitopes is mixed, with some CBRs making identical contacts to the native epitope and others promoting binding to the Ab paratope by entirely different mechanisms. It has been our observation that peptides that mimic discontinuous epitopes almost always require constraints imposed by disulfide bridging. In contrast, linear epitope mimics require structural constraints found in the cognate epitope (e.g., beta-turn structures) and may not require constraints by disulfide bridging (see Ref. [2]). Thus, the mechanism of protein–epitope mimicry varies, depending on the type of epitope an Ab binds and the type of library screened.

Peptides can be functional, and sometimes structural, mimics of CHO and other nonprotein Agns. Examples of functional, but probably not structural, CHO mimicry were described by Harris et al. [8], who screened 11 RPLs with three very similar MAbs against the cell wall polysaccharide of group A streptococcus. Mapping of the cognate CHO epitope with a panel of synthetic oligosaccharides showed that the

MAbs recognize the same CHO epitope; moreover, these MAbs have closely related gene sequences. Despite the similarity of the MAbs, the consensus sequences of the peptides selected by each MAb were distinct, with most peptides being specific for only the selecting MAb and rarely cross-reactive with the other, closely related MAbs. These results suggest that the mechanisms by which the MAbs bind to cross-reactive peptides and to CHO differ: peptide binding is primarily determined by unique features of each Ab, whereas CHO binding is determined by features that are shared among the MAbs. This concept has been further supported by structural studies by Vyas et al. [9,10].

The fundamental uses of phage display libraries with Abs are to identify the cognate protein Agn for an Ab, to map an epitope (if the Agn is protein), or to simply identify ligands that cross-react with an Ab. These ligands can be used as tools for probing an Ab and/or its cognate epitope, as diagnostics that will recognize the presence of Abs against the cognate Agn, and as vaccine leads that will elicit Abs having the same specificity as the screening Ab. Different types of library are suited for each purpose. Whole-Agn libraries are best for identifying the cognate protein Agn for an Ab. AFLs are used to map linear epitopes on a protein, or discontinuous epitopes that are located within a small, folded domain. RPLs are useful for identifying ligands that cross-react with Agn; these ligands may or may not be mimics of the native epitope. In Section 6.2, we review in more detail the general uses and optimal application of each library type, focusing on the kinds of information one can expect to obtain from each library type.

6.2 PHAGE-DISPLAY LIBRARIES AS TOOLS FOR EPITOPE DISCOVERY

6.2.1 Types of Display

The proteins and peptides displayed by filamentous phage libraries are typically fused to the gene 3 minor coat protein, pIII, or the gene 8 major coat protein, pVIII. Three types of display have been described for applications ranging from whole protein to peptide display [11]. One type of display comprises the so-called Type 3 and Type 8 vectors, which fuse the library amino acid sequence to the N-terminus of all copies of pIII or pVIII, respectively. Thus, gene 3 or gene 8 in the phage genome is modified to encode a library. There is little restriction to the variety of proteins and peptides that can be displayed by Type 3 vectors, probably due to the relatively low production of pIII and the low copy number of this molecule on the virion. In contrast, Type 8 vectors display peptides fused to each copy of pVIII along the entire length of the phage body, such that pVIII fusions are expressed at much higher levels and comprise the entire length of the virion. As a consequence, Type 8 libraries are restricted to peptides of five residues or less.

In contrast, the Type 33/88 and Type 3+3/8+8 systems produce "hybrid" virions that bear two forms of the pIII or pVIII protein: recombinant coat protein fusion and wild-type (WT) coat protein. There are advantages to both pIII- and pVIII-based hybrid systems. Type 3+3 phagemid systems are especially useful for cloning more complex proteins (e.g., for Fab display), whereas Type 8+8 or 88 systems allow longer peptides to be displayed. Type 33 and Type 88 vectors encode both recombinant and WT copies of the coat protein gene on the viral genome, whereas the Type 8+8 and 3+3 systems use phagemid to express coat protein fusions and helper phage to express the WT coat protein. For example, a Type 88 vector, such as f88–4 [12], is a phage that carries on its genome a WT gene 8 along with an expression cassette for recombinant gene 8 ligated to DNA encoding a library. This produces phage with coats that contain both WT pVIII and recombinant pVIII fusion, and the levels of recombinant pVIII incorporation depend upon the sequence of the displayed peptide and the phage display system used. For Type 3+3 and 8+8 systems, phagemid is used, which carries an f1 phage origin of replication and an expression cassette encoding the library protein or peptide fused to coat protein (pIII or pVIII); it also contains a selectable marker such as beta-lactamase. Phagemid DNA is used to transform *Escherichia coli*, which are then super-infected with helper phage to provide the other proteins necessary for phage assembly, including WT pVIII or pIII. Thus, phagemid DNA is selectively packaged into virions comprising library-pVIII fusions expressed by the phagemid and all other proteins provided by helper phage [11]. For an extensive review of phage display vectors, library construction, and screening, refer to Barbas et al. [13]. The type of display can affect the screening outcome, as discussed at the end of this section.

6.2.2 Whole-Agn Libraries

Phage libraries made from cDNA display the protein repertoire of an organism or cell line in full-length form. Such whole-protein Agns are more likely to be displayed in their correctly folded state if they do not require posttranslational modifications, such as glycosylation; such proteins are typically difficult to express in *E. coli*. Crameri and Suter [14] first described the pJuFo vector for a whole-Agn library. The pJuFo vector system was designed to overcome problems associated with the presence of stop codons at the 3′ ends of cDNAs. Since the C-terminus of pIII is believed to be buried in the viral capsid, displayed proteins must be fused to the N-terminus of pIII for efficient display. Thus, for display on pIII, the 3′ end of a cDNA must be ligated to the 5′ end of the gene 3 coding region. This presents two problems in making a cDNA display library; the presence of stop codons at the 3′ ends of cDNAs will prevent protein fusions from being made, and it is not possible to predict where open reading frames (ORFs) start for all cDNAs, especially those encoding uncharacterized proteins. Expression of the cDNA as a protein, independent from pIII expression, circumvents this complication. To construct libraries expressed by the pJuFo phagemid vector, whole proteins encoded by cDNA fragments are fused to the C-terminus of the leucine zipper region of the Fos transcription factor. The same vector also encodes the leucine zipper region of Jun fused to the N-terminus of pIII. Phage assembly results in heterodimerization between the Fos and Jun leucine zippers, and cysteines that have been added to the N- and C-termini of the Fos and Jun leucine zippers form a covalent bond that allows display of the cDNA gene product. Helper phages are required for phage packaging and virion assembly. Alternative systems for cDNA display have been described; for example, fusion to the C-terminus of the pVI minor protein overcomes some of the difficulties described here [15,16]. Others have created systems for fusing cDNA to both pVIII and pIII [17].

Whole-Agn libraries have multiple applications, with a major one being in allergy research. Immunoglobulin E (IgE) from allergic patients is used to identify protein allergens from cDNA libraries made from the mRNA of whole organisms. Once identified, these proteins can then be used in clinical diagnostic allergy tests. Whole-Agn libraries are generally useful for the identification of unknown Agn. This type of library has also been used to identify tumor-specific Agns from cell lines (see Section 6.3.1) [18,19]. Since this type of display uses a bacterial expression system, problems with proper glycosylation and protein folding can arise. To circumvent these problems, whole-Agn display on yeast or mammalian cells could be considered.

6.2.3 Agn Fragment Libraries

Also referred to as gene fragment libraries or natural peptide libraries, AFLs are encoded by randomly sheared fragments of cDNA rather than complete cDNAs. AFLs can be made from the whole genomes of lower organisms (e.g., a viral genome), the cDNA repertoire of a cell line, or the cDNA or gene of a single protein. AFLs are constructed by digesting a genome (from a lower organism that lacks introns), cDNA from total RNA of a cell or organism, or cDNA from a gene of interest, with a nonspecific endonuclease such as DNase I [20]. After addition of linker DNA to fragment ends, the fragments are ligated into an appropriate phage vector, and proteins related to the gene fragments are displayed on the phage surface. Digestion can control the size of the fragments and hence the length of expressed AFs. However, most of the clones in a library do not encode the parental protein sequence. Since each fragment can be ligated in the wrong reading frame at either end, or it may be inverted, only 1 out of 18 possible insertions will encode the parent protein sequence. Zacchi et al. [21] addressed this problem by devising a system to preselect library inserts for ORFs before cloning them into a phage display vector. The pPAO2 vector was designed to ligate gene fragments upstream from a beta-lactamase gene flanked by lox-recombination sites, which is upstream of the phage gene 3. Gene fragments are cloned as fusions to the beta-lactamase gene, and functional fusions are selected with ampicillin. Phagemid DNA from ampicillin-resistant clones is then used to transform cells expressing the Cre recombinase; this removes the sequences between the gene fragment and the gene 3, causing ORFs to be joined in frame to the 5′ end of the gene 3 coding region. Superinfection of these cells with helper phage allows production of the library with clones displaying expressed ORF sequences. The authors used this system to make a library and reported that all the clones they analyzed contained ORFs, of which 83% were localized to known genes.

There are multiple applications for AFLs. They are well suited to mapping epitopes for antiprotein MAbs or PCAbs. The amino acid sequences of clones isolated from an AFL by a MAb should align to identify the Mab-binding epitope. PCAbs

against a protein or set of proteins may isolate sets of aligning peptides corresponding to multiple epitopes. This method is most suitable for the isolation of linear
epitopes; however, discontinuous epitopes may be isolated if the library contains
a folded domain bearing CBRs required for MAb binding. Since the origin of the
epitope is the native protein, optimization by building sublibraries is not necessary;
however, doped libraries may be used to optimize binding and to characterize the
epitope in more detail (see Section 6.2.6).

6.2.4 Random Peptide Libraries

The idea of RPLs displayed by filamentous phage was first proposed by Parmley
and Smith [22] and then reported in 1990 [23–25]. RPLs consist of "randomized"
amino acids fused to pIII or pVIII, and are usually expressed by Type 3, or Type
88/8+8 vectors, respectively. They range in size from 10^8 to 10^{11} clones and cover a
range of peptide lengths; the peptides can be "linear" or "constrained" by the introduction of fixed disulfide bridges. RPLs are encoded by degenerate oligonucleotides
that are synthesized using a degenerate codon strategy (NNK or NNS, in which N
represents an equimolar mixture of all four nucleotides, K is an equimolar mixture
of G and T, and S is an equimolar mixture of G and C). NNK and NNS codons each
comprise 32 codons that encode all 20 amino acids and one amber stop codon. Since
there are 32 possible codons, not all of the amino acids are equally represented in
the mixture, producing a bias toward certain residues. There are three codons for
Arg, Leu, and Ser, and two codons for Val, Pro, Thr, Ala, and Gly; the remaining
amino acids are each encoded by a single codon. This bias can be overcome by
codon-based oligonucleotide synthesis strategies (Glen Research, Sterling, Virginia).
This approach uses mixed trinucleotide phosphoramidites, instead of single nucleotide phosphoramidites, to synthesize degenerate oligonucleotides that will encode
a library [26]. There are several advantages to this. First, stop codons are omitted,
allowing more clones in a library to express and display long amino acid sequences.
Second, since single codons can be mixed together in any proportion, codon bias
can be completely controlled. Last, codons that are optimized for expression can
potentially be used for library construction. The methods described here for producing degenerate oligonucleotides for RPLs are also useful for peptide optimization, as
described in Section 6.2.6.

RPLs can be used to identify peptide ligands for MAbs and PCAbs, regardless of
the type of Agn they recognize (i.e., CHO, DNA, or protein). RPLs have an advantage
over AFLs in that the same RPL can be used for different applications, whereas an
AFL is only specific to one protein or organism. As with AFLs, RPLs can be used to
map epitopes on protein Agns. An RPL would be screened with an antiprotein MAb,
and the sequences of peptides from MAb-selected clones aligned. Peptide sequences
selected with a MAb against a linear epitope often can be aligned into consensus
sequences. The consensus residues, having been selected by the MAb, are potential
CBRs, and they can be confirmed as CBRs by amino acid replacement studies. If the
protein is known, its gene can be searched for matches to the consensus sequence
to identify the epitope. If the gene for the protein is unknown, consensus sequences

can be used to perform BLAST searches, whose purpose is to identify potential Agn targets for the Ab. This latter application has been useful, particularly in identifying Agns and the organisms expressing them that may be involved in immune responses. However, this method is unreliable. It can identify leads, but they must be tested further for MAb binding and other characteristics.

6.2.5 Library Screening to Isolate High-Affinity Ligands or a Broad Range of Ligands

Screening conditions affect the type of ligands isolated from a library. These include the type of display used (pVIII vs. pIII), the panning strategy (in-solution vs. solid-phase), the concentration of the target molecule, and preadsorption to remove phage that binds to plastic or reagents used in screening rather than the desired target.

Solid-phase panning uses immobilized Ab to capture phage. The Ab can be immobilized by adsorption to a plate or to beads, or the Ab can be captured by immobilized protein A; the Ab can also be biotinylated and then captured on immobilized streptavidin. A phage library is allowed to bind to the plate, and nonbinding phage is removed by washing. Bound phage can be eluted by Ab denaturation in acid, and then the eluates can be neutralized and used to infect *E. coli* cells that will amplify the eluted phage. Alternatively, cells can be added to the plate-bound phage and infected, particularly if the phage vector is a Type 8+8 or Type 88. In solid-phase panning, phages are captured by multiple Abs in the well of a microtiter plate or on beads; multivalent binding produces an avidity effect that can effectively capture phage, bearing even relatively weak-binding peptides. Thus, phages bearing peptides covering a whole range of affinities are captured by solid phase panning. Importantly, this is the most efficient method of capturing phages and is generally used in the first round of panning, when each clone in the library is available to be captured by Abs.

In contrast, in-solution panning allows more stringent selection, because binding of each Ab (especially if it is in Fab form) to a phage-borne peptide is an independent event. This approach is especially useful in screening sublibraries, as a means of optimizing or further characterizing a peptide ligand. In-solution panning begins by mixing Abs and phages, and allowing the reactions to reach equilibrium. Subsequently, Ab–phage complexes are captured by brief exposure to immobilized streptavidin or protein A. As Ab concentration drives the binding of Abs to peptide, the number of Abs bound to a given phage will depend upon the K_d between the displayed peptide and the Ab. Thus, at low Ab concentrations (near or below the K_d), a phage bearing tighter-binding peptides will bind proportionally more Ab molecules than a phage bearing weaker-binding peptides; the former, having more bound Abs per phage, will be more efficiently captured by immobilized protein A or streptavidin. Thus, clones having tighter-binding peptides will be favored in selections using low Ab concentrations. The disadvantage of this strategy is that capture of phages out of solution is inefficient and results in lower yields [27]. Thus, it is best used on phages that have been enriched by an initial round of solid-phase panning. A range of Ab concentrations should be used for in-solution pannings as it is difficult to estimate the best concentration of Abs to use beforehand [27]. Since libraries typically consist of millions to billions of phage clones, one should also consider practices that will reduce phage binding to all facets of the screening system, including reagents (streptavidin, protein A), beads, tubes, or plates in which the pannings are carried out, and parts of the Ab that are not involved in Agn binding. To decrease background caused by such phages, preadsorption of the library on these reagents is also recommended.

Before screening, it should be decided if the goal of the experiment is to obtain high-affinity clones or clones covering a broad range of affinities and sequences. In-solution screening with MAb or Fab is more likely to isolate clones bearing high-affinity peptides, whereas solid-phase screening PCAb is best used to map multiple epitopes on a protein. Peptides displayed by Type 3 vectors, and probably by Type 8+8/88 ones, are aligned closely enough to allow bivalent binding to a single IgG [28]. Multivalent binding to phages can produce avidity effects that mask intrinsic affinity differences between peptides, and thus decrease affinity discrimination; this effect can be overcome by using Fab for screening. If PCAbs are being used to identify multiple peptide ligands, solid-phase screenings (or in-solution screenings using relatively high Ab concentrations) should be performed to isolate a broad range of phage encompassing both tight and weak binders. Moreover, PCAbs, being derived from serum, can include Abs with specificities for Agns other than the target antigen. To avoid Abs that are not specific for Agn, it is prudent to screen libraries with PCAbs that have been affinity purified on Agn. This will ensure that the ligands selected from a library will very likely cross-react with Agn.

6.2.6 Sublibraries for Optimization of Peptides and Agn Fragments

Phage display libraries can be used to optimize ligands for Ab binding by two general approaches. The first is most applicable to ligands isolated from RPLs whose sequences align to produce a consensus. Sublibraries can then be designed in which the consensus residues are fixed and all other residues surrounding them randomized [29]. Alternatively, sublibraries can be produced that extend a consensus sequence by placing additional randomized residues on either side of it. This creates longer peptides and thus potentially increases binding affinity by increasing binding contacts with the Ab [30–32].

If a single ligand is isolated from an RPL, the CBRs will likely be unknown. A doped library can be made to identify CBRs and optimize binding. This approach can also be used for peptides derived from AFLs to define CBRs, confirm epitope length, and optimize binding. Doped libraries of three types can be constructed. In a nucleotide-doped library, each nucleotide used to encode the peptide sequence is mixed with an N nucleotide mixture; this results in a degenerate oligonucleotide that is biased toward the sequence encoding the peptide ligand, but depending on the level of N-nucleotide doping, each nucleotide in the sequence will have a chance of being replaced by a different nucleotide. The drawback of these doped libraries is that each N-nucleotide–doped codon is biased to express a different subset of amino acids. This makes it difficult to discriminate between amino acids selected as a result of improved affinity from those that are overrepresented in the library. An NNK-codon–doped library circumvents this problem but is more difficult to prepare, as it requires oligonucleotide synthesis on parallel columns [33]. During synthesis of each codon that is to be doped, the resin on which the oligonucleotide is being synthesized is split between two columns. On one column, the codon from the parental peptide sequence is synthesized; on the other, an NNK codon is synthesized, and after synthesis, the resin from the two columns is combined. This procedure is repeated for all residues that are to be doped. More recently, with the commercial availability of trinucleotide phosphoramidites (Glen Research), codon-doped libraries of any type can be made, and probably more easily than multiple-column synthesis. Specific trinucleotide phosphoramidites can be mixed with NNK ones or a limited mixture of trinucleotide phosphoramidites to produce degenerate oligonucleotides [26]. A doped library is screened with the initial MAb under conditions that will stringently select relatively tight binders, yielding more in-depth binding and sequence information on the consensus sequence.

6.3 DIAGNOSTICS

Section 6.2 reviewed Ab–Agn interactions and discussed several types of phage libraries as a source of Ab-binding ligands. In this section, we review the application of phage libraries to diagnostics, focusing on methods that detect the presence of specific Abs in the blood. Many diagnostic assays use Agn or epitopes to detect the presence of specific Abs in serum, indicating the presence of a pathogen and/or the existence of a disease state. For example, a serum sample assayed for the presence of specific IgE against a particular allergen indicates whether an individual has an allergic sensitivity. Such assays can also be used to identify Abs that bind viral or bacterial proteins, indicating infection. Theoretically, even noninfectious states may be assessed with assays developed to specifically detect Abs associated with diseases such as autoimmune disorders and cancer.

Phage display libraries are useful as a source of Agn for diagnostic assays. A primary use of phage display libraries in diagnostics is to identify protein Agns that react with Abs against a known organism (with the exception of membrane bound proteins, since they are insoluble). For example, a whole-Agn library produced from genomic or cDNA from an allergenic organism can be screened with reactive sera to identify specific allergen proteins. These allergens can then be produced in their recombinant form and used in clinical diagnostics to test other individuals for reactivity to the same allergen. Similarly, whole-Agn libraries can be screened to identify protein ligands for infections caused by a known pathogen. For example, human immunodeficiency virus type-1 (HIV-1) is diagnosed by the presence of serum Abs against the viral protein p24. In circumstances in which the diagnostic agent is unstable, rare, expensive, or difficult to purify, an Agn fragment or peptide may be a viable alternative. RPLs can also be applied to situations in which one wants to distinguish between two closely related species. For example, sera from individuals infected with *Chlamydophila pneumoniae* can cross-react with *C. trachomatis* and

C. psittaci [34]. Identification of peptides that differentiate between these species can improve diagnosis of an infectious disease. Specific Abs are often produced in auto-immune disorders and cancer, and perhaps in some idiopathic chronic disorders (i.e., chronic fatigue syndrome). Peptide libraries can be screened with sera from affected individuals to identify ligands associated with disease-related Abs; however, it is difficult to determine with certainty whether particular peptides are associated with a specific disease state. The following reviews the applications of phage libraries to these problems.

6.3.1 Whole-Agn Libraries for Agn Identification

Soon after describing the pJuFo vector (see Section 6.2.2), Crameri et al. [35] applied it to construct a whole-Agn library from the cDNA of the fungus *Aspergillus fumigatus*. They isolated single proteins from the library using IgE from individuals with *A. fumigatus* allergies. Later, they identified one of these proteins as a manga-nese superoxide dismutase that reacted in skin tests in allergic individuals [36]. More recently, this group used high-throughput methods to screen a cDNA library from the allergenic fungus *Cladosporium herbarum* with IgE from 13 sensitized individ-uals. They identified the allergen as a hydrophobic component of the cell wall; this is the first report of a fungal cell wall protein as being a clinically relevant allergen [37]. Allergenic proteins have been identified by screening whole-Agn libraries pro-duced from several organisms including *Alternaria alternata*, *Candida albicans*, wheat germ, peanut, and *E. coli* [38]. Significantly, this approach identifies allergens that can be expressed recombinantly; other allergen preparations may be compara-tively difficult to produce.

Whole-Agn libraries have been used to identify tumor-specific Agns. The cDNA for such libraries has been produced from a cancer cell line, such as breast cancer lines T47D and MCF-7 [19] or the colorectal cancer cell line HT-29 [18]. Library screening with sera from cancer patients, including those who had been immunized with autologous tumor cells, isolated specific clones, whose sequences identified novel tumor-associated Agns. These examples used pVI display described by Jespers et al. [15] and, more recently, Hufton et al. [16]. The tumor Agns identified may serve as candidates for tumor vaccination, sero-diagnosis of cancer, prognostic markers, or as probes for monitoring tumor cell-based vaccination trials. Other groups have used whole-Agn libraries to identify auto-Agns [39].

6.3.2 RPLs for Identifying Diagnostic Peptides for Known Pathogens

Folgori et al. [40] first described the screening of RPLs with serum PCAbs from pathogen-infected individuals to identify peptides for use in disease diagno-sis. Success is dependent on identifying peptides that bind all or most sera from infected individuals, while being unreactive with sera from uninfected individuals. Kouzmitcheva et al. [41] used this technique to identify peptides that indicate Lyme disease. They used affinity-purified PCAbs from eight individuals infected with *Borrelia burgdorferi* in a screening strategy that selected each clone with IgG from at least two individuals. They screened a panel of 12 RPLs with various constraints and isolated 17 peptides that were recognized by all 10 sera from infected individuals whose sera had not been used in screening. None of the consensus sequences isolated matched *B. burgdorferi* proteins. A similar technique was used by Birch-Machin et al. [42] to identify peptides that bind to equine herpes virus (EHV) antisera; however, they reported homology of consensus sequences to several EHV proteins. For these peptides to be useful for diagnostics, they must be further studied and developed. Ultimately, they must react with at least 75% of the infected population to be considered for commercial assays [41].

Studies identifying diagnostic leads are best performed with PCAbs rather than MAbs, and preferably with PCAbs from more than one individual. There are several reasons for this. First, different individuals or species responding to the same Agn may not make the same Abs, or may not make Abs against the same epitope. Second, if a MAb is made against a nonimmunodominant epitope, then individuals may have only a small concentration of that MAb in their sera, which may be difficult to detect. The work of Benguric et al. [43] illustrates this point. Their goal was to identify peptides and Abs that would be specific for a MAb against *Mycoplasma capricolum* subsp. *capripneumoniae* (*MCCP*). They used murine MAb 4.52 to isolate two peptides from a 17-mer RPL. Mab 4.52 was also used to immunize chickens; the resulting chicken PCAbs were adsorbed on mouse IgG to remove anti-isotypic and antiallotypic Abs, and then affinity purified on MAb 4.52 to isolate specific Abs against the Ab combining site. (These are called anti-idiotype [Id] Abs or anti-Id Abs; see Section 6.5.2 for more details about Ids.) Both peptides and the purified anti-Id Ab blocked MAb 4.52 binding to *MCCP* cell lysate in a competition ELISA. However, when goat antisera against inactivated *MCCP* was tested for binding to both the peptides and anti-Id Abs, no binding was detected. Thus, the peptides and anti-Id Ab were specific for the murine MAb, but not for Abs from a different species that were presumably against the same Agn (though this was not shown). The authors might have had better luck if they had used a PCAb rather than a MAb for library screening, and perhaps an AFL rather than an RPL as a source of peptides.

6.3.3 AFLs and RPLs for Auto-Agn Identification and Idiopathic Disease Diagnosis

AFLs and RPLs can be applied to the identification of auto-Agns for autoimmune disorders and may eventually lead to diagnostic tools for specific autoimmune disorders. As an example of this, Fierabracci et al. [44] used RPLs to identify a new auto-Agn that may be associated with the autoimmune disease, insulin-dependent diabetes mellitus (IDDM, also known as type II or juvenile-onset diabetes). They screened two 9-mer peptide libraries with a single serum sample from a patient with IDDM and then screened the enriched phage pools with two additional IDDM+ serum samples, isolating peptides that bound to Abs from all three serum samples. To ensure that the peptides identified during the screening were IDDM specific, the enriched phage pools were further screened with three additional IDDM+ sera and counterscreened with sera from eight healthy people. Five unique IDDM-related peptides were identified by testing 70 phage clones for binding to serum samples from newly diagnosed and long-term IDDM patients, and for lack of binding to sera from healthy individuals and patients with other autoimmune disorders. One peptide (CH1p) that showed no homology to any human protein was detected by 70% of newly diagnosed IDDM patient sera compared with 10% of healthy control sera.

The authors used the CH1p peptide to identify a putative, IDDM-associated auto-Agn. To further test the ability of the CH1p peptide to behave as an immunogenic mimic of an epitope on the auto-Agn, the CH1p peptide was used to immunize rabbits. The resulting anti-CH1p serum (R-anti-CH1p) was used to screen a cDNA expression library from human pancreatic islet cells. Three clones were identified and all shared 99% homology with human osteopontin, a marker for late osteoblast differentiation [45]. A number of experiments were then performed to assess whether the CH1p peptide is an immunogenic mimic of osteopontin. SDS-PAGE of human islet cell preparations followed by western blotting with R-anti-CH1p and antiosteopontin PCAbs indicated that both Ab preparations bound to a protein with a molecular weight corresponding to that of osteopontin. R-anti-CH1p also bound to purified osteopontin in a dot blot, and binding could be blocked by preincubation with CH1p peptide. Immunohistochemical staining of paraffin-embedded pancreatic sections with both R-anti-CH1p and IDDM⁺/CH1p⁺ sera produced similar binding patterns to somatostatin-producing alpha cells. However, the R-anti-CH1p, but not the IDDM⁺/CH1p⁺ sera, was blocked from binding to these cells when preincubated with CH1p peptide. It is not clear whether the IDDM⁺/CH1p⁺ sera that were tested also had osteopontin-binding activity (if they did not, then it is perhaps understandable why the peptide did not block their binding to the cells), nor whether binding of any of these Abs would be blocked by osteopontin. One can conclude that that the CH1p peptide can elicit osteopontin-binding Abs in rabbits, and that such reactivities occur in CH1p⁺ sera from IDDM patients. However, the relationship between CH1p-binding and osteopontin-binding reactivities in IDDM⁺ sera is not clear.

The frequency of osteopontin reactivity in IDDM⁺ patient sera was further analyzed by radioimmunoassay (RIA) using recombinant human osteopontin. Serum samples from 146 people were tested: 64 were from newly diagnosed IDDM patients, four were from patients with long-term IDDM, and the remaining serum samples were controls from healthy individuals and patients with autoimmune disorders. Although, 70% of newly diagnosed IDDM sera reacted with the CH1p peptide in ELISA, only 8% of the tested sera bound to osteopontin by RIA. If the CH1p peptide is supposed to be a marker for IDDM-related Abs against the putative auto-Agn, osteopontin, it is not clear why so many IDDM patients with CH1p⁺ sera are osteopontin⁻. A central question from these studies that remains unanswered is whether Abs from different patients that bind to CH1p also bind specifically to osteopontin. Thus, these intriguing findings do not clearly show that osteopontin is an auto-Agn associated with IDDM. Moreover, the frequency of osteopontin reactive Abs in patient sera indicates that, even if osteopontin is an auto-Agn, it may not be a very common one. Interestingly, independent studies from this work have shown that osteopontin is involved in the vascular calcification that accompanies diabetes of all types (reviewed in Ref. [46]).

Fierabracci et al. [44] developed a novel approach to the discovery of putative auto-Agns simply by finding a peptide that commonly reacted with serum Abs from diseased patients. Under the assumption that this peptide could behave as an immunogenic mimic of a corresponding auto-Agn, they used the peptide to obtain antipeptide sera and, in turn, used this antiserum to identify a putative cognate auto-Agn. However, to prove the connection between a disease state, an Ab-binding peptide, and its putative cognate Agn, several criteria should be met. First, serum Abs from diseased patients should be affinity-purified on the peptide and then tested for binding to the cognate Agn to show that disease-specific Abs cross-react with both Agns. Second, sera that react with a disease-specific peptide should also react with the cognate Agn. If this is not the case (i.e., if more sera bind peptide than cognate Agn), then there may be confounding reactivities in the patient sera. For example, the peptide reactivity may initially arise via one Agn but then "spread" to the putative cognate Agn; this could account for sera that react with peptide but not the putative cognate Agn. Third, if a putative cognate Agn is truly an auto-Agn, one would expect to find signs of "epitope spreading" on the auto-Agn; more disease-specific sera should react with the cognate Agn than with its corresponding disease-specific peptide. Last, as this approach depends on serum Ab responses, they should be tested at every step for nonspecific reactivity, which can confound experimental results. One way to remove some, but not all, nonspecific reactivities in a serum is to incubate it with a complex antigenic mixture, like a bacterial lysate, then to test treated and untreated sera for binding to the Agns of interest (disease-specific peptide, putative cognate Agn, and unrelated protein controls). One should see a drop in binding to the unrelated controls, but not in binding to the targeted peptides and cognate Agns. Given these caveats, this approach, using RPLs to discover disease-specific Ab reactivities, may be useful in identifying new Agns involved in a variety of disease states, including autoimmune diseases and cancer. The hope is that the pathology of some idiopathic diseases may be clarified through the identification of disease-associated Agns.

RPLs have also been used to identify ligands for the diagnosis of autoimmune disorders such as Crohn's disease (CD). By screening an RPL displaying nonamer peptides with sera from CD patients, Saito et al. [47] identified five CD-related peptides; only two of them shared similar sequences and binding characteristics. Multiple antigenic peptides (MAPs) made from four of the peptide sequences were recognized by 52 of 92 (56.4%) patient sera tested but reacted with only 6.2% of negative controls. Others have used a similar approach to identify peptides that specifically bind an anticardiolipin MAb in studying homologous, disease-associated anticardiolipin Abs in patients with antiphospholipid syndrome [48].

There are many diseases for which there is no known associated pathogen. While the examples in this section describe discovery of disease-related ligands, this approach is not always successful. Our lab has screened RPLs with PCAbs from patients diagnosed with Kawasaki disease, a disorder that occurs in young children and is characterized by sudden onset of a high fever followed by severe vasculitis and cardiac aneurysm. We screened a panel of RPLs with sera from children with Kawasaki's disease who had not received IVIG treatment, but found no peptides that bound patient sera but not that of controls. Another group screened a panel of RPLs with purified IgG from 23 patients with chronic fatigue syndrome and identified specific peptides associated with this illness; however, no difference in phage clones was seen between patient and control sera (Smith, personal communication, 2003).

Diagnostic applications of phage libraries include the identification of whole Agn, the identification of AFs when a disease-causing organism is known, or the identification of peptides associated with autoimmune or chronic disorders and other idiopathic diseases. The approach of screening whole-Agn libraries to identify specific allergens has yielded much success. Other applications have met with limited success, such as the identification of peptide ligands specific for Abs associated with idiopathic diseases. Still, there is the promise that the diagnosis of, or clues to the pathogenesis of, these latter diseases may be revealed by novel approaches using RPLs. In Section 6.4, we discuss epitope mapping, one of the more common uses for phage display libraries.

6.4 PHAGE LIBRARIES FOR EPITOPE MAPPING

It is useful to describe traditional epitope mapping to begin with, and the work of Jin et al. [49] is an elegant example of this. They used 21 MAbs to fine-map 19 discontinuous and two linear epitopes on human growth hormone (huGH). First, gross epitope regions were mapped using 12 mutants of huGH that each had a different region replaced with the corresponding region from a homologous protein. Binding by 19 of the MAbs was decreased by more than one mutant, indicating that it involved more than one region on huGH. Moreover, for each of the 19 MAbs, these regions mapped to localized patches on the surface of huGH, further supporting the idea that they form a discontinuous epitope. Only two linear epitopes were identified (i.e., MAbs whose binding was affected by replacement of only one region), reflecting the dominance of discontinuous epitopes in the murine Ab response.

Second, fine epitope mapping was completed using Ala replacement of individual residues within localized regions identified in the homolog mapping studies. As expected, the CBRs of epitopes were spatially clustered and mostly surface-exposed residues; but surprisingly, they were restricted to only a few types of amino acid: Arg, Pro, Glu, Asp, Phe, and Ile. This elegant example of a detailed epitope mapping study serves as a starting point from which to analyze epitope mapping work performed with phage libraries.

AFLs and RPLs have been used to identify epitopes on a protein Agn for MAbs and PCAbs. Linear epitopes are relatively easy to identify for two reasons. First, they are typically restricted to a short sequence, making it likely that peptides containing fragments of the appropriate length, or longer, will be present in an AFL or RPL. Second, linear epitopes usually comprise three to five CBRs, most of which are residues that are shared with the cognate protein Agn. Thus, a consensus sequence has a good chance of aligning with the cognate protein's sequence, even if it originates from an RPL. The advantage of an RPL is that CBRs in an epitope are more likely to be identified from the sequences of Ab-binding clones (through the identification of consensus residues), whereas the advantage of an AFL is that Ab-binding clones contain most or all of the cognate epitope, whose boundaries are defined by alignment of sequences from multiple clones.

Discontinuous epitopes are more difficult to identify using either type of library. Compared to linear epitopes, discontinuous epitopes are less likely to be present in an AFL because of their distantly spaced CBRs and/or their dependence on the folding of a larger protein domain, Hence, they typically require a longer protein sequence. Nevertheless, AFLs are best for identifying a discontinuous epitope, given the restrictions that (1) the targeted epitope must be reproduced by a relatively short region or small domain from the cognate Agn, (2) an AF containing the discontinuous epitope will most likely be considerably larger than the epitope itself, and (3) CBRs will not be identified by this approach. The smallest Ab-binding AF could be used to produce a doped library, and Ab-binding clones coming from this sublibrary may reveal CBRs.

Alternatively, ligands for Abs against discontinuous epitopes can be identified from RPLs. If CBRs are identified from clones isolated from an RPL, they can be used to map a discontinuous epitope. However, many of the CBRs may serve to promote the unique structure of the peptide, without contacting the Ab directly. In this instance, only a few CBRs may match between a peptide ligand and its corresponding discontinuous epitope, making it difficult to identify the epitope by sequence alignment with the cognate Agn. Peptides that mimic discontinuous epitopes typically have more CBRs than peptides that mimic linear continuous epitopes, and they are typically less abundant in a library. Thus, it can be difficult to isolate multiple, independent clones sharing a consensus sequence by a typical library panning procedure. A rare, discontinuous-epitope mimic derived from an RPL may require, in addition to the initial RPL screening, the effort of producing and screening sublibraries to reveal its CBRs and to produce acceptable affinity. For these reasons, discontinuous epitopes can be difficult to map regardless of the type of library used for Ab screening.

PCAbs have been used to map multiple epitopes on a protein or organism [2,3,42,50]. It is not possible to predict the Abs in a PCAb mixture that will cross-react with peptides from either an AFL or RPL, and thus a cross-reactive peptide may not be selected for a single epitope of interest that is bound by a PCAb. PCAbs usually isolate a number of peptides corresponding to different Ab subspecificities; thus, they can provide a peptide "signature" that reflects, in part, the breadth of the Ab response. Moreover, such peptides can be used to compare Ab responses from different individuals and from such comparisons, to identify immunodominant epitopes that are common to multiple sera [51,52]. The use of multiple libraries in PCAb screenings [3,41,53] can provide a wider range of selected clones.

Epitope mapping can be used to define epitopes for MAbs with known biological function. By using such MAbs for screening, AFLs and RPLs have been used to identify binding sites for neutralizing Abs [50,54,55] or immunodominant regions on Agn [50,56–58]. Moreover, peptides have potential use in the design of vaccines that specifically target a response against a cognate epitope on an antigen, including pathogens and toxins (see Section 6.5). Thus, RPLs and AFLs have the potential for producing vaccines that target particular Abs and thus specific epitopes.

6.4.1 AFLs for Epitope Mapping

AFLs are used to identify regions of a protein that bear a cognate epitope. Very likely, the whole epitope must be present on a fragment for binding to occur. Once an AFL is screened, sequences from binding clones can be aligned to reveal the epitope, deduced as the shortest region shared by all binding clones. As the screening of an AFL may not always yield ligands (especially if a MAb binds a discontinuous epitope), RPLs may be screened along with the AFL to further ensure that clones bearing ligand peptides will be isolated. Fack et al. [59] used four MAbs (each against a different protein) to screen two RPLs and four AFLs (one AFL for each protein). Two of the MAbs (215 and Bp53–11) were known to bind linear epitopes, and the other two MAbs (GDO5 and L13F3) were also thought to bind linear sequences, but they had not been previously mapped. The RPLs yielded consensus sequences for only two MAbs (Bp53–11 and GDO5). In contrast, all four MAbs isolated ligands from their respective AFLs. Alignment of the sequences isolated by each MAb determined the region that was shared between clones, and this shared sequence was used to define the cognate epitope on the protein. The two MAbs with known epitopes (215 and Bp53–11) selected several short fragments from their AFLs, confirming their previously identified linear epitopes. One of the MAbs with an unknown epitope (GDO5) isolated three overlapping AFs, which identified a 10-residue region containing a putative linear epitope. All putative epitopes were confirmed by testing Ab-binding to overlapping peptides that spanned the region identified by the AFs or RPLs. In contrast, it was concluded that the fourth MAb (L13F3) binds a discontinuous epitope, since it isolated two partially overlapping AFs that spanned 50 residues; it did not isolate any binding peptides from the RPLs. Taken together, these results indicate that AFLs are suitable for identifying linear epitopes and perhaps discontinuous ones.

This example indicates that AFLs are a better choice for mapping linear epitopes, but draws no conclusions about which library type is better for mapping discontinuous epitopes. It is not surprising that a discontinuous-epitope binding MAb did not isolate binding phage from the RPLs used by Fack et al. [59], as neither of the libraries they used contained disulfide constraints. It has been our experience that this type of Ab always selects constrained peptides. Other examples that use AFLs for epitope mapping include the work of Bentley et al. [50], who mapped several immunodominant antigenic regions on the outer capsid protein of African horsesickness virus using chicken and horse PCAbs, and the work of Holzem et al. [60], who identified a seven-residue epitope for a MAb against tobacco mosaic virus.

In a unique approach, Huang et al. [61] used AFLs to map discontinuous epitopes on porcine rotavirus. They produced AFLs expressed from the VP7 gene; cDNA was self-ligated (presumably into multimers), digested with DNa-seI, and then blunt-end–ligated to limiting amounts of phage vector. This produced clones containing short stretches of VP7 sequence as well as clones containing discontinuous VP7 sequences ligated together, which might function as discontinuous epitopes. The library was screened with seven virus-neutralizing MAbs thought to bind discontinuous VP7 epitopes; however, no binding clones were isolated. The library was also screened with PCAbs from mice, rabbits, and convalescing pigs, and sequences of the isolated clones mapped to portions of VP7. The PCAbs selected clones containing mostly single fragments as well as some in-frame fragments that were flanked by out-of-frame ones; only 2 out of the 23 analyzed clones bore in-frame fragments from two regions of the protein. One composite clone, isolated by mouse PCAbs, contained fragments from two antigenic regions of VP7; however, there is no evidence to suggest that a single Ab bound to both regions. It appears that this library did not produce mimics of discontinuous epitopes, as none of the seven discontinuous-epitope binding MAbs selected any clones; this probably occurred because only a small proportion of this library expressed the desired composite sequences in-frame. Perhaps an approach like this would succeed with additional work. For instance, a DNA shuffling may improve the mixing of and variation within fragments [62], and in addition, ligated fragments could be selected for ORFs [21] before cloning DNA fragments into phage display vectors.

6.4.2 RPLs for Epitope Mapping

RPLs have been used to identify ligands for antiprotein Abs against linear and discontinuous epitopes. As explained in Section 6.2, peptides that cross-react with linear epitopes are most dependably obtained, since they usually do not require a large number of CBRs to bind Ab. Linear epitopes are also easier to map once the CBRs are known. Discontinuous epitopes are more difficult to identify, since CBRs on the sequence may relate to different parts of the protein, and many CBRs may be needed to structure a peptide ligand. Thus, peptides that cross-react with discontinuous epitopes are more difficult to obtain and may require several stages of development.

Work from our laboratory with human MAb b12 illustrates this point. MAb b12 neutralizes a broad range of HIV-1 isolates by binding to the envelope protein gp120 at a large discontinuous epitope, which overlaps with the CD4-binding site [63]. Screening of a gp120-based AFL with MAb b12 yielded no binding peptides (Parren, personal communication, 2003). Bonnycastle et al. [53] used MAb b12 to screen a panel of 11 RPLs and two clones were identified that share a five-residue sequence. Based on this consensus, Zwick et al. [29] produced and screened two sublibraries, and the best-binding peptide (B2.1) shared a significant amount of sequence homology with the D loop of gp120. Later, Ala replacement studies showed that most of the homologous residues were CBRs (Ollman-Saphire E, Montero M, Menendez A, van Houten NE, Irving MB, Zwick MB, Parren PWHI, Burton DR, Scott JK, Wilson IA, manuscript in preparation). In support of this homology with the gp120 D loop, the crystal structure of MAb b12 bound to the peptide showed that, on a gross structural level, the peptide binds MAb b12 at the same paratope subsite as that deduced for the D loop [63]. Surprisingly, five of the seven CBRs in the peptide did not contact the Ab and thus were required for the correct structure of the peptide (Ollman-Saphire E, Montero M, Menendez A, van Houten NE, Irving MB, Zwick MB, Parren PWHI,

Burton DR, Scott JK, Wilson IA, manuscript in preparation). Of the two CBRs that directly contacted the Ab, only one matched with a CBR on the D loop, and replacement of this Asp residue with Ala ablates MAb b12 binding to gp120. Thus, except for a common Asp residue, the B2.1 peptide and the D loop of gp120 do not appear to share a common mechanism for binding to MAb b12.

In an ambitious study, Bresson et al. [57] identified CBRs derived from screening phage libraries and aligned them to regions on the auto-Agn, thyroperoxidase (TPO), to localize regions that contribute to an immunodominant, discontinuous epitope. Four phage libraries were screened with a MAb that binds a discontinuous epitope on TPO and competes for binding with most anti-TPO sera. Three consensus motifs were identified, and Ala replacement was used to determine the CBRs of these peptides. Consensus sequences were aligned to TPO, identifying five regions that potentially contribute to MAb binding. In many cases, the alignment was questionable, since spaces had to be added to the CBRs from the peptides to match "homologous" residues on TPO. For each of the five regions identified, four to ten residues were replaced with the amino acid sequence from the corresponding region of a homologous protein. This was meant to change the sequences in these regions without affecting the overall folding of TPO. These TPO mutants were tested by western blot for binding to three MAbs and a rabbit PCAb, and by ELISA for binding to sera from patients with the autoimmune thyroid disorders, Grave's disease and Hashimoto's thyroiditis. Patient sera showed reduced or no binding to four of the TPO mutants, and the MAbs did not bind to two of these mutants. The authors concluded from these results that four regions contribute to the immunodominant epitope on TPO.

It is unclear whether the loss in binding was due to mutations in the immunodominant region of TPO or to denaturation, since the authors did not determine whether the amino acid replacements affected global folding of TPO. Other studies, using homolog replacement (see Ref. [49]), showed decreased MAb binding to certain homolog replacement mutants. When MAbs were tested for binding to single Ala replacement mutants in these regions, there was no effect on the degree of MAb binding, and this indicated that homolog scans may introduce some disruptive effects to protein folding. The study by Bresson et al. [57] presents an interesting method for identification of immunodominant epitopes. However, further studies could strengthen the conclusions; for example, the putative immunodominant epitope could be subjected to Ala replacement scans.

Several laboratories have used MAb-binding motifs identified from RPLs to map a linear epitope and determine its secondary protein structure. Stern et al. [64] isolated two MAbs (GV1A8 and GV4D3) from a mouse immunized with HIV-1 gp120. These were shown to bind to the N-terminal portion of gp120 between residues 1 and 142 by probing protease-digested, reduced gp120 on a western blot with both MAbs. To map the MAb-binding site on gp120, a 20-mer RPL was constructed and screened with MAb GV1A8. A single phage clone (ϕ35) was identified, but the sequence did not match gp120, and therefore, shared amino acids could not be identified. To obtain more clones, the authors rescreened the library under less stringent conditions, and they isolated an additional 18 clones that shared a common dipeptide motif (Leu/Ile-Trp). Alignment of these clones to the N-terminal region of the gp120 sequence, based on the dipeptide motif, revealed other residues that were shared with gp120, but were not shared among all the phage clones; this identified the general region of the MAb GV1A8 epitope. Each of the 19 clones was analyzed based on its homology with gp120, and clone φ66 exhibited the highest homology. The gp120 epitope encompassed residues 108–113, as revealed by alignment with clone φ66, and confirmed by MAb binding to overlapping peptides that covered this region. However, even though clone φ66 shared the greatest homology with the gp120 epitope, it exhibited lower affinity for the MAb GV1A8 than did five other binding clones. This led the authors to continue to look for determinants of homology with gp120.

The alignment of all 19 phage clones with gp120 revealed a pattern of conserved residues at four positions along the gp120 sequence (HxxIxxLW). Only one clone (φ35) bore the full sequence motif, and this clone exhibited the highest affinity for MAb GV1A8. The analysis indicated that the full binding motif (HxxIxxLW) was consistent with two turns of an alpha-helix, and it was hypothesized that the C1-region of HIV-1 had an alpha-helical secondary structure. A computer model of residues 89–117 of gp120, configured as an alpha-helix, placed the CBRs for MAb GV1A8 on a contiguous surface, whereas this was not the case when this region on gp120 was modeled as a beta strand. To further confirm the analysis, it was postulated that the second MAb (GV4D3), which binds to gp120 in the same region as MAb GV1A8, would also select a series of phage clones with sequences consistent with an alpha-helical motif. This was confirmed by screening the same 20-mer RPL with MAb GV4D3. Seven phage clones were isolated and analyzed by alignment with gp120. The authors concluded that this region also bore a helical motif. However, fewer clones were isolated for this MAb, and the peptides had overall lower homology scores than those isolated for MAb GV1A8. The authors did not confirm their structural hypothesis with physical analyses, such as circular dichroism. However, it may not be surprising that a putative alpha-helical motif was identified, given that helicity is a common structure among proteins. The computationally derived hypothesis, that residues 93–112 of HIV-1 gp120 form an alpha-helix, was later confirmed by the crystallographic structure of gp120 [65].

In a similar vein, work by Ferrieres et al. [66] used RPLs to show that residues in a linear epitope that are not CBRs can contribute to Ab affinity, probably by conferring structural stability to the peptides. The authors first used Ala replacement to identify the consensus sequence, YXTEPH, for a linear epitope on troponin I; replacement of residues flanking this region did not appear to affect MAb binding. They next screened an RPL with the MAb, yielding peptides bearing the YXTEPH consensus; however, synthetic versions of the phage-displayed peptides were found to have lower affinities than synthetic peptides bearing the native epitope. To improve affinity of the phage-selected sequences, the authors replaced sequences flanking the consensus on a phage-selected sequence with those flanking the consensus on native troponin I. Although Ala replacement of these sequences did not affect MAb binding, transfer of either the N-terminal or the C-terminal region from troponin I to flank the consensus sequence of a phage-selected peptide improved affinity for immobilized MAb. In some cases, improved affinity correlated with improved structural stability, as shown by circular dichroism studies. Thus, longer sequences, which apparently do not contain CBRs, can affect binding affinity. As with the study of Stern et al. [64], the results of Ferrieres et al. [66] indicate a global, multiresidue effect that is consistent with structural stabilization, which in some cases could be detected.

Our laboratory has produced further evidence of sequences that confer increased affinity due to global effects that do not depend on discrete CBRs. Working with a MAb that recognizes a linear epitope on the membrane proximal region of the HIV-1 envelope protein gp41, Menendez et al. [67] constructed and screened RP sublibraries that extended the peptide on either side of the core epitope (DKW). Two Ala residues were placed N-terminal to the DKW and Ser was placed C-terminal to DKW in the sublibraries so that the high-affinity epitope, ELDKWA, could not be produced; this would make selection of higher-affinity peptides dependent upon the N- and C-terminal sequences from the sublibrary. Thus, two libraries, X_{12}AADKWS and AADKWSX$_{12}$, were constructed and screened with the MAb under stringent conditions that would select the tightest-binding clones. No such clones were selected from the X_{12}AADKWS library, but three relatively tight-binding clones were selected from the AADKWSX$_{12}$ library. To perform in-solution affinity studies using a BIAcore instrument, the sequences of the displayed peptides were transferred to the N-terminus of the maltose-binding protein of *E. coli*, which allowed monovalent display in the context of a fusion protein [68]. The authors showed that the affinity of the peptides could be increased to levels similar to the affinity of gp41 by replacing the two Ala residues preceding the DKW motif with Asp and Leu; this supports the conclusion that there is overlap between the sites on the MAb that the peptides and native epitope bind. The affinities were two orders of magnitude greater than that of a synthetic peptide bearing the ELDKWAS epitope flanked by Gly residues.

Ala replacement studies on the phage-displayed version of two of the sequences showed little effect of singly changing any of the last four residues. Yet, deletion of the last three residues of the same peptides decreased binding of the Fab version of the MAb significantly by >50% in one case and >80% in the other. (As Fab binds phage-borne peptides monovalently, its apparent affinity is less affected by peptide polyvalency than IgG, which is bivalent. Thus, it is more sensitive to differences in affinity.) These studies now await crystallographic study to determine the role of these C-terminal sequences in enhancing the affinity of binding to the MAb. Fusion of the peptides to smaller proteins should allow structural studies (NMR, circular dichroism) that may reveal the degree to which structural stabilization plays a role in Ab affinity.

Thus, the works of Stern et al. [64], Ferrieres et al. [66], and Menendez et al. [67] have elucidated three different approaches to analyzing the role of non-CBRs in promoting Ab affinity. There remains much to be done in revealing the connection between affinity and structural stability in all of these cases.

In a recent study, Enshell-Seijffers et al. [69] describe a novel computational approach to mapping discontinuous epitopes on the surfaces of proteins with known tertiary structures, using the sequences of MAb-binding peptides isolated from RPLs. The algorithm is based upon two general assumptions: (1) Ab-selected peptides contain amino acid pairs that occur at a higher frequency than the amino acid pairs in a RPL, and (2) those selected amino acid pairs will occur in close proximity in the cognate protein epitope. The analysis was applied to three MAbs that recognized overlapping epitopes on the gp120 envelope protein of HIV-1. The epitope of one of the MAbs (17b) was previously shown to comprise four beta-strands of the HIV-1 envelope protein gp120, based on a cocrystal structure of MAb 17b bound to gp120 [65]. A constrained 12-mer RPL was screened with MAb 17b, and 11 peptides, having no homology to gp120, were selected and shown to bind by dot blot. The frequency of each amino acid pair in the peptide sequences was calculated. The most frequently occurring pairs were assumed to represent either continuous residues on the cognate epitope or noncontinuous amino acids that are brought together by protein folding. The authors identified all amino acid pairs that are within a short distance of each other on the Agn surface and matched the amino acid pairs identified from the RPL screening to them. Twenty-six pairs from the RPL screening mapped into a region that overlapped with the crystallographically defined MAb 17b epitope. The region comprised four discontinuous regions, containing seven of the eight known segments that were observed to contact the MAb, some of which spanned up to eight contact residues. Another MAb (CG10), which competes with MAb 17b, was thought to bind an epitope on gp120 that overlaps with the MAb 17b epitope. Twenty-eight peptides were selected from the library screening and analyzed as described for MAb 17b. The identified epitope contained five residues known to affect MAb CG10 binding and, as suspected, was located near the MAb 17b epitope.

A discontinuous epitope-mimic peptide was constructed by fusing the four gp120 regions identified in the analysis into reconstructed epitope bound to MAb CG10 and competed with gp120. The authors did not determine its affinity, the regions involved in binding, nor the CBRs within them. Ala replacement studies would further clarify the mechanism of binding, as would a crystal structure of MAb CG10 bound to the peptide mimic; the latter would also show the degree to which the peptide mimics the cognate epitope on gp120. This approach shows promise as an alternative to homolog replacement studies for identifying discontinuous epitopes for MAbs that bind proteins with known structures. Moreover, it may serve as a viable approach to the structure-based design of epitope-mimic peptides for vaccine and diagnostic applications.

The main promise of epitope mapping is its application in the design of vaccines that elicit the production of Abs against specific epitopes. There are a number of reports of peptides that, when used as immunogens, will elicit Abs that cross-react with a cognate Agn, presumably at a single, targeted epitope. Typically, the sequences of these peptides are based on a cognate linear epitope. AFLs are a superior source of immunogenic mimics of protein epitopes. Being derived from the cognate Agn, AFs are more likely to elicit Abs that will cross-react with cognate Agn. However, as most Abs recognize discontinuous epitopes, Ab-binding peptides may not be present in a given AFL. Alternate approaches may be required if AFs are not found for a specific Ab. For example, Enshell-Seijffers et al. [69] used sequence data from clones selected from an RPL to identify the structure of a discontinuous epitope, and then designed a peptide mimic of that epitope based on that structure. This construct was never tested as an immunogenic mimic, but may prove to be one.

Another approach to epitope-targeted HIV-1 vaccines is being explored by Pantophlet et al. [70]. Their goal is to produce a vaccine that will elicit Abs having the properties of MAb b12, a human MAb that neutralizes a broad range of HIV-1 primary isolates by binding to the CD4-binding site and blocking the virus–receptor interaction. Instead of using a small domain to mimic a discontinuous epitope on gp120, this group engineered gp120 to block Ab binding to unrelated, immunodominant epitopes and to bind strongly to MAb b12, but not to the nonneutralizing MAb b6, which also binds the CD4-binding site. Immunodominant epitopes on gp120 were blocked by the addition of glycosylation sites, and the CD4-binding site was engineered to decrease binding to MAb b6 and to bind more tightly to MAb b12. This engineered gp120 thus does not bind CD4, the MAb b6, or a number of other gp120-binding Abs, but binds well to MAb b12, and immunization studies with it are underway. The engineering of gp120 was based on work by Ollman Saphire et al. [63], who produced a computationally derived model of the MAb b12 structure docked onto the gp120 structure of Kwong et al. [65], as well as extensive gp120 mutagenesis studies that differentiated residues on gp120 that are important for MAb b12 binding from those that are important for MAb b6 binding [71].

Both the approaches of Pantophlet et al. [71] and Enshell-Seijffers et al. [69] are structure-based epitope-targeted approaches, designed to create immunogens that elicit Abs against a specific epitope. It may also be possible to target the production of a specific Ab (as opposed to Abs against a specific epitope) by using a prime-boost approach (see Section 6.6.2.1). After priming with cognate Agn, production of the desired Ab is selected and amplified during the boost with an Ab-specific peptide. For this approach to work, the desired Abs must be elicited during the priming immunization. One caveat is that complex Abs, such as those having extensive somatic mutation, and/or those that have rare specificities, may be difficult to target by any of the approaches described. Section 6.5 will briefly introduce the topic of vaccines and follow it with an examination of the use of peptides, derived from AFLs and RPLs, as immunogens for targeting the production of Abs of predetermined specificity.

6.5 PHAGE DISPLAY LIBRARIES FOR VACCINE DEVELOPMENT

Since Jenner and Pasteur, vaccines have evolved from poorly defined "bacterial soup" mixtures into molecularly well-defined formulae [72]. Some vaccines comprise attenuated or whole killed organisms (e.g., cholera and pertussis) and others "detoxified" exotoxin (e.g., tetanus toxoid, TT). More recently, vaccines have evolved into purer formulae comprising partially purified subunits (e.g., acellular pertussis and influenza virus vaccines), recombinant whole Agn (e.g., hepatitis B Agn), and polysaccharide conjugate vaccines (e.g., pneumococcal capsular polysaccharide conjugated to TT) [72,73]. All vaccines today elicit neutralizing Abs as a part of their protective action; some, like the antitoxin vaccines, rely completely on neutralizing Abs for protection. Molecular vaccines target the Ab response against a specific Agn molecule (i.e., induce the formation of Agn-specific Abs) and theoretically could target the production of Abs against a specific epitope. As discussed in Section 6.4, epitope mapping can provide leads for targeting the production of specific Abs against an epitope during immunization. In the following, we review the application of epitope mapping to vaccine design. Specifically, we explore how different types of phage libraries may be applied to find vaccine leads that target epitopes and their corresponding Abs.

6.5.1 Phage Libraries for Vaccine Design

The primary application of phage libraries to vaccine design is in identifying ligands that target biologically active Abs and/or the epitopes they recognize. An Ab-targeted vaccine elicits Abs that have the functional properties of a biologically active Ab (i.e., neutralization), whereas an epitope-targeted vaccine elicits Abs against a specific, biologically relevant epitope. Both approaches are viable and perhaps often reflect different aspects of the same process. Peptides have great potential for targeting both specific epitopes and Abs.

The idea of Ab-targeted vaccines evolved from the idiotypic network theory developed by Jerne and others. Jerne [74] defined an Id as "a set of epitopes displayed by the variable regions of a set of antibody molecules." Abs differ from one another by virtue of their variable region sequences. These form the Ab combining site, which to some degree overlaps with the Ab paratope (the regions of the Ab that contact the Agn's epitope). The Ab combining site also overlaps with the Id (the regions of the Ab that are immunologically distinct and can elicit an Ab response in syngenic animals). On a functional level, the Id is serologically defined by the Abs produced against it, called anti-Id Abs. Some anti-Id Abs can be thought of as carrying a "mirror image" of the Ab's paratope, and thus mimicking the corresponding epitope on the cognate Agn; these Abs will compete with Agn for binding to the Id Ab. Other anti-Id Abs may bind to regions of the Id that are immunologically distinct from the paratope and will not compete with Agn for binding to the Id Ab.

By immunizing with a single Id Ab (i.e., a MAb or highly restricted PCAb response), it is theorized that at least some anti-Id Abs will mimic the Agn in binding the paratope. By immunizing with anti-Id Abs that can compete with Agn, some of the resulting anti–anti-Id Abs may behave as the original Id Ab and bind to Agn as well as the original Id. Such anti–anti-Id Abs are thought to be similar to the original Id Ab [75,76]. Thus, several groups have sought to use anti-Id Abs to target the production of specific Abs. This has been especially attractive as an alternative way of producing Abs against T cell-independent, weakly immunogenic Agns like polysaccharide [77] and more recently DNA [78], as well as self-Agn targets for cancer vaccines [79,80]. For reviews, see Refs. [81,82].

Until very recently, there had been little effort to use anti-Id Abs to target production of antiprotein Abs. In an impressive piece of work, Goldbaum et al. [75] demonstrated that immunization with a single Id Ab, followed by immunization with a PC anti-Id preparation, can target the production of specific anti–anti-Id Abs against a discontinuous protein epitope. Rabbits were hyperimmunized with Id MAb D1.3, which binds a discontinuous epitope on hen eggwhite lysozyme (HEL). The resulting anti-Id IgGs were affinity-purified on MAb D1.3 and extensively absorbed with mouse IgG to remove any anti-isotypic or antiallotypic Abs. Mice were immunized with these rabbit anti-D1.3 Id PCAbs and the resulting anti–anti-Id sera were tested for binding to HEL and to the D1.3-anti-Id MAb E5.2, which is known to structurally mimic the D1.3 epitope on HEL [83]. The anti–anti-Id sera bound to HEL but bound better to MAb E5.2. Hybridomas were produced from immune mice, and two anti–anti-Id MAbs (AF14 and AF52) were isolated that bind both HEL and MAb E5.2. Interestingly, the amino acid sequences of MAbs AF14 and AF52 differed from D1.3 by 10 residues at most, indicating they were probably derived from the same germ line genes and shared the same type of V gene rearrangements [75]. These results indicate that it is possible to target the production of specific Abs using anti-Id Abs as structural mimics of a cognate discontinuous epitope. However, there are problems associated with using anti-Id Abs in vaccines. The affinity of anti–anti-Id MAbs, AF14 and AF52, was greater for the anti-Id MAb E5.2 than for HEL. This was especially true for MAb AF14, whose association constant (K_a) for MAb E5.2 was three orders of magnitude greater than that for HEL. Also, immunizing with whole Abs can be expensive, and as Abs are highly conserved proteins, not all individuals will mount a strong response against an Ab-based vaccine. Finally, there is the chance that such a vaccine would elicit an autoimmune response.

Despite these drawbacks, targeting a specific Ab is a desirable way to approach vaccine design for some purposes. Because of tolerance, the immunogenic regions on an anti-Id Ab are limited to its Ab combining site; thus, the Ab response is limited to a relatively small region on a very large immunogen. This is reminiscent of the work of Pantophlet et al. [70], who blocked the immunodominant sites on gp120, limiting it to a single epitope; perhaps epitopes as restricted as this would act as anti-Id Abs and elicit a highly restricted Ab response, with MAb b12-like reactivity being a major component of it. As with the approach of Enshell-Seijffers et al. [69], peptides and AFs represent an alternative to large molecules with restricted immunogenicity as components for Ab-targeted vaccines. In this case, a peptide or AF is more like a hapten, in being small and having only a few epitopes. As compared to a larger, more complex Agn, a hapten-like peptide or AF would elicit only a limited number of reactivities. Sections 6.5 and 6.6 discuss how the polypeptides and peptides identified from AFLs and RPLs can be used to target the production of Abs having specific functions.

The success of an Ab-targeted vaccine relies on choosing an Ab that has a desired biological activity. Biological activity has many forms; for example, in the case of HIV-1, some Abs block infection by neutralizing the virus, in vitro, or protect against infection, in vivo. An in vitro-neutralizing Ab may block infection by binding viral receptors. However, in vivo protection is mediated by a number of factors. For example, Abs may opsonize the surface of a bacterium, thus triggering complement-mediated lysis, or phagocytosis by macrophages. Abs such as these would be useful to target with a vaccine. Second, the targeted Ab must be producible in the first place; if an Ab has extensive somatic mutation and/or uncommon variable-gene usage, then this approach may not work.

Ab- and epitope-targeted vaccination is especially suited to situations in which immunization with whole Agn produces a negative or blocking effect. For example, immunization with cognate Agn may induce Abs that interfere with a neutralizing Ab response [32] (see Section 6.5.4), or, in the case of anticancer vaccines, some MAbs may stimulate cell division when they bind a surface receptor whereas others may block it [55]. Ab- and epitope-targeted vaccine approaches can avoid the production of such unwanted Abs and focus the immune system to produce effective specificities.

Most current vaccines are designed to prevent infection by certain bacteria or viruses; however, Ab- and epitope-targeted vaccines may be used against diseases whose origin is not infectious. For example, some vaccines are being designed to treat cancer [55] or to block autoimmune diseases [84]. Other studies are aimed at producing tolerance against specific allergens, thus preventing allergic reactions [85]. New work also suggests that vaccines can prevent degenerative diseases like Alzheimer's disease [86]. The methods discussed in Section 6.5.2 can be applied to these diverse ailments as well as to infectious diseases. We discuss how AFLs and RPLs can be used in Ab- and epitope-targeted vaccine design.

6.5.2 AFLs and Epitope-Targeted Vaccines

AFs, selected by specific Abs, are effective immunogens that can be used to target the production of Abs against specific epitopes. AFLs yield fragments that are more likely to elicit Abs that cross-react with cognate Agn than peptides found in RPLs. This was demonstrated by Matthews et al. [3] using T4 phage as a model pathogen in a murine system. Mice were immunized six times with T4 phage, and their purified IgGs were affinity-purified on T4 phage. These IgGs were used to screen 12 phage libraries comprising a set of 11 RPLs and one T4 genome AFL. Of the 35 unique clones isolated from the RPLs, only two unique peptides contained 5-mer sequences that mapped directly to T4 proteins. Consensus sequences that did not map to T4 proteins were observed in four sets of clones derived from different libraries. Screening of the AFL with the anti-T4 IgG resulted in 16 unique clones ranging from 15 to 97 residues in length, and they were all in-frame fragments of T4 ORFs.

Thirteen phage-displayed peptides from RPLs and eight from AFLs were used to immunize mice. Sera were tested after three immunizations for both antipeptide and anti-T4 Abs. Cross-reactivity with T4 was defined by the percent drop in Ab titer after serum Abs were adsorbed with whole T4 compared to mock-adsorbed sera. Since intact T4 phage particles were used, Abs against internal T4 proteins may not have been removed from sera, if they were present. Sera from only two RP-immunized groups cross-reacted with T4 at a 10-fold dilution; this level of cross-reactivity was not considered significant. Only four AFs elicited antipeptide titers, and of these, two elicited 40%–100% cross-reactivity with T4 phage. The low anti-peptide Ab response elicited by AFs was likely due to their low copy number on phage. To increase anti-T4 cross-reactive Ab titers, six AFs (including the two that elicited cross-reactivity) and one RP were fused to the N terminus of an internal T4 protein as a carrier for immunization. Three AFs elicited high-titer Abs against T4 phage, and the percent cross-reactivity elicited by the RP was lower than those elicited by the AFs. This study demonstrated that, although peptides derived from RPLs are immunogenic and elicit antipeptide Abs, they are less likely than AFs to elicit Abs that cross-react with the cognate Agn. This approach should be valuable in defining epitopes for subunit vaccines. AFLs should be considered for identifying vaccine leads prior to using RPLs as a source of ligands for specific Abs.

MAbs against the self-oncoprotein ErbB2 can either stimulate or inhibit tumor growth depending on the epitope they bind. Yip et al. [55] used an ErbB2 AFL to identify an anticancer vaccine-lead peptide. Seven anti-ErbB2 MAbs were used to select AFs from two ErbB2 libraries, one of which was constrained by flanking Cys residues; the AFs ranged from 16 to 50 residues in length. Four of the MAbs did not select phage from the AFLs; most likely, they bound to discontinuous epitopes not present in the AFLs. Three murine MAbs (N12, N28, and L87) isolated AFs, and these MAbs were tested for their ability to effect the growth of several breast cancer cell lines. MAb N12 inhibited the growth of the four breast cancer cell lines tested, whereas MAb N28 inhibited the growth of only two and enhanced the growth of the other two. Thus, MAbs N12 and N28 affected cell growth, presumably, by binding ErbB2.

MAb N12 probably bound to a discontinuous epitope, since it repeatedly selected AFs comprising the same 55-residue sequence corresponding to residues 531–586 of ErbB2, and it did not bind to any overlapping 15-mer synthetic peptides that covered this region. The AFs selected by MAbs N28 and L87 shared a 20-mer overlapping epitope region; however, their alignment was not shown. MAb L87 also bound to one peptide from a panel of 15-mer overlapping synthetic peptides, and MAb N28 did not, indicating that these MAbs bind overlapping, but nonidentical epitopes. This is consistent with MAb N28 having biological activity, but not MAb L87. A competition ELISA between these two MAbs would further confirm whether they share identical or overlapping epitopes, but this was not mentioned. Three RPLs were screened to identify CBRs for the MAbs but no peptides were selected.

Mice were immunized five times with GST fusions to either the 55-residue AF selected by MAb N12 or to a 20-residue AF selected by MAbs L87 and N28. Sera were analyzed by ELISA after three and five immunizations for anti-AF Abs and anti-ErbB2 Abs. Only the 55-residue AF elicited Abs that cross-reacted with ErbB2, even though the Ab response was six- to sevenfold lower than that against the AF. Cross-reactive, protein A affinity-purified IgGs were tested for their ability to inhibit growth of one of the breast cancer tumor cell line, BT474. IgGs from the 55-residue AF-immunized mice inhibited 85% of cell growth compared to 65% inhibition by the positive control MAb 4D5, 15% inhibition by negative control anti-GST IgGs, and 45% inhibition by MAb N12, the Ab used to select the AF. The approach discussed here targets the production of Abs against a folded protein subdomain, as opposed to a single epitope or the production of one specific Ab. However, these results illustrate that using AFs as vaccine leads may avoid the production of Abs against undesirable epitopes. In addition, there are instances in which MAbs do not select fragments from AFLs. RPLs may provide ligands for these MAbs, as discussed in Section 6.5.3.

6.5.3 RPLs and Ab-Targeted Vaccines

AFs are a good source of Ab-specific ligands. However, not all Abs will select AFs, especially those that bind discontinuous epitopes, whose CBRs are distant on the primary protein structure. In these situations, RPLs may be a better source of

Ab-specific ligands. Peptides derived from RPLs interact with an Ab paratope by mechanisms that are not necessarily identical to the mechanism of binding to the cognate epitope, but they can be very specific to that Ab; their specificity can make them a good choice for targeting particular Abs. Although peptides from RPLs may be a good alternative to anti-Id vaccines, in targeting specific Abs against a discontinuous epitope, peptides derived from RPLs may cover only a small portion of the paratope in comparison to an anti-Id Ab, which can cover the entire paratope. Building sublibraries to extend the coverage of the paratope by peptide ligand is one way to overcome this problem [30–32].

There are several examples that demonstrate that immunization with peptides derived from RPLs results in protection against infection. Yu et al. [56] demonstrated protection from infection with an encephalopathogenic strain of murine hepatitis virus (E-MHV) after immunization with peptides isolated from RPLs. Three MAbs (5B170, 5B19, and 7–10A) that neutralized E-MHV in vitro and protected mice in vivo against intracerebral challenge with E-MHV were used to screen a panel of 13 RPLs. MAbs 5B170 and 5B19 are known to bind to linear epitopes on the E-MHV S_2-glycoprotein subunit, and MAb7-10A is predicted to bind a discontinuous epitope on the same antigen. MAb 5B170 selected five unique clones and a six-residue consensus motif was identified. MAb 5B19 selected one peptide that contained the same six-residue motif as the peptides selected by MAb 5B170. This was assumed to be the consensus sequence for MAb 5B19, even though only one binding sequence was identified. An ELISA, in which phage clones were titrated onto plate-bound MAb, showed that both MAbs 5B170 and 5B19 bound to the same phage clones. The S_2-glycoprotein was examined for homology to the consensus sequence isolated by MAb 5B170 and MAb 5B19, and a high degree of similarity was found for several different MHV strains, indicating that these are conserved residues on the virus. Thus, MAbs 5B19 and 5B170 compete for binding to E-MHV and probably share a conserved, linear epitope.

The putative discontinuous epitope-binding MAb 7–10A isolated 10 unique phage clones from three libraries. A consensus sequence for MAb 7–10A was not easily identified, since some peptides contained only two residues and others had common residues spaced apart from one another. One peptide, which did not share consensus to other peptides, shared six out of nine residues with the S_2-glycoprotein. Phages bearing these peptides were tested for binding to MAb 710-A in the presence of a reducing agent; disulfide reduction ablated binding to three peptides, consistent with binding to a discontinuous epitope.

C57BL/6 and BALB/c mice were immunized four times with 10 selected phage-displayed peptides (five for MAb 7–10A, one for MAb 5B19, and four for MAb 5B170) at 14-day intervals. Mice were given an intracerebral viral challenge with 10_{LD50} MHV-A59, which induces 100% mortality after 5 days. Only three of six C57BL/6 mice immunized with clone 9.1 (selected by MAb 5B170) survived the viral challenge. BALB/c mice immunized with clone 9.1 and all other phage-immunized groups died before 10 days. Serum analysis by western blotting revealed that only mice immunized with clone 9.1 produced serum Abs that bound to recombinant S_2-glycoprotein; this was also true for BALB/c mice even though they were not protected from infection. To determine the mechanism of protection, immunized mouse sera were used in neutralizing antibody-dependent cell-mediated cytolysis and antibody-dependent complement-mediated lysis assays. No activity was detected in any of these assays at 50-fold dilutions. Thus, this study showed that only one phage-borne peptide, which mimicked a linear epitope on E-MHV, elicited partial protection from viral challenge, although the mechanism of action was not clear. The protective response was also strain dependent.

Working with sera from HIV-1-infected people, Scala et al. [51] theorized that specific Abs protected long-term infected (LTI) patients from progression to AIDS, and that peptides that bound to these Abs could be used as a vaccine lead to target their production. They used serum IgGs from disease-free LTI patients to isolate peptides from two RPLs. Five peptides that bound with high frequency to LTI patient sera and low frequency to AIDS patient sera were chosen for further study; the peptides also bound sera from SHIV-infected macaques. The peptide sequences were compared to the envelope proteins, gp120 and gp41. One clone (p217) was mapped to the gp120 C2 region, whereas another (p197) mapped to the cluster I epitope of gp41. The p195 peptide was questionably mapped to the V1 region of gp120. The remaining two peptides (p287 and p335) did not bear homology to gp120 or gp41. Serum Abs from an LTI individual were affinity purified with each of the phage clones and were then used in western blots to identify whether they bound to gp160 and/or gp120. Abs purified on all of the clones bound both gp120 and gp160, with the exception of clone p197, which bound to neither. Purified phage clones were used to immunize mice, and the purified immune IgGs were shown to neutralize three isolates of HIV-1 in vitro (two were lab-adapted strains, and one was a primary HIV-1 isolate). It is puzzling that immune Ab against a cluster I epitope mimic (p197) would neutralize infection by an HIV-1 primary isolate, since strong Ab responses are typically present in HIV-1 infected people at all levels of progression (LTIs, rapid, and normal progressors) and are not associated with viral neutralization. Nevertheless, this work clearly demonstrates the identification of peptides that are frequently recognized by the sera of HIV-1-infected people, and produced preliminary data indicating that they could be used to produce Abs that cross-react with HIV-1 envelope and, perhaps, neutralize the virus.

In an extension of the work by Scala et al. [51], Chen et al. [87] tested the five phage clones for their ability to protect rhesus macaques from infection with SHIV89.6PD, a pathogenic SIV–HIV chimera. Five macaques were immunized five times, at 10-week intervals, with pools of the five phage-displayed peptides, and four negative-control macaques were immunized with WT phage. Ab titers against synthetic versions of each phage-displayed peptide in the experimental group ranged from 800 to 24,000, and anti-gp120 Ab titers ranged from 100 to several thousands. One phage-peptide-immunized monkey produced low antiphage-peptide Ab titers that did not cross-react with gp120. At 44 weeks, naive, WT-phage-immunized and peptide-phage-immunized macaques were challenged with SHIV-89.6PD. All groups became infected by the virus, as determined by a peak in viremia at 12 days, which coincided with a drop in CD4$^+$ T-cell counts. However, four of the peptide-immunized macaques showed a significant increase in CD4$^+$ T cells and lower virus levels compared to naive or WT-phage-immunized controls. Analysis of the postchallenge Ab response showed that five of the peptide-immunized macaques elicited an anti-gp140–89.6 Ab response with titers in the 100,000 range. In contrast to the macaques immunized with the phage-peptides, one macaque in each of the control groups (naive and WT-phage immunized animals) developed an anti-gp120–89.6 Ab response with significantly lower titers (100–3000). All of the naive and WT-phage-immunized macaques developed severe AIDS-like illness 6 weeks postchallenge. Among the phage-peptide-immunized macaques, one developed AIDS-like symptoms that progressed at similar rates to control macaques; this was the same macaque that did not show any prechallenge Ab response to gp120. Another macaque, in the peptide-immunized group, became ill with acute glomerulonephritis, but disease pathology was unrelated to SHIV infection; this animal most likely died from an unrelated preexisting condition. The three remaining peptide-immunized macaques were alive and healthy at 270 days postchallenge [87]. Thus, immunization with a mixture of phage displaying five different peptides elicited Ab responses that bound synthetic peptides representing the five phage-displayed peptides, cross-reacted with gp140, and appeared to reduce the disease caused by infection with SHIV 89.6PD.

The examples discussed in this section demonstrate that peptides from RPLs have potential as vaccine leads but require significant optimization before a functional vaccine could be developed. For example, the peptides used by Chen et al. [87] elicited immune responses that prevented development of an AIDS-like disease in SHIV-infected macaques, but were unable to prevent infection. Furthermore, only half of the mice immunized with a peptide isolated by Yu et al. [56] were protected from MHV infection. It may be that these peptides would more effectively target the production of the desired Abs if used in an alternative immunization approach, such as the prime-boost strategy outlined in Section 6.6.2.

6.5.4 Peptide Ligands for Anti-CHO Abs and Their Use in Anti-CHO Vaccines

Vaccines against CHOs present a very different set of problems from those directed against proteins, and some of these may be addressed with peptides from RPLs. The best defense against bacteria, such as *Haemophilus* spp. and *Streptococcus pneumoniae*, is to produce Abs that will opsonize their polysaccharide coat. Unfortunately, polysaccharides are weakly immunogenic, comprising repeating units having limited epitope diversity. Thus, anti-CHO Ab responses are often weak in infants, the elderly, and immunocompromised individuals. Polysaccharides are T-independent Agns and therefore form weak to no immunological memory in these age groups [73]. Several recently released vaccines comprise polysaccharide that is chemically conjugated to the immunogenic carrier protein, TT. The immune response against such conjugates is similar to the hapten-carrier response (described in Section 6.6.1) in that it elicits a T cell-dependent immune response against TT and creates immunological memory in both Agn binding B and T cells. Since B cells recognize both the TT carrier and the CHO conjugated to it, B-cell memory is formed against both Agns. Thus, stronger and longer-lasting Ab responses are elicited in response to multiple vaccinations due to the T cell-dependent secondary response. Notwithstanding these advances, there continue to be problems with the development of some CHO vaccines. First, it is not guaranteed that the antipolysaccharide Ab within this population is increasing proportionally to the anti-TT Abs, and sometimes anti-CHO responses may remain weak. Second, it can be difficult to impossible to synthesize complex CHO Agns, and sometimes native CHO Agns are difficult to isolate in a pure form. Third, some viable CHO targets, such as oligosaccharide moieties on LPS, are difficult to separate from larger toxic molecules. Finally, in some cases, only some of the Abs elicited against a CHO Agn will confer the desired biological activity. This can be due to the production of Abs having the wrong isotype (e.g., nonopsonizing Abs) or to Abs with the wrong specificity (i.e., they do not neutralize the targeted pathogen).

A number of alternative strategies have been explored to develop vaccines with strong and protective anti-CHO responses (reviewed by Lesinski and Westerink [88]). The earliest strategy explored was the use of anti-Id Abs as vaccines that target CHO Agns. Immunization with peptides from RPLs represents a more recent twist to this approach, since it removes the difficulties associated with developing, producing, and immunizing with Id Abs while targeting the Id Ab. Several studies have used peptides for CHO vaccines, as they may be optimized for desired characteristics that enhance structural and/or functional mimicry of the cognate CHO epitope (e.g., see Ref. [32]).

The nature of CHO mimicry by peptides that bind anti-CHO Abs has been explored in the structural studies of Vyas et al. [10] using the MAb SYA/J6, which is specific for the LPS O-Agn polysaccharide of *Shigella flexneri* Y. They determined the crystallographic structures of Fab SYA/J6 bound to either an eight-residue peptide [8] or to a synthetic oligosaccharide analog of the cognate CHO epitope [9]. The structures were quite similar on a gross structural level, with each Agn occupying the same site in the Ab combining site. The peptide made six hydrogen bonds and 126 van der Waals contacts with the Fab, whereas the pentasaccharide made eight hydrogen bonds and 74 van der Waals contacts. For both ligands, 57% of the contacts were made with the light chain. However, for the peptide, most of these contacts were with the CDR3 loop, which forms a shallow hydrophobic cavity; this interaction is not shared by the pentasaccharide. Furthermore, only half of the 37 contacts that are shared between the peptide and pentasaccharide are similar in type (i.e., polar–polar, nonpolar–nonpolar, polar–nonpolar). Nine Fab residues made contact solely with the peptide compared to two solely with the pentasaccharide. The peptide also interacted with a greater number of water molecules, 14 compared to 2 for the pentasaccharide [10]. The authors concluded that the peptide had better shape complementarity for the Fab than the pentasaccharide (as reflected by the larger number of contacts with the Fab), and that the peptide is thus not a structural mimic of the pentasaccharide. These differences in structural interaction may explain our preliminary immunizations with the peptide (as a synthetic peptide conjugate and as a phage-displayed peptide), which do not elicit Abs that cross-react with the LPS (our unpublished data). It is possible that by using this peptide in a prime-boost immunization strategy, as described in Section 6.6.2.1, Abs will be elicited that behave like SYA/J6, in cross-reacting with the bacterial LPS and the peptide.

The first example of the immunogenic mimicry of a CHO Agn by a phage-displayed peptide was reported by Phalipon et al. [89]. They used two protective IgA MAbs that bind to *Shigella* serotype 5a LPS to screen two RPLs (the peptides in one of the libraries were constrained by flanking Cys residues). MAb C5 selected 13 peptides and MAb I3 selected 6. Five of the peptides selected by MAb I3 also reacted with MAb C5. This type of cross-reactivity, between MAbs that bind the same CHO, has been observed by other groups [90] and may reflect a restricted Ab response against the CHO. BALB/c mice were immunized six times with 1 of the 19 phage-displayed peptides, and sera were tested for binding to LPS by western blot. Only two peptides elicited cross-reactivity with the LPS: p100c (selected by MAb I3) and p115 (which was selected by MAb C5 and cross-reacted with MAb I3). The cross-reactivity was relatively weak, as the anti-LPS titers were measured at 100-fold dilutions, whereas the antiphage titers were measured at 10,000-fold dilutions. The serum Abs are serotype specific, since Abs bound to LPS from *Shigella* serotype 5a bacteria but not to serotype 2a bacteria [89]. No in vitro neutralization or in vivo challenge studies were performed to analyze the protective properties of the Abs elicited by the peptides. Nevertheless, the authors clearly showed that immunizations with phage-displayed peptides can elicit Abs that specifically cross-react with an LPS.

Immunization with whole *Cryptococcus neoformans* glucuronoxylomannan (GXM), a component of the capsular polysaccharide, elicits both protective and nonprotective Abs. Nonprotective Abs can block the function of protective Abs; moreover, the Abs produced by immunization with the GXM do not protect against infection. However, some anti-GXM MAbs do protect against infection. Beenhouwer et al. [32] used an Ab-targeted, prime-boost immunization strategy to elicit the production of protective Abs against GXM. Protective MAb 2H1 was used to screen phage libraries for ligands, and peptide PA1 was identified. It did not elicit anti-GXM Abs when used as an immunogen on its own. To improve the binding of MAb 2H1 to PA1, a sublibrary was built around the PA1 core residues by adding one random residue at the center of the sequence and six random residues at the C- and N-termini. Six rounds of screening isolated peptide P206.1, which had an estimated K_d of 4 nM compared to 300 nM for the original PA1 peptide. Direct and competition ELISAs showed that P206.1 also bound to other anti-GXM protective MAbs, whereas it did not bind to a nonprotective MAb. The P206.1 peptide was used to make a number of different immunogens. It was produced as a synthetic peptide and conjugated to TT (P206.1-TT), as a multiple antigenic peptide (P206.1-MAP) in which multiple copies of the peptide were attached to a polylysine backbone, and as a free peptide. Immunizations with P206.1-TT, P206.1 alone, and P206 1-MAP did not elicit anti-GXM titers above those seen in control mice immunized with TT alone.

It was hypothesized that P206.1 could specifically stimulate B cells producing protective Abs from a preexisting population of B cells producing Abs against both protective and nonprotective epitopes. Thus, mice were first immunized with a low dose of GXM conjugated to TT (GXM-TT). This, in theory, activated B cells that produce Abs against the protective epitopes, as well as B clones whose Abs recognize nonprotective epitopes. No anti-GXM Abs were detected after the prime. To amplify the production of Abs against only the protective epitopes, mice were boosted with either P206.1-TT, P206.1 alone, P206.1-MAP, or TT alone, and anti-GXM titers were determined after 14 days. Only the group that was primed with GXM-TT and boosted with P206.1-TT developed anti-GXM titers that were significantly higher than the TT boosted control group. These anti-GXM serum Abs cross-reacted with both GXM and peptide, as shown by competition ELISA, and did not bind to de-O-acetylated GXM, which is preferentially bound by nonprotective Abs. Immune mice were not challenged with *C. neoformans* to assess protection. This elegant work clearly shows that prime-boost immunizations can significantly augment the production of peptide-targeted Abs.

A study by Pincus et al. [90] demonstrated immunogenic mimicry by a peptide selected by a protective MAb (S9) against the type-III capsular polysaccharide (type-III CPS) of group B *Streptococcus* (GBS). MAb S9 isolated two peptides from an RPL, and the sera of mice infected with GBS bound to a synthetic version of one of the peptides. This peptide was conjugated to ovalbumin (OVA), bovine serum albumin (BSA), and keyhole limpet hemocyanin (KLH), and the conjugates were used to immunize mice in complete Freund's adjuvant (CFA). Immune sera, diluted 1000-fold, appeared to cross-react with whole GBS and with the type-III CPS, compared to preimmune sera. However, the analysis was inconclusive, since titers were not calculated, and control sera from carrier-only immunizations were not analyzed. Such controls are important, as glutaraldehyde-conjugated KLH and unconjugated KLH have been shown to elicit nonspecific CHO-binding Abs [91]. Thus, the carrier proteins may have produced false-positive reactivity; data from carrier-only immune sera would clarify this. In vitro or in vivo challenge studies were not performed to confirm biological activity of the immune sera. Recent NMR studies have analyzed the structure of the peptide in solution [92], and ultimately, these may help to reveal the structure of the corresponding epitope on the type-III CPS.

The most conclusive study to date showing protective Ab production by an immunogenic-mimic peptide is described in the work of Grothaus et al. [93]. They produced hybridomas specific for meningococcal group A polysaccharide (MGAPS), and MAb 9C10 was used to screen an RPL. After three rounds of selection, the sequences of 60 phage clones were analyzed and a dominant sequence representing 38 of the clones was identified. Direct ELISA showed that human hyperimmune sera against group C *Meningococcus* bound to phage displaying the peptide almost as well as to MGAPS, and the sera did not bind to control phage. A synthetic analog of the peptide was incorporated into proteasomes prepared from synthetic lipopeptides and outer membrane complex vesicles from group B meningococci. The first immunizations comprised a dose response study where mice were immunized three times with either 1, 5, 10, 50 µg peptide/proteosome complex, with proteosome alone or with 5-µg MGAPS. A 5-µg dose of the peptide/proteosome complex was shown, by titration ELISA, to give the highest anti-MGAPS Ab titers. These Ab titers were much higher than those for the group immunized with MGAPS alone. The second immunization regimen comprised a prime-boost strategy, in which mice were primed with 1, 5, 10, or 50-μg peptide–proteasome complex and then boosted at day 28 with 5-μg MGAPS. This part of the study did not include an MGAPS control. Although titers were calculated in two different ways (the dose response study used 50% of maximal titers and the prime-boost study used 25% of maximal titers), the prime-boost approach for the 5-μg dose group appeared to elicit similar to higher anti-MGAPS Ab titers than did three immunizations with the peptide–proteasome complex alone or with MGAPS alone. Serial dilutions of immunized mouse sera were tested for in vitro bactericidal activity in the presence of complement, with data reported as the dilution of serum required to kill 50% of the bacteria. Sera from groups immunized twice with the peptide–proteasome complex killed 50% of bacteria at a dilution of 25 compared to a dilution of 23 for sera produced 1 week after a single MGAPS immunization. Mice that were primed with peptide–proteasome complex and boosted with MGAPs killed 50% of bacteria at a significantly greater dilution of 150.

Thus, several studies have shown that RPLs are a viable source of peptides that can serve as both functional mimics and, in some cases, immunogenic mimics of CHOs; however, the structural mechanisms behind this cross-reactivity are unclear. Several studies have demonstrated that peptide immunization can elicit Abs that cross-react with cognate CHO [89,90,93]. Others were only successful when a CHO-prime and peptide-boost were used [32]. Despite these successes, there is little evidence to suggest that cross-reactive Abs protect in vivo. Grothaus et al. [93] performed in vitro studies confirming the antibiotic capabilities of the peptide-elicited Abs. The Ab response to CHOs is sometimes restricted and may contribute to the success of peptide immunogens for anti-CHO, Ab-targeted vaccines. Perhaps this restriction enabled prime-boost immunizations, such as the one conducted by Beenhouwer et al. [32], to succeed. It remains to be seen if the prime-boost approach will work for specific antiprotein MAbs.

In general, vaccines made from whole-protein immunogens are preferable to subunit vaccines. However, if the whole Agn is difficult to produce, or if a non-immunodominant epitope is to be targeted, then AFLs may be a useful source of immunogens that will target specific epitopes on the Agn. There are instances in which an AF cannot be isolated, especially for some MAbs directed against discontinuous epitopes. In these circumstances, and for CHO-binding MAbs, RPLs may yield Ab targeting peptides. Very likely, epitope- and Ab-targeting AFs and peptides from RPLs require modification to become strongly immunogenic. In Section 6.6, we discuss considerations for immunization strategies with peptides derived from these libraries, their formulation as vaccine components, and finally, the analysis of immune responses to vaccination with them.

6.6 DEVELOPING IMMUNOGENS FROM PEPTIDE LEADS

The studies described in Sections 6.4 and 6.5 focus on the isolation and analysis of peptide ligands for MAbs and PCAbs that bind protein and carbohydrate Agns. In the majority of these examples, peptides isolated from AFLs and RPLs were used for mapping epitopes to proteins or for immunization with the purpose of eliciting Abs that cross-react with the cognate Agn. The desired outcome of this approach is to develop vaccines that elicit biologically relevant Abs when used as immunogens. One caveat of peptides as immunogens is that, by themselves, they are rarely immunogenic. The following reviews methods for incorporating vaccine-lead peptides into effective immunogens. Since this chapter focuses mostly on targeting Abs, we will discuss methods for enhancing the humoral response, rather than cellular immunity, even though both aspects of the immune response should be considered when designing a vaccine.

The process of formulating an Ab-binding peptide into a vaccine involves numerous steps. The first step, after isolating the peptide, is to test the immunogenicity of the peptide. The peptide must be incorporated into a carrier for immunization, and other components of the vaccine must be decided on. These include dose, adjuvant, immunization route, and timing between immunizations. Each of these components may require optimization to enhance antipeptide response and cross-reactivity with the cognate Agn. The second step is thorough analysis of the immunized sera. This involves measuring antipeptide, anticognate Agn, and anticarrier Ab titers by direct ELISA. Cross-reactive Abs that bind both peptide and cognate Agn are determined by direct ELISA and should be confirmed by competition ELISA. Results from these assays indicate the requirement for further optimization of the vaccination strategy or specific vaccine components.

Cross-reactivity with cognate Agn by peptide-elicited Abs is the first indication that a vaccination strategy is producing a desired outcome. Cross-reactivity is only significant if the peptide-elicited Abs have the appropriate biological effect. The later stage of vaccine-lead development requires stringent testing of the biological activity of the Abs elicited by the peptide vaccine. In vitro neutralization assays followed by in vivo challenge studies confirm that Abs being targeted by a peptide vaccine share the same biological activity as those Abs elicited after immunization with the vaccine. These criteria must be met if a vaccine-lead peptide is to be considered for further development into a vaccine.

6.6.1 Carrier Proteins and the Induction of T-Cell Help

A humoral immunogen requires both B-cell epitopes (BCE) and T_H-cell epitopes (TCEs) to trigger Ab responses that include affinity maturation and immunological memory. The peptides we have discussed in Section 6.6 are functional mimics of BCEs given that they bind the Ab paratope. These peptides may serve as a replacement for a cognate BCE in a vaccine whose purpose is to target the production of specific Abs; however, additional components are required to make BCE peptides immunogenic. Carrier molecules are proteins that can serve as sources of TCEs and, when attached to small-molecule BCEs, like haptens and peptides, can make them immunogenic. Traditionally, a hapten or peptide is chemically coupled to a carrier protein, and immunization with the conjugate produces an Ab response against the carrier and the molecules coupled to it. More recently, carriers can be a recombinant protein (including a phage) to which a BCE peptide is fused. Choices for carrier proteins include traditional immunogens such as OVA, BSA, TT, and KLH. The most effective carrier proteins are those that contain many TCEs, as they will elicit Ab responses in a variety of species and strains. However, as mentioned, carrier proteins also contain BCEs that elicit the production of Abs. These may interfere with analysis of immune sera. For example, May et al. [91] detected Abs that cross-reacted with GXM from *C. neoformans* in sera from mice immunized with KLH. These Abs preferentially bound to KLH over GXM in competition ELISAs and, when used to passively immunize mice, did not protect them from challenge with *C. neoformans*.

Self-proteins are typically nonimmunogenic; however, they can be made immunogenic by the addition of exogenous or foreign TCEs. For example, the highly conserved protein ubiquitin fused to an exogenous TCE was able to elicit an Ab response specifically against a V3-based peptide from HIV-1 (which served as the targeted BCE) [94]. Exogenous TCE coupled to self-protein is critical to this approach, as the former drives helper T-cell responses, whereas B-cell clones that would recognize the self-protein are probably deleted. Thus, carriers comprising an immunogenic TCE and a self-protein avoid the production of Abs and allow the Ab response to be focused against exogenous B-cell epitopes.

Vaccine carriers are not exclusively limited to proteins, so long as they incorporate immunogenic TCEs. Other methods for vaccine delivery include liposomes, comprising single or multilamellar bilayer membrane vesicles where Agn is membrane bound or within the intermembrane spaces. The immunomodulation effects of liposomes depend on the composition of the micelle and the types of proteins and adjuvants incorporated into the lipid bilayer. They are particulate, and they target APCs and create Agn depots that allow longer exposure of the immune system to the Agn [95]. Liposomes (called virosomes) are approved for human use in influenza vaccines [96].

MAPs can also serve as carriers, if they include TCE peptides in their structure. MAPs comprise multiple synthetic peptides, of single or multiple specificity, which are coupled to a polylysine backbone [97]. Like self-proteins, they overcome problems associated with immunodominant BCEs on carrier proteins and can be suitable for focusing Ab responses against a BCE peptide. Microparticles, made of biodegradable polymers, may also serve as carriers; these are suitable for single-dose administration since they create long-term Agn depots and do not require repeated boosts. Synthetic peptide leads can be spray-dried onto the microspheres [98].

The topic of vaccine carriers covers a large field, which will not be discussed further. Since this chapter focuses on phage-displayed peptides for vaccine development, we turn to the advantages and disadvantages of filamentous phage as carriers for BCE peptides and AFs.

The use of filamentous phage as carriers for peptides is well established. In 1988, de la Cruz et al. [99] first demonstrated the use of phage for eliciting Ab responses against a displayed peptide. They immunized rabbits with phage displaying a peptide from *Plasmodium falciparum* fused to pIII and discovered that they were immunogenic, even in animals immunized without adjuvant. Later, Meola et al. [100] compared immune response to several peptides derived from screening RPLs with antihepatitis B surface Agn (HbsAgn) MAbs. Peptides were fused to pVIII or pIII, made into MAPs, and conjugated to hepatitis B virus core peptide (HBV core) and to human ferritin. Phage immunogens were administered without adjuvant, whereas adjuvant was used in immunizations with the HBV core and ferritin conjugates and the MAPs. It was found that pVIII-displayed peptides consistently elicited higher anti-HbsAgn Ab titers than the other carrier proteins, with pIII display coming in a close second. All of the recombinant carriers elicited higher Ab titers against recombinant HbsAgn than the synthetic peptide MAP.

Bastien et al. [101] demonstrated that a protective immune response could be induced by phage displaying a peptide, derived from respiratory syncytial virus (RSV), and was known to protect from infection after immunization. Mice were immunized with phage displaying the RSV peptide on pIII. They were then challenged with RSV, and their lungs were checked for the presence of virus 5 days post-challenge. All of the phage-immunized mice had cleared the virus from their lungs, whereas controls had not. These earlier successes have led many groups to use phage as carriers for vaccine-lead peptides derived from phage libraries (Section 6.5).

More recently, Grabowska et al. [102] used MAb H5 against herpes simplex virus (HSV) glycoprotein G to isolate three peptides from an RPL. Consensus motifs deduced from the three peptides aligned to a linear epitope on the native glycoprotein G amino acid sequence. Prior to immunization, phage clones were either treated to remove LPS or left untreated. BALB/c mice were immunized with various doses of the pooled phage clones. After two immunizations, serum Ab responses against both phage and glycoprotein G were measured in ELISA by comparing OD_{490} of sera at 1000-fold dilution. Ab titers were not determined. Groups of mice immunized with untreated phage clones produced antiphage and antiglycoprotein G Ab responses in a dose-dependent manner, whereas mice immunized with treated phage had much lower Ab responses. This did not affect the protective effect of the Abs in vivo, since mice immunized with high doses (100 and 70 μg) of either treated or untreated phage were completely protected from lethal challenge with HSV-2. Three of the groups immunized with lower doses of phage (50 and 10 μg) had only a 30% survival rate. The exception was the fourth low-dose group, which received 50 μg of treated phage and had a 65% survival rate. The level of serum Ab did not correlate with survival rates in this study, indicating that aspects of the immune response other than serum Ab were responsible for protection. Although LPS did not increase the ability of mice to survive viral challenge, its presence did enhance both antiphage and antiglycoprotein-G Ab responses. LPS is a B-cell mitogen and confers this advantage to the immunogenic effects of phage as a carrier for BCE peptides. Further advantages are discussed in Section 6.6.1.1.

6.6.1.1 Advantages of Using Phage as a Carrier Protein

Although filamentous phages are unlikely to be used in commercial vaccine formulations, there are several advantages to using them as carriers for BCE peptides. Phages are easily amplified to large quantities, they can be rapidly purified by PEG precipitation, and they can be further purified by CsCl density-gradient centrifugation. This makes the production of phage immunogens a rapid, inexpensive approach to testing peptide immunogenicity, compared to synthesizing peptides and chemically conjugating them to carrier proteins. Filamentous phage particles have the general dimensions of a virus but are noninfectious to animal tissues; this makes them good pathogen mimics. Also, as the phage surface comprises mostly the outer 10 residues of the major coat protein, pVIII, phage coats are homogeneous and restricted to a few B-cell epitopes [103]. This can result in lower titers against the phage carrier compared to more complex carriers, such as OVA (van Houten NE, Zwick MB, Menendez A, Scott JK, manuscript in preparation).

As mentioned in Section 6.1.1, carrier proteins enhance the immune response by providing TCEs. There is evidence that phage bearing exogenous TCEs can target either the cellular or humoral branches of the immune system, or both. Willis et al. [104] reported that the humoral response to phage-displayed peptides is helper T cell-dependent. They immunized nude, heterozygous, and BALB/c mice with phage displaying a peptide, and tested by ELISA whether class switching occurred in antipeptide Abs (i.e., whether IgGs were produced). Nude mice, lacking T cells necessary for B-cell class switching, produced only IgM, whereas BALB/c mice produced IgG, and the heterozygous mice produced a mixture of IgM and IgG. They also demonstrated that T cells were recruited by WT phage in the presence or absence of an adjuvant.

Furthermore, phage can be genetically manipulated to enhance T-cell help and recruit cytoxic T-cell responses. De Berardinis et al. [105] engineered phage to simultaneously display a helper T-cell epitope (p23) and a cytotoxic T-cell (CTL) epitope (RT2) derived from HIV-1 reverse transcriptase. Phage displaying both epitopes elicited cytotoxic activity in human T-cell lines (as determined by ^{51}Cr release), whereas phage displaying the RT2 CTL epitope alone did not induce cytotoxic activity. For other examples, see Refs. [106,107].

Phages are effective carrier proteins for synthetic peptides. Our laboratory engineered an additional lysine residue near the N-terminus of pVIII to be used as an accessible primary amine for amine-reactive cross-linking agents (Zwick, unpublished data). This allowed the conjugation of a peptide to every two to three pVIII molecules on the phage, and this corresponds to 1000–2000 peptides per phage. After three subcutaneous (SC) immunizations, the antipeptide Ab response exceeded that of the antiphage Ab response by twofold (van Houten NE, Zwick MB, Menendez A, Scott JK, manuscript in preparation).

6.6.1.2 Disadvantages of Using Phage as a Carrier Protein

The copy number of a displayed peptide on the phage coat is a major determinant controlling the Ab response against a given peptide. Peptides displayed by pIII (via Type 3 vectors) have a copy number fixed at three to five copies per phage and produce relatively weak antipeptide Ab responses. Type 8 vectors produce phages displaying 2500–3000 copies of peptide per phage, and in theory, this should make them excellent for peptide-targeted immunization. However, pVIII cannot tolerate more than four or five additional residues without blocking viral assembly [28,108,109]. This severely limits the type of peptide that can be used for immunization with Type

8 vectors. The Type 88 vector, f88.4, produces hybrid phages having a displayed peptide copy number varying from 1% to 15% of the total pVIII, with other such vectors [110] producing peptide copy numbers reaching 30% (personal communication, Cesareni). Variation in the copy number of a peptide displayed in a Type 88 system depends on the amino acid composition and length of the peptide and is probably related to the efficiency with which peptide–pVIII fusions assemble, along with WT pVIII, onto the virion as it emerges from the inner membrane of the bacterium. Thus, longer fragments, such as those found in AFLs, are displayed at lower frequencies and, although immunogenic, may not induce a detectible Ab response due to low copy number [3]. Peptide–pVIII fusions that assemble poorly onto the phage result in virions bearing peptide at low copy numbers, making them poor immunogens compared to phages that display peptides in high copy numbers.

Analysis of the copy number of peptide–pVIII fusions is unreliable; SDS-PAGE and amino acid analysis has been used for this purpose, but each has its shortcomings. Probably the best method is separation of peptide–pVIII fusions from WT pVIII by SDS-PAGE [111]. The goal is to separate the two types of protein and to determine the relative concentration of protein in each band. This is done by running side-by-side dilutions of the phage and identifying peptide fusion and WT pVIII bands in different dilutions that have comparable intensities. The percentage of peptide–pVIII fusion is then calculated using the dilution factor between these two samples. However, in some instances, the recombinant and WT forms of pVIII will not resolve from each other, making this approach impossible. Amino acid analysis can be used to calculate peptide copy number, based on the fact that pVIII forms most of the protein mass of the virion. As pVIII is 50 residues long and has a restricted amino acid composition (e.g., Cys, His, and Arg are absent), the peptide copy number can be calculated by comparing the molar amounts of two categories of amino acids in the analysis: (1) the amount of an amino acid that is present in the peptide, but absent from pVIII, or the amount of an amino acid that is present in pVIII and absent from the peptide; and (2) the amount of an amino acid that is present in both pVIII and the peptide.

Given a peptide of interest that is displayed at a low copy number, in our hands, it has been very difficult to improve display density on phage. We have attempted this for several clones by using different vectors and different promoters with only limited success (unpublished data). Thus, it is likely that a low peptide copy number cannot be raised on a routine basis. This makes the use of synthetic peptides attractive; however, this approach comes with a different set of problems. In our experience, the affinity of a peptide selected from an RPL is typically much lower as a synthetic peptide than for its recombinant-fusion counterpart. This problem has occurred with several synthetic peptides that we have derived from RPLs. Most likely, these peptides adopt a more stable conformation in the context of a fusion protein, whereas their free-peptide counterparts lack persistent structure [112].

Several studies demonstrate that phage may not be an ideal carrier protein in all circumstances, and various options should be investigated to find the ideal immunogen. For example, Yip et al. [113] tested the immunogenicity of a 55-residue peptide that binds to an anti-ErbB2 MAb. The authors compared peptide fusions to the N-termini of pIII and pVIII, and to the C-terminus of GST, which produced three to five copies per Type 3 phage, an unknown copy number per Type 88 phage, and two copies per GST dimer. The GST-peptide fusion elicited the highest antipeptide and anti-ErbB2 Ab titers. The low titers produced by the phage probably reflected a low copy number from both vectors: pIII display is low, and the low titers elicited by the Type 88 vector are likely due to the length of the peptide. Rubinchik and Chow [114] also compared different carrier proteins expressing residues 47–64 from the Staphylococcal superAgn, toxic shock syndrome toxin-1 (TSST-1). They compared fusions to GST, the outer membrane porin protein (OprF) of *E. coli*, pIII, and its synthetic counterpart conjugated to BSA (a relatively weak carrier protein). All carriers, with the exception of pIII, induced high Ab titers against TSST-1; however, only Abs from the GST immunized group inhibited binding of the synthetic peptide to MHC class II and inhibited TSST-1-induced T-cell proliferation.

As mentioned in Section 6.6.1.1, Willis et al. [104] showed that phages elicit T-cell responses. However, our unpublished observations have led us to question the strength of the T-cell response against phages, as compared to commonly used carrier proteins, since this was not done in the earlier study. Thus, we immunized Balb/c mice with a synthetic peptide conjugated to phage or to the carrier protein, OVA. The Ab response against the peptide was far stronger in the group immunized with the peptide–OVA conjugate. Moreover, Ab response against the phage carrier plateaued after five immunizations, whereas the response against the OVA carrier was still on the increase after seven immunizations. From this, we deduced that phages elicit a weaker helper T-cell response than OVA. Thus, phages can produce good Ab responses against a displayed foreign peptide or AF, but have the drawbacks that the copy number of the displayed peptide or AF can be low, and that they may not elicit very strong T-cell responses, as compared to more traditional carrier proteins.

6.6.2 Vaccine Formulation

Choosing a carrier molecule is the first step in formulating a vaccine for a lead peptide or AF. This is followed by developing an immunization strategy. A complete vaccine comprises adjuvants, timing of boosts, boosting immunogens, dose, and injection route. Section 6.6.2.1 gives a brief overview of how these different components are incorporated into a vaccine.

6.6.2.1 Immunization Strategy

The immunization strategy should be one of the first considerations in vaccine formulation. "Straight" immunization, in which a single Agn is used for several immunizations, is commonly used. This elicits an immune response focused on immunodominant regions of the immunogen. This approach works well if a peptide or AF immunogen elicits Abs that cross-react with cognate Agn, or if immunodominant Abs are being targeted. However, certain peptides do not elicit Abs that cross-react with cognate Agn after straight immunization. To elicit Abs that bind to such peptides and cross-react with cognate Agn, a prime-boost approach could be considered (Figure 6.1).

A prime-boost immunization strategy is a method for enhancing the production of targeted Abs that recognize a specific epitope or region on a selected Agn. This approach comprises a priming immunization with the cognate Agn followed by a boosting immunization with a peptide or AF that specifically binds the targeted Abs. Thus, the priming immunization activates and expands a B-cell population

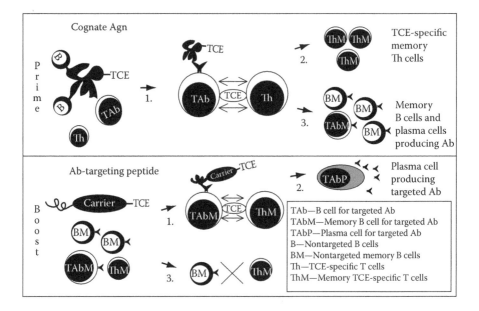

Figure 6.1 Strategy for targeting the production of specific Abs using a prime-boost immunization approach. Prime: The first immunization comprises cognate-Agn conjugated to a TCE-containing carrier protein (e.g., TT) or to a foreign TCE peptide. This elicits a PC response from B cells (B) that bind to numerous epitopes on the cognate Agn, as well as a small proportion of B cells that produce the Ab that is being targeted (TAb). (1) The foreign TCE, presented by B cells (both TAb and B, only TAb is shown), activates a TCE-specific population of helper T cells (Th) that signal to B, Tab, and other Th cells to differentiate and mature. (2) Memory Th cells (ThM), specific for the foreign TCE, are elicited during the prime immunization and await subsequent encounters with the same TCE. (3) Both B and TAb differentiate into memory B cells (BM and TAbM, respectively) that may be selectively amplified by the boosting immunogen. Plasma cells that secrete Ab for B and TAb are also produced (not shown). Boost: The second immunization comprises a TAb-specific peptide immunogen. It may be expressed as a recombinant fusion protein or chemically conjugated to the same carrier protein used in the prime. (1) The boosting immunogen contains the same foreign TCE as the priming immunogen and stimulates ThM cells elicited during the prime. The peptide portion of the immunogen amplifies TAb-specific memory B cells (TAbM) from the preexisting population of memory B cells. (2) The stimulated TAbM cells differentiate into Ab-producing plasma cells (TAbP), increasing the specific Ab titer in serum. (3) The preexisting population of BM cells that produce the nontargeted Ab will not be amplified since there is no protein epitope for them to bind to in the boosting immunogen.

that includes those that produce the targeted Abs. Also in the prime, helper T cells provide signals that drive the differentiation of naive B cells, which recognize the cognate Agn, into memory B cells and plasma cells. The boosting immunogen, comprising an Ab-specific peptide or epitope-specific AF coupled to a carrier protein, amplifies only the memory B cells that produce the targeted Ab. To produce a strong response, the boost should be driven by memory T cells that were formed after the prime; this requires that the priming and boosting immunogens carry a strong, identical TCE. Such a strategy has been used by Beenhouwer et al. [32] to elicit the production of specific Abs against the capsular polysaccharide of *C. neoformans*. Animals were primed with capsular polysaccharide coupled to TT and then boosted with an Ab-specific peptide coupled to TT to produce the targeted Ab.

6.6.2.2 *Adjuvants*

Adjuvants enhance the immune response and can selectively activate specific branches of the immune system (i.e., humoral vs. cellular). Cox and Coulter [95] have classified adjuvants into two broad categories: particulate and nonparticulate adjuvants. Others have categorized adjuvants by their source of origin [96]. Particulate adjuvants create Agn depots that stimulate Agn presenting cells (APCs) and provide short- to long-term exposure of the immunogen to the immune system. They include aluminum salts (e.g., alum), water-in-oil emulsions (e.g., Freund's complete and incomplete adjuvants; CFA and IFA, respectively), liposomes, and biodegradable microparticles. Particulate adjuvants can be further modified by immunostimulatory effects. For example, IFA is modified to become CFA by the addition of heat-killed *Mycobacterium tuberculosis*, which acts as an irritant and enhances the activity of macrophages [1]. Other particulate adjuvants, such as alum, contribute an innate immunostimulatory effect. Alum stimulates T_H2 responses resulting in increased IgG production [96].

Nonparticulate adjuvants have a direct effect on the cellular components of the immune system. These include cytokines, block copolymers, and bacterial components such as monophosphoryl lipid A (MPL) and CpG oligodeoxynucleotides [95,96]. Cytokines target specific immune cells; for example, IL-12 activates T_H1-dependent cell-mediated immunity, granulocyte-macrophage colony stimulating factor (GM-CSF) activates dendritic cells, and IL-4 stimulates B cells [115]. Block copolymers such as those included in TiterMax™ target APCs [95].

Only aluminum salts (e.g., alum), virosomes, and MF59 are for use in humans, but many other adjuvants are available for research [96]. Adjuvants commonly used in research include CFA and IFA, alum, TiterMax, quil A (a saponin-based adjuvant that induces both humoral and cellular responses), and Ribi™ (which contains MPL and induces a strong T_H1 response) [95,116–118]. Certain adjuvants, like CFA, are commonly used but are associated with severe side effects, including granulomas that can ulcerate into abscesses [116]. While CFA and IFA can elicit responses against weak immunogens, caution and restraint should be used because of their severe side effects. Furthermore, Kenney et al. [118] discovered that Abs elicited by CFA bound to denatured epitopes, whereas Abs induced by other adjuvants were more selective for the folded protein Agn.

The most appropriate adjuvant for any given application should be empirically determined. Bennett et al. [116] compared TiterMax, alum, Ribi, and CFA and found that Ab titers induced by TiterMax were comparable to those of CFA, but resulted in fewer side effects. Kenney et al. [118] compared quil A, Ribi, CFA, alum, and SAF-1 and discovered that quil A and Ribi produced a more suitable immune response for making hybridoma, even though CFA produced higher Ab titers. In our hands, alum has induced higher Ab titers in Balb/c mice to the model Agn HEL compared to CFA or TiterMax (unpublished data).

6.6.2.3 *Immunization Route, Dose, Timing, and Number of Boosts*

Since pathogens use different entry routes, so must protective vaccines. Immunization routes should be considered so that the protective response is focused in areas most likely to encounter pathogens. Vaccines elicit systemic immune responses by subcutaneous (SC), intramuscular (IM), or intraperitoneal (IP) delivery, whereas vaginal, rectal, oral, or nasal routes confer mucosal protection. Phages have been successfully used in multiple routes including IP, SC, oral (PO), or intranasal (IN) [119,120]. In our experience with phage immunizations in mice, we have had the greatest success with SC immunizations as compared to IP for eliciting systemic Ab titers; this may be due to more effective access to dendritic cells (DCs). Several immunization routes may be tried to optimize immune response, and adjuvants should be chosen according to route; for example, cholera toxin is often used to aid the formation of a mucosal immune response (reviewed in Ref. [121]).

Other considerations for vaccine development include dose, timing of boost, and the number of boosts. These typically need to be empirically determined and are Agn dependent; however, there are some general points that may be considered. For example, a lower dose of Agn will selectively activate B cells producing high-affinity Ab, compared to a higher dose that activates a greater breadth of response. This is an important consideration when targeting specific Abs, since a discrete immune response may be preferred. Timing of the boost may be critical if a prime-boost immunization strategy is being considered. One method of assaying the success of a prime-boost immunization strategy is to measure Ab levels against the cognate Agn before and after the boosting immunization to determine if there has been an increase in titer. This can only be done if the immune response against the cognate Agn has peaked and then dropped. If the boost occurs too soon after the prime, then an increase in Abs against the cognate Agn may be due to the natural kinetics of the priming immunization, rather than the boosting immunogen. Furthermore, several boosting immunizations may be required to amplify the targeted Abs.

6.6.3 Assays

The primary goal of a vaccine is to elicit a protective immune response against a specific disease or disorder. Experimentally, there are several ways to track the immune response to indicate that a protective response is being formed. Protection correlates to levels of specific Abs in blood; thus, measuring specific Ab response by serum ELISA is the first and fastest method of analysis. Immunization with a vaccine-lead peptide or AF will elicit populations of Abs against all components of the vaccine including the carrier protein. The peptide itself will elicit Abs that bind only to the peptide, as well as Abs that cross-react with peptide and cognate Agn. The latter are preferred, since they represent the protective response. However, Abs produced against the carrier protein are likely to dominate the response and may also cross-react with cognate Agn [91]. Serum ELISAs can reveal the proportion of the Ab response that is being diverted from the desired target. Calculation of the antipeptide to anticarrier ratio can indicate skewing of the Ab response toward an immunodominant or more prevalent component of a vaccine. Thus, optimizing a vaccine to reduce response against unwanted components of an immunogen may produce a "cleaner" protective response.

After ELISA, Abs produced by a vaccine should be analyzed for in vitro neutralization activity or other relevant biological activity. Since Abs may prevent or reduce infection by different mechanisms, such as blocking of viral receptors or complement-mediated lysis, different assays are used to confirm such activity. The ability of an Ab to block cellular infection by virus can be determined by plaque assay. For example, Burton et al. [122] preincubated titrated Ab with HIV-1 virus that was then added to uninfected MT-2 cells. Cells were topped with agarose and the plates centrifuged to form cell monolayers. After several days of incubation, the plates were labeled with propidium iodide and fluorescent plaques were counted. Neutralizing titers were defined as the concentration of Ab required to give 50%–90% reduction in plaque numbers compared to controls with no Ab.

A more recent example of this type of assay was reported by Richman et al. [123], who used two different expression vectors to generate HIV-1 virus-like particles (VLPs). The first vector comprised the HIV-1 genome with the envelope gene replaced with a luciferase gene. The second vector bore an expression cassette for intact *env* (which encodes gp160) from primary HIV-1 isolates. Cells that are coinfected with both vectors make noninfectious VLPs bearing gp160 from the primary isolate and an HIV-1 genome that expresses luciferase instead of Env. These VLPs were then used in neutralization assays with sera from HIV-1-infected patients (either from the same patient that donated the envelope gene [autologous sera] or from a different patient [heterologous sera]). Titrated sera and VLPs were coincubated with a cell line that expresses the HIV-1 receptor CD4 and the HIV-1 coreceptors CCR5 and CXCR4. Neutralization activity was measured as a decrease in luciferase activity in the cells after infection, as compared to a control.

In vitro assays are also used to test the ability of an Ab or sera to induce complement-mediated lysis of bacteria. Prinz et al. [124] immunized mice with proteosomes bearing a peptide that binds to MAb that protects from infection with a *Neisseria meningitidis* serogroup C (for a similar study, see Ref. [81]). A discrete cell number of *N. meningitidis* spirochets were mixed with baby rabbit serum as a source of complement. Dilutions of sera from the immunized mice were mixed with the cell/complement mixture and incubated for 1 h. Aliquots of the reactions were plated at time 0 and after 1 h, and colonies were counted after overnight incubation.

The serum bactericidal titer was considered as the lowest reciprocal dilution that was required for killing 50% of the bacteria.

In vitro assays are the first way to test if a vaccine elicits Abs with biological activity. However, protection assays are necessary for testing the ability of a vaccine to prevent infection in vivo. These assays follow immunization with a vaccine-lead peptide or AF and culminate in direct exposure of the immunized animal to the pathogen. The results of the challenge may be analyzed in several ways. The simplest is to calculate percent survival [56,91,102]. Other methods include checking for viral clearance [101,125] or slowed disease progression [87]. Depending on the pathogen, challenge studies may be conducted with live bacteria [91,124] or virus [56,87,101,102].

For example, May et al. [91] used a challenge assay to demonstrate that Abs, elicited by gluteraldehyde-treated KLH (gKLH), did not protect mice from infection with *C. neoformans*, even though the Abs bound to the coat polysaccharide, GXM. Mice, twice immunized with gKLH, were given an IV challenge comprising a lethal dose of *C. neoformans* strain 24067. The endpoint was the number of surviving animals; none of the mice survived. Unfortunately, positive controls were not included to confirm the efficacy of the challenge model. Another example of a bacterial challenge model was that used by Prinz et al. [124] to test the efficacy of their *N. meningitidis* serogroup C vaccine. Mice that were immunized three times with either peptide/proteosome complex or meningococcal serogroup C polysaccharide (MCPS) were prepared for challenge by IP injection of iron dextran, which enhances susceptibility to bacteria. After 7 days, mice were given an IP challenge with 10 times the lethal dose of *N. meningitidis* serogroup C. The endpoint was measured as percent survival over time. After 96 h, mice immunized with peptide/proteosome complex and MCPS showed 80% and 100% survival, respectively, whereas the proteosome immunized group had zero survival after 36 h. The endpoint in challenge studies is not always measured by survival. For example, challenge studies that test vaccine-induced immune response against RSV monitor viral clearance rather than survival [101,125]. In both studies, mice immunized with vaccine-lead peptides were given IN challenge with RSV. Mice were sacrificed 4 or 5 days postchallenge. Lung tissue was harvested, and viral clearance was determined with plaque assays.

Passive transfer studies are also used to analyze the effect of Abs in systems other than the ones in which they were elicited. This has been a common approach to test the neutralizing capability of human MAbs against HIV-1 or SHIV. This has been done in monkeys [126–128] and hu-PBL-SCID mice [129]. These types of studies may be adapted to test sera from vaccine-lead immunized animals in other models.

Section 6.5 described assays at several stages of analysis. A "vaccine-lead" peptide or AF should satisfy all of the parameters discussed in this section to be considered for further development as a vaccine component. At the Ab level, there should exist a high level of Abs that cross-react with both peptide or AF and cognate Agn. These Abs should demonstrate biological activity in vitro (e.g., in the form of neutralization or complement mediated lysis). However, since other mechanisms may contribute to the protective effects of an Ab in vivo, challenge studies should also be performed. These results, taken together, comprise a "gold standard" for assessing vaccine-lead peptides.

6.7 SUMMARY

This chapter has reviewed the technology of phage display and its application to diagnostics, epitope mapping, and vaccine design. The technology and its applications have been reviewed in the context of the three major types of libraries: whole-Agn libraries, AFLs, and RPLs. It has been the purpose of this review to outline, as well as critique, the methods and analysis used to develop diagnostic or vaccines leads from phage-displayed libraries. The specific focus has been on developing leads for Ab-targeted vaccines using ligands from AFLs and RPLs. As a result, there has not been an extensive discussion on the use of whole-Agn libraries, despite the importance of this technology. In the following, the three library types discussed in this chapter will be summarized in the context of their limitations and alternatives. Whole-Agn libraries are mentioned, but the focus will be on AFLs and RPLs.

Whole-Agn libraries, made from the cDNA of a whole organism or cell line, are used to identify diagnostic or vaccine-lead proteins. The primary applications have been for identifying allergens or therapeutic targets [37,38,130], for isolating and identifying tumor-specific Agns [18,19,131], and for identifying Agns for autoimmune diseases such as lupus [39]. The whole-Agn approach has several advantages. For example, the candidate protein is physically associated with its gene, and furthermore, it is possible to screen libraries comprising many different protein Agns using high throughput methods [37].

There are limitations to this method. For instance, cDNAs contain stop codons that prevent the expression of open reading frames; this can be overcome with suitable vectors [14]. Moreover, bacterial expression affects translation of eukaryotic proteins, specifically, protein folding and glycosylation. Yeast [132] or mammalian display systems should be considered as alternatives for expression of eukaryotic cDNA.

AFLs, comprising cDNA fragments as opposed to whole cDNAs, are made from genes for single proteins [59–61] or from whole genomes [3]. They have been used for mapping linear epitopes [50,59,60], for identifying vaccine-lead protein fragments [3,55], and for mapping immunodominant regions on a protein using PCAbs [50,61]. AFLs are a better source of ligands for PCAbs or MAbs than RPLs since fragments are derived from cognate protein. The exception is for MAbs that bind to discontinuous epitopes on regions of the cognate protein that are distantly spaced or nonprotein Agns such as DNA or CHO. One study has shown that AFs are more frequently immunogenic than peptides derived from RPLs [3]. However, AFs isolated by MAbs do not always elicit Abs that cross-react with cognate protein [55]. The primary applications for AFLs are epitope mapping and the identification of ligands for epitope or Ab-targeted vaccines.

Like whole-Agn libraries, AFLs are also affected by *E. coli* codon usage and post-translational modifications when eukaryotic proteins are expressed. Furthermore, each application requires the construction of a new library, and this is a time-consuming task. One of the primary uses of AFLs is epitope mapping. AFLs are well suited to mapping linear epitopes; however, overlapping fragments need to be isolated to identify the discrete epitope rather than just the general region of an epitope. There are greater problems associated with using AFLs for mapping epitopes of MAbs that bind to discontinuous epitopes. AFLs do not often produce fragments that contain discontinuous epitope regions on the same discrete fragment. This is in spite of attempts by Huang et al. [61] to create an AFL with distant AFs ligated together. Homolog scanning, such as that done by Jin et al. [49], is an alternative method for mapping discontinuous epitopes. Other options include using RPLs to define CBRs that are mapped to the cognate Agn [57,69].

RPLs are the most versatile of the library types discussed in this chapter since they are sources of ligands for MAbs that bind to discontinuous epitopes or nonprotein Agns such as CHOs and DNA. They can be made in different lengths, with or without constraints, and one library may be reused for many applications. RPL technology is applicable to diagnostics, epitope mapping, and vaccine design. For example, diagnostic panels of peptides, identified with serum Abs from infected patients, characterize the "footprint" of a specific disease and may be used for diagnosis [40,41]. This same approach is used for idiopathic diseases or autoimmune disorders [44,47] since RPL technology does not depend on identification of the cognate Agn.

As mentioned in Section 6.2.4, RPLs are sources of ligands for MAbs that bind to discontinuous epitopes. Moreover, peptides isolated from RPLs by such MAbs may be used to identify the putative epitope on the cognate protein. This approach has been explored with computer-modeled epitopes for MAbs that bind RPs [57,69]. These studies show promise in the field of discontinuous epitope mapping but require Ab-binding studies with mutations to the putative CBRs.

Peptides derived from RPLs are commonly used in vaccine development; however, they are not always optimal or do not elicit Abs that cross-react with cognate Agn [3,32,56,89]. For example, Matthews et al. [3] showed that peptides derived from RPLs are less immunogenic than those derived from AFLs. In addition, numerous peptides need to be tested to find one that elicits Abs that cross-react with cognate Agn [56,89]. However, the use of RPL-derived peptides may be unavoidable, especially if a MAb binds a discontinuous epitope or a CHO Agn.

6.8 CONCLUSION

The primary aim of this chapter has been to explore the application of phage libraries to Ab-targeted vaccine design. Specific Abs or a group of Abs against a specific epitope may be targeted by immunization with peptides or AFs derived from RPLs and AFLs. The targeted Abs are elicited by straight immunization with peptide or AF, or by a prime-boost immunization, where the cognate Agn is used as the priming immunogen and the peptide or AF is used as the boosting immunogen.

Analysis of the immune response follows immunization to determine whether the putative vaccine lead is successful or if alternative approaches should be explored. In the following, the discussion is focused on the criteria that should be evaluated before declaring a vaccine lead a success.

We propose that several criteria must be analyzed to measure the quality of Abs elicited by an Ab-targeting vaccine-lead peptide or AF. First, immunization with the vaccine-lead peptide should elicit high titers of Abs that cross-react with the cognate Agn; these represent the proportion of Abs that are "protective." Second, these Abs should share the mechanism of biological activity with the parental Ab. For example, they should block against viral infection or activate ADCC; this is demonstrated by in vitro neutralization assays and in vivo challenge studies. Third, the Abs should be genetically identical or very similar to the parental Ab (i.e., they should have similar gene usage and levels of somatic mutation).

The majority of the studies discussed in this chapter focus on different aspects of Ab-targeted vaccine design. Some studies focus on in-depth biochemical analysis of vaccine-lead peptides whereas others focus on the immunological effect of peptide immunization. For example, Beenhouwer et al. [32] used a very well characterized vaccine-lead peptide and a sophisticated immunization strategy that successfully elicited high Ab titers against the cognate Agn. However, it was not determined if these Abs protected from infection either in vivo or in vitro. In contrast, the peptides isolated by Scala et al. [51] were not optimized and were poorly characterized. Nonetheless, immunization with these peptides resulted in Abs that neutralized HIV-1 in vitro and protected macaques from disease progression after infection with SHIV-89.6PD [87]. A more stringent biochemical analysis of these peptides would contribute to the knowledge of how they elicited partial protection, and optimization of these peptides may lead to better immunogens.

In conclusion, a vaccine lead should be considered successful when it meets both biochemical and immunological expectations. In-depth biochemical analysis of a peptide lead is an excellent precursor to immunization studies but a poor predictor of how that peptide will focus the immune response against a pathogen. Conversely, a peptide that elicits a protective immune response must be well characterized to ensure that it is optimized. To date, there are no commercial peptide vaccines. Building an effective peptide vaccine will require the expertise of both biochemical and immunological disciplines and a great deal of intense study.

REFERENCES

1. Janeway CA, Travers P, Walport M, Shlomchik M. *Immuno-Biology: The Immune System in Health and Disease*, 5th ed. New York: Garland Publishing, 2001.
2. Craig L, Sanschagrin PC, Rozek A, Lackie S, Kuhn LA, Scott JK. The role of structure in antibody cross-reactivity between peptides and folded proteins. *J Mol Biol* 1998; 281:183–201.
3. Matthews LJ, Davis R, Smith GP. Immunogenically fit subunit vaccine components via epitope discovery from natural peptide libraries. *J Immunol* 2002; 169:837–846.

4. Wells JA. Binding in the growth hormone receptor complex. *Proc Natl Acad Sci U S A* 1996; 93:1–6.

5. Barlow DJ, Edwards MS, Thornton JM. Continuous and discontinuous protein antigenic determinants. *Nature* 1986; 322:747–748.

6. Geysen HM, Tainer JA, Rodda SJ, Mason TJ, Alexander H, Getzoff ED, Lerner RA. Chemistry of antibody binding to a protein. *Science* 1987; 235:1184–1190.

7. Davies DR, Cohen GH. Interactions of protein antigens with antibodies. *Proc Natl Acad Sci U S A* 1996; 93:7–12.

8. Harris SL, Craig L, Mehroke JS, Rashed M, Zwick MB, Kenar K, Toone EJ et al. Exploring the basis of peptide-carbohydrate crossreactivity: Evidence for discrimination by peptides between closely related anti-carbohydrate antibodies. *Proc Natl Acad Sci U S A* 1997; 94:2454–2459.

9. Vyas NK, Vyas MN, Chervenak MC, Johnson MA, Pinto BM, Bundle DR, Quiocho FA. Molecular recognition of oligosaccharide epitopes by a monoclonal Fab specific for Shigella flexneri Y lipopolysaccharide: X-ray structures and thermodynamics. *Biochemistry* 2002; 41:13575–13586.

10. Vyas NK, Vyas MN, Chervenak MC, Bundle DR, Pinto BM, Quiocho FA. Structural basis of peptide-carbohydrate mimicry in an antibody-combining site. *Proc Natl Acad Sci U S A* 2003; 100:15023–15028.

11. Smith GP, Petrenko VA. Phage display. *Chem Rev* 1997; 97:391–410.

12. Zhong G. Conformational mimicry through random constraints plus affinity selection. *Methods Mol Biol* 1998; 87:165–173.

13. Barbas CF III, Burton DR, Scott JK, Silverman GJ. *Phage Display: A Laboratory Manual.* Cold Spring Harbor, NY: Cold Spring Harbor Laboratory Press, 2001.

14. Crameri R, Suter M. Display of biologically active proteins on the surface of filamentous phages: A cDNA cloning system for selection of functional gene products linked to the genetic information responsible for their production. *Gene* 1993; 137:69–75.

15. Jespers LS, Messens JH, De Keyser A, Eeckhout D, Van den Brande I, Gansemans YG, Lauwereys MJ et al. Surface expression and ligand-based selection of cDNAs fused to filamentous phage gene VI. *Biotechnology (N Y)* 1995; 13:378–382.

16. Hufton SE, Moerkerk PT, Meulemans EV, de Bruine A, Arends JW, Hoogenboom HR. Phage display of cDNA repertoires: The pVI display system and its applications for the selection of immunogenic ligands. *J Immunol Methods* 1999; 231:39–51.

17. Jacobsson K, Frykberg L. Gene VIII-based, phage-display vectors for selection against complex mixtures of ligands. *Biotechniques* 1998; 24:294–301.

18. Somers VA, Brandwijk RJ, Joosten B, Moerkerk PT, Arends JW, Menheere P, Pieterse WO et al. A panel of candidate tumor antigens in colorectal cancer revealed by the serological selection of a phage displayed cDNA expression library. *J Immunol* 2002; 169:2772–2780.

19. Sioud M, Hansen MH. Profiling the immune response in patients with breast cancer by phage-displayed cDNA libraries. *Eur J Immunol* 2001; 31:716–725.

20. van Zonneveld AJ, van den Berg BM, van Meijer M, Pannekoek H. Identification of functional interaction sites on proteins using bacteriophage-displayed random epitope libraries. *Gene* 1995; 167:49–52.

21. Zacchi P, Sblattero D, Florian F, Marzari R, Bradbury AR. Selecting open reading frames from DNA. *Genome Res* 2003; 13:980–990.

22. Parmley SF, Smith GP. Antibody-selectable filamentous fd phage vectors: Affinity purification of target genes. *Gene* 1988; 73:305–318.

23. Cwirla SE, Peters EA, Barrett RW, Dower WJ. Peptides on phage: A vast library of peptides for identifying ligands. *Proc Natl Acad Sci U S A* 1990; 87:6378–6382.

24. Devlin JJ, Panganiban LC, Devlin PE. Random peptide libraries: A source of specific protein binding molecules. *Science* 1990; 249:404–406.

25. Scott JK, Smith GP. Searching for peptide ligands with an epitope library. *Science* 1990; 249:386–390.

26. Virnekaes B, Ge L, Plückthun A, Schneider KC, Wellnhofer G, Moroney SE. Trinucleotide phosphoramidites: Ideal reagents for the synthesis of mixed oligonucleotides for random mutagenesis. *Nucleic Acids Res* 1994; 22:5600–5607.

27. Menendez A, Bonnycastle LL, Pan OCC, Scott JK. Screening peptide libraries. In: Barbas CF III, Burton DR, Scott JK, Silverman GJ, eds. *Phage Display: A Laboratory Manual.* Cold Spring Harbor, NY: Cold Spring Harbor Laboratory Press, 2001:17.11–17.32.

28. Kishchenko G, Batliwala H, Makowski L. Structure of a foreign peptide displayed on the surface of bacteriophage M13. *J Mol Biol* 1994; 241:208–213.

29. Zwick MB, Bonnycastle LL, Menendez A, Irving MB, Barbas CF III, Parren PW, Burton DR et al. Identification and characterization of a peptide that specifically binds the human, broadly neutralizing anti-human immunodeficiency virus type 1 antibody b12. *J Virol* 2001; 75:6692–6699.

30. Wrighton NC, Farrell FX, Chang R, Kashyap AK, Barbone FP, Mulcahy LS, Johnson DL et al. Small peptides as potent mimetics of the protein hormone erythropoietin. *Science* 1996; 273:458–464.

31. Zhu ZY, Minenkova O, Bellintani F, De Tomassi A, Urbanelli L, Felici F, Monaci P. "In vitro evolution" of ligands for HCV-specific serum antibodies. *Biol Chem* 2000; 381:245–254.

32. Beenhouwer DO, May RJ, Valadon P, Scharff MD. High affinity mimotope of the polysaccharide capsule of *Cryptococcus neoformans* identified from an evolutionary phage peptide library. *J Immunol* 2002; 169:6992–6999.

33. Glaser SM, Yelton DE, Huse WD. Antibody engineering by codon-based mutagenesis in a filamentous phage vector system. *J Immunol* 1992; 149:3903–3913.

34. Marston EL, James AV, Parker JT, Hart JC, Brown TM, Messmer TO, Jue DL et al. Newly characterized species-specific immunogenic *Chlamydophila pneumoniae* peptide reactive with murine monoclonal and human serum antibodies. *Clin Diagn Lab Immunol* 2002; 9:446–452.

35. Crameri R, Jaussi R, Menz G, Blaser K. Display of expression products of cDNA libraries on phage surfaces. A versatile screening system for selective isolation of genes by specific gene-product/ligand interaction. *Eur J Biochem* 1994; 226:53–58.

36. Crameri R, Faith A, Hemmann S, Jaussi R, Ismail C, Menz G, Blaser K. Humoral and cell-mediated autoimmunity in allergy to *Aspergillus fumigatus*. *J Exp Med* 1996; 184:265–270.

37. Weichel M, Schmid-Grendelmeier P, Rhyner C, Achatz G, Blaser K, Crameri R. Immunoglobulin E-binding and skin test reactivity to hydrophobin HCh-1 from *Cladosporium herbarum*, the first allergenic cell wall component of fungi. *Clin Exp Allergy* 2003; 33:72–77.

38. Rhyner C, Kodzius R, Crameri R. Direct selection of cDNAs from filamentous phage surface display libraries: Potential and limitations. *Curr Pharm Biotechnol* 2002; 3:13–21.

39. Kemp EH, Herd LM, Waterman EA, Wilson AG, Weetman AP, Watson PP. Immunoscreening of phage-displayed cDNA-encoded polypeptides identifies B-cell targets in autoimmune disease. *Biochem Biophys Res Commun* 2002; 298:169–177.

40. Folgori A, Tafi R, Meola A, Felici F, Galfre G, Cortese R, Monaci P et al. A general strategy to identify mimotopes of pathological antigens using only random peptide libraries and human sera. *EMBO J* 1994; 13:2236–2243.
41. Kouzmitcheva GA, Petrenko VA, Smith GP. Identifying diagnostic peptides for lyme disease through epitope discovery. *Clin Diagn Lab Immunol* 2001; 8:150–160.
42. Birch-Machin I, Ryder S, Taylor L, Iniguez P, Marault M, Ceglie L, Zientara S et al. Utilisation of bacteriophage display libraries to identify peptide sequences recognised by equine herpesvirus type 1 specific equine sera. *J Virol Methods* 2000; 88:89–104.
43. Benguric DR, Dungu B, Thiaucourt F, du Plessis DH. Phage displayed peptides and anti-idiotype antibodies recognised by a monoclonal antibody directed against a diagnostic antigen of *Mycoplasma capricolum* subsp. *capripneumoniae*. *Vet Microbiol* 2001; 81:165–179.
44. Fierabracci A, Biro PA, Yiangou Y, Mennuni C, Luzzago A, Ludvigsson J, Cortese R et al. Osteopontin is an autoantigen of the somatostatin cells in human islets: Identification by screening random peptide libraries with sera of patients with insulin-dependent diabetes mellitus. *Vaccine* 1999; 18:342–354.
45. Yamagishi S, Fujimori H, Yonekura H, Tanaka N, Yamamoto H. Advanced glycation endproducts accelerate calcification in microvascular pericytes. *Biochem Biophys Res Commun* 1999; 258:353–357.
46. Chen NX, Moe SM. Arterial calcification in diabetes. *Curr Diab Rep* 2003; 3:28–32.
47. Saito H, Fukuda Y, Katsuragi K, Tanaka M, Satomi M, Shimoyama T, Saito T et al. Isolation of peptides useful for differential diagnosis of Crohn's disease and ulcerative colitis. *Gut* 2003; 52:535–540.
48. Visvanathan S, Scott JK, Hwang KK, Banares M, Grossman JM, Merrill JT, FitzGerald J et al. Identification and characterization of a peptide mimetic that may detect a species of disease-associated anticardiolipin antibodies in patients with the antiphospholipid syndrome. *Arthritis Rheum* 2003; 48:737–745.
49. Jin L, Fendly BM, Wells JA. High resolution functional analysis of antibody–antigen interactions. *J Mol Biol* 226:851–865.
50. Bentley L, Fehrsen J, Jordaan F, Huismans H, du Plessis DH. Identification of antigenic regions on VP2 of African horsesickness virus serotype 3 by using phage-displayed epitope libraries. *J Gen Virol* 2000; 81:993–1000.
51. Scala G, Chen X, Liu W, Telles JN, Cohen OJ, Vaccarezza M, Igarashi T et al. Selection of HIV-specific immunogenic epitopes by screening random peptide libraries with HIV-1-positive sera. *J Immunol* 1999; 162:6155–6161.
52. Enshell-Seijffers D, Smelyanski L, Vardinon N, Yust I, Gershoni JM. Dissection of the humoral immune response toward an immunodominant epitope of HIV: A model for the analysis of antibody diversity in HIV+ individuals. *FASEB J* 2001; 15:2112–2120.
53. Bonnycastle LL, Mehroke JS, Rashed M, Gong X, Scott JK. Probing the basis of antibody reactivity with a panel of constrained peptide libraries displayed by filamentous phage. *J Mol Biol* 1996; 258:747–762.
54. Zhu Z, Ming Y, Sun B. Identification of epitopes of trichosanthin by phage peptide library. *Biochem Biophys Res Commun* 2001; 282:921–927.
55. Yip YL, Smith G, Koch J, Dubel S, Ward RL. Identification of epitope regions recognized by tumor inhibitory and stimulatory anti-ErbB-2 monoclonal antibodies: Implications for vaccine design. *J Immunol* 2001; 166:5271–5278.
56. Yu MW, Scott JK, Fournier A, Talbot PJ. Characterization of murine coronavirus neutralization epitopes with phage-displayed peptides. *Virology* 2000; 271:182–196.

57. Bresson D, Cerutti M, Devauchelle G, Pugniere M, Roquet F, Bes C, Bossard C et al. Localization of the discontinuous immunodominant region recognized by human anti-thyroperoxidase autoantibodies in autoimmune thyroid diseases. *J Biol Chem* 2003; 278:9560–9569.

58. Parhami-Seren B, Keel T, Reed GL. Sequences of antigenic epitopes of strepto-kinase identified via random peptide libraries displayed on phage. *J Mol Biol* 1997; 271:333–341.

59. Fack F, Hugle-Dorr B, Song D, Queitsch I, Petersen G, Bautz EK. Epitope mapping by phage display: Random versus gene-fragment libraries. *J Immunol Methods* 1997; 206:43–52.

60. Holzem A, Nahring JM, Fischer R. Rapid identification of a tobacco mosaic virus epi-tope by using a coat protein gene-fragment–pVIII fusion library. *J Gen Virol* 2001; 82:9–15.

61. Huang JA, Wang L, Firth S, Phelps A, Reeves P, Holmes I. Rotavirus VP7 epitope map-ping using fragments of VP7 displayed on phages. *Vaccine* 2000; 18:2257–2265.

62. Stemmer WP. Rapid evolution of a protein in vitro by DNA shuffling. *Nature* 1994; 370:389–391.

63. Ollman Saphire E, Parren PW, Pantophlet R, Zwick MB, Morris GM, Rudd PM, Dwek RA et al. Crystal structure of a neutralizing human IGG against HIV-1: A template for vaccine design. *Science* 2001; 293:1155–1159.

64. Stern B, Denisova G, Buyaner D, Raviv D, Gershoni JM. Helical epitopes determined by low-stringency antibody screening of a combinatorial peptide library. *FASEB J* 1997; 11:147–153.

65. Kwong PD, Wyatt R, Robinson J, Sweet RW, Sodroski J, Hendrickson WA. Structure of an HIV gp120 envelope glycoprotein in complex with the CD4 receptor and a neutral-izing human antibody. *Nature* 1998; 393:648–659.

66. Ferrieres G, Villard S, Pugniere M, Mani JC, Navarro-Teulon I, Rharbaoui F, Laune D et al. Affinity for the cognate monoclonal antibody of synthetic peptides derived from selection by phage display. Role of sequences flanking the binding motif. *Eur J Biochem* 2000; 267:1819–1829.

67. Menendez A, Chow KC, Pan O, Scott JK. Human immunodeficiency virus type 1-neu-tralizing monoclonal antibody 2F5 is multispecific for sequences flanking the DKW core epitope. *J Mol Biol* 2004; 338:311–327.

68. Zwick MB, Bonnycastle LL, Noren KA, Venturini S, Leong E, Barbas CF III, Noren CJ et al. The maltose-binding protein as a scaffold for monovalent display of peptides derived from phage libraries. *Anal Biochem* 1998; 264:87–97.

69. Enshell-Seijffers D, Denisov D, Groisman B, Smelyanski L, Meyuhas R, Gross G, Denisova G et al. The mapping and reconstitution of a conformational discontinuous B-cell epitope of HIV-1. *J Mol Biol* 2003; 334:87–101.

70. Pantophlet R, Wilson IA, Burton DR. Hyperglycosylated mutants of human immunode-ficiency virus (HIV) type 1 monomeric gp120 as novel antigens for HIV vaccine design. *J Virol* 2003; 77:5889–5901.

71. Pantophlet R, Ollmann Saphire E, Poignard P, Parren PW, Wilson IA, Burton DR. Fine mapping of the interaction of neutralizing and nonneutralizing monoclonal antibodies with the CD4 binding site of human immunodeficiency virus type 1 gp120. *J Virol* 2003; 77:642–658.

72. Plotkin SA. Vaccines, vaccination, and vaccinology. *J Infect Dis* 2003; 187:1349–1359.

73. Moylett EH, Hanson IC. 29. Immunization. *J Allergy Clin Immunol* 2003; 111:S754–S765.

74. Jerne NK. Towards a network theory of the immune system. *Ann Immunol (Paris)* 1974; 125C:373–389.

75. Goldbaum FA, Velikovsky CA, Dall'Acqua W, Fossati CA, Fields BA, Braden BC, Poljak RJ et al. Characterization of anti–anti-idiotypic antibodies that bind antigen and an anti-idiotype. *Proc Natl Acad Sci U S A* 1997; 94:8697–8701.

76. Hutchins WA, Adkins AR, Kieber-Emmons T, Westerink MA. Molecular characterization of a monoclonal antibody produced in response to a group C meningococcal polysaccharide peptide mimic. *Mol Immunol* 1996; 33:503–510.

77. Westerink MA, Campagnari AA, Wirth MA, Apicella MA. Development and characterization of an anti-idiotype antibody to the capsular polysaccharide of *Neisseria meningitidis* serogroup C. *Infect Immun* 1988; 56:1120–1127.

78. Caton M, Diamond B. Using peptide mimetopes to elucidate anti-polysaccharide and anti-nucleic acid humoral responses. *Cell Mol Biol (Noisy-le-grand)* 2003; 49:255–262.

79. Benvenuti F, Cesco-Gaspere M, Burrone OR. Anti-idiotypic DNA vaccines for B-cell lymphoma therapy. *Front Biosci* 2002; 7:d228–d234.

80. Bendandi M. Anti-idiotype vaccines for human follicular lymphoma. *Leukemia* 2000; 14:1333–1339.

81. Westerink MA, Smithson SL, Hutchins WA, Widera G. Development and characterization of anti-idiotype based peptide and DNA vaccines which mimic the capsular polysaccharide of *Neisseria meningitidis* serogroup C. *Int Rev Immunol* 2001; 20:251–261.

82. Kieber-Emmons T, Getzoff E, Kohler H. Perspectives on antigenicity and idiotypy. *Int Rev Immunol* 1987; 2:339–356.

83. Fields BA, Goldbaum FA, Ysern X, Poljak RJ, Mariuzza RA. Molecular basis of antigen mimicry by an anti-idiotope. *Nature* 1995; 374:739–742.

84. Putterman C, Deocharan B, Diamond B. Molecular analysis of the autoantibody response in peptide-induced autoimmunity. *J Immunol* 2000; 164:2542–2549.

85. Ganglberger E, Grunberger K, Sponer B, Radauer C, Breiteneder H, Boltz-Nitulescu G, Scheiner O et al. Allergen mimotopes for 3-dimensional epitope search and induction of antibodies inhibiting human IgE. *FASEB J* 2000; 14:2177–2184.

86. Frenkel D, Katz O, Solomon B. Immunization against Alzheimer's beta-amyloid plaques via EFRH phage administration. *Proc Natl Acad Sci U S A* 2000; 97:11455–11459.

87. Chen X, Scala G, Quinto I, Liu W, Chun TW, Justement JS, Cohen OJ et al. Protection of rhesus macaques against disease progression from pathogenic SHIV-89.6PD by vaccination with phage-displayed HIV-1 epitopes. *Nat Med* 2001; 7:1225–1231.

88. Lesinski GB, Westerink MA. Novel vaccine strategies to T-independent antigens. *J Microbiol Methods* 2001; 47:135–149.

89. Phalipon A, Folgori A, Arondel J, Sgaramella G, Fortugno P, Cortese R, Sansonetti PJ et al. Induction of anti-carbohydrate antibodies by phage library-selected peptide mimics. *Eur J Immunol* 1997; 27:2620–2625.

90. Pincus SH, Smith MJ, Jennings HJ, Burritt JB, Glee PM. Peptides that mimic the group B streptococcal type III capsular polysaccharide antigen. *J Immunol* 1998; 160:293–298.

91. May RJ, Beenhouwer DO, Scharff MD. Antibodies to keyhole limpet hemocyanin cross-react with an epitope on the polysaccharide capsule of *Cryptococcus neoformans* and other carbohydrates: Implications for vaccine development. *J Immunol* 2003; 171:4905–4912.

92. Johnson MA, Jaseja M, Zou W, Jennings HJ, Copie V, Pinto BM, Pincus SH. NMR studies of carbohydrates and carbohydrate-mimetic peptides recognized by an anti-group B Streptococcus antibody. *J Biol Chem* 2003; 278:24740–24752.

93. Grothaus MC, Srivastava N, Smithson SL, Kieber-Emmons T, Williams DB, Carlone GM, Westerink MA. Selection of an immunogenic peptide mimic of the capsular polysaccharide of *Neisseria meningitidis* serogroup A using a peptide display library. *Vaccine* 2000; 18:1253–1263.

94. Lohnas GL, Roberts SF, Pilon A, Tramontano A. Epitope-specific antibody and suppression of autoantibody responses against a hybrid self protein. *J Immunol* 1998; 161:6518–6525.

95. Cox JC, Coulter AR. Adjuvants—A classification and review of their modes of action. *Vaccine* 1997; 15:248–256.

96. Edelman R. The development and use of vaccine adjuvants. *Mol Biotechnol* 2002; 21:129–148.

97. Tam JP. Synthetic peptide vaccine design: Synthesis and properties of a high-density multiple antigenic peptide system. *Proc Natl Acad Sci U S A* 1988; 85:5409–5413.

98. Coombes AG, Lavelle EC, Jenkins PG, Davis SS. Single dose, polymeric, microparticle-based vaccines: The influence of formulation conditions on the magnitude and duration of the immune response to a protein antigen. *Vaccine* 1996; 14:1429–1438.

99. de la Cruz VF, Lal AA, McCutchan TF. Immunogenicity and epitope mapping of foreign sequences via genetically engineered filamentous phage. *J Biol Chem* 1988; 263:4318–4322.

100. Meola A, Delmastro P, Monaci P, Luzzago A, Nicosia A, Felici F, Cortese R et al. Derivation of vaccines from mimotopes. Immunologic properties of human hepatitis B virus surface antigen mimotopes displayed on filamentous phage. *J Immunol* 1995; 154:3162–3172.

101. Bastien N, Trudel M, Simard C. Protective immune responses induced by the immunization of mice with a recombinant bacteriophage displaying an epitope of the human respiratory syncytial virus. *Virology* 1997; 234:118–122.

102. Grabowska AM, Jennings R, Laing P, Darsley M, Jameson CL, Swift L, Irving WL. Immunisation with phage displaying peptides representing single epitopes of the glycoprotein G can give rise to partial protective immunity to HSV-2. *Virology* 2000; 269:47–53.

103. Kneissel S, Queitsch I, Petersen G, Behrsing O, Micheel B, Dubel S. Epitope structures recognised by antibodies against the major coat protein (g8p) of filamentous bacteriophage fd (Inoviridae). *J Mol Biol* 1999; 288:21–28.

104. Willis AE, Perham RN, Wraith D. Immunological properties of foreign peptides in multiple display on a filamentous bacteriophage. *Gene* 1993; 128:79–83.

105. De Berardinis P, Sartorius R, Fanutti C, Perham RN, Del Pozzo G, Guardiola J. Phage display of peptide epitopes from HIV-1 elicits strong cytolytic responses. *Nat Biotechnol* 2000; 18:873–876.

106. De Berardinis P, D'Apice L, Prisco A, Ombra MN, Barba P, Del Pozzo G, Petukhov S et al. Recognition of HIV-derived B and T-cell epitopes displayed on filamentous phages. *Vaccine* 1999; 17:1434–1441.

107. Manoutcharian K, Terrazas LI, Gevorkian G, Acero G, Petrossian P, Rodriguez M, Govezensky T. Phage-displayed T-cell epitope grafted into immunoglobulin heavy-chain complementarity-determining regions: An effective vaccine design tested in murine cysticercosis. *Infect Immun* 1999; 67:4764–4770.

108. Petrenko VA, Smith GP, Gong X, Quinn T. A library of organic landscapes on filamentous phage. *Protein Eng* 1996; 9:797–801.

109. Makowski L. Structural constraints on the display of foreign peptides on filamentous bacteriophages. *Gene* 1993; 128:5–11.

110. Felici F, Castagnoli L, Musacchio A, Jappelli R, Cesareni G. Selection of antibody ligands from a large library of oligopeptides expressed on a multivalent exposition vector. *J Mol Biol* 1991; 222:301–310.

111. Zwick MB, Shen J, Scott JK. Homodimeric peptides displayed by the major coat protein of filamentous phage. *J Mol Biol* 2000; 300:307–320.

112. Monette M, Opella SJ, Greenwood J, Willis AE, Perham RN. Structure of a malaria parasite antigenic determinant displayed on filamentous bacteriophage determined by NMR spectroscopy: Implications for the structure of continuous peptide epitopes of proteins. *Protein Sci* 2001; 10:1150–1159.

113. Yip YL, Smith G, Ward RL. Comparison of phage pIII, pVIII and GST as carrier proteins for peptide immunisation in Balb/c mice. *Immunol Lett* 2001; 79:197–202.

114. Rubinchik E, Chow AW. Recombinant expression and neutralizing activity of an MHC class II binding epitope of toxic shock syndrome toxin-1. *Vaccine* 2000; 18:2312–2320.

115. Kato H, Bukawa H, Hagiwara E, Xin KQ, Hamajima K, Kawamoto S, Sugiyama M et al. Rectal and vaginal immunization with a macromolecular multicomponent peptide vaccine candidate for HIV-1 infection induces HIV-specific protective immune responses. *Vaccine* 2000; 18:1151–1160.

116. Bennett B, Check IJ, Olsen MR, Hunter RL. A comparison of commercially available adjuvants for use in research. *J Immunol Methods* 1992; 153:31–40.

117. Johnston BA, Eisen H, Fry D. An evaluation of several adjuvant emulsion regimens for the production of polyclonal antisera in rabbits. *Lab Anim Sci* 1991; 41:15–21.

118. Kenney JS, Hughes BW, Masada MP, Allison AC. Influence of adjuvants on the quantity, affinity, isotype and epitope specificity of murine antibodies. *J Immunol Methods* 1989; 121:157–166.

119. Zuercher AW, Miescher SM, Vogel M, Rudolf MP, Stadler MB, Stadler BM. Oral anti-IgE immunization with epitope-displaying phage. *Eur J Immunol* 30:128–135.

120. Delmastro P, Meola A, Monaci P, Cortese R, Galfre G. Immunogenicity of filamentous phage displaying peptide mimotopes after oral administration. *Vaccine* 1997; 15:1276–1285.

121. Holmgren J, Czerkinsky C, Eriksson K, Mharandi A. Mucosal immunisation and adjuvants: A brief overview of recent advances and challenges. *Vaccine* 2003; 21(suppl 2): S89–S95.

122. Burton DR, Pyati J, Koduri R, Sharp SJ, Thornton GB, Parren PW, Sawyer LS et al. Efficient neutralization of primary isolates of HIV-1 by a recombinant human monoclonal antibody. *Science* 1994; 266:1024–1027.

123. Richman DD, Wrin T, Little SJ, Petropoulos CJ. Rapid evolution of the neutralizing antibody response to HIV type 1 infection. *Proc Natl Acad Sci U S A* 2003; 100:4144–4149.

124. Prinz DM, Smithson SL, Westerink MA. Two different methods result in the selection of peptides that induce a protective antibody response to *Neisseria meningitidis* serogroup C. *J Immunol Methods* 2004; 285:1–14.

125. Chargelegue D, Obeid OE, Hsu SC, Shaw MD, Denbury AN, Taylor G, Steward MW. A peptide mimic of a protective epitope of respiratory syncytial virus selected from a combinatorial library induces virus-neutralizing antibodies and reduces viral load in vivo. *J Virol* 1998; 72:2040–2046.

126. Mascola JR, Stiegler G, VanCott TC, Katinger H, Carpenter CB, Hanson CE, Beary H et al. Protection of macaques against vaginal transmission of a pathogenic HIV-1/SIV chimeric virus by passive infusion of neutralizing antibodies. *Nat Med* 2000; 6:207–210.

127. Parren PW, Marx PA, Hessell AJ, Luckay A, Harouse J, Cheng-Mayer C, Moore JP et al. Antibody protects macaques against vaginal challenge with a pathogenic R5 simian/human immunodeficiency virus at serum levels giving complete neutralization in vitro. *J Virol* 2001; 75:8340–8347.

128. Baba TW, Liska V, Hofmann-Lehmann R, Vlasak J, Xu W, Ayehunie S, Cavacini LA et al. Human neutralizing monoclonal antibodies of the IgG1 subtype protect against mucosal simian-human immunodeficiency virus infection. *Nat Med* 2000; 6:200–206.

129. Gauduin MC, Parren PW, Weir R, Barbas CF, Burton DR, Koup RA. Passive immunization with a human monoclonal antibody protects hu-PBL-SCID mice against challenge by primary isolates of HIV-1. *Nat Med* 1997; 3:1389–1393.

130. Crameri R, Blaser K. Cloning Aspergillus fumigatus allergens by the pJuFo filamentous phage display system. *Int Arch Allergy Immunol* 1996; 110:41–45.

131. Rotblat B, Enshell-Seijffers D, Gershoni JM, Schuster S, Avni A. Identification of an essential component of the elicitation active site of the EIX protein elicitor. *Plant J* 2002; 32:1049–1055.

132. Boder ET, Wittrup KD. Yeast surface display for screening combinatorial polypeptide libraries. *Nat Biotechnol* 1997; 15:553–557.

CHAPTER 7

Exploring Protein–Protein Interactions Using Peptide Libraries Displayed on Phage

Kurt Deshayes and Jacob Corn

CONTENTS

7.1 INTRODUCTION

It has become evident that protein–protein interactions play a central role in signal transduction, and are thus key regulators of cell function. It has also become clear that the identification of therapeutically important protein–protein interactions from among the thousands of contenders requires rapid and robust screening methodologies. Fortunately, the display of naive peptide libraries on phage has proven an effective tool for the exploration of binding surfaces and the discovery of novel binding partners [1]. The utility of phage display is demonstrated by the repeated success of phage sorting in yielding potent binders against proteins for which other methods fail to discover specific ligands.

7.2 EXTRACELLULAR PROTEIN–PROTEIN INTERACTIONS

Small molecule screening efforts against extracellular protein–protein interactions generally fail to identify antagonists, presumably because extracellular protein binding sites are presented over large areas without significant invagination [2]. In contrast, phage-displayed peptide libraries consistently yield binders that recognize features presented over large surface areas.

The ability of phage display to discover potent peptide reagents has enabled the detailed elucidation of biological recognition in several systems. As illustrated in Figure 7.1, peptide binders selected against extracellular targets consistently contain internal disulfide bonds. The constraints imposed by disulfide bonds stabilize peptide

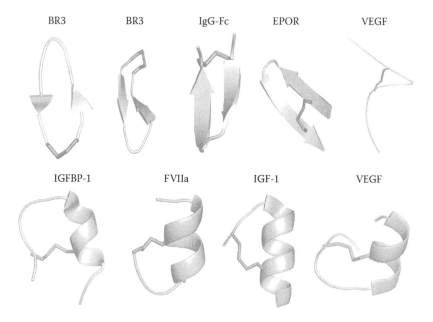

Figure 7.1   Selected phage-derived peptide antagonists and agonists for which structures have been determined bound to, or free of, their target proteins. The name of the target protein is shown above each peptide. Secondary structural elements are shown as ribbons along with the disulfide bonds. The literature references are as follows: BAFF receptor (BR3) [3], crystallizable fragment of human immuoglobulin G (IgG-Fc) [13], erythropoietin receptor (EPOR) [10], vascular endothelial growth factor (VEGF) [15,16], insulin-like growth factor-binding protein 1 (IGFBP-1) [4], FVIIa [5], and insulin-like growth factor 1 (IGF-1) [9].

structure and organize the binding surfaces. The consistent selection of structured peptides [3–9] that bind to extracellular proteins suggests that preorganization of the binding surface is required for high-affinity binding. Peptides that bind to a variety of targets frequently have either β-hairpin or turn-helix structures, but other folds are also observed (Figure 7.1) [4–15].

The interactions between phage-derived peptide and receptor often lead to observable changes in function. Frequently, the phage-derived peptide competes with the natural ligand for overlapping binding sites [4–14,16]. This observation leads one to surmise that proteins have evolved regions that function as definitive binding sites. The characteristics of these binding sites are an important issue when evaluating potential binding partners or evaluating the possibility of small molecule therapeutics. This chapter presents examples that illustrate how phage-displayed peptide libraries can be used to assess the physical characteristics of protein–protein interactions.

7.2.1 Erythropoietin Receptor

An important example of efficacy of phage display is the discovery of peptides that either agonize or antagonize the primary pathway for red blood cell proliferation [10,11]. Polyvalent phage display methods were used to identify the 20-residue peptide epithelial membrane protein (EMP1), which binds tightly to EPOR (Figure 7.2) [10].

Not only does EMP1 contain an internal disulfide bond that stabilizes the β-hairpin structure, it also exists as a homodimer in solution, with a 320 Å^2 interface between the monomers. EMP1 blocks binding of erythropoietin to EPOR with an IC$_{50}$ of 200 nM. Structural analysis shows that the EMP1 dimer forms a 1720 Å^2 interface that simultaneously contacts two EPOR molecules (Figure 7.3).

Binding of the EMP1 dimer activates the erythropoietin-mediated signaling pathway, as demonstrated by the proliferation of erythroid cells [10]. The initial EMP1 results agree with the model in which dimerization of isolated EPOR molecules brings together associated kinase domains residing across the cellular membrane [17]. Interactions between the kinase domains initiate a series of events that

Figure 7.2   Solution structure of the EMP1 dimer.

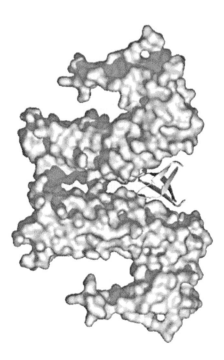

Figure 7.3 Structure of the complex between the EMP1 dimer and the erythropoietin receptor dimer. The peptide is shown as a ribbon.

transmit a signal into the nucleus. Amazingly, substitution of the tyrosine residue in
EMP1 with 3,5-dibromotyrosine converts the agonist into an antagonist (EMP33)
[11]. Structural analysis reveals that EMP33 is also a dimer, structurally identical to
EMP1 except for deviations at both termini [18]. These variations in peptide structure are sufficient to reorient the two EPOR molecules within the complex, preventing the requisite interaction between associated kinases (Figure 7.3).

Comparison of the EMP1–EPOR and EMP33–EPOR structures shows that
subtle changes in EPOR orientation distinguish an inactive structure. Subsequent
studies provided evidence that two EPORs exist in an inactive preformed dimer,
and signaling through EPOR occurs upon alignment of two preassociated EPOR
molecules into an activated state [18,19]. This is distinct from the mechanisms of
other cytokines, such as human growth hormone, in which dimerization of two
independent receptors is required for signaling [17]. The relative generality of these
two mechanisms has not been established, but the observation that the mechanism
of cytokine signal transduction varies among different pathways has far-reaching
mechanistic implications.

7.2.2 Insulin-Like Growth Factor

The recent solution structure of IGF-1 is a good example of how peptide
phage display can provide reagents that can be used to provide crucial structural

information unavailable through structural analysis of the receptor alone [9,20]. IGF-1 is a 70-residue peptide hormone that regulates both mitogenic and metabolic functions, primarily via binding to the cell surface IGF-1 receptor (IGFR), a cell surface receptor related to the insulin receptor (IR) [21]. IGF-1 activity is tightly regulated, existing in vivo primarily bound to one of six soluble IGFBPs that seques-ter the hormone within high-affinity complexes, thereby controlling IGF-1 biological availability and plasma half-life [22].

Nuclear magnetic resonance studies of IGF-1 in solution gave poor results due to self-association and unfavorable internal dynamics [23,24]. The situation changes markedly upon the addition of the phage-derived IGF-1 ligand F1-1. F1-1 is a 17-residue peptide that contains a single disulfide bond and inhibits IGF-1 binding to four recep-tors with IC_{50} values in the low micromolar range [9]. Interestingly, IGF-1 becomes highly structured upon binding F1-1 (Figure 7.4), adopting a conformation similar to that of insulin [20]. This result clearly demonstrates that IGF-1 only adopts a stable conformation when bound to an ancillary molecule [20,25,26].

Comparison of the solution structure of the IGF-1 F1-1 complex with crystal structures of IGF-1 bound to either a fragment of IGFBP-5 or a detergent molecule reveals that the same hydrophobic patch on IGF-1 makes contact with each ligand [20]. The differences in IGF-1 structure in complexes are due to changes in helix 3 conformation that change the orientation of important contact residues. For example, when IGF-1 binds a detergent molecule, helix 3 begins as a 3_{10} helix but ends as a regular α-helix [27]. In contrast, when F1-1 is bound to IGF-1, helix 3 adopts the 3_{10} helix conformation exclusively [20]. The helix 3 conformation of IGF-1 is intermit-tent between the other structures when in complex with the IGFBP-5 fragment [26]. The variations in helix 3 conformation have a pronounced affect on the location of Leu54, a key residue in the interactions with every ligand. Changes in residue orientation upon binding may account for the ability of a small peptide hormone to

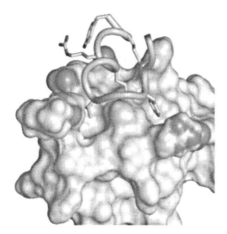

Figure 7.4 Structure of the complex between the phage-derived peptide F1-1 and IGF-1. The peptide is shown as a worm.

recognize six binding proteins and two receptors. The dynamic characteristics that make IGF-1 difficult to characterize in solution may also permit the hormone to efficiently adjust conformation in order to recognize dissimilar receptors.

7.2.3 Vascular Endothelial Growth Factor

Peptide phage display has been used to generate compelling evidence that proteins contain discrete sites that dominate the binding interactions. An important example of the conservation of binding sites is the study of complexes between VEGF and four unrelated ligands. VEGF is a primary modulator of physiological angiogenesis that exists in solution as a homodimer and functions by inducing dimerization of its tyrosine kinase receptors, *fms*-like tyrosine kinase-1 (Flt-1) and kinase-insert domain containing receptor (KDR) [28]. The crystal structure of VEGF in complex with the second Ig-like domain of Flt-1 (Flt-1$_{D2}$) revealed symmetric binding sites at the poles of the dimeric VEGF receptor–binding domain [29] that are similar to the KDR-binding sites [30–33] revealed by alanine-scanning mutagenesis. The binding sites for antibodies that neutralize VEGF have also been shown to overlap with the KDR- and Flt-1-binding sites [31,34,35].

A library of disulfide-constrained peptides was sorted against VEGF and yielded three peptide classes that bound to VEGF and blocked KDR binding [16]. Representatives of two of these peptide classes, v108 and v107, have subsequently had their structures determined in complex with VEGF [15,36]. The structures reveal that the binding sites for both peptides overlap significantly with each other and the receptor- and antibody-binding sites. The contact between v108 and VEGF involves primarily main-chain-mediated hydrogen bonds (Figure 7.5a), while in contrast, v107 makes extensive hydrophobic side-chain-mediated contacts (Figure 7.5b).

Each peptide complex resembles another structurally characterized VEGF complex. The binding mode of v108 most closely resembles that of an antigen-binding fragment (Fab) from the neutralizing anti-VEGF antibody Avastin (Figure 7.5c) [34,35]. The interaction of v107 with VEGF is most similar to that observed for Flt-1$_{D2}$ (Figure 7.5d) [29].

Although the modes of interaction are similar, the dissociation constant (K_d) values of the phage-derived peptides (0.6 and 2.2 μM for v107 and v108, respectively) [15,16] are significantly weaker than the 2–10 nM reported for Flt-1$_{D2}$ and 13 and 0.11 nM reported for the Avastin Fab and its affinity-optimized version, respectively [15,29,35]. The 19-residue peptide v107 buries 1167 Å^2 of hydrophobic binding surface, while Flt-1$_{D2}$ uses 101 residues to bury 1672 Å^2 of surface area. All the VEGF residues that contact v107 in the peptide complex also contact Flt-1$_{D2}$. The 20-residue peptide v108 buries a total surface area of 1350 Å^2 upon binding to VEGF, while the receptor-blocking Fabs bury 1750–1800 Å^2. Up to 13 intermolecular hydrogen bonds are observed in the interface of the v108 complex, while 10–12 are found in the Fab complexes. All VEGF atoms involved in hydrogen bonds in the v108 complex also hydrogen-bond to the Fabs.

That phage-derived peptides adopt VEGF-binding interactions similar to those observed for protein ligands is not likely due to random chance; rather, it suggests

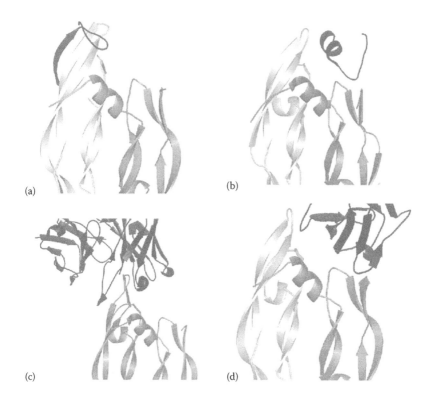

Figure 7.5 Structures of the interfaces within complexes of VEGF with (a) peptide v108, (b) peptide v107, (c) the neutralizing Fab Avastin, and (d) Flt-1$_{D2}$.

that only limited regions of the VEGF surface support high-affinity binding interactions. An illuminating comparison can be made using data obtained from crystal structure analysis of protein–protein complexes. Examination of 32 protein dimer structures revealed that the size of the protein–protein interfaces ranged from 368 to 4761 Å [37]. A similar analysis of protease inhibitor complexes and antibody–protein antigen complexes showed a range of 1600 ± 350 Å [38]. VEGF-binding peptides present interaction surfaces on reduced scaffolds, but the buried surface areas in these interfaces (~1000 Å^2) fall within the observed norms for protein–protein interactions. Thus, in order to effectively bind to VEGF, it is necessary to utilize a major part of the optimal binding region.

7.2.4 Immunoglobulin G

Another phage-based study that points to the existence of optimal binding sites is the probing of the binding surface of IgG-Fc. It has been shown that the same hinge region of the IgG-Fc interacts with four different proteins: protein A [39], protein G [40], neonatal factor [41], and rheumatoid factor (Figure 7.6) [42].

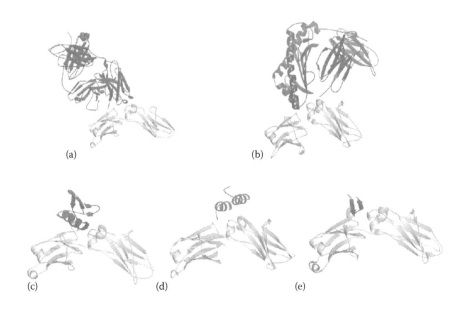

(a)   (b)

(c)   (d)   (e)

Figure 7.6 Structures of the IgG-Fc in complex with (a) rheumatoid factor, (b) neonatal factor domain, (c) domain C2 of protein G, (d) domain B1 of protein A, and (e) the phage-derived peptide Fc-III.

Amazingly, four proteins recognize the same binding surface using four drastically different folds. Is it possible that the proteins bind to that region due to functional considerations and not because the intrinsic properties of the site make it optimal for binding? In order to answer these questions, a library of disulfide-constrained peptides was panned against the IgG-Fc. A single peptide, Fc-III, was selected and further optimized to produce a 13-residue peptide that inhibits protein A binding with a K_i of 25 nM. Structural analysis revealed that although Fc-III adopts a β-hairpin conformation distinct from the four proteins that bind to the hinge region, the peptide also binds to the same location (Figure 7.6e) [13]. Since there is no functional bias involved in the selection of peptides displayed on phage, this result indicates that the Fc region of immunoglobulin G has physical properties that make it the most suitable binding site for protein ligands. Although Fc-III is much smaller than the natural ligands, the area of interaction (650 Å²) is close to that of the natural ligands (~740 Å²). It appears that the consensus binding patch on the IgG-Fc is an area of approximately 700 Å² that is more adaptive and hydrophobic than other surfaces of the protein.

The advantages of using phage display to guide the reengineering of proteins was demonstrated in a paper in which one of the helices from the B region of protein A was removed in order to create a minimized binding domain (Figure 7.7) [39]. Although the binding interface remained intact, the affinity went from a K_d of 10 nM for protein A to greater than 1 mM for the minimized protein. Using results from phage libraries that explored three regions of the two-helix structures, it was found that 12 mutations dramatically increased the α-helical structure of the miniprotein, thereby preorganizing the binding surface. In effect, the mutations discovered using phage display carry

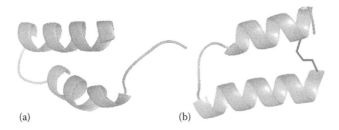

Figure 7.7 Phage-derived peptides that mimic the IgG-Fc–binding surface of protein A. (a) The original optimized structure and (b) the structure of the 34-residue peptide containing a stabilizing disulfide bond, shown in dark gray.

out the same function as the third helix and restore the affinity for the IgF-Fc (K_d = 43 nM). In subsequent work, the preorganization of the binding surface was increased by adding a disulfide bond between the two α-helices, which eventually yielded a 34-residue variant (Figure 7.7b) that binds to the Fc with a ninefold increase in affinity [43]. Thus, affinity maturation yielded a peptide that contains half the number of residues found in the native domain and yet mimics both its structure and function.

7.2.5 Factor VII

Many enzymes are regulated via binding to so-called exosites, sites distant from the active site. Inhibition via exosite binding can occur either by rearrangement of the enzyme into a less active conformation or by the physical blocking of the interaction between enzyme and substrate. An important application of phage display is shown in the work of Dennis et al. [5], which produced a novel 20-residue peptide (E-76) that acts as an exosite enzyme regulator of factor VII activity (Figure 7.8).

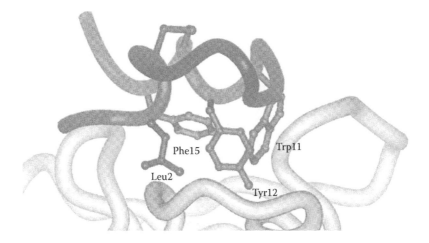

Figure 7.8 Structure of E-76 (dark gray) bound to factor VII. Key peptide side chains that make contact with factor VII are shown and labeled.

Factor VII is a serine protease that is activated by tissue factor upon vascular damage and serves as a key component in the formation of fibrin clots in the coagulation cascade [44]. The coagulation cascade is a cluster of serine proteases that control the balance between hemostasis, blood vessel repair, and thrombosis—the formation of artery-blocking clots. There is evidence that selective inhibition of the coagulation cascade will maintain the essential balance between hemostasis and thrombosis and thereby avert further cardiovascular degeneration [44].

The sorting of the peptide libraries was carried out against factor VII bound to tissue factor, so that peptides that compete with tissue factor binding could not be selected. E-76 binds tightly to factor VII (K_d = 8.5 nM) and inhibits the factor VII-mediated activation of factor X, the next component in the coagulation cascade, with an IC_{50} of 1 nM. No binding was detected between E-76 and other members of the coagulation cascade. E-76 contains a disulfide bond that maintains a well-defined solution structure consisting of a distorted type I reverse turn and a type I helix (Figure 7.8). This structure has been shown to be maintained upon binding to factor VII [5].

Structural and functional data indicate that E-76 makes contact with a 660 $Å^2$ site distant from both the enzyme active site and the tissue factor-binding site. Mutagenesis data suggest that four hydrophobic residues (Leu2, Trp11, Tyr12, and Phe15) contribute an inordinate amount to the binding energy, indicating that the binding surface is made up of hydrophobic contacts between the peptide and the protein.

E-76 displays a mixed inhibitor mechanism, indicating that it does not directly compete with factor X for binding to factor VII. This result differs from other known exosite inhibitors. For example, the dodecapeptide hirugen, derived from the leech protein hirudin, binds at the fibrinogen-binding site and acts as a competitive inhibitor, unlike E-76, which displays a mixed inhibitor mechanism [45,46]. It has been suggested that the formation of a binding surface to accommodate E-76 rearranges the protein hydrogen-binding network of factor VII so that the "oxyanion hole" is disrupted, thereby removing a crucial feature of the serine protease mechanism [5]. It is possible that the E-76 epitope has no equivalent in nature and the inhibition mechanism is unique to E-76. One can speculate that blocking the tissue factor-binding site has biased the selection against the tightest binders, and the unique nature of E-76 is a product of the selection process. This result highlights an interesting application of phage display; by varying the panning conditions, it may be possible to obtain diverse binders to a single protein. For example, once the primary binding site is occupied, can other binders be identified that recognize other regions of the protein? How many times can this process be repeated?

7.2.6 Summary of Extracellular Protein–Protein Interactions

The inability of phage display to identify small molecule–binding sites on extracellular proteins is not due to a lack of potential binding epitopes being presented to the target protein. Increases in library diversity from 10^7 to greater than 10^{11} individual members have not changed the types of peptides selected [1]. It appears that many, if not most, extracellular proteins preferentially bind epitopes that resemble "miniproteins." There are exceptions (e.g., Eph receptor/ephrin [47], BAFF/Blys [3] in which extracellular proteins have evolved binding sites for small epitopes, but the majority of extracellular proteins studied to date have given results similar to the VEGF example. Thus, we conclude that if diverse peptide libraries exclusively produce large, extended binding surfaces, the protein is unlikely to be a good target for small molecule therapeutics.

7.3 INTRACELLULAR PROTEIN–PROTEIN INTERACTIONS

The accumulated evidence suggests that the majority of extracellular proteins contain widely dispersed binding sites. This generality does not appear to carry over to intracellular protein–protein interactions. Small, continuous epitopes contained within large proteins are often recognized by intracellular receptors. For example, proline-rich sequences are recognized by two different domains: Src homology 3 (SH3) domains [48] and WW domains (named for two highly conserved tryptophan residues found in the consensus sequence) [49]. Another example is phosphotyrosine, which is recognized by both the Src homology 2 (SH2) [50] and phosphotyrosine-binding (PTB) domains [51].

7.3.1 PDZ Domains

A novel binding module, the PDZ domain—named after the first proteins in which it was observed: postsynaptic density-95, discs large, and zonula occludens-1—that recognizes the C-terminal sequences, has attracted a lot of attention [52]. The PDZ domain is a compact module consisting of approximately 90 amino acids that are found imbedded in a larger protein, often with other PDZ modules or other module types. Multiple modules can act in concert to bind and localize multiple partners and construct intracellular architecture. In many cases, short peptides of four to eight residues are specific, potent ligands. The structure of a phage-optimized ligand bound to the PDZ domain of the protein Erbin is shown in Figure 7.9 and is representative of how these receptors recognize specific C-termini [53]. The peptide main chain has antiparallel β-sheet interactions with the PDZ domain main chain, and the terminal carboxylate is inserted into a "carboxylate-binding loop." Essentially all PDZ domains maintain these interactions.

The unique nature of PDZ domain-mediated recognition makes it possible to use phage display to explore ligand specificity of domain families, understand the structure–function relationships, and predict and validate putative natural protein–protein interactions. PDZ domains assemble proteins into functional complexes localized at specific subcellular sites within eukaryotic cells: epithelial tight junctions or neuronal synaptic densities, for example [54,55]. Novel C-terminally displayed peptide phage libraries were used to investigate the binding specificities of two of the six PDZ domains from a membrane-associated guanylate kinase (MAGI-3) [56]. Each domain bound specifically to small peptides, and the binding specificities were

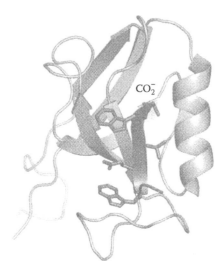

Figure 7.9 Structure of a phage-optimized peptide (WETWV) bound to the Erbin PDZ domain. The C-terminus of the peptide is labeled, and peptide side chains are shown.

very different from each other, suggesting that the two PDZ domains bind different ligands. Like many PDZ-containing proteins, MAGI-3 contains multiple PDZ domains, and thus, a single MAGI-3 protein likely acts as a multiligand scaffold to assemble multicomponent protein complexes.

More recently, C-terminal peptide phage libraries were used to study the binding specificity of the single PDZ domain of Erbin [57], a protein that was originally identified as a putative ligand for ErbB2 in a yeast two-hybrid screen [58]. ErbB2 is an epidermal growth factor receptor-related tyrosine kinase that is a causal factor in the development of some cancers [59]. The intent was to discover high-affinity ligands for the Erbin PDZ domain that could be used to disrupt its interaction with ErbB2. Surprisingly, phage display revealed a binding consensus for Erbin PDZ ([D/E][T/S]WV$_{COOH}$) that differed significantly from the C-terminal sequence of ErbB2 (DVPV$_{COOH}$) [56]. Furthermore, searches of genomic databases revealed that the phage-derived consensus closely matched the C-termini of δ-catenin and two related homologs (ARVCF and p0071), which all terminate in an identical sequence (DSWV$_{COOH}$). Since these catenins are also mediators of intracellular signaling [60,61], it is possible that the interaction with Erbin is physiologically relevant. Subsequent in vitro and in vivo experiments clearly demonstrated that Erbin binds to δ-catenin and its homologs with high affinity and specificity, while its affinity for ErbB2 is significantly lower. Subsequent work demonstrated that neither Erbin nor the C-terminus of ErbB2 interacts in the proposed manner [62]. Furthermore, the in vivo interaction between Erbin and ARVCF was successfully disrupted by intracellular delivery of phage-derived high-affinity peptides [57], thus demonstrating the utility of peptide ligands for intracellular target validation.

In studying the relationships between PDZ domain structure and function, we have made extensive use of in vitro affinity assays with synthetic peptides to accurately map the determinants of affinity and specificity. Our results suggest that PDZ domains can use up to five side chains at the C-termini of proteins to bind with high affinity to their cognate ligands while excluding other closely related sequences. This point was illustrated by comparing the binding specificity of the Erbin PDZ domain to that of MAGI-3 PDZ2. Peptide phage libraries revealed that these two domains recognize C-terminal consensus sequences that differ at only one site in the last four positions ([D/E][T/S]WV$_{COOH}$ vs. [C/V/I][T/S]WV$_{COOH}$ for Erbin PDZ and MAGI-3 PDZ2, respectively). However, this single difference was sufficient to alter affinity by at least two orders of magnitude; a peptide bearing a glutamate side chain (TGWETWV$_{COOH}$) interacted exclusively with the Erbin PDZ domain, while a peptide in which glutamate was replaced with isoleucine (TGWITWV$_{COOH}$) interacted only with MAGI-3 PDZ2 [57].

The type of ligand side chains accepted by a PDZ domain depends on the binding surface defined by the side chain residues that line the peptide-binding groove. Analysis of these interactions with phage-derived peptides gives deep insights into the manner in which PDZ domains recruit specific targets [53]. Furthermore, the observation that PDZ domains bind short linear peptides with high affinity and specificity suggests that they may be valid small molecule targets.

7.3.2 Inhibitors of Apoptosis

A second class of receptor proteins that recognize short linear amino acid epitopes is the inhibitor of apoptosis proteins (IAPs), which provide protection for cells against diverse proapoptotic stimuli [63]. The IAPs were first identified in the baculovirus, where they suppress the cell death response during viral infection [64,65], and they were subsequently detected in both invertebrates and vertebrates [58,66–74]. The IAP family is characterized by the presence of the baculovirus IAP repeat (BIR) motif. BIRs are ~70-residue zinc-binding domains that bind to, and thereby inhibit, the caspase proteases that mediate apoptosis.

The ubiquitously expressed X-chromosome–linked IAP (X-IAP) contains three BIR domains; the second BIR domain (BIR2), together with the immediately preceding linker region, inhibits active caspase-3 and caspase-7 [75–79], while the third BIR domain (BIR3) is a specific inhibitor of caspase-9 activation [77,80,81]. Another interesting member of the IAP family is melanoma IAP (ML-IAP), a protein that is not detectable in most normal adult tissues but is strongly overexpressed in melanoma cells [82,83]. ML-IAP contains a single BIR domain and has also been shown to be a strong inhibitor of apoptosis [58,82–84]. Caspase-9 binds to BIR3 of X-IAP largely through interactions involving the N-terminus of the small subunit of caspase-9 (AVPT) [84–86]. Importantly, the exposed N-terminus of the small subunit of caspase-9 is homologous to the processed N-termini of natural antagonists of the IAPs, including the mammalian proteins Smac/DIABLO (AVPI) and HtrA2/Omi (AVPF). IAP antagonists promote apoptosis by releasing caspases from the BIR domains and thereby allowing the apoptotic cascade to proceed [63].

Phage display was used to investigate the sequence diversities that bind to BIR domains. Peptide libraries were sorted against two BIR domains from X-IAP and the BIR domain from ML-IAP [87]. Only the four extreme N-terminal positions show sequence consensus, indicating that only these positions are essential for recognition by the BIR domain. Significant observations are that alanine was exclusively selected at the N-terminus, and proline is highly favored in the third position for X-IAP BIR3 and ML-IAP BIR but not for BIR2. Another difference is the preference for glutamate in the second position of X-IAP BIR2 ligands, whereas the other BIR domains prefer small hydrophobic residues such as valine or isoleucine. The last difference is that ML-IAP and X-IAP BIR3 select aromatic amino acids in position 4, while X-IAP BIR2 prefers small hydrophobic residues in this position.

Several peptide sequences derived from the phage results were synthesized and assayed against three BIR domains with fascinating results; both ML-IAP and X-IAP BIR2 can tolerate an acidic group in the second position, while ML-IAP BIR cannot [87]. A combination of x-ray crystal structure analysis and homology modeling was used to explain the differences in selectivity between X-IAP BIR3 and ML-IAP BIR (Figure 7.10).

X-IAP BIR3 has an aspartate residue proximal to the position 2 binding site, which precludes the placement of a negatively charged residue in this site (Figure 7.10a). The aspartate is replaced by a serine in ML-IAP BIR, and this residue is capable of hydrogen bonding to a carboxylate group residing at position 2 of the ligand (Figure 7.10b). Another interesting observation is that ML-IAP BIR can more readily tolerate β-branched amino acids such as valine in position 3 than X-IAP BIR3. This difference is explained by the change of a phenylalanine in ML-IAP BIR to the more sterically demanding tyrosine in X-IAP BIR3.

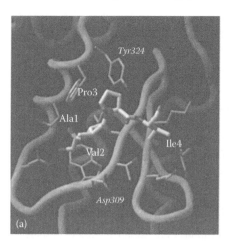 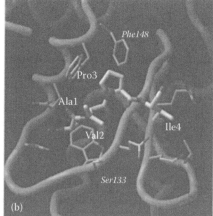

Figure 7.10   Structures of a peptide (AVPI) bound to (a) the BIR3 domain of X-IAP and (b) the BIR domain of ML-IAP. Note the two key differences in sequence that contribute to differences in specificity; Asp and Tyr residues in X-IAP BIR3 are substituted by Ser and Phe, respectively, in ML-IAP BIR. Peptide side chains are shown and labeled. Side chains on the BIR domains are labeled in italics.

The final position for selectivity pointed out by the phage display experiment is that both ML-IAP BIR and X-IAP BIR3 have deep, adaptive, binding pockets at position 4 and can therefore accommodate large aromatic amino acids. In contrast, position 4 of X-IAP BIR2 is better suited for small hydrophobic amino acids such as valine and isoleucine.

The exceptional feature of this application of peptide phage display is that one experiment reveals the roots of specificity between three closely related domains. Subsequently, binding data were combined with structural analysis to construct a detailed model of how BIR domains recognize contiguous linear epitopes. It is difficult to imagine another methodology that can so rapidly elucidate so many key features of a binding interface.

7.4 STRUCTURED LIBRARIES AND SMALL PROTEINS

Peptide phage display has been used to engineer binding interfaces within structured scaffolds, which has proven to be a productive approach for antagonizing protein–protein interfaces [88]. In contrast to naive peptide phage libraries that contain large fractions of unstructured or poorly structured peptides, libraries of well-characterized scaffolds organize the randomized amino acids of the library into well-defined structures. Although an organized binding surface will not be complementary to all targets, this approach yields peptide libraries that exhaustively probe the binding potential of specific shapes. Since the precise conformation of the binding surface is molded by amino acid composition, one anticipates that an adaptive scaffold may recognize a range of surfaces with high binding affinity.

7.5 BICYCLES—CHEMICALLY TETHERED PEPTIDE LOOPS

It has long been appreciated that high-affinity peptides selected to bind extracellular targets frequently contain an internal disulfide bond, presumably prepaying the entropic cost of stabilizing a flexible peptide into a single binding configuration. Libraries have been designed to explicitly take advantage of this property, with a pair of invariant cysteines encoded at fixed positions within peptide libraries of variable length. This strategy has been quite successful, yielding potent binders with diverse structures against a multitude of targets (Figure 7.1).

A recent approach, termed *bicycles*, extends the rigidification concept into chemical biology by physically linking three cysteines via a small molecule, tris-(bromomethyl)benzene (TBMB) [89]. In this approach, the three cysteines were first separated by two stretches of six variable positions ("6 × 6" libraries), affording two "loops" of diversity (Figure 7.11a). Since each loop is tethered to the mesitylene core, there is a potential for selection to generate either independent or coupled organization within these loops, as well as direct interactions with the target.

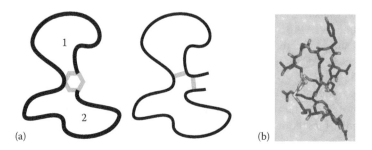

Figure 7.11   (a) Bicycles are covalently constrained peptides, with two loops extending beyond the cross-linking points. The cross-link can be formed via three cysteines and a small molecule, such as TBMB (left), or four cysteines forming two pairs of disulfides (right). (b) A crystal structure of a bicycle bound to urokinase-type plasminogen activator revealed two independent loops (black), each making extensive contacts to the target (gray) and centrally constrained by the small molecule core (white).

Selection and affinity maturation of the first generation of bicycle libraries yielded relatively potent and specific inhibitors of cathepsin G, plasma kallikrein, and urokinase-type plasminogen activator (uPA) [89,90]. These phage-selected peptides could be readily synthesized and cyclized with TBMB, and one such molecule was even crystallized in complex with uPA [90]. This structure revealed an extensive peptide–protease interface (~700 Å), with one loop blocking the uPA S1 pocket (Figure 7.11b). While each loop was physically tethered to the small molecule core, there were no other interactions with the mesitylene moiety [90], and NMR data showed that bicycles tend to be unstructured in solution [89]. Reselection of uPA-binding bicycles with alternate small molecule cores containing hydrogen bond donors and acceptors led to greatly enhanced intramolecular interactions but no profound increases in affinity. Alternate enthalpic contacts with uPA complicate a detailed analysis, but this result may suggest that the physical tethering of the peptide via TBMB is sufficient to constrain configurational space to prepay the majority of the entropic cost of binding.

Subsequent generations of bicycle libraries substantially varied the lengths of the inter–cross-link loops. One such library, optimized to inhibit human plasma kallikrein by shortening the loops to 3-mers or 5-mers ("3 × 3" or "5 × 5"), achieved cross-species nanomolar inhibition while sparing related proteases [91]. A more thorough exploration of bicycle loop lengths explored most combinations of $i \times j$ libraries, where i and j were loop combinations of 3, 4, 5, or 6 amino acids in length (e.g., 5 × 3 or 4 × 6), while selecting for affinity to uPA. This yielded greater sequence diversity among the selected clones, with a consensus apparent even between different loop lengths [92]. The extreme variety afforded by varying both loop length and composition suggests that the bicycle approach is quite robust, though it remains to be seen whether this format will be as successful in isolating binders to targets other than proteases.

A major drawback of the traditional bicycles is the use of the TBMB reagent for cyclization. This molecule's reactivity reduces phage infectivity, both directly via off-target lysine cross-linking and by necessitating the use of a nonoptimal disulfide-free g3 phage coat protein [89]. Incorporation of the TBMB reaction into a phage workflow is also cumbersome, requiring purification of phage and a carefully optimized reaction for each round of selection. To avoid these complications, the most recent generation of bicycles combined three constant cysteines with four distinct variable regions, relying on the lower infectivity of phage with an unpaired cysteine to preferentially evolve a fourth cysteine and form two disulfides [93].

This approach also explored variable loop lengths, and selecting against uPA yielded molecules with paratopes very similar to those observed with TBMB-linked bicycles. While the ease of selection sans TBMB cross-linking affords substantial advantages, alternate disulfide pairing can complicate the isolation of disulfide-linked bicycles. Since the evolved fourth cysteine can pair with any of the existing cysteines, each binder can potentially exist in three disulfide topologies. Local folding of the peptide may favor one topology over the others, but care must always be taken in purifying and characterizing disulfide-rich molecules.

7.6 SMALL PROTEINS—LIBRARIES BASED ON THE Z-DOMAIN SCAFFOLD

An early example of altering the binding specificity of a naturally occurring small protein is the randomization of the binding surface of the 58-amino-acid, three-helix Z-domain from staphylococcal protein A (introduced in Section 7.2.4, "Immunoglobulin G"). Structural analysis led to the selection of 13 crucial amino acids that make up the binding surface that runs across helices 1 and 2 for the construction of Z-domain phage library using NN (G/T) codons (Figure 7.12a).

Z-domain-based peptides with low nanomolar and subnanomolar affinities were obtained for targets such as taq DNA polymerase and the Her2 extracellular domain (ECD) [94]. Interestingly, analysis of x-ray structures of complexes show binding

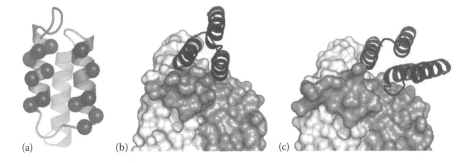

Figure 7.12 (a) Structure of the Z-domain of protein A with the 13 randomized positions shown as solid spheres. (b) Structure of the three helices of the Z-Domain bound to VEGF. (c) Structure of the dimer of two-helix "mini-Z" helices bound to VEGF. Note that one helix from each peptide makes contact with the binding site of VEGF. (The disulfide bound has been omitted for clarity.) (From Fedorova, A., *Chem. Biol.*, 18, 2011.)

surfaces in the range of 800–900 Å^2, which are primarily made up of hydrophobic amino acids. These results are in line with typical protein–protein interactions [95]. Combining two Z-domains in head-to-tail dimers led to a decrease in the off rate and higher affinity. Interestingly, the increase in affinity for Her2-expressing cells was more pronounced than isolated HER2 ECD due to avidity between the dimer and the multiple copies of Her2 expressed on the cell surface [96]. Further optimization of the Her2 binding Z-domain led to a monomer that binds to the Her2 ECD with subnanomolar affinity. These interesting reagents can be made via solid-phase peptide synthesis (SPPS) and have been modified for application, ranging from the specific delivery of chemotherapeutics to HER2-expressing tumors to radionuclide diagnostics [97]. A similar approach was taken in which the Her2 binding interface was grafted on the two-helix version of the Z-domain (mini-Z) introduced in Section 7.2.4, "Immunoglobulin G" [98]. Affinity was improved to a K_d of 5 nM through the selective inclusion of unnatural amino acids. Good-resolution images of Her2-expressing tumors were obtained using positron emission tomography (PET).

A similar approach was taken in the development of radionuclide imaging agents for VEGF-expressing tumors [99]. As previously described in Section 7.2.3, investigators at Genentech obtained two VEGF-binding peptides, v107 and v108, from sorting naive peptide libraries against VEGF. Although these peptides specifically bound to VEGF, the affinity was not sufficient for imaging applications. In order to obtain peptide agents with higher affinity, a phage library containing the 58-residue Z-domain scaffold with 9 residues associated with the predicted binding surface was sorted against VEGF, while an analogous selection was concurrently undertaken with the 38-residue two-helix version developed by Braisted and Wells (see Section 7.2.4, "Immunoglobulin G").

The 58-residue Z-domain library yielded binders that mutated on phage to 59-residue three-helix structures with K_d values in the low nanomolar range (Figure 7.12b). It has been postulated this is due to a subtle change in the registry of the first helix. X-ray crystal structure analyses show that the Z-domain-derived peptides bind to the same area as the natural receptors and v107 and v108, albeit with improved affinities compared to the other peptides. The naive libraries contain much smaller peptides that are apparently not capable of organizing residues into a binding surface as effective as the 744 Å^2 binding site produced on the Z-domain. The constant regions of the Z-domain constrain residues within the binding surface into a shape that can accommodate the VEGF-binding site. The facility with which the Z-domain can be muted to recognize a diverse set of targets is a strong statement to the utility of applying structured scaffolds for ligand discovery.

The 38-residue two-helix library yielded peptides that bound to the same binding site as the three-helix Z-domain but could only bind to VEGF as a dimer of two-helix mini-Z peptides. It has been postulated that the disulfide bond crucial for structuring the two helices of the mini-Z peptide does not allow the structure to adapt to the shape of VEGF-binding site, so that two mini-Z molecules must dimerize in order to arrange the helices into the VEGF-binding site (Figure 7.12c). In order to accomplish this, a new dimerization interface was selected that effectively places one helix from each monomer into the VEGF-binding pocket.

This result demonstrates the limits of preorganized binding interfaces. When amino acids are held in positions that are uncomplementary to the target binding site, it is implausible that any potent binders can be selected without a major change in scaffold structure. There must be a sufficient correspondence between binding surfaces to achieve serviceable potency. Where it may be possible to obtain binders, albeit with limited potency, against a wide range of targets with unstructured libraries, highly organized ligands will bind a smaller range of targets but potentially with higher affinity. It is a testament to the power of combinatorial biology that phage display found potent solutions to VEGF binding by altering properties of the original Z-domain in ways that are quite distinct from the less successful initial design.

7.7 SCAFFOLD LIBRARIES AGAINST D-PROTEINS

Peptide phage display was crucial in identifying VEGF-binding peptides based on the 56-amino-acid GB1 scaffold from streptococcal protein G that are constructed from unnatural amino acids. A GB1 library was constructed with 15 contiguous residues randomized using KHT codons that allows Y, A, D, S, F, and V. This GB1 library was sorted against VEGF synthesized from D amino acids using native chemical ligation [100]. This example highlights the use of limited-diversity libraries to cover large surface areas that would be fatally underrepresented if randomization were performed using a library that uses all 20 amino acids at each position. First introduced in antibody libraries by Fellouse and Sidhu [101], this approach has now been validated for peptide libraries. Once a lead sequence is obtained, affinity optimization is carried out to give potent binders.

The mirror image of the high-potency binder RFX001 found via phage display was synthesized from D amino acids using SPPS (Figure 7.13). It is important to stress that two peptides made with the amino acids of opposite stereochemistry will be mirror images.

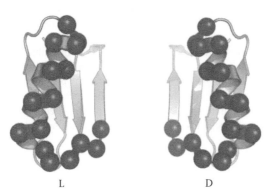

L   D

Figure 7.13 Structures of L-RFX001 (L) and D-RFX001 (D) shown with the randomized residues depicted as solid spheres. Note the mirror-image relationship of the peptides and binding sites.

As predicted by the diastereomeric relationship between the complexes, the affinities of D-RFX001 to the L-VEGF (K_d = 85 ± 12 nM) are analogous to L-RFX001 bound to D-VEGF (K_d = 95 ± 8 nM). The D enantiomers of RFX001 and VEGF bind only to the L isomer of VEGF and RFX001, respectively. The D ligand is predicted to be nonimmunogenic and resistant to proteolysis. If D-peptides have significantly improved stability and clearance compared to the corresponding L isomers, their potential as therapeutic agents will be greatly enhanced.

7.8 CONCLUSIONS

This chapter has highlighted the utility of phage-displayed peptide libraries for investigating protein–protein interactions. The advances in our ability to display peptide libraries on phage are accelerating in concert with the exploration of protein–protein interactions. These gains in technology will help us more efficiently harvest the information generated by the ongoing genomics and proteomics efforts. Phage display not only enables the rapid analysis of many of the thousands of proposed protein–protein interactions but also allows us to assess the potential of using small molecules to control these interactions.

REFERENCES

1. Sidhu SS, Fairbrother WJ, Deshayes K. Exploring protein–protein interactions with phage display. *ChemBioChem* 2003; 4(1):14–25.
2. Cochran AG. Antagonists of protein–protein interactions. *Chem Biol* 2000; 9:R85–R94.
3. Gordon NC et al. BAFF/BLyS receptor 3 comprises a minimal TNF receptor-like module that encodes a highly focused ligand-binding site. *Biochemistry* 2003; 42(20):5977–5983.
4. Lowman HB et al. Molecular mimics of insulin-like growth factor 1 (IGF-1) for inhibiting IGF-1: IGF-binding protein interactions. *Biochemistry* 1998; 37(25):8870–8878.
5. Dennis MS et al. Peptide exosite inhibitors of factor VIIa as anticoagulants. *Nature* 2000; 404(6777):465–470.
6. Nakamura GR et al. A novel family of hairpin peptides that inhibit IgE activity by binding to the high-affinity IgE receptor. *Biochemistry* 2001; 40(33):9828–9835.
7. Nakamura GR et al. Stable "zeta" peptides that act as potent antagonists of the high-affinity IgE receptor. *Proc Natl Acad Sci U S A* 2002; 99:1303–1308.
8. Skelton NJ et al. Amino acid determinants of β-hairpin conformation in erythropoietin receptor agonist peptides derived from a phage display library. *J Mol Biol* 2002; 316:1111–1125.
9. Deshayes K et al. Rapid identification of small binding motifs with high-throughput phage display: Discovery of peptidic antagonists of IGF-1 function. *Chem Biol* 2002; 9:495–505.
10. Livnah O et al. Functional mimicry of a protein hormone by a peptide agonist: The EPO receptor complex at 2.8 Å. *Science* 1996; 273(5274):464–471.
11. Livnah O et al. An antagonist peptide–EPO receptor complex suggests that receptor dimerization is not sufficient for activation. *Nat Struct Biol* 1998; 5:993–1003.

12. Eckert DM et al. Inhibiting HIV-1 entry: Discovery of D-peptide inhibitors that target the gp41 coiled-coil pocket. *Cell* 1999; 99(1):103–115.
13. DeLano WL et al. Convergent solutions to the binding at a protein–protein interface. *Science* 2000; 287:1279–1283.
14. Scherf T et al. A β-hairpin structure in a 13-mer peptide that binds α-bungarotoxin with high affinity and neutralizes its toxicity. *Proc Natl Acad Sci U S A* 2001; 98:6629–6634.
15. Pan B et al. Solution structure of a phage-derived peptide antagonist in complex with vascular endothelial growth factor. *J Mol Biol* 2002; 316:769–787.
16. Fairbrother WJ et al. Novel peptides selected to bind vascular endothelial growth factor target the receptor-binding site. *Biochemistry* 1998; 37(51):17754–17764.
17. de Vos AM, Ultsch MH, Kossiakoff AA. Human growth hormone and extracellular domain of its receptor: Crystal structure of the complex. *Science* 1992; 255:306–312.
18. Livnah O et al. Crystallographic evidence for preformed dimers of erythropoietin receptor before ligand activation. *Science* 1999; 283(5404):987–990.
19. Remy I, Wilson IA, Michnick SW. Erythropoietin receptor activation by a ligand-induced conformation change. *Science* 1999; 283(5404):990–993.
20. Schaffer ML et al. Complex with a phage display-derived peptide provides insight into the function of insulin-like growth factor I. *Biochemistry* 2003; 42:9324–9334.
21. De Meyts P. The structural basis of insulin and insulin-like growth factor-I receptor binding and negative co-operativity, and its relevance to mitogenic versus metabolic signalling. *Diabetologia* 1994; 37(suppl 2):S135–S148.
22. Ballard FJ et al. Effects of interactions between IGFBPs and IGFs on the plasma clearance and in vivo biological activities of IGFs and IGF analogs. *Growth Regul* 1993; 3(1):40–44.
23. Sato A et al. Three-dimensional structure of human insulin-like growth factor-I (IGF-I) determined by 1H-NMR and distance geometry. *Int J Peptide Protein Res* 1993; 41(5):433–440.
24. Cooke RM, Harvey TS, Campbell ID. Solution structure of human insulin-like growth factor 1: A nuclear magnetic resonance and restrained molecular dynamics study. *Biochemistry* 1991; 30(22):5484–5491.
25. Vajdos FF et al. Comprehensive functional maps of the antigen-binding site of an anti-ErbB2 antibody obtained with shotgun scanning mutagenesis. *J Mol Biol* 2002; 320:415–428.
26. Zeslawski W et al. The interaction of insulin-like growth factor-I with the N-terminal domain of IGFBP-5. *EMBO J* 2001; 20(14):3638–3644.
27. Vajdos FF et al. Crystal structure of human insulin-like growth factor-1: Detergent binding inhibits binding protein interactions. *Biochemistry* 2001; 40(37):11022–11029.
28. Folkman J. Angiogenesis in cancer, vascular, rheumatoid and other disease. *Nat Med* 1995; 1:27–31.
29. Wiesmann C et al. Crystal structure at 1.7 Å resolution of VEGF in complex with domain 2 of the Flt-1 receptor. *Cell* 1997; 91(5):695–704.
30. Keyt BA et al. The carboxyl-terminal domain (111–165) of vascular endothelial growth factor is critical for its mitogenic potency. *J Biol Chem* 1996; 271(13):7788–7795.
31. Muller YA et al. Vascular endothelial growth factor: Crystal structure and functional mapping of the kinase domain receptor binding site. *Proc Natl Acad Sci U S A* 1997; 94(14):7192–7197.
32. Fuh G et al. Requirements for binding and signaling of the kinase domain receptor for vascular endothelial growth factor. *J Biol Chem* 1998; 273(18):11197–11204.
33. Li B et al. Receptor-selective variants of human vascular endothelial growth factor. Generation and characterization. *J Biol Chem* 2000; 275(38):29823–29828.

34. Muller YA et al. VEGF and the Fab fragment of a humanized neutralizing antibody: Crystal structure of the complex at 2.4 Å resolution and mutational analysis of the interface. *Structure* 1998; 6(9):1153–1167.

35. Chen Y et al. Selection and analysis of an optimized anti-VEGF antibody: Crystal structure of an affinity-matured Fab in complex with antigen. *J Mol Biol* 1999; 293(4):865–881.

36. Wiesmann C et al. Crystal structure of the complex between VEGF and a receptor-blocking peptide. *Biochemistry* 1998; 37(51):17765–17772.

37. Jones S, Thornton JM. Protein–protein interactions: A review of protein dimer structures. *Prog Biophys Mol Biol* 1995; 63:31–65.

38. Davies DR, Padlan EA, Sheriff S. Antibody–antigen complexes. *Annu Rev Biochem* 1990; 59:439–473.

39. Deisenhofer J. Crystallographic refinement and atomic models of a human Fc fragment and its complex with fragment B of protein A from *Staphylococcus aureus* at 2.9- and 2.8-Å resolution. *Biochemistry* 1981; 20(9):2361–2370.

40. Sauer-Eriksson AE et al. Crystal structure of the C2 fragment of streptococcal protein G in complex with the Fc domain of human IgG. *Structure* 1995; 3(3):265–278.

41. Burmeister WP, Huber AH, Bjorkman PJ. Crystal structure of the complex of rat neonatal Fc receptor with Fc. *Nature* 1994; 372(6504):379–383.

42. Corper AL et al. Structure of human IgM rheumatoid factor Fab bound to its autoantigen IgG Fc reveals a novel topology of antibody–antigen interaction. *Nat Struct Biol* 1997; 4(5):374–381.

43. Starovasnik MA, Braisted AC, Wells JA. Structural mimicry of a native protein by a minimized binding domain. *Proc Natl Acad Sci U S A* 1997; 94(19):10080–10085.

44. Nemerson Y. Tissue factor and hemostasis. *Blood* 1988; 71(1):1–8.

45. Skrzypczak-Jankun E. Structure of hirugen and hirulog 1 complexes of alpha-thrombin. *J Mol Biol* 1991; 221:1379–1393.

46. Stubbs MT, Bode WA. A player of many parts: The spotlight falls on thrombin's structure. *Thrombin Res* 1993; 69:1–58.

47. Himanen JP et al. Crystal structure of an Eph receptor–ephrin complex. *Nature* 2001; 414(6866):933–938.

48. Ren R et al. Identification of a ten-amino acid proline-rich SH3 binding site. *Science* 1993; 259:1157–1161.

49. Chen HI, Sudol M. The WW domain of Yes-associated protein binds a proline-rich ligand that differs from the consensus established for Src homology 3-binding modules. *Proc Natl Acad Sci U S A* 1995; 92:7819–7823.

50. Songyang Z et al. SH2 domains recognize specific phosphopeptide sequences. *Cell* 1993; 72:767–778.

51. Kavanaugh WM, Williams LT. An alternative to SH2 domains for binding tyrosine-phosphorylated proteins. *Science* 1994; 266:1862–1865.

52. Cowburn D. Peptide recognition by PTB and PDZ domains. *Curr Opin Struct Biol* 1997; 7:835–838.

53. Skelton NJ et al. Origins of PDZ domain ligand specificity. Structure determination and mutagenesis of the Erbin PDZ domain. *J Biol Chem* 2003; 278(9):7645–7654.

54. Tomita S, Nicoll RA, Bredt DS. PDZ protein interactions regulating glutamate receptor function and plasticity. *J Cell Biol* 2001; 153:F19–F23.

55. Fanning AS, Anderson JM. Protein modules as organizers of membrane structure. *Curr Opin Cell Biol* 1999; 11:432–439.

56. Fuh G et al. Analysis of PDZ domain–ligand interactions using carboxyl-terminal phage display. *J Biol Chem* 2000; 275:21486–21491.

57. Laura RP et al. The Erbin PDZ domain binds with high affinity and specificity to the carboxyl termini of δ-catenin and ARVCF. *J Biol Chem* 2002; 277:12906–12914.

58. Borg J-P et al. ERBIN: A basolateral PDZ protein that interacts with the mammalian ERBB2/HER2 receptor. *Nat Cell Biol* 2000; 2:407–414.

59. Klapper LN et al. Biochemical and clinical implications of the ErbB/HER signaling network of growth factor receptors. *Adv Cancer Res* 2000; 77:25–79.

60. Lu Q et al. Brain armadillo protein delta-catenin interacts with Abl tyrosine kinase and modulates cellular morphogenesis in response to growth factors. *J Neurosci Res* 2002; 67:618–624.

61. Fraser PE et al. Presenelin function: Connections to Alzheimer's disease and signal transduction. *Biochem Soc Symp* 2001; 67:89–100.

62. Shelly M et al. Polar expression of ErbB-2/HER2 in epithelia: Bimodal regulation by Lin-7. *Dev Cell* 2003; 5:475–486.

63. Salvesen GS, Duckett CS. IAP proteins: Blocking the road to death's door. *Nat Rev Mol Cell Biol* 2002; 3(6):401–410.

64. Crook NE, Clem RJ, Miller LK. An apoptosis-inhibiting baculovirus gene with a zinc finger-like motif. *J Virol* 1993; 67(4):2168–2174.

65. Birnbaum MJ, Clem RJ, Miller LK. An apoptosis-inhibiting gene from a nuclear poly-hedrosis virus encoding a polypeptide with Cys/His sequence motifs. *J Virol* 1994; 68(4):2521–2528.

66. Rothe M et al. The TNFR2–TRAF signaling complex contains two novel proteins related to baculoviral inhibitor of apoptosis proteins. *Cell* 1995; 83(7):1243–1252.

67. Roy N et al. The gene for neuronal apoptosis inhibitory protein is partially deleted in individuals with spinal muscular atrophy. *Cell* 1995; 80(1):167–178.

68. Hay BA, Wassarman DA, Rubin GM. *Drosophila* homologs of baculovirus inhibitor of apoptosis proteins function to block cell death. *Cell* 1995; 33(7):1253–1262.

69. Duckett CS et al. A conserved family of cellular genes related to the baculovirus *iap* gene and encoding apoptosis inhibitors. *EMBO J* 1996; 15(11):2685–2694.

70. Liston P et al. Suppression of apoptosis in mammalian cells by NAIP and a related family of IAP genes. *Nature* 1996; 379(6563):349–353.

71. Uren AG et al. Cloning and expression of apoptosis inhibitory protein homologs that function to inhibit apoptosis and/or bind tumor necrosis factor receptor-associated factors. *Proc Natl Acad Sci U S A* 1996; 93(10):4974–4978.

72. Ambrosini G, Adida C, Altieri DC. Activation-dependent exposure of the inter-EGF sequence Leu83–Leu88 in factor Xa mediates ligand binding to effector cell protease receptor-1. *J Biol Chem* 1997; 272(13):8340–8345.

73. Fraser AG et al. *Caenorhabditis elegans* inhibitor of apoptosis protein (IAP) homologue BIR-1 plays a conserved role in cytokinesis. *Curr Biol* 1999; 9(6):292–301.

74. Uren AG et al. Role for yeast inhibitor of apoptosis (IAP)-like proteins in cell division. *Proc Natl Acad Sci U S A* 1999; 96(18):10170–10175.

75. Takahashi R et al. A single BIR domain of XIAP sufficient for inhibiting caspases. *J Biol Chem* 1998; 273(14):7787–7790.

76. Sun C et al. NMR structure and mutagenesis of the inhibitor-of-apoptosis protein XIAP. *Nature* 1999; 401(6755):818–822.

77. Chai J et al. Structural basis of caspase-7 inhibition by XIAP. *Cell* 2001; 104(5):769–780.

78. Huang Y et al. Structural basis of caspase inhibition by XIAP: Differential roles of the linker versus the BIR domain. *Cell* 2001; 104(5):781–790.

79. Riedl SJ et al. Structural basis for the inhibition of caspase-3 by XIAP. *Cell* 2001; 104(5):791–800.

80. Deveraux QL et al. Cleavage of human inhibitor of apoptosis protein XIAP results in fragments with distinct specificities for caspases. *EMBO J* 1999; 18(19):5242–5251.

81. Sun C et al. NMR structure and mutagenesis of the third Bir domain of the inhibitor of apoptosis protein XIAP. *J Biol Chem* 2000; 275(43):33777–33781.

82. Vucic D et al. ML-IAP, a novel inhibitor of apoptosis that is preferentially expressed in human melanomas. *Curr Biol* 2000; 10(12):1359–1366.

83. Kasof GM, Gomes BC. Livin, a novel inhibitor of apoptosis protein family member. *J Biol Chem* 2001; 276(5):3238–3246.

84. Vucic D et al. SMAC negatively regulates the anti-apoptotic activity of melanoma inhibitor of apoptosis (ML-IAP). *J Biol Chem* 2002; 277(14):12275–12279.

85. Srinivasula SM et al. A conserved XIAP-interaction motif in caspase-9 and Smac/DIABLO regulates caspase activity and apoptosis. *Nature* 2001; 410(6824):112–116.

86. Shiozaki EN et al. Mechanism of XIAP-mediated inhibition of caspase-9. *Mol Cell* 2003; 11(2):519–527.

87. Franklin MC et al. Structure and function analysis of peptide antagonists of melanoma inhibitor of apoptosis (ML-IAP). *Biochemistry* 2003; 42(27):8223–8231.

88. Weidle U et al. The emerging role of new protein scaffold-based agents for treatment of cancer. *Cancer Genomics Proteomics* 2013; 10:155–168.

89. Heinis C et al. Phage-encoded combinatorial chemical libraries based on bicyclic peptides. *Nat Chem Biol* 2009; 5:502–507.

90. Angelini A et al. Bicyclic peptide inhibitor reveals large contact interface with a protease target. *ACS Chem Biol* 2012; 7:817–821.

91. Baeriswyl V et al. Bicyclic peptides with optimized ring size inhibit human plasma kallikrein and its orthologues while sparing paralogous proteases. *ChemMedChem* 2012; 7:1173–1176.

92. Rebollo IR, Angelini A, Heinis C. Phage display libraries of differently sized bicyclic peptides. *MedChemComm* 2013; 4:145–150.

93. Rentero Rebollo I et al. Bicyclic peptide ligands pulled out of cysteine-rich peptide libraries. *J Am Chem Soc* 2013; 135:6562–6569.

94. Nord K et al. Ligands selected from combinatorial libraries of protein A for use in affinity capture of apolipoprotein A-1 M and Taq DNA polymerase. *J Biotech* 2000; 80:45–54.

95. Nygren P. Alternative binding proteins: Affibody binding proteins developed from a small three-helix bundle scaffold. *FEBS J* 2008; 275:2668–2676.

96. Friedman M et al. Phage display selection of Affibody molecules with specific binding to the extracellular domain of the epidermal growth factor receptor. *Protein Eng Des Sel* 2007; 20:189–199.

97. Orlova A et al. Tumor imaging using a picomolar affinity HER2 binding affibody molecule. *Cancer Res* 2006; 66:4339–4348.

98. Webster J et al. Engineered two-helix small proteins for molecular recognition. *ChemBioChem* 2009; 10:1293–1296.

99. Fedorova A et al. The development of peptide-based tools for the analysis of angiogenesis. *Chem Biol* 2011; 18:839–845.

100. Mandal K et al. Chemical synthesis and X-ray structure of a heterochiral {D-protein antagonist plus vascular endothelial growth factor} protein complex by racemic crystallography. *Proc Natl Acad Sci U S A* 2012; 109:14779–14784.

101. Fellouse F, Weisemann C, Sidhu S. Synthetic antibodies from a four-amino-acid code: A dominant role for tyrosine in antigen recognition. *Proc Natl Acad Sci U S A* 2004; 101:12467–12472.

CHAPTER 8

Substrate Phage Display

Shuichi Ohkubo

CONTENTS

8.1 OVERVIEW

"Substrate phage display" is a powerful application that makes it possible to screen substrate sequences of enzymes from large and diverse collections of randomized sequences without any initial substrate data. The sensitivity and versatility of this technique have been clearly established through the discrimination of the substrate specificity of closely related proteases or protein tyrosine kinases. The information obtained from substrate phage display has substantially improved our understanding of substrate recognition in catalysis and signal transduction. This chapter highlights recent advances in the field of substrate phage display and illustrates its utility in cancer research.

8.2 INTRODUCTION

Dysregulation of specific enzymes, such as proteases, protein kinases, and protein phosphatases, has been implicated in various pathological conditions, especially malignancies. Therefore, much effort has been directed towards developing potent inhibitors of specific enzymes for cancer chemoprevention. However, many of these compounds are broad-spectrum inhibitors that show undesirable side effects among these closely related enzymes. Thus, there is still a considerable need for more selective inhibitors and for new and high-quality treatments. A better understanding of substrate specificities of these enzymes may significantly improve our overall knowledge about these enzymes, and this will facilitate the design and optimization of potent and selective inhibitors. In recent years, high-throughput screening (HTS), where hundreds of thousands of compounds can be tested for activity during a short period, has been increasingly used to discover novel lead candidate molecules. A key step in establishing an HTS format is to identify a highly selective substrate for use in enzyme assays. A basic understanding of substrate preferences allows further clarification of the physiological roles of these enzymes.

Traditionally, the substrate specificities of enzymes have been studied using synthetic peptides corresponding to sequences derived from known substrate proteins. However, in the cases of proteases and protein tyrosine kinases it has been shown that the sequences derived from natural substrates are not always optimal for these enzymes when the catalytic activities are studied in vitro with synthetic peptides [1–4]. Moreover, this approach does not permit the identification of novel substrate sequences and thereby new target molecules, which remain largely unknown.

"Substrate phage display" is a powerful application of the phage display technique, which enables us to screen substrate sequences of enzymes from large, diverse collections of randomized sequences without any initial substrate data [5–7]. In contrast to other combinatorial approaches, substrate phage display has an added advantage of being able to generate a vast number of possible combinations, thereby enabling rapid library construction and substrate optimization in a cost-effective manner [5,6,8–10]. Substrate phage display can be used to identify the substrate sequences of proteases [5], protein serine/threonine kinases [11], and protein tyrosine kinases [10]. In this chapter, we summarize the basic concepts of substrate phage display and discuss how they might be used to facilitate cancer research.

8.3 THE CONCEPT OF SUBSTRATE PHAGE DISPLAY

The concept of substrate phage display has been aptly described by Gram [7], as follows:

> In general terms, the substrate phage concept is based upon the principle that a phage library of potential substrate peptides is subjected to modification by an enzyme, followed by selection of those phage displaying a modified or catalytically processed peptide.

Substrate phage display was first introduced by Matthews and Wells [5] to identify substrates for various proteases, including subtilisin BPN′ mutant, factor Xa, and HIV protease. Since then, this technique has been successfully used to identify substrates for several types of proteases and protein kinases, as described in Sections 8.3.1 and 8.3.2.

8.3.1 Screen of Substrate Sequences for Proteases

Substrate phage display has been used to identify substrates for several classes of endopeptidases, including serine protease, aspartyl protease, and metalloprotease. Table 8.1 summarizes the various protease substrates and their cleavage sites that have been identified by substrate phage display to date [2–5,9,12–30].

Two strategies, monovalent and multivalent display, have been used for the selection and optimization of substrates for proteolytic enzymes [5,9]. In either case, gene *III* of filamentous phage is modified such that a randomized sequence is fused to the N-terminus of protein III (pIII) and an additional affinity tag sequence that enables attachment of the phage to an immobile phase is fused to N-terminus of the random sequence (Figure 8.1). As shown in Figure 8.2, this phage library is incubated with a protease, and uncleaved phage can be excluded from the solution by affinity binding. Uncaptured phage carry a cleavable substrate site within the randomized sequence that can easily be recovered and subjected to repeated selection and amplification cycles. In another approach, the phage library is first allowed to bind to a solid support via the affinity sequence. The phage are then subjected to proteolysis to release those phage that express peptide sequences that are susceptible to the enzyme. The released phage are then amplified, bound, and cleaved again.

Protein or peptide epitopes, or histidine or FLAG tags, have been successfully used as affinity tags to identify protease substrate sequences [5,9,16,18,27]. In order to reduce the frequency of false-positive clones during the selection, a high binding affinity between the tag and its corresponding matrix is desired; for example, a phage that included a histidine tag sequence at the N-terminus of pIII and exhibited high-affinity binding for nickel-nitrilotriacetic acid (Ni-NTA), with an equilibrium binding constant (K_D) of 12 nM [21].

8.3.2 Screen of Substrate Sequences for Protein Kinases

Westendorf et al. [11] used a different phage display approach to identify substrates of protein kinases. These investigators isolated phage clones that can be phosphorylated by partially purified protein serine/threonine kinases. This approach has also been used to determine the substrate specificity of protein tyrosine kinases [10]. Table 8.2 provides a summary of the various kinase substrates and their phosphorylation sites that have been identified to date [10,11,31,32].

The insertion of randomized sequences into pIII of filamentous phage is the most common method of making phage display libraries [6]. In this screen, gene *III* is also modified such that a randomized sequence is expressed at the N-terminus of pIII. A schematic overview of a selection for kinase substrates is shown in Figure 8.3. To identify and characterize the substrate sequences of protein kinases, a phage library

Table 8.1   Substrate Preferences of Proteases Identified Using Substrate Phage
Display

	Protease	Substrate Sequence	References
Serine protease	Subtilisin BPN′ mutant S24C/H64A/E156S/G166A/G169A/Y217L	TSM↓HT	[5]
	Subtilisin BPN′ mutant N62D/G166D	GNLMRK↓G	[12]
	Human factor Xa	(G/A/T/F)R↓	[5]
	Mouse furin	(L/P)RRF(K/R)↓RP	[13]
	Human tissue-type plasminogen activator (t-PA)	GGSGPFGR↓SALVPEE	[2]
		FRGR↓K[a]	[14]
	Human urokinase-type plasminogen activator (u-PA)	GSGK↓S[b]	[15]
	HSV-1 protease	LVLA↓SSSF	[16]
	Mouse tryptase	SLSSR↓QSP	[17]
	Rat granzyme B	IEXD↓XG	[18]
	Human prostate specific antigen (PSA)	SS(Y/F)Y↓S(G/S)	[3]
	Human elastase mutant H57A	MEHV↓VY	[19]
	Human plasmin	GIYR↓SR	[4]
	Human membrane-type serine protease 1	(R/K)XSR↓A	[20]
		X(R/K)SR↓A	
	Human α-thrombin	PR↓G	[21]
		GR↓	
		R↓G	
	Human kallikrein 2 (hK2)	LR↓SRA	[22]
	Rat mast cell protease 4(rMCP-4)	LVWF↓RG	[23]
	Staphylococcus aureus signal peptidase SpsB	LPASLPSF	[24]
Aspartyl protease	HIV protease	SQNYPIVQ	[5]
		GSGIF↓LETSL	[25]
Metalloprotease	Humangelatinase A(MMP-2)	PXX↓X$_{Hy}$	[26]
		(L/I)XX↓X$_{Hy}$	
		X$_{Hy}$SX↓L	
		HXX↓X$_{Hy}$	
	Stromelysin 1 (MMP-3)	PFE↓LRA	[9]
	Matrilysin (MMP-7)	PLE↓LRA	[9]
	Human gelatinase B (MMP-9)	PR(S/T)↓X$_{Hy}$(S/T)	[27]
	Human collagenase 3 (MMP-13)	GPLG↓MRGL	[28]
	Human membrane type-1 matrix metalloproteinase (MT1-MMP, MMP-14)	PX(G/P)↓L	[29]
		RIGF↓LRTA[c]	[30]

Note: ↓: site of digestion; X$_{Hy}$: hydrophobic residue.
[a]   Minimized, t-PA selective sequence.
[b]   Minimized, u-PA selective sequence.
[c]   One of the highly selective sequences.

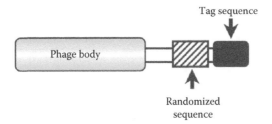

Tag sequence

Phage body

Randomized
sequence

Figure 8.1 Schematic representation of phage particles used to construct a substrate phage library to identify substrates for proteases. For simplicity, this schematic representation of the phage particle shows only one pIII on the phage body. The substrate phage library consists of the affinity tag sequence and the protease target randomized sequence at the N-terminus of pIII.

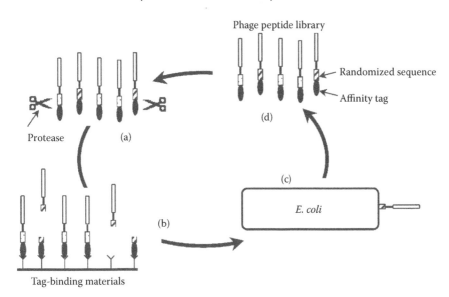

Phage peptide library

Randomized sequence

Affinity tag

(d)

Protease   (a)

(c)

E. coli

(b)

Tag-binding materials

Figure 8.2 Schematic illustration of the use of substrate phage selection to select substrates for proteases. (a) Phage displaying randomized substrates and an affinity tag fused to pIII are incubated with the protease of interest. (b) The uncleaved (non-substrate) phage is captured by tag-binding materials. (c) The phage displaying substrates susceptible to protease cleavage are recovered and amplified in *Escherichia coli*. (d) Either individual clones are subjected to DNA sequencing or the amplified phage are used in a subsequent round of selection.

is incubated in the presence of protein kinases. Phages expressing peptides that can be efficiently phosphorylated are selected by incubation with an antibody specific for epitopes containing a certain phosphoamino acid [11] or with antiphosphotyrosine antibodies [10,31,32], followed by multiple rounds of conventional selection and amplification. After the final selection, individual clones are isolated, and the substrate peptide sequences are determined by sequencing the relevant portion of the phage DNA.

Table 8.2 Substrate Preferences of Protein Kinases Identified Using Substrate Phage Display

Kinase		Substrate Sequence	References
Serine/threonine kinase	Human partial purified M-phase kinases[a]	LTPLK[b]	[11]
Tyrosine kinase	Bovine Fyn	E(X_{Hy}/T) **Y** GXX_{Hy}	[31]
	Human c-Src	(D/E)X(I/L) **Y**(G/W)X(F/W)X	[10]
	Human Blk	XXI **Y** (D/E)XLP	[10]
	Human Lyn	(D/E)X(I/L) **Y** (D/E)XLP	[10]
	Human Syk	XXD **Y** EXXX	[10]
	Human Tie-2	RLVA **Y** EGWV	[32]

Note: **T**: phosphorylated threonine residue; **X**_{Hy}: hydrophobic residue. **Y**: phosphorylated tyrosine residue.
[a] Partial purified kinase-containing fraction prepared from M-phase arrested HeLa cells.
[b] Selected by MPM2 antibody, which can recognize a phospho amino acid-containing epitope of M-phase phosphoproteins.

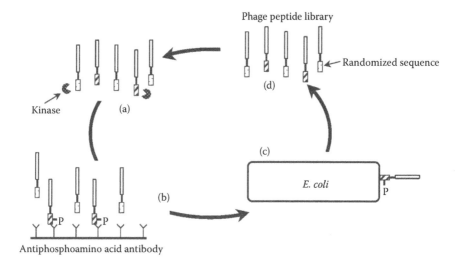

Figure 8.3 Schematic illustration of the use of substrate phage selection to select substrates for protein kinases. (a) Phage displaying randomized substrates are incubated with the protein kinase of interest. (b) The phosphorylated phage is captured by an antiphosphoamino acid antibody, such as antiphosphotyrosine. (c) The phage displaying specific substrates are recovered and amplified in *Escherichia coli*. (d) Individual clones are either subjected to DNA sequencing or the amplified phage is used for a subsequent round of selection.

8.3.3 Characterization of Proteolysis Reaction by Analysis of Substrate Phage

Matthews et al. [13] developed a method to identify the true positive clones among the selected phage and to compare the relative rates at which the isolated sequences are hydrolyzed. Fusion proteins were prepared that contained an affinity peptide tag sequence, the substrate sequence obtained from the phage screen, and an alkaline phosphatase. These fusion proteins were then immobilized onto the affinity tag-binding protein, and the time course of substrate hydrolysis was followed by monitoring the activity of alkaline phosphatase released after incubation with protease. In a similar procedure, Cloutier et al. [22] used cyan fluorescent protein (CFP), a variant of the green fluorescent protein, in place of alkaline phosphatase, and the time course of substrate hydrolysis was followed by monitoring the released fluorescence.

Smith et al. [9] developed another simple and rapid method to compare relative rates of hydrolysis of the isolated sequences. Isolated phage clones were incubated in a solution containing the protease of interest, and the mixture was then spotted onto nitrocellulose membranes. The membranes were then probed with an antibody directed against the affinity tag. If the tag had been hydrolytically removed from the phage, the antigen would not be retained on the membrane. In this way, the loss of the tag sequence from the phage by proteolytic digestion could be monitored, and the efficacy of cleavage of the substrate sequence could thereby be determined. Using this dot blot assay, Smith et al. were able to compare quantitatively the cleavage rates of the selected sequences. By monitoring the time-dependent loss of the tag sequence from the phage during proteolytic digestion, the relative concentration of the phage retaining the tag sequence {P} was determined at different time points. A set of slopes corresponding to first-order decay rates was obtained by plotting log{P} versus time for each of the isolated clones. Thus, these investigators concluded that the relative *kcat/Km* values of the individual clones could be determined by comparing the decay rates.

Sharkov et al. [33] monitored the hydrolysis of individual phage substrates by an ELISA. Isolated phage were incubated with protease and placed in microtiter plate wells, which were coated with an antitag antibody. The captured phage were then detected with an antiphage antibody. Sharkov et al. characterized the reaction kinetics of the stromelysin protease in this way. The phage can also be first immobilized in microtiter plate wells and then cleaved by the protease, and the remaining tag sequence can be detected by an antitag antibody [27]. In general, the enzyme is present in excess as compared to the substrate during proteolytic digestion of substrate phage [6,33], resulting in a single-turnover reaction, in which the equations for steady-state kinetics are not considered to be valid. Sharkov et al. also showed that the proteolysis of substrate phage was a single-exponential process, and single-exponential rate constants appeared linear with respect to the enzyme concentration throughout the range examined. Thus, these investigators concluded that the reaction between an enzyme and its substrate phage obeys the rules of pseudo-first-order kinetics, suggesting that time of incubation and the amount of enzyme are critical factors in designing substrate phage selection experiments. They further suggested that few substrates would be found if the selection conditions were too stringent, that is, if the incubation time was too short or the enzyme concentration too low. On the other hand, this technique would offer little discrimination among good substrates if the conditions were too relaxed [33]. Thus, it is critical to optimize the conditions of substrate phage selection when using this approach to identify specific substrates.

8.3.4 Subtracted Substrate Phage Library

Tissue-type plasminogen activator (t-PA) and urokinase-type plasminogen activator (u-PA) are members of the chymotrypsin gene family that share a high degree of structural similarity [34,35]. A highly efficient substrate sequence for t-PA was identified by substrate phage display and was found to be cleaved up to 5300 times more efficiently by t-PA than was the amino acid sequence of the actual cleavage site present in plasminogen [2,14]. However, this newly identified sequence was also cleaved efficiently by u-PA, and the selectivity of this substrate was only 4.7-fold higher for t-PA than for u-PA [2,14]. To select substrates specific for t-PA, Ke and colleagues [14] developed a novel protocol, which included a subtraction step during the phage selection, that removed substrates that were cleaved efficiently by u-PA. As a first step, a random substrate phage library was subjected to high stringency selection with t-PA, to generate an intermediate library enriched in phage that are efficient substrates of t-PA. This intermediate library was then digested at low stringency with u-PA to remove phage that were moderate or good substrates for u-PA. The phage that remained undigested with u-PA were defined as a subtracted substrate phage library, and the phage specific for t-PA were recovered by digesting the subtracted substrate phage library with t-PA. One of the highly selective substrates isolated from the subtracted substrate phage library showed 47-fold higher catalytic efficiency for t-PA than that for u-PA (Q-R-G-R-K-A at the P4–P2′ sites*). Comparison of the amino acid sequences of substrates, derived from the subtracted substrate phage library and the standard substrate phage library, suggested that the P3 and P4 residues play a critical role in determining the specificity between t-PA and u-PA for a given substrate. Whereas plasminogen activator inhibitor type I (PAI-I) is a primary physiological inhibitor of both enzymes, mutation of valine to glutamine at P4 and serine to arginine at P3 enhanced the specificity of PAI-I for t-PA by approximately 600-fold [14].

8.4 APPLICATION OF SUBSTRATE PHAGE DISPLAY TO CANCER RESEARCH

8.4.1 Angiogenesis

Angiogenesis is the biological process by which new capillaries are formed from preexisting blood vessels [36]. This occurs under both physiological and pathological conditions. For example, the transition from an avascular to a vascularized state is considered to be a critical turning point in the development of tumors. Vascularization provides oxygen and essential nutrients to the tumor and increases the proliferation rate of cancer cells. It has been established that a tumor mass cannot exceed a size of ~1 mm^3 in an avascular state [37,38]. Even though virtually every

---

* The nomenclature for the substrate amino acid preference is Pn, Pn − 1,..., P2, P1, P1′, P2′,..., Pm − 1′, Pm′. Amide bond hydrolysis occurs between P1 and P1′.

cancer exhibits a slightly different phenotype, the tumor endothelium is relatively uniform in all solid tumors. Thus, angiogenesis is regarded as a common and key target for cancer chemopreventive agents.

Tumor angiogenesis depends mainly on the release of specific growth factors from neoplastic cells. These growth factors bind to receptor tyrosine kinases (RTKs) expressed on the endothelial cell surface. Binding of growth factors leads to the phosphorylation and activation of RTK, which eventually results in endothelial cell recruitment and proliferation [36,39]. Expression of vascular endothelial growth factor (VEGF) is widely induced during angiogenesis, rendering it a prime target for antivascular therapy [40]. VEGF has five isoforms that exist as homodimers and bind to the fms-like tyrosine kinase (flt-1) and fetal liver kinase (flk-1) receptors. The major known physiological function of VEGF is to promote angiogenesis in response to hypoxia. Several inhibitors of VEGF and VEGF receptors have now reached the stage of clinical trials [36,39]. In addition, Tie-2 (tyrosine kinase with immunoglobulin and epidermal growth factor homology domain) is an endothelium-specific RTK that binds the angiopoietin ligands and plays an indispensable role in vascular remodeling and maturation [41]. Angiopoietins seem to function in a complementary and coordinated fashion with VEGF [42]. It has been reported that increased Tie-2 expression is associated with the growth of several types of tumors [43–45], and the growth of experimental tumors could be inhibited by blocking the Tie-2 pathway [46–48]. These data suggest that inhibitors of Tie-2 also may be useful as antiangiogenic cancer drugs.

Using substrate phage display, Deng et al. [32] have identified the substrate sequences for Tie-2 to develop high-throughput screens for Tie-2 inhibitors. The phage libraries consisted of sequences in the X-X-X-Y-X-X-X-X or X-X-X-X-Y-X-X-X-X motifs, where Y represents tyrosine and X represents a random mixture of all 20 amino acids. The peptide library was incubated with the catalytic domain of Tie-2 and phosphorylated phage particles were captured with an antiphosphotyrosine antibody, and the peptide sequences were analyzed. Amongst four identified substrate sequences, R-L-V-A-Y-E-G-W-V exhibited the best catalytic efficiency, with a $kcat/Km$ of 5.9×10^4 M^{-1} s^{-1}. This activity was sufficient to develop a number of HTS assays, including dissociation-enhanced lanthanide fluoroimmunoassays (DELFIAs), radioactive plate binding (RPB), and time-resolved fluorescent resonance energy transfer (TR-FRET). Automated DELFIA and TR-FRET assays were later developed, which have been successfully used to screen a combined set of >600,000 small organic molecules against Tie-2 [32].

8.4.2 Prognostic Markers for Prostate Cancer

The human kallikrein family of serine proteases now includes 15 family members that share significant homology [49,50]. Human glandular kallikrein 3 (hK3), or prostate-specific antigen (PSA), is considered to be a prognostic marker for prostate cancer, and serum PSA levels are used for screening and early detection of prostate cancer [51–53]. In addition to PSA, human kallikrein 2 (hK2) has recently emerged as a complementary marker for detecting prostate cancer, especially at low PSA values [54–56]. PSA and hK2 are the most highly homologous members of the human kallikrein family, with 78% and 80% identity at the amino acid and DNA levels, respectively [50]. Whereas hK2 has a trypsin-like specificity for arginine and lysine, PSA more closely resembles other tissue kallikreins, exhibiting a chymotrypsin-like preference for tyrosine and leucine [57,58]. PSA hydrolyzes semenogelin I, semenogelin II, and fibronectin in vivo and plays a role of semen liquefaction, a biological process that immediately follows ejaculation [59–62]. A number of other potential substrates for PSA have been identified, including transforming growth factor beta (TGF-β) [63], parathyroid hormone-related protein [64] and insulin-like growth factor binding proteins (IGF-BPs) [65]. hK2 has also been shown to play a possible role in the early stage of semen liquefaction and to have hydrolyzing activity towards fibronectin and semenogelins [66]. In addition, hK2 can activate u-PA [67], inactivate PAI-I [68], and cleave IGF-BPs [65]. However, a role of these enzymes in cancer development has not yet been clearly defined. Characterization of the substrate specificity of PSA and hK2 may shed some light on their roles during tumor progression and on additional physiological functions of these enzymes. Moreover, sensitive activity-based assays of these enzymes are useful for monitoring prostate cancer diagnosis.

By using two independent approaches, substrate phage display and iterative optimization of the synthetic peptides derived from native substrate sequences of semenogelin, a consensus substrate sequence for PSA was defined as S-S-(Y/F)-Y-S-G at the P4–P2′ sites [3]. These sequences were cleaved by PSA with catalytic efficiencies ($kcat/Km$) as high as 2200–3100 M^{-1} s^{-1}, as compared with values of 2–46 M^{-1} s^{-1} for peptides derived from the physiological target sequences of semenogelin. The optimized consensus substrate sequence (S-S-Y-Y-S-G at the P4–P2′ sites) does not fit into the known structural model for PSA, which has been deposited in the Protein Data Bank (1PFA.PDB). Thus, a new three-dimensional model for PSA was reconstructed based on the known structure of porcine tissue kallikrein [3]. This new model indicates that a tyrosine residue at P1 is preferred and suggests that the P2 residue of a substrate or inhibitor can be docked into a pocket formed by a large insertion loop.

The substrate specificity of hK2 was also characterized using a random pentapeptide phage library [22]. After eight rounds of selection, genes encoding the random substrate were subcloned into an expression vector, to generate CFP-XXXXX-HHHHHH fusion proteins. These fusion proteins were immobilized onto Ni-NTA beads, treated with hK2, and the time course of substrate hydrolysis was followed by monitoring the fluorescence released from the beads.

Thirty peptides were selected by phage screen with catalytic efficiencies ($kcat/Km$) for cleavage by hK2 ranging from 1.7×10^4 M^{-1} s^{-1} for L-R-S-R-A at the P2–P3′ sites to 9.9×10^1 M^{-1} s^{-1} for E-R-V-S-P at the P2–P3′ sites. The data show that hK2 cleaves quite selectively after arginine residues, which is consistent with previous reports [69,70], and that hK2 specificity is further enhanced by a serine in the P1′ position. A Swiss-Prot database search with selected sequences identified three putative substrates: a disintegrin-like and metalloprotease domain with a disintegrin and metalloproteinase with thrombospondin motifs 8 (ADAM-TS8) precursor, a cadherin-related tumor suppressor homolog, and a collagen α (IX) chain precursor.

It is plausible that the cleavage of these proteins by hK2 could play a role in cancer progression [71–74].

8.4.3 Tumor Invasion and Metastasis

Matrix metalloproteinases (MMPs) are a family of zinc endopeptidases capable of degrading extracellular matrix (ECM). These enzymes are essential for embryonic development, morphogenesis, reproduction, tissue resorption, and remodeling [75–78]. MMPs also participate in various pathological processes, including rheumatoid arthritis, osteoarthritis, periodontitis, autoimmune blistering disorders of the skin, and tumor invasion and metastasis [75–78]. To date, the MMP family includes 21 known enzymes that are categorized by their domain structure and by their preferences for macromolecular substrates (Table 8.3) [75–78]. The catalytic domain of MMPs is comprised of a five-stranded β-sheet, three α-helices, and bridging loops, which form a backbone structure that is highly conserved among MMP family members [79–85]. X-ray crystallographic analysis showed that the S1′ subsite in MMPs, which is the most well-defined binding area, forms a hydrophobic pocket [79,82,83]. The substrate specificities of MMPs have been studied using collagen sequence-based synthetic peptides, which are natural substrates of MMPs. This topic was reviewed by Nagase and Fields [86].

Substrate phage display has also been used to characterize the substrate specificities of MMPs, including MMP-2, -3, -7, -9, -13, and -14 (membrane type-1 MMP [MT1-MMP]) [9,21,26–30]. Smith et al. [9] compared the substrate specificity between stromelysin 1 (MMP-3) and matrilysin (MMP-7) by using a random hexamer phage library. While both enzymes favored proline at the P3 position, which is frequently found in substrates for other MMPs [86], differences were observed in the preferences of stromelysin 1 and matrilysin at the P2 and P1′ positions of the substrates. The substitution of leucine for methionine at the P1′ resulted in a modest increase in stromelysin 1 activity but decreased matrilysin activity by eightfold. The substitution of leucine for phenylalanine at P2 resulted in a threefold decrease in the catalytic efficiency of stromelysin 1 but a nearly threefold increase in matrilysin activity.

Highly selective and efficient substrates of human collagenase 3 (MMP-13) were identified by the substrate phage technique [28]. A consensus substrate sequence for collagenase 3 at the P3–P3′ sites (P-L-G-M-R-G) was deduced based on the preferred residue in each subsite position from 35 selected phage clones. A synthetic peptide corresponding to the consensus sequence (G-P-L-G-M-R-G-L) exhibited a higher $kcat/Km$ value (4.2×10^6 M^{-1} s^{-1}) for hydrolysis by collagenase 3 and was a more efficient substrate than the previously reported substrates, such as McaPChaGNva-HADpa-NH$_2$ and McaPLGLDpaAR-NH$_2$ [87]. The catalytic efficiency of collagenase 3 for hydrolysis of the consensus sequence was 1344-, 11-, and 820-fold higher than those for stromelysin 1 (MMP-3), gelatinase B (MMP-9), and collagenase 1 (MMP-1), respectively. Substitution at the P3 residue revealed that collagenase 3 also favors proline as a P3 residue. A search of the Swiss-Prot and translated the European Molecular Biology Laboratory (EMBL) protein databases with selected

Table 8.3   Human Matrix Metalloproteinase Family

Enzyme		Known Substrates

Collagenases

MMP-1	Collagenase-1	Collagen types, I, II, III, VII, VIII, X, aggrecan, entactin, α1-PI, IL-1β, TFPI, IGF-BP3, SAA, AFPs, proMMP-2
MMP-8	Collagenase-2	Collagen types I, II, III, aggrecan, substance-P, TFPI
MMP-13	Collagenase-3	Collagen types I, II, III, IV, IX, X, XIV, aggrecan, FN, gelatin, casein, tenascin, proMMP-9, osteonectin

Gelatinases

MMP-2	Gelatinase A (72 kDa)	Collagens types I, IV, V, X, IX, XI, proMMP-9, 13, Eph B1, aggrecan, VN, SAA, AFPs, APP, galectin-3 tenascin, IL-1β, decorin, osteonectin
MMP-9	Gelatinase B (92 kDa)	Collagens types I, III, IV, V, XI, XIV, XVII, entactin, α1-PI, galectin-3, substance-P, VN, aggrecan, IL-1β, osteonectin, elastin, plasminogen, TFPI, gelatin, myelin basic protein

Stromelysins

MMP-3	Stromelysin-1	ProMMP-1, 8, 9, 13, collagen types I, II, III, IV, IX, aggrecan, α1-PI, IGF- BP3, FN, laminin, casein, gelatin, tenascin, osteopontin, transferring, α2-macroglobulin, VN, osteonectin, fibrinogen, fibrin, IL-1β, decorin, elastin, PAI-I, scu-PA, SAA, AFPs
MMP-10	Stromelysin-2	Collagen type IV, proMMP-1,-7, -8,-9, proteoglycan, gelatin, casein, aggrecan

Matrilysins

MMP-7	Matrilysin	Aggrecan, entactin, αI-PI, proMMP-2, decorin, TFPI, VN, carboxymethylated transferring, osteopontin, tenascin, casein, collagens types I, IV, XVIII, E-cadherin, FN, fibulin, osteonectin, beta4-integrin, proTNF-α, plasminogen, fibrinogen, fibrin, FasL
MMP-26	Matrilysin-2	Collagen type IV, FN, fibrinogen, gelatin, α1-PI: proMMP9

Membrane-Type MMPs

MMP-14	MTI-MMP	ProMMP2, proMMP-13, FN, tenascin, nidogen, perlecan, collagen types I, II, III, VN, laminin, tTG, α2-microglobulin, aggrecan, fibrinogen, fibrin
MMP-15	MT2-MMP	ProMMP2, tTG, FN, tenascin, nidogen, aggrecan, perlecan, laminin
MMP-16	MT3-MMP	ProMMP2, collagen type III, gelatin, laminin-1, aggrecan, FN, VN, α1-PI, α2-macroglobulin, tTG
MMP-17	MT4-MMP	Gelatin, pro-TNF-α, FN, fibrin
MMP-24	MT5-MMP	ProMMP-2, chondroitin sulphate proteoglycan, FN, dematin sulfate proteoglycan
MMP-25	MT6-MMP	ProMMP-2, collagen type IV, gelatin, FN, fibrin

Table 8.3 (Continued)   Human Matrix Metalloproteinase Family

Enzyme		Known Substrates
		Others
MMP-11	Stromelysin-3	α1-PI, serine protease inhibitor, IGFBP-1, weak activity for ECM proteins
MMP-12	Metalloelastase	Elastin, collagen types I, IV, V, osteonectin, FN, VN, gelatin, laminin, pro-TNF-α, TFPI, myelin basic protein, α1-antitrypsin
MMP-19	RASI-1	Collagen type IV, laminin, nidogen, fibronectin, gelatin, COMP, aggrecan
MMP-20	Enamelysin	Amelogenin, aggrecan, COMP
MMP-23	CA-MMP	Unknown
MMP-28	Epilysin	Casein

Note: AFPs: AA amyloid fibril proteins; APP: amyloid protein precursor; COMP: cartilage oligomeric matrix protein; ECM: extracellular matrix; FN: fibronectin; PAI-I: plasminogen activator inhibitor I; SAA: acute-phase serum amyloid A; scu-PA: single-chain urokinase-type plasminogen activator; TEPI: tissue factor pathway inhibitor; tTG: tissue transglutaminase; VN: vitronectin; α1-PI: α1-proteinase inhibitor.

phage sequences identified potential collagenase 3 cleavage sites in type IV collagen, biglycan, and the latency-associated peptide of TGF-β3. This match is consistent with the role of MMPs in proteolytic activation of TGF-β3 [88].

The MMPs have been implicated in the process of tumor growth, invasion and metastasis [89–92]. Among MMPs, the gelatinases (MMP-2 and MMP-9), which can both degrade components of basement membranes, have been most consistently detected in malignant tissues and associated with tumor aggressiveness, metastatic potential, and a poor prognosis [78,89,93–95]. Most MMPs are secreted as latent precursors (zymogens) and are subsequently activated by proteolysis. MT1-MMP has been cloned as an activator of MMP-2 [96]. It has been reported that MT1-MMP is also overexpressed in several types of tumors [97–103], suggesting that the activation of MMP-2 by MT1-MMP may play a critical role in cancer cell invasion and metastasis. On the other hand, MT1-MMP itself can digest many types of ECM proteins, such as interstitial collagens, gelatin, and proteoglycans [104–106]. These results suggest that MT1-MMP plays a dual role in the digestion of ECM, both by directly cleaving the substrate and by activating MMP-2. Since the relationships between MMPs and cancer invasion and metastasis have been delineated, several MMP inhibitors (MMPIs) have been developed and are expected to represent a new approach to cancer treatment, in addition to the traditional cytotoxic drugs [78,95]. However, many of these MMPIs are broad-spectrum inhibitors, and exhibit undesirable side effects [78,95]. The principal side effect of these drugs is musculoskeletal pain, suggesting that for cancer treatment, it will be important to develop selective gelatinase inhibitors with limited activity against collagenases involved in the maintenance of normal joint function.

The substrate sequence specificity of MT1-MMP was analyzed using a random hexamer phage library [21,29]. By aligning the selected clones, the consensus substrate sequence for MT1-MMP was defined as P-X-(G/P)-(L/I) at the P3–P1' sites. The deduced consensus substrate sequence resembles the canonical collagen-like P-X-X-L motif, which has been previously established for MMPs [86]. In fact, the P-X-G-L/I sequence was reported to be present in human type I collagen (α1)(P-Q-G-I), type I collagen (α2)(P-Q-G-L), type II collagen (P-Q-G-L), type III collagen (P-L-G-I), and α_2-macroglobulin (P-E-G-L). These proteins were susceptible to digestion by MT1-MMP in vitro [104]. The synthetic peptide prepared based on the consensus sequence (G-P-L-G-L-R-S-W from P4 to P4') was cleaved efficiently by MT1-MMP; however, this peptide was also efficiently hydrolyzed by MMP-2 and MMP-9.

Highly selective substrate sequences for MMP-2, MMP-9, and MT1-MMP were identified in a series of substrate phage studies by Smith and coworkers [26,27,30]. In each screen, the canonical MMP substrate sequence, P-X-X-X_{Hy} (where X_{Hy} is a hydrophobic residue) at the P3–P1' sites emerged as a consensus sequence. However, other specific substrate sequences for individual MMPs were also identified from the selected clones. Substrate phage display was used to identify substrate sequences of MMP-2, and these were categorized into four distinct groups, based on their sequence similarities [26]. Whereas the substrates containing the P-X-X-X_{Hy} canonical motif lacked selectivity, the other three groups contained a unique consensus sequence and showed higher selectivity for MMP-2 over the other MMPs tested. These substrates contained consensus motifs of (L/I)-X-X-X_{Hy}, X_{Hy}-S-X-L, and H-X-X-X_{Hy} at the P3–P1' sites. Among these sequences, the (L/I)-X-X-X_{Hy} peptides exhibited the most selectivity, and one of the highly selective substrates (S-G-R-S-L-S-R-L-T-A at the P7–P3' sites) was cleaved 200-fold more efficiently by MMP-2 than by its closely related homolog, MMP-9. Though this motif does not contain a proline at the P3 position, which frequently occurs in MMP substrates, substitution at the P3 position of (L/I)-X-X-X_{Hy} peptide revealed that the absence of proline at this position is not the sole determinant for MMP-2 selectivity. Rather, the P2 position plays a major role in determining specificity between MMP-2 and MMP-9. Using substrate phage display, the consensus substrate sequence for MMP-9 was characterized as P-R-(S/T)-X_{Hy}-(S/T) at the P3–P2' sites, and the peptides containing arginine at the P2 position were cleaved most efficiently by MMP-9 [27]. Since arginine was rarely present at P2 in the substrates selective for MMP-2, arginine was substituted at this position within the MMP-2–selective peptides, and catalytic activities were determined. Interestingly, an arginine to serine amino acid substitution at the P2 position dramatically increased hydrolysis by MMP-9, and significantly decreased hydrolysis by MMP-2. These observations suggest that the interaction between the P2 position of the substrate and the S2 position within the catalytic cleft of MMP plays a key role in distinguishing substrate recognition by MMP-2 and MMP-9. Indeed, analysis of the structure of the S2 subsite within MMP-2 and MMP-9 identified a potential structural basis for this distinction in substrate recognition at the P2 position [26].

The MT1-MMP-selective phage were isolated from substrate phage clones by comparing the catalytic activity of MT1-MMP, MMP-2, and MMP-9 for individual phage [30]. No consensus sequence could be defined from the MT1-MMP-selective clones; however, the residues that include long side chains, particularly arginine, were favored at the P4 position. In fact, the substitution of alanine for arginine at P4 resulted in substrates that were poorly cleaved by MT1-MMP. Like the canonical substrate sequences for MMPs, a hydrophobic residue was preferred at the P1' position in MT1-MMP-selective substrates. Interestingly, proline, which is the favored residue at the P3 position among the various substrates of MMPs, is absent from these sequences. The peptide derived from phage clones (S-G-R-S-E-N-I-R-T-A at the P6–P4' sites) was hydrolyzed 83-fold more efficiently by MT1-MMP than by MMP-9. Hence, the substitution of serine for proline at the P3 position of this sequence converted a substrate selective for MT1-MMP to a substrate that is recognized equally well by both enzymes. The idea that MT1-MMP recognizes substrates in two distinct modes arose from these observations [30]. One mode makes use of the P3 and P1' positions as dominant contact points that bind to nonselective substrates. In the other mode, the P4 and P1' subsites appear to form contacts that are critical for recognizing specific substrates. Indeed, a three-dimensional modeling of the selective and nonselective substrates bound to the catalytic pocket of MT1-MMP could explain two separate binding modes. These new insights can be used to help design highly selective inhibitors of individual MMPs and may also provide additional clues for understanding of the physiological roles of MMPs.

8.5 CONCLUSIONS

As we have discussed in this chapter, substrate phage display has enormous potential for delineating the substrate specificities of proteases and protein kinases. This information may facilitate the development of potent and selective inhibitors for improved therapeutics. Knowledge of the sequence specificities of enzymes, especially of closely related enzymes, might serve as a template for understanding structure–activity relationships and for the rational design of new drugs targeting these enzymes. The selective substrates can be converted into probes to be used in HTS assays. In addition, the results of substrate phage display can be used for other applications, such as activity-based measurement of enzymes and prediction of the physiological and pathophysiological substrates for enzymes. Indeed, several novel potential protease substrates have been successfully identified using substrate phage display (Table 8.4).

Though substrate phage display has been applied to screen for substrates of proteases and kinases, one would predict that this technique could also be applied to identify substrates for other enzymes. Enzymatic modifications of proteins that can be carried out in vitro, including acetylation, ubiquitination, and glycosylation, should be good candidates for applying substrate phage display. It is also conceivable that,

Table 8.4   Novel Potential Substrate Proteins Identified by Substrate Phage Display

Protease	Substrate Sequence Identified	Potential Substrate and Target Sequence	References
Rat granzyme B	IEXD↓XG	Poly (ADP-ribose) polymerase (PARP) (VDPD↓SG, LEID↓YG)	[18]
		Pro-caspase 3 (IETD↓SG)	
		Pro-caspase 7 (IQAD↓SG)	
Human membrane-type serine protease 1	(R/K)XSR↓A	Protease-activated receptor (PAR) 2 (SKGR↓S)	[20]
	X(R/K)SR↓A	Single-chain urokinase-type plasminogen activator (sc-uPA) (PREK↓)	
Human collagenase 3 (MMP-13)	GPLG↓MRGL	Biglycan (PKG↓VFS)	[28]
		TGF-β3 (PKG↓ITS)	
Human gelatinase B (MMP-9)	PR(S/T)↓ X$_{Hy}$(S/T)	Kallikrein 14 (PRT↓IT)	[27]
		Ladinin 1 (PRT↓IS)	
		Endoglin (PRT↓VT)	
		Endothelin receptor (PRT↓IS)	
		Laminin α3 chain (PRS↓LT)	
		Phosphate regulating neutral endopeptidase (PRS↓LS)	
		ADAM 2 (PRT↓IS)	
		Desmoglein 3 (PRS↓LT)	
		Integlin β$_5$ (PRS↓IT)	
Rat mast cell protease 4 (rMCP-4)	LVWF↓RG	Protein C precursor (VVFF↓RG)	[23]
		Procollagen C-proteinase enhancer protein (LLWY↓SG)	
		Coagulation factor V (VMYF↓NG)	
		TGF-β receptor type III (VVYY↓NS)	
		Cystein-rich secretory protein-3 (VVWY↓SS)	
		Plasminogen activator inhibitor-1 (ALYF↓NG)	
		Low affinity Igγ Fc region receptor III (LVWF↓HA)	
Human kallikrein 2 (hK2)	LR↓SRA	ADAM-TS 8 precursor (RGR↓SE)	[22]
		Cadherin-related tumor suppressor homologue precursor (GVFR↓S)	
		Collagen α (IX) chain precursor (PGR↓AP)	
Human gelatinase A (MMP-2)	X$_{Hy}$SX↓L	Eph B1 tyrosine kinase receptor (YKSE↓LRE)	[26]

Note: ↓: site of digestion; X$_{Hy}$: hydrophobic residue.

with modification, a similar strategy could be applied to screen for substrates of protein phosphatases. Dente et al. [31] have shown that by extending the kinase reaction time, the sequence specificity is weakened, and practically any tyrosine-containing sequence can become phosphorylated. Thus, it is possible to make a modified phage peptide library, where all the tyrosine residues are converted to phosphotyrosine in vitro. Such a modified phage library could then be used to identify specific substrates for protein tyrosine phosphatases.

As of August 2013, 1023 human known and putative peptidases were deposited in the MEROPS peptidase database [107]. It is expected that some of these will be considered as new drug targets [108]. It is now possible to quickly determine the activities and substrate specificities of these proteases by substrate phage display. This technique will undoubtedly contribute to molecular medicine and the development of novel therapeutic strategies in the future.

REFERENCES

1. Zhou S, Cantley LC. Recognition and specificity in protein tyrosine kinase-mediated signalling. *Trends Biochem Sci* 1995; 20:470–475.
2. Ding L, Coombs GS, Strandberg L, Navre M, Corey DR, Madison EL. Origins of the specificity of tissue-type plasminogen activator. *Proc Natl Acad Sci U S A* 1995; 92:7627–7631.
3. Coombs GS, Bergstrom RC, Pellequer JL, Baker SI, Navre M, Smith MM, Tainer JA et al. Substrate specificity of prostate-specific antigen (PSA). *Chem Biol* 1998; 5:475–488.
4. Hervio LS, Coombs GS, Bergstrom RC, Trivedi K, Corey DR, Madison EL. Negative selectivity and the evolution of protease cascades: The specificity of plasmin for peptide and protein substrates. *Chem Biol* 2000; 7:443–453.
5. Matthews DJ, Wells JA. Substrate phage: Selection of protease substrates by monovalent phage display. *Science* 1993; 260:1113–1117.
6. Kay BK, Winter J, McCafferty J. *Phage Display of Peptides and Proteins*. London: Academic Press, 1996.
7. Gram H. Phage display in proteolysis and signal transduction. *Comb Chem High Throughput Screen* 1999; 2:19–28.
8. Smith GP, Scott JK. Libraries of peptides and proteins displayed on filamentous phage. *Methods Enzymol* 1993; 217:228–257.
9. Smith MM, Shi L, Navre M. Rapid identification of highly active and selective substrates for stromelysin and matrilysin using bacteriophage peptide display libraries. *J Biol Chem* 1995; 270:6440–6449.
10. Schmitz R, Baumann G, Gram H. Catalytic specificity of phosphotyrosine kinases Blk, Lyn, c-Src and Syk as assessed by phage display. *J Mol Biol* 1996; 260:664–677.
11. Westendorf JM, Rao PN, Gerace L. Cloning of cDNAs for M-phase phosphoproteins recognized by the MPM2 monoclonal antibody and determination of the phosphorylated epitope. *Proc Natl Acad Sci U S A* 1994; 91:714–718.
12. Ballinger MD, Tom J, Wells JA. Designing subtilisin BPN' to cleave substrates containing dibasic residues. *Biochemistry* 1995; 34:13312–13319.

13. Matthews DJ, Goodman LJ, Gorman CM, Wells JA. A survey of furin substrate specificity using substrate phage display. *Protein Sci* 1994; 3:1197–1205.

14. Ke SH, Coombs GS, Tachias K, Navre M, Corey DR, Madison EL. Distinguishing the specificities of closely related proteases. Role of P3 in substrate and inhibitor discrimination between tissue-type plasminogen activator and urokinase. *J Biol Chem* 1997; 272:16603–16609.

15. Ke SH, Coombs GS, Tachias K, Corey DR, Madison EL. Optimal subsite occupancy and design of a selective inhibitor of urokinase. *J Biol Chem* 1997; 272:20456–20462.

16. O'Boyle DR II, Pokornowski KA, McCann PJ III, Weinheimer SP. Identification of a novel peptide substrate of HSV-1 protease using substrate phage display. *Virology* 1997; 236:338–347.

17. Huang C, Wong GW, Ghildyal N, Gurish MF, Sali A, Matsumoto R, Qiu WT et al. The tryptase, mouse mast cell protease 7, exhibits anticoagulant activity in vivo and in vitro due to its ability to degrade fibrinogen in the presence of the diverse array of protease inhibitors in plasma. *J Biol Chem* 1997; 272:31885–31893.

18. Harris JL, Peterson EP, Hudig D, Thornberry NA, Craik CS. Definition and redesign of the extended substrate specificity of granzyme B. *J Biol Chem* 1998; 273: 27364–27373.

19. Dall'Acqua W, Halin C, Rodrigues ML, Carter P. Elastase substrate specificity tailored through substrate-assisted catalysis and phage display. *Protein Eng* 1999; 12:981–987.

20. Takeuchi T, Harris JL, Huang W, Yan KW, Coughlin SR, Craik CS. Cellular localization of membrane-type serine protease 1 and identification of protease-activated receptor-2 and single-chain urokinase-type plasminogen activator as substrates. *J Biol Chem* 2000; 275:26333–26342.

21. Ohkubo S, Miyadera K, Sugimoto Y, Matsuo K, Wierzba K, Yamada Y. Substrate phage as a tool to identify novel substrate sequences of proteases. *Comb Chem High Throughput Screen* 2001; 4:573–583.

22. Cloutier SM, Chagas JR, Mach JP, Gygi CM, Leisinger HJ, Deperthes D. Substrate specificity of human kallikrein 2 (hK2) as determined by phage display technology. *Eur J Biochem* 2002; 269:2747–2754.

23. Karlson U, Pejler G, Froman G, Hellman L. Rat mast cell protease 4 is a beta-chymase with unusually stringent substrate recognition profile. *J Biol Chem* 2002; 277:18579–18585.

24. Sharkov NA, Cai D. Discovery of substrate for type I signal peptidase SpsB from Staphylococcus aureus. *J Biol Chem* 2002; 277:5796–5803.

25. Beck ZQ, Hervio L, Dawson PE, Elder JH, Madison EL. Identification of efficiently cleaved substrates for HIV-1 protease using a phage display library and use in inhibitor development. *Virology* 2000; 274:391–401.

26. Chen EI, Kridel SJ, Howard EW, Li W, Godzik A, Smith JW. A unique substrate recognition profile for matrix metalloproteinase-2. *J Biol Chem* 2002; 277:4485–4491.

27. Kridel SJ, Chen E, Kotra LP, Howard EW, Mobashery S, Smith JW. Substrate hydrolysis by matrix metalloproteinase-9. *J Biol Chem* 2001; 276:20572–20578.

28. Deng SJ, Bickett DM, Mitchell JL, Lambert MH, Blackburn RK, Carter HL III, Neugebauer J et al. Substrate specificity of human collagenase 3 assessed using a phage-displayed peptide library. *J Biol Chem* 2000; 275:31422–31427.

29. Ohkubo S, Miyadera K, Sugimoto Y, Matsuo K, Wierzba K, Yamada Y. Identification of substrate sequences for membrane type-1 matrix metalloproteinase using bacteriophage peptide display library. *Biochem Biophys Res Commun* 1999; 266:308–313.

30. Kridel SJ, Sawai H, Ratnikov BI, Chen EI, Li W, Godzik A, Strongin AY et al. A unique substrate binding mode discriminates membrane type-1 matrix metalloproteinase from other matrix metalloproteinases. *J Biol Chem* 2002; 277:23788–23793.

31. Dente L, Vetriani C, Zucconi A, Pelicci G, Lanfrancone L, Pelicci PG, Cesareni G. Modified phage peptide libraries as a tool to study specificity of phosphorylation and recognition of tyrosine containing peptides. *J Mol Biol* 1997; 269:694–703.

32. Deng SJ, Liu W, Simmons CA, Moore JT, Tian G. Identifying substrates for endothelium-specific Tie-2 receptor tyrosine kinase from phage-displayed peptide libraries for high throughput screening. *Comb Chem High Throughput Screen* 2001; 4:525–533.

33. Sharkov NA, Davis RM, Reidhaar-Olson JF, Navre M, Cai D. Reaction kinetics of protease with substrate phage. Kinetic model developed using stromelysin. *J Biol Chem* 2001; 276:10788–10793.

34. Spraggon G, Phillips C, Nowak UK, Ponting CP, Saunders D, Dobson CM, Stuart DI et al. The crystal structure of the catalytic domain of human urokinase-type plasminogen activator. *Structure* 1995; 3:681–691.

35. Lamba D, Bauer M, Huber R, Fischer S, Rudolph R, Kohnert U, Bode W. The 2.3 A crystal structure of the catalytic domain of recombinant two-chain human tissue-type plasminogen activator. *J Mol Biol* 1996; 258:117–135.

36. Ribatti D, Vacca A, Nico B, De Falco G, Giuseppe Montaldo P, Ponzoni M. Angiogenesis and anti-angiogenesis in neuroblastoma. *Eur J Cancer* 2002; 38:750–757.

37. Woodhouse EC, Chuaqui RF, Liotta LA. General mechanisms of metastasis. *Cancer* 1997; 80:1529–1537.

38. Ferrara N, Alitalo K. Clinical applications of angiogenic growth factors and their inhibitors. *Nat Med* 1999; 5:1359–1364.

39. Dy GK, Adjei AA. Novel targets for lung cancer therapy: Part II. *J Clin Oncol* 2002; 20:3016–3028.

40. Toi M, Matsumoto T, Bando H. Vascular endothelial growth factor: Its prognostic, predictive, and therapeutic implications. *Lancet Oncol* 2001; 2:667–673.

41. Lauren J, Gunji Y, Alitalo K. Is angiopoietin-2 necessary for the initiation of tumor angiogenesis? *Am J Pathol* 1998; 153:1333–1339.

42. Holash J, Wiegand SJ, Yancopoulos GD. New model of tumor angiogenesis: Dynamic balance between vessel regression and growth mediated by angiopoietins and VEGF. *Oncogene* 1999; 18:5356–5362.

43. Stratmann A, Risau W, Plate KH. Cell type-specific expression of angiopoietin-1 and angiopoietin-2 suggests a role in glioblastoma angiogenesis. *Am J Pathol* 1998; 153:1459–1466.

44. Takahama M, Tsutsumi M, Tsujiuchi T, Nezu K, Kushibe K, Taniguchi S, Kotake Y et al. Enhanced expression of Tie2, its ligand angiopoietin-1, vascular endothelial growth factor, and CD31 in human non-small cell lung carcinomas. *Clin Cancer Res* 1999; 5:2506–2510.

45. Etoh T, Inoue H, Tanaka S, Barnard GF, Kitano S, Mori M. Angiopoietin-2 is related to tumor angiogenesis in gastric carcinoma: Possible in vivo regulation via induction of proteases. *Cancer Res* 2001; 61:2145–2153.

46. Lin P, Polverini P, Dewhirst M, Shan S, Rao PS, Peters K. Inhibition of tumor angiogenesis using a soluble receptor establishes a role for Tie2 in pathologic vascular growth. *J Clin Invest* 1997; 100:2072–2078.

47. Siemeister G, Schirner M, Weindel K, Reusch P, Menrad A, Marme D, Martiny-Baron G. Two independent mechanisms essential for tumor angiogenesis: Inhibition of human melanoma xenograft growth by interfering with either the vascular endothelial growth factor receptor pathway or the Tie-2 pathway. *Cancer Res* 1999; 59:3185–3191.

48. Lin P, Buxton JA, Acheson A, Radziejewski C, Maisonpierre PC, Yancopoulos GD, Channon KM et al. Antiangiogenic gene therapy targeting the endothelium-specific receptor tyrosine kinase Tie2. *Proc Natl Acad Sci U S A* 1998; 95:8829–8834.

49. Diamandis EP, Yousef GM, Clements J, Ashworth LK, Yoshida S, Egelrud T, Nelson PS et al. New nomenclature for the human tissue kallikrein gene family. *Clin Chem* 2000; 46:1855–1858.

50. Yousef GM, Diamandis EP. The new human tissue kallikrein gene family: Structure, function, and association to disease. *Endocr Rev* 2001; 22:184–204.

51. Catalona WJ, Smith DS, Ratliff TL, Dodds KM, Coplen DE, Yuan JJ, Petros JA et al. Measurement of prostatespecific antigen in serum as a screening test for prostate cancer. *N Engl J Med* 1991; 324:1156–1161.

52. Oesterling JE. Prostate specific antigen: A critical assessment of the most useful tumor marker for adenocarcinoma of the prostate. *J Urol* 1991; 145:907–923.

53. Labrie F, Dupont A, Suburu R, Cusan L, Tremblay M, Gomez JL, Emond J. Serum prostate specific antigen as pre-screening test for prostate cancer. *J Urol* 1999; 147:846–851; discussion 851–852.

54. Kwiatkowski MK, Recker F, Piironen T, Pettersson K, Otto T, Wernli M, Tscholl R. In prostatism patients the ratio of human glandular kallikrein to free PSA improves the discrimination between prostate cancer and benign hyperplasia within the diagnostic "gray zone" of total PSA 4 to 10 ng/mL. *Urology* 1998; 52:360–365.

55. Partin AW, Catalona WJ, Finlay JA, Darte C, Tindall DJ, Young CY, Klee GG et al. Use of human glandular kallikrein 2 for the detection of prostate cancer: Preliminary analysis. *Urology* 1999; 54:839–845.

56. Nam RK, Diamandis EP, Toi A, Trachtenberg J, Magklara A, Scorilas A, Papnastasiou PA et al. Serum human glandular kallikrein-2 protease levels predict the presence of prostate cancer among men with elevated prostate-specific antigen. *J Clin Oncol* 2000; 18:1036–1042.

57. Watt KW, Lee PJ, M'Timkulu T, Chan WP, Loor R. Human prostate-specific antigen: Structural and functional similarity with serine proteases. *Proc Natl Acad Sci U S A* 1986; 83:3166–3170.

58. Akiyama K, Nakamura T, Iwanaga S, Hara M. The chymotrypsin-like activity of human prostate-specific antigen, gamma-seminoprotein. *FEBS Lett* 1987; 225:168–172.

59. Lilja H. A kallikrein-like serine protease in prostatic fluid cleaves the predominant seminal vesicle protein. *J Clin Invest* 1985; 76:1899–1903.

60. Lilja H, Oldbring J, Rannevik G, Laurell CB. Seminal vesicle-secreted proteins and their reactions during gelation and liquefaction of human semen. *J Clin Invest* 1987; 80:281–285.

61. Lilja H. Structure and function of prostatic- and seminal vesicle-secreted proteins involved in the gelation and liquefaction of human semen. *Scand J Clin Lab Invest Suppl* 1988; 191:13–20.

62. Robert M, Gagnon C. Semenogelin I: A coagulum forming, multifunctional seminal vesicle protein. *Cell Mol Life Sci* 1999; 55:944–960.

63. Killian CS, Corral DA, Kawinski E, Constantine RI. Mitogenic response of osteoblast cells to prostate-specific antigen suggests an activation of latent TGF-beta and a proteolytic modulation of cell adhesion receptors. *Biochem Biophys Res Commun* 1993; 192:940–947.

64. Iwamura M, Hellman J, Cockett AT, Lilja H, Gershagen S. Alteration of the hormonal bioactivity of parathyroid hormone-related protein (PTHrP) as a result of limited proteolysis by prostate-specific antigen. *Urology* 1996; 48:317–325.

65. Rehault S, Monget P, Mazerbourg S, Tremblay R, Gutman N, Gauthier F, Moreau T. Insulin-like growth factor binding proteins (IGFBPs) as potential physiological substrates for human kallikreins hK2 and hK3. *Eur J Biochem* 2001; 268:2960–2968.

66. Deperthes D, Frenette G, Brillard-Bourdet M, Bourgeois L, Gauthier F, Tremblay RR, Dube JY. Potential involvement of kallikrein hK2 in the hydrolysis of the human seminal vesicle proteins after ejaculation. *J Androl* 1996; 17:659–665.

67. Frenette G, Tremblay RR, Lazure C, Dube JY. Prostatic kallikrein hK2, but not prostate-specific antigen (hK3), activates single-chain urokinase-type plasminogen activator. *Int J Cancer* 1997; 71:897–899.

68. Mikolajczyk SD, Millar LS, Kumar A, Saedi MS. Prostatic human kallikrein 2 inactivates and complexes with plasminogen activator inhibitor-1. *Int J Cancer* 1999; 81:438–442.

69. Bourgeois L, Brillard-Bourdet M, Deperthes D, Juliano MA, Juliano L, Tremblay RR, Dube JY et al. Serpin-derived peptide substrates for investigating the substrate specificity of human tissue kallikreins hK1 and hK2. *J Biol Chem* 1997; 272:29590–29595.

70. Mikolajczyk SD, Millar LS, Kumar A, Saedi MS. Human glandular kallikrein, hK2, shows arginine-restricted specificity and forms complexes with plasma protease inhibitors. *Prostate* 1998; 34:44–50.

71. Dunne J, Hanby AM, Poulsom R, Jones TA, Sheer D, Chin WG, Da SM et al. Molecular cloning and tissue expression of FAT, the human homologue of the Drosophila fat gene that is located on chromosome 4q34–q35 and encodes a putative adhesion molecule. *Genomics* 1995; 30:207–223.

72. Georgiadis KE, Hirohata S, Seldin MF, Apte SS. ADAM-TS8, a novel metalloprotease of the ADAM-TS family located on mouse chromosome 9 and human chromosome 11. *Genomics* 1999; 62:312–315.

73. Wang SS, Virmani A, Gazdar AF, Minna JD, Evans GA. Refined mapping of two regions of loss of heterozygosity on chromosome band 11q23 in lung cancer. *Genes Chromosomes Cancer* 1999; 25:154–159.

74. Muragaki Y, Kimura T, Ninomiya Y, Olsen BR. The complete primary structure of two distinct forms of human alpha 1 (IX) collagen chains. *Eur J Biochem* 1990; 192:703–708.

75. Johnson LL, Dyer R, Hupe DJ. Matrix metalloproteinases. *Curr Opin Chem Biol* 1998; 2:466–471.

76. Shapiro SD. Matrix metalloproteinase degradation of extracellular matrix: Biological consequences. *Curr Opin Cell Biol* 1998; 10:602–608.

77. Nagase H, Woessner JF Jr. Matrix metalloproteinases. *J Biol Chem* 1999; 274:21491–21494.

78. Vihinen P, Kahari VM. Matrix metalloproteinases in cancer: Prognostic markers and therapeutic targets. *Int J Cancer* 2002; 99:157–166.

79. Bode W, Reinemer P, Huber R, Kleine T, Schnierer S, Tschesche H. The X-ray crystal structure of the catalytic domain of human neutrophil collagenase inhibited by a substrate analogue reveals the essentials for catalysis and specificity. *EMBO J* 1994; 13:1263–1269.

80. Lovejoy B, Cleasby A, Hassell AM, Longley K, Luther MA, Weigl D, McGeehan G et al. Structure of the catalytic domain of fibroblast collagenase complexed with an inhibitor. *Science* 1994; 263:375–377.

81. Stocker W, Grams F, Baumann U, Reinemer P, Gomis-Ruth FX, McKay DB, Bode W. The metzincins—topological and sequential relations between the astacins, adamalysins, serralysins, and matrixins (collagenases) define a superfamily of zinc-peptidases. *Protein Sci* 1995; 4:823–840.

82. Grams F, Reinemer P, Powers JC, Kleine T, Pieper M, Tschesche H, Huber R et al. X-ray structures of human neutrophil collagenase complexed with peptide hydroxamate and peptide thiol inhibitors. Implications for substrate binding and rational drug design. *Eur J Biochem* 1995; 228:830–841.

83. Dhanaraj V, Ye QZ, Johnson LL, Hupe DJ, Ortwine DF, Dunbar JB Jr, Rubin JR et al. X-ray structure of a hydroxamate inhibitor complex of stromelysin catalytic domain and its comparison with members of the zinc metalloproteinase superfamily. *Structure* 1996; 4:375–386.

84. Gomis-Ruth FX, Maskos K, Betz M, Bergner A, Huber R, Suzuki K, Yoshida N et al. Mechanism of inhibition of the human matrix metalloproteinase stromelysin-1 by TIMP-1. *Nature* 1997; 389:77–81.

85. Morgunova E, Tuuttila A, Bergmann U, Isupov M, Lindqvist Y, Schneider G, Tryggvason K. Structure of human pro-matrix metalloproteinase-2: Activation mechanism revealed. *Science* 1999; 284:1667–1670.

86. Nagase H, Fields GB. Human matrix metalloproteinase specificity studies using collagen sequence-based synthetic peptides. *Biopolymers* 1996; 40:399–416.

87. Knauper V, Lopez-Otin C, Smith B, Knight G, Murphy G. Biochemical characterization of human collagenase-3. *J Biol Chem* 1996; 271:1544–1550.

88. Yu Q, Stamenkovic I. Cell surface-localized matrix metalloproteinase-9 proteolytically activates TGF-beta and promotes tumor invasion and angiogenesis. *Genes Dev* 2000; 14:163–176.

89. Stetler-Stevenson WG, Aznavoorian S, Liotta LA. Tumor cell interactions with the extracellular matrix during invasion and metastasis. *Annu Rev Cell Biol* 1993; 9:541–573.

90. Chambers AF, Matrisian LM. Changing views of the role of matrix metalloproteinases in metastasis. *J Natl Cancer Inst* 1997; 89:1260–1270.

91. Kahari VM, Saarialho-Kere U. Matrix metalloproteinases and their inhibitors in tumour growth and invasion. *Ann Med* 1999; 31:34–45.

92. Kleiner DE, Stetler-Stevenson WG. Matrix metalloproteinases and metastasis. *Cancer Chemother Pharmacol* 1999; 43(suppl):S42–S51.

93. Noel A, Emonard H, Polette M, Birembaut P, Foidart JM. Role of matrix, fibroblasts and type IV collagenases in tumor progression and invasion. *Pathol Res Pract* 1994; 190:934–941.

94. Ray JM, Stetler-Stevenson WG. The role of matrix metalloproteases and their inhibitors in tumour invasion, metastasis and angiogenesis. *Eur Respir J* 1994; 7:2062–2072.

95. Hidalgo M, Eckhardt SG. Development of matrix metalloproteinase inhibitors in cancer therapy. *J Natl Cancer Inst* 2001; 93:178–193.

96. Sato H, Takino T, Okada Y, Cao J, Shinagawa A, Yamamoto E, Seiki M. A matrix metalloproteinase expressed on the surface of invasive tumour cells. *Nature* 1994; 370:61–65.

97. Okada A, Bellocq JP, Rouyer N, Chenard MP, Rio MC, Chambon P, Basset P. Membrane-type matrix metalloproteinase (MT-MMP) gene is expressed in stromal cells of human colon, breast, and head and neck carcinomas. *Proc Natl Acad Sci U S A* 1995; 92:2730–2734.

98. Nomura H, Sato H, Seiki M, Mai M, Okada Y. Expression of membrane-type matrix metalloproteinase in human gastric carcinomas. *Cancer Res* 1995; 55:3263–3266.

99. Tokuraku M, Sato H, Murakami S, Okada Y, Watanabe Y, Seiki M. Activation of the precursor of gelatinase A/72 kDa type IV collagenase/MMP-2 in lung carcinomas correlates with the expression of membrane-type matrix metalloproteinase (MT-MMP) and with lymph node metastasis. *Int J Cancer* 1995; 64:355–359.

100. Gilles C, Polette M, Piette J, Munaut C, Thompson EW, Birembaut P, Foidart JM. High level of MT-MMP expression is associated with invasiveness of cervical cancer cells. *Int J Cancer* 1996; 65:209–213.

101. Yamamoto M, Mohanam S, Sawaya R, Fuller GN, Seiki M, Sato H, Gokaslan ZL et al. Differential expression of membrane-type matrix metalloproteinase and its correlation with gelatinase A activation in human malignant brain tumors in vivo and in vitro. *Cancer Res* 1996; 56:384–392.

102. Ohtani H, Motohashi H, Sato H, Seiki M, Nagura H. Dual over-expression pattern of membrane-type metalloproteinase-1 in cancer and stromal cells in human gastrointestinal carcinoma revealed by in situ hybridization and immunoelectron microscopy. *Int J Cancer* 1996; 68:565–570.

103. Ellenrieder V, Alber B, Lacher U, Hendler SF, Menke A, Boeck W, Wagner M et al. Role of MT-MMPs and MMP-2 in pancreatic cancer progression. *Int J Cancer* 2000; 85:14–20.

104. Ohuchi E, Imai K, Fujii Y, Sato H, Seiki M, Okada Y. Membrane type 1 matrix metalloproteinase digests interstitial collagens and other extracellular matrix macromolecules. *J Biol Chem* 1997; 272:2446–2451.

105. d'Ortho MP, Will H, Atkinson S, Butler G, Messent A, Gavrilovic J, Smith B et al. Membrane-type matrix metalloproteinases 1 and 2 exhibit broad-spectrum proteolytic capacities comparable to many matrix metalloproteinases. *Eur J Biochem* 1997; 250:751–757.

106. Fosang AJ, Last K, Fujii Y, Seiki M, Okada Y. Membrane-type 1 MMP (MMP-14) cleaves at three sites in the aggrecan interglobular domain. *FEBS Lett* 1998; 430:186–190.

107. Rawlings ND, Waller M, Barrett AJ, Bateman A. *MEROPS*: The database of proteolytic enzymes, their substrates and inhibitors. *Nucleic Acids Res* 2014; 42:D503–D509.

108. Southan C. A genomic perspective on human proteases as drug targets. *Drug Discov Today* 2001; 6:681–688.

CHAPTER 9

Mapping Intracellular Protein Networks

Malgorzata E. Kokoszka, Zhaozhong Han, Ece Karatan, and Brian K. Kay

CONTENTS

9.1 INTRODUCTION

In recent years, it has become well recognized that eukaryotic cells utilize protein–protein interactions for a number of cellular processes, such as assembly of the cytoskeleton, transference of signals during signal transduction, and to compartmentalize proteins. Across different eukaryotic organisms, the predicted number of protein–protein interactions ranges from tens to hundreds of thousands [1]. Protein interaction modules often serve to mediate such protein–protein interactions [2] and have the following properties: they are typically 60–140 amino acids in length, fold autonomously within the context of the protein that contains the modular domain, and bind segments of other proteins known as short linear motifs (SLiMs), ranging from 2 to 11 amino acids in length [3,4]. Examples include the Eps15 homology (EH) domain, the phosphotyrosine-binding (PTB) domain, the postsynaptic density/disc-large/ZO1 (PDZ) domain, the Src homology (SH) 2 and 3 domains, and the WW domain (Figure 9.1).

Since protein interaction modules bind SLiMs, a fruitful approach of identifying their optimal peptide ligands has been to screen combinatorial peptide libraries. Such libraries can be synthesized on a solid support either with mixtures of amino acids or by the "split-mix" method [5,6]. Alternatively, phage-displayed combinatorial peptide libraries [7], commonly generated via mutagenesis [8] can be screened with a modular domain, and its ligand preferences can be deduced from the primary structures of the selected peptides [9]. Thus, by either approach, one can deduce the optimal ligand preferences of a protein interaction module in a few weeks. Not only is this information useful in understanding how the specificity of modules varies from one another; there is often excellent correspondence between the primary structures of the peptide ligands and regions within known interacting proteins. We have termed this phenomenon "convergent evolution" [10]. Hence, a productive process for mapping protein–protein interactions within a proteome is to identify the optimal peptide ligands for a protein interaction module, predict the interacting proteins via computational methods, and test those hypothetical interactions that make intuitive biological sense. Recent efforts in this direction are summarized in Sections 9.2 and 9.3.

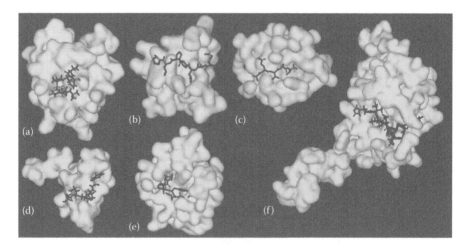

Figure 9.1   Three-dimensional structure of peptides complexed with several protein interaction modules. Short peptides (stick diagrams) are shown complexed with modules (surface view). (a) STNPFL peptide complexed with the middle EH domain of Eps15 (PDB accession number 1FF1), (b) RALPPLPRY peptide complexed with the N-terminal SH3 domain of Crk (1CKA), (c) AQTSV peptide complexed with third PDZ domain of PSD95 (1BE9), (d) GTPPPPYTVG peptide complexed with the WW domain of YAP (1JMQ), (e) pYEEI peptide complexed with the Src SH2 domain (1SPS), and (f) HIIENPNpYFSDA peptide complexed with the PTB domain of Shc (1SHC). (pY represents phosphotyrosine.) For clarity's sake, only shorter versions (underlined) of the peptides are shown complexed to the domains.

9.2 DOMAIN-MEDIATED INTERACTIONS

9.2.1 EH Domains

One domain, present in a number of proteins involved in the transport and sorting of molecules within the cell, is the EH domain [11]. This domain is 100 amino acids in length, and several three-dimensional structures have been determined by nuclear magnetic resonance spectroscopy [12–15]. To examine the molecular recognition properties of this domain, a phage-displayed combinatorial 9-mer peptide library was screened by affinity selection with glutathione-S-transferase (GST) fusions to the three EH domains present in the endocytic protein, Eps15. Interestingly, every one of the selected peptides contained the tripeptide motif NPF [16]. This motif was demonstrated to be biologically relevant in several ways: (1) It occurred multiple times in Eps15-interacting proteins; (2) mutation of any of the NPF residues destroyed binding to the EH domain; (3) the NPF residues contact the most conserved residues on the surface of the EH domain [14,17]; and (4) two NPF motifs of the same Eps15-interacting protein, such as stonin2, can simultaneously bind the same EH domain, where one of the motifs comes in contact with a noncanonical binding pocket on the surface of the domain, increasing the observed affinity [14].

Phage display has been used to determine the specificity of EH domains present in other proteins [16,18]. In each case, the EH domains selected peptides with the NPF motif, although certain residues predominated in the flanking regions, suggesting that the sequence context of the motif contributes to the interaction. For instance, the binding of phage-displayed peptide to the EH domains in intersectin is enhanced if the NPF motif is conformationally constrained by flanking cysteine residues that form intramolecular disulfide bonds [18]. While the elucidation of the EH domain's ligand preferences has proven important in mapping the interaction of EH domain-containing proteins, the motif occurs too frequently in proteomes to predict a manageable number of candidate interacting proteins for testing.

9.2.2 SH3 Domains

Src homology 3 (SH3) domains are roughly 60 amino acids in length and present in a wide array of membrane-associated and cytoskeletal protein, as well as proteins with enzymatic activity and adaptor proteins without catalytic activity [19]. Almost 300 SH3 domains, present in over 200 proteins, have been identified in the human genome [20]. In general, SH3 domains bind short (i.e., ~7 amino acid) peptides, usually proline rich, with the core motif of PxxP, where x is any of the 20 amino acids. Next to yeast two-hybrid screens and structural studies, phage-display technology functions as an integral part in the process of characterizing SH3 domain interactions [21]. Phage display has proven invaluable in determining the peptide ligand specificity of SH3 domains and predicting the cellular interactions of a large number of proteins.

For example, the N-terminal SH3 domain of the adaptor protein, Crk, selected peptides from a phage-displayed library with the motif PxLPxK (Figure 9.2). Searching GenBank with the motif demonstrated that this motif is present in Abl

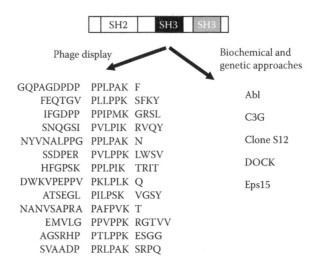

Figure 9.2 Selection of peptide ligands to the central SH3 domain of Crk. A GST fusion to the N-terminal SH3 domain of human Crk was used to screen an x_6PxxPx_6 combinatorial peptide library, displayed at the N-terminus of mature protein III of bacteriophage M13. The primary structures of the selected peptides are shown aligned to highlight the consensus motif, PxLPxK. Through a combination of co-immunoprecipitation, filter lift, pull-down, and yeast two-hybrid experiments, the Crk SH3 domain has been shown to bind to Abl, C3G, Clone S12, DOCK, and Eps15.

(PLLPTK), C3G (PALPPK), clone S12 (PGLPSK), DOCK 180 (PPLPLK), and Eps15 (PALPPK), all of which are previously identified, through a series of bio-chemical experiments (i.e., immunoprecipitation, far Western blotting, and screen-ing complementary DNA [cDNA] expression libraries), as Crk-interacting proteins. Therefore, several of the cellular ligands of the Crk SH3 domain could have been predicted directly through phage-display experiments.

The knowledge about specificity of individual SH3 domains, determined through phage display, provides useful insight into how this module mediates specific protein–protein interactions in the cell. For example, the motif [22], which was selected for binding to the SH3 domain of Src (RxLPxLP), occurs in the potassium channel Kv1.5 (RPLPxxP) and synapsin (RPQPPPP), while the motif [22] selected for the SH3 domain of cortactin (+PPxPxKPxWL) occurs in CBP90 (KPPVPPKPKMK) and Shank (KPPVPPKPKLK). All of these molecular interactions have been shown to be SH3 domain-mediated [23–26], and one can verify the importance of the match-ing peptide sequences through deletion and mutagenesis experiments. (In addition, one can use these motifs to predict that the alpha 1 subunit of the type E neuronal Ca^{2+} channel [RQLPPVP] and the faciogenital dysplasia protein [KPQVPPKPSYL] likely interact with the SH3 domains of Src and cortactin, respectively.) Phage-display experiments have also proven invaluable in mapping the SH3 domain-mediated interactions of endophilin and amphiphysin with synaptojanin [27], Eps8 with Abi-1 [28], and Abp1p with the Ser/Thr kinases Prk1p and Ark1p [29].

The availability of sequenced genomes has rendered large-scale proteome-wide analysis of protein–protein interactions plausible. It is now possible to construct protein interaction networks consisting of all of the protein interaction modules and their interaction partners in the proteome of an organism. For instance, Tong et al. [30] applied phage-display technology in a proteome-wide approach to create an interaction network of all the SH3 domain-containing proteins and their binding partners in *Saccharomyces cerevisiae*. The authors identified 28 SH3 domains in the yeast proteome and constructed GST fusions to each of the domains, 24 of which could be expressed as soluble proteins in *Escherichia coli*. After three rounds of selection, ligands were identified for all but four of the SH3 domains, which the authors suspect may not bind short peptides with sufficient (i.e., micromolar) affinity. For the remaining 20 SH3 domains, consensus sequences were identified and used to scan the yeast proteome by computer for potential binding partners: The search resulted in 206 proteins with 394 potential interactions. The result set is displayed in Figure 9.3 as a network consisting of nodes and lines, representing the individual proteins and interactions between the proteins, respectively.

To confirm the SH3 domain-mediated interaction predicted through phage display, Tong et al. [30] have utilized yeast two-hybrid screens [31,32]. Eighteen of the SH3 domains were used as bait in two-hybrid screens of either individual open reading frames or libraries of fragmented cDNAs fused to the Gal4 activation domain, resulting in the identification of 145 interacting proteins with 233 potential interactions [30]. Comparison of the protein–protein interactions predicted by phage display and yeast two-hybrid analyses yielded 39 proteins with 59 common interactions. To test the physiological relevance of the overlapping data sets, Las17p, a yeast protein involved in actin assembly [33], was selected for further scrutiny. The two data sets predicted that 10 different SH3 domain-containing proteins bind Las17p, of which 3 were known binding partners, 4 have been identified in other two-hybrid screens, and 3 are previously unidentified binding partners. Coimmunoprecipitation analysis of yeast cells, which were transformed with hemagglutinin (HA) epitope-tagged Las17p and c-myc epitope-tagged SH3 domain-containing proteins, demonstrated that many of the predicted protein interactions occurred in vivo. To map the sites of interaction within the Las17p protein, five proline-rich regions of Las17p were displayed separately as fusions to capsid protein D of bacteriophage lambda and assayed for binding to the SH3 domains in an enzyme-linked immunosorbent assay (ELISA). The experimental binding sites agreed in 9 out of 10 cases with the binding sites inferred from the peptide ligands selected by phage display.

Over the years, different SH3 domains of interest were frequently subject to screenings against phage-displayed peptide libraries with the common goal to isolate peptide ligands that would later be a starting point in predicting cellular interactions. On the other hand, a "reverse" approach can be also used to study SH3 domain–ligand interactions. In the study by Karkkainen et al. [20], a phage-displayed library of 281 human SH3 domains has been generated as a fusion to the pVIII gene of the M13 phagemid vector. Among the three targets used in the study were human immunodeficiency virus-1 Karkkainen Nef, p21-activated kinase 2 (PAK2), and ADAM15, all

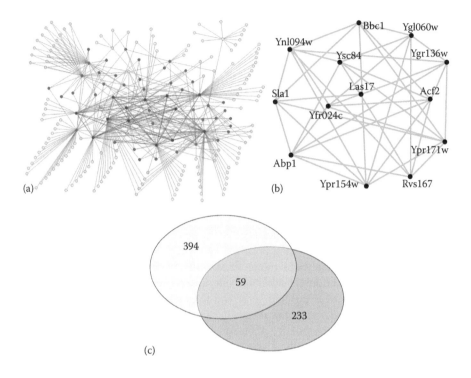

Figure 9.3   Interactions among yeast proteins predicted to be mediated by SH3 domains. (a) Diagram showing all the interactions predicted by phage-display experiments. The nodes and lines represent proteins and protein–protein interactions, respectively. (b) Enlargement of the proteins with the most SH3 domain–mediated interactions. (Panels a and b are modified from Tong AH et al., *Science*, 295:321–24, 2002.) (c) Venn diagram comparing the number of SH3 domain–mediated interactions predicted by phage display (394, light gray) with yeast two-hybrid screening (233, dark gray), where 59 of the predicted interactions overlap.

established ligands for specific SH3 domain interactions. The results of this study not only confirmed already-known SH3 domain–ligand interactions but also identified some novel interacting proteins such as POSH2, PAK2-binding adaptor protein, and SNX30, an ADAM15-binding partner. SH3 domain–ligand interactions are usually in micromolar range, but in this study, when the targets were screened against the full set of SH3 domains, several nanomolar–range interactions were isolated. For instance, measured K_d values for POSH2–PAK2 and SNX30–ADAM15 are 75 nM and 130 nM, respectively.

9.2.3 PDZ Domains

PSD95/discs large/ZO1 (PDZ) domains are 80–90 amino acids in length and frequently found in proteins associated with the cellular membrane, where they coordinate the assembly of transmembrane and cytosolic components into multipro-tein complexes [34,35].

The PDZ domain has an overall structure very much like the PTB domain, even though they are unrelated in function. It contains a core of five or six β-sheets (β1–β6) and two α-helices (α1 and α2), and peptide ligands fit into a hydrophobic pocket formed by α2, β2, and the conserved GLGF loop that connects the β1 and β2 strands [35]. Unlike other protein interaction modules, PDZ domains recognize the free carboxy-termini (C-termini) of target proteins, along with the penultimate three or four residues. Depending on the consensus sequence of its preferred ligands, PDZ domains can be grouped into different classes: Type I domains recognize C-terminal peptides with the consensus sequence (T/S)x(V/I/L)-COOH, and type II domains recognize (Ψ)x(Ψ)-COOH, where Ψ is a hydrophobic residue (i.e., V, Y, F, L, I) [36,37].

To learn what the ~2000 PDZ domains in sequenced genomes bind to, a use-ful approach has been to affinity-select ligands from combinatorial peptide libraries and use them to predict the candidate interacting proteins. PDZ domain specificity has been studied extensively with repertoires of chemically synthesized combinato-rial peptides [37] and a variety of display systems. By fusing coding regions to the C-terminus of the Lac repressor [38], it has been possible to define the specificity of the PDZ domain in neuronal nitric oxide synthase (nNOS) [39]. Guided by the DxV-COOH consensus in the selected peptides, several candidate interacting pro-teins have been identified for nNOS, including the glutamate and melatonin recep-tors. With a library displaying combinatorial peptides at the C-terminus of the capsid D protein of bacteriophage λ, Vaccaro et al. [40] have defined the ligand specific-ity of the seven PDZ domains of the human INADL protein. From the consensus sequences of the selected peptides, six of the PDZ domains correspond to either type I or II, with the seventh belonging to a new type characterized by the presence of an acidic residue at the C-terminal position of the peptide ligands.

While the carboxylate group makes an important contribution of ligand binding to many PDZ domains, it does not for all. When an M13 phage library display-ing 12-mer combinatorial peptides displayed at the N-terminus of protein III was affinity-selected with the PDZ domain of α-syntrophin, a component of the dys-trophin protein complex, three peptides with intramolecular disulfides were recov-ered [41]. The peptides shared residues with several known cellular ligands for the α-syntrophin PDZ domain but without a C-terminal carboxylate group, suggesting that conformationally constrained peptide ligands could mimic ligands for certain PDZ domains. The significance of this observation became clearer with the subse-quent discovery that β-hairpin peptide "fingers" could fit into the groove of PDZ domains that normally binds the C-termini of cellular ligands [42].

It is also possible to display peptides at the C-termini of proteins III or VIII of bacteriophage M13 [43]. In a modified phagemid display vector, C-terminal fusions result in display levels comparable to those achieved with conventional N-terminal fusions to either capsid protein. With a library of combinatorial peptides fused to the C-terminus of protein VIII, peptide ligands have been selected for a variety of PDZ domains. Affinity selection experiments with the PDZ2 and PDZ3 domains of MAGI 3, a membrane-associated guanylate kinase, selected peptides with the consensus Xxx-Xxx-Xxx-Cys/Val-Ser/Thr-Trp-Val-COOH, while the PDZ3 domain selected only a single 7-mer peptide sequence (i.e., TRWWFDI-COOH). When synthetic peptides corresponding to the PDZ2 consensus were tested for binding to the domain, they were found to bind stronger than a peptide corresponding to the C-terminal sequence (i.e., HTQITKV-COOH) of a natural PDZ2 ligand, the tumor suppressor PTEN/MMAC [43]; examination of the peptides suggested that the Trp residue (underlined) in the Cys/Val-Ser/Thr-<u>Trp</u>-Val-COOH sequence contributed to the stronger binding. Protein database searches with the motif of the peptide ligands of PDZ2 suggest that that δ-catenin, an epithelial cell junction protein, which contains such a sequence at its C-terminus, might interact with MAGI 3 via its PDZ2 domain. Affinity selection experiments of a library of combinatorial peptides displayed at the C-terminus of pVIII with the PDZ domain of ERBIN, a basolateral protein that interacts with the mammalian ERBB2/HER2 receptor, yielded type 1 ligands that bound strongly and specifically [44]. The selected peptides closely resembled the C-terminal sequence of three p120-like catenins (δ-catenin, ARVCF, and p0071), in addition to ERBB2/HER2. To test the significance of these predicted interactions with the ERBIN PDZ domain, both δ-catenin and ARVCF were confirmed to interact in vitro and in vivo, suggesting that ERBIN may play a role of integrating cytoskeletal functions with epithelial cell morphology and polarization.

A recent large-scale study of human and worm PDZ domain–ligand interactions, using combinatorial peptide libraries, revealed 16 distinct specificity classes, thus expanding the canonical type I/II classification system. Different classes are defined by up to six C-terminal positions of the binding site (i.e., 0, -1, -2, -3, -4, -5). Interestingly, each of the analyzed positions showed several distinct specificities, thus showing that different PDZ domains can bind diverse sequence motifs. For example, a hydrophobic residue is always preferred at position 0; however, depending on the specificity class, position 0 can be occupied by C, L, I, F, V, or W. Also, position -1, originally defined as "any residue," shows seven different specificities, such as W, D, R/K, S/R, or L [45].

9.2.4 WW Domains

Another protein interaction module to discuss is the WW domain, named for the presence of two invariable tryptophan residues present in the 40-amino-acid module [46]. Across the human genome there are over 50 proteins containing ~100 WW domains, often occurring multiple times in a single protein, ranging from one to four copies [47]. Characterization of two proteins, WBP1 and WBP2, discovered in a screen of a λ-cDNA expression library with a radiolabeled form of the WW domain of Yes-associated protein (YAP), revealed a common PPPPY peptide ligand sequence [48]. In addition, a 10-mer peptide encompassing this motif was able to bind the WW domain of YAP, and alanine scanning of the peptide sequence revealed that the motif PPxY, where x is any amino acid, was necessary for this interaction. Based on this motif, a variety of proteins have been proposed and confirmed to interact with WW domain-containing proteins, such the cytosolic domain of epithelial amiloride-sensitive sodium channels (ENaC). Genetic evidence supports the biological importance of the interaction of the ENaC with Nedd4, an ubiquitin ligase with three WW domains, as mutations in the PPPY motif in the channel cause Liddle's syndrome [49], a form of hypertension that is due to the extended cellular half-life of the channel [50].

The peptide ligand specificity of a variety of WW domains has been defined with phage-displayed combinatorial peptides, and this information then has been used to predict their cellular interaction partners. Selection of peptides with the YAP WW domain yielded the core sequence PPPPYP [51], which is present in the interacting proteins WBP1 and WBP2. Computer analysis also identified this motif within the p53 binding protein-2 (p53BP2), and this protein–protein interaction was later confirmed in a yeast two-hybrid screen [52]. Another matching protein was the p45 subunit of the NF-E2 transcription factor, which was confirmed by biochemical [53] and mutational [54] experiments. Screens of phage libraries with a dozen other WW domains, related in primary structure to the YAP WW domain, revealed an almost absolutely conserved core motif of Pro-Xxx-Tyr among the peptides selected, with different WW domains preferring varying residues N- and C-terminal of this core [55]. Computer-aided searches of the protein database revealed several candidate interactions, including the homologous to the E6-AP C-terminus (HECT) domain, which is involved in the ubiquitination of proteins.

Sometimes, screens of phage-displayed combinatorial peptide libraries with certain WW domains fail to yield peptide motifs. While screens with the WW domain of the cytoskeletal protein, utrophin, were initially negative, a larger protein segment (including flanking EF hands and a ZZ domain) yielded binding phage that displayed the motif PPxY [56]. This motif is present near the C-terminus of β-dystroglycan, a membrane protein known to bind utrophin and link the actin cytoskeleton to the extracellular basal lamina. Thus, the ability of the utrophin WW domain to bind peptides and cellular ligands is dependent on its structural context. Support for this conclusion comes from biochemical and structural analyses of the WW domain in a related protein, dystrophin: The WW domain of dystrophin requires its flanking EF hands to bind to a C-terminal segment of β-dystroglycan [57], and the β-dystroglycan peptide binds a composite surface formed by the WW domain and its flanking regions [58]. Thus, when phage library screens are negative, larger segments of the target proteins may be necessary to promote folding or provide additional contacts for peptide ligand binding. Screens with the WW domain of the Pin1 protein also failed to yield peptide ligands [55]. This negative result is to be expected, as the Pin1 WW domain recognizes a posttranslationally modified peptide motif, pSer/pThr-Pro, where pSer and pThr represent phosphoserine and phosphothreonine, respectively [59].

9.2.5 PTB and SH2 Domains

Another protein interaction module is the PTB domain, a 100-amino-acid-long domain that binds to specific phosphotyrosine peptide sequences within proteins. In general, the specificity of PTB domains is mediated by amino acids N-terminal to the phosphotyrosine in the cellular ligand [60]. While chemically synthetic peptide libraries have been quite useful in mapping the ligand preferences of individual PTB domains [61], it also has been possible to use phage-displayed peptide libraries after incubating them with a protein tyrosine kinase and then affinity-selecting phage with the PTB domain. For example, incubation of phage displaying combinatorial 9-mer peptides with the protein tyrosine kinase Fyn permitted the selection of phage that bound to the PTB domain of the adaptor protein Shc [62]. Among the 18 different isolates, there was the consensus (F/Y)xNPTpYxx(Y/W), where x and pY correspond to any amino acid and phosphotyrosine, respectively. This phage-derived motif compares favorably with the motifs (F/Y)xNPxpY and NPxpY, which were identified from screens with synthetic combinatorial peptide libraries [63] and inspection of the primary structures of known cellular ligands of the PTB domain of Shc, respectively.

Combinatorial peptide libraries have also played a useful role in mapping the specificity of the Src homology 2 (SH2) domain, another module that binds to specific phosphotyrosine peptide sequences within proteins. Unlike PTB domains, the specificity of SH2 domains is mediated by amino acids C-terminal to the phosphotyrosine in the cellular ligand. The ligand preferences of a large number of SH2 domains have been defined with synthetic peptides [61] and phage-displayed combinatorial peptide libraries, which have been preincubated with protein tyrosine kinases and then affinity-selected phage with an SH2 domain. For example, experiments in which phage were incubated with a mixture of protein tyrosine kinases (c-Src, Blk, and Syk) and selected with the SH2 domain of growth factor receptor binding protein 2 (Grb2) revealed that this SH2 domain recognizes the peptide sequence pY(M/E)NW [64]. By a similar approach, the peptide ligand preferences of the SH2 domains of Grb2, Shc, Sli, and Rai have also been defined [62].

From phage display of cDNA fragments, it is also possible to identify candidate interacting proteins for SH2 domains. For example, a library of phage displaying fragments of leukocyte cDNA was incubated with protein tyrosine kinase Fyn and selected with the tandem SH2 domains of SHP-2, a cytoplasmic tyrosine phosphatase. This approach led to the selection of clones encoding the cytoplasmic domain of PECAM-1, a known cellular ligand for SHP-2 [65]. As more phage-display (i.e., lambda, T7) cDNA libraries become available, it should be possible to identify cellular ligands of a large variety of different PTB and SH2 domains by this approach [66].

Even though SH2 domains typically bind to specific phosphotyrosine-containing peptide sequences within cellular proteins, they have also been reported to bind proteins in a phosphotyrosine-independent manner. Two research groups have selected nonphosphorylated peptide ligands from phage-displayed combinatorial peptide libraries that can bind to the Grb2 SH2 domain. One group isolated a single peptide sequence, CELYENVGMYC, that bound selectively to the SH2 domain of Grb2 and, when synthesized in a cyclized, disulfide form, competed the binding of natural phosphopeptide ligands, with an inhibitory concentration for 50% activity (IC_{50}) value of 15.5 μM [67]. Replacement of either the tyrosine or asparagine residues in this peptide led to a loss of binding, demonstrating that the YxN motif contributes to binding to the Grb2 SH2 domain, even without tyrosine phosphorylation. A second research group also screened three phage-displayed combinatorial peptide libraries with a GST fusion to the Grb2 SH2 domain and isolated peptides that contained the YxN motif [68]. While the selected peptides were preceded and followed by charged and hydrophobic residues, respectively, only a subset of peptides contained pairs of cysteine residues. Synthetic forms of the peptides were competent to compete with the recognition of the phosphorylated epidermal growth factor (EGF-R) by the Grb2 SH2 domain with an IC_{50} of 2–60 μM. Interestingly, synthesis of the phage-selected YxN peptides with a phosphotyrosine in place of the tyrosine residue increased their affinity 10- to 100-fold for the Grb2 SH2 domain [68]. Thus, the primary and secondary structural contexts of the YxN motif contribute significantly in positioning peptides into the SH2 domain pocket.

Nonphosphorylated peptide ligands have also been selected for two other SH2 domains. The C-terminal SH2 domain within the 85 kDa subunit (p85) of the phosphatidylinositol 3-kinase (PI 3-kinase) has been used to affinity-select peptides from a phage-displayed combinatorial peptide library [69]. The selected peptides shared the motif (L/A)A(R/K)IR, and a search of GenBank matched the serine/threonine kinase A-Raf, which contained several examples of the motif. Coimmunoprecipitation experiments confirmed that these two matching proteins indeed form a protein complex in a mouse cell line. The SH2 domain of Grb7, an adaptor-type signaling protein, was used to screen an $x_4Cx_{10}Cx_4$ combinatorial peptide library and selected peptides with the motif Y(A/D/E/G)N in the central 10-mer region [70]. From a second library randomizing this motif, the investigators isolated peptides that bound the SH2 domain of Grb7 and not that of Grb2 or Grb14. Synthetic, cyclized forms of these evolved peptides were able to compete the binding of Grb7 to the phosphorylated form of the ErbB3 receptor, with an IC_{50} of ~20 μM. While the sequence of such peptides does not exactly match the sequences of Grb7-interacting proteins, these peptides may prove useful in future drug discovery efforts (see Section 9.6).

9.2.6 Chromo Shadow Domain

In chromatin, heterochromatin-associated protein 1 (HP1) is thought to regulate heterochromatin structure through interactions with other proteins. To define the basis for selective protein interactions, the C-terminal chromo shadow domain of the *Drosophila melanogaster* HP1 protein was used to select peptide ligands from a phage-displayed combinatorial peptide library. Many of the peptides contained the consensus sequence P-R/W/Y-V-L/M/V-L/M/V [71]. Interestingly, not only is this pentapeptide sequence present in the primary structure of many reported HP1-associated proteins, but it is also occurs in the shadow domain, suggesting that HP1 dimerization may occur through binding this motif as well. This conclusion is supported by the observation that synthetic forms of the peptides, which bind the shadow domain, disrupt HP1 dimerization.

9.3 NONDOMAIN-MEDIATED PROTEIN–PROTEIN INTERACTIONS

In a number of cases, screens of phage-displayed combinatorial peptide libraries have also proven useful in mapping and predicting protein–protein interactions. Peptides selected for binding to protein VanR, the two-component signal transduction response regulator that controls expression of vancomycin resistance in *Enterococcus faecium*, yielded a 12-mer peptide that could be aligned with an 18-mer sequence of the catalytic center dimerization domain of VanS, a sequence with which VanR also normally interacts [72]. Amino acid replacement experiments support a model in which the selected peptide mimics the VanS phosphorylatable sequence with which the regulatory domain of VanR interacts and thus functions as a "minimalist" analogue of VanS. Selection of peptides that bind to tumor necrosis factor β (TNF-β) yielded a peptide sequence, RKEMGGGGGPGWSENLFQ, which contained two short motifs (RKEM, WSENLFQ) that identically matched two regions spaced 25 amino acids apart in the primary structure of tumor necrosis factor receptor 1 (TNFR1) [73]. Selection of peptides that bind to the two HMG boxes of the high mobility group protein 1 (HMGB1) yielded a large collection of diverse sequences, several of which matched regions within a number of nuclear proteins and transcription factors [74]. GST fusion protein pull-down experiments confirmed that a number of the candidate proteins could interact with HMGB1 in vitro.

9.4 SOFTWARE FOR IDENTIFYING CANDIDATE INTERACTING PARTNERS

Several different resources, such as algorithms, web servers, and databases, are available for analyzing mimotopes and predicting interacting proteins in the proteomes, based on matching the motif of peptide ligands identified through phage display. In the review by Huang et al. [75], five special databases for mimotope data (e.g., ASPD, MOTIF, MimoDB) and 18 algorithms, programs, and web servers and their applications for phage display studies have been described. For instance, the MimoDB database (http://immunet.cn/mimodb/) contains affinity selection results from combinatorial peptide libraries, with peptide sequences less than 40 amino acids long [76]. The current version is a collection of 20,572 unique peptide sequences to 1448 targets from 1143 original publications.

On the other hand, with the SeqIt protein database mining tool, a web-based program available from Array Genetics (http://www.arraygenetics.com), one can rapidly search a protein database with a motif and identify all proteins containing that amino acid sequence of interest. Conversely, with the Scansite algorithm (http://scansite.mit.edu), one can identify protein sequences likely to bind to certain protein interaction modules, such as SH2 or PDZ, or likely to be phosphorylated by specific protein kinases [77]. For SH3 domains, the SH3-SPOT algorithm can be used to predict cellular ligands [78].

With the increasing frequency of publications of particular protein–protein interactions, a number of protein interaction databases have been created. The Molecular

Interaction (MINT) database (http://cbm.bio.uniroma2.it/mint/), in addition to cataloging interactions, also stores other types of functional interactions, including enzymatic modifications of one of the partners [79]. Presently, MINT contains over 240,000 interactions. The Database of Interacting Proteins (DIP) is a database (http://dip.doe-mbi.ucla.edu) that documents experimentally determined protein–protein interactions. The DIP presently catalogs more than 77,000 interactions among over 26,700 proteins from >665 organisms. The Biomolecular Object Network Databank (BOND, http://bond.unleashedinformatics.com/) is a data bank that integrates several existing databases, such as Small Molecule Interaction Database (SMID) or Biomolecular Interaction Network Database (BIND), and contains information on small molecule and protein interactions. For instance, the SMID is a small molecule–protein domain interaction database and is based on structural data from the Protein Data Bank (PDB) [80]. The BIND database records the interaction between two objects (i.e., protein, DNA, RNA, ligand, molecular complex). The database also describes the cellular location, experimental conditions used to observe the interaction, conserved sequence, molecular location, chemical action, kinetics, thermodynamics, and chemical state [81].

9.5 ANALYZING PREDICTED INTERACTIONS

Once the ligand preferences for a protein interaction module or a target protein have been defined, candidate interacting proteins can be tested for their ability to interact with the module (or target protein) by a variety of methods such as affinity chromatography [82], cross-linking, filter lifts [83], mass spectrometry [84], and two-hybrid screening [85] experiments. However, for a number of reasons, it is reasonable to expect that some of the interactions predicted from the phage-display data may not actually occur in the cell or organism. First, it is possible that the two proteins are not expressed in the same cell type or at the same time, or occur in different cellular compartments. Second, it is possible that due to predominance of one protein–protein interaction, other protein–protein interactions occur infrequently in the cell. Third, it is also possible that another motif, not identified through phage display, is responsible for molecular interactions. Fourth, it is also possible that peptide ligand sequence is not accessible within the candidate protein either due to being buried within the tertiary structure or the presence of another protein that sterically interferes with binding. Fifth, flanking resides (or posttranslational modifications) may influence the strength of binding of the candidate protein with the target. Thus, while phage display has the potential to predict both physiologically relevant and nonrelevant interactions, there are many experimental options available for investigators to sort through them quickly.

It should also be noted that there are several limitations and disadvantages of phage-displayed peptide libraries. For example, the adequate size of the library can be sometimes unachievable to cover all the possible recombinants in the library. Also, runs of positively charged residues, such as arginine, lead to defective phage production, thus drastically decreasing the titer and resulting in underrepresentation of those library individuals [86]. A position-dependent amino acid bias in the composition of combinatorial peptide libraries has also been observed [87]. Finally, the selection process favors even numbers of cysteines [88] and is biased toward hydrophobic residues, thus possibly compromising the likelihood of correctly predicting the protein–protein interactions [89].

To minimize such discrepancies, one can synthesize large numbers of peptides corresponding to all the regions within a proteome that resemble the phage-display motif [27,90], monitor the actual binding of the peptides to the target in vitro, and then rank-order the list of candidate interacting proteins for further testing based on their strength of binding. Also recently, a proteomic library encoding human and viral peptide sequences has been employed for mapping protein–protein interactions of several PDZ domains [91]. In this study, custom-printed microarrays with oligonucleotides, encoding over 50,000 human and 10,000 viral 7-mer C-terminal protein sequences, were synthesized and the encoded sequences PCR-amplified for subsequent library construction. The resulting peptide libraries were displayed as C-terminal fusions to the major coat protein, pVIII, using a phagemid vector; affinity selections confirmed several known ligands of PDZ domains, and the selected peptides were less hydrophobic than the ones isolated from traditional combinatorial peptide libraries. Therefore, one can use this approach to assess systematically any proteome of choice for its potential to interact with a particular target.

9.6 RELEVANCE TO BIOTECHNOLOGY AND DRUG DISCOVERY

Once peptide ligands have been discovered and used to predict protein interactions, one can consider using the peptides in three different manners for drug discovery. First, the peptide can be injected, overexpressed, or linked to a membrane translocating sequence for delivery into cells [92], where it binds to the target protein and prevents a particular protein–protein interaction. If the cell requires that interaction for a critical cellular function, then potentially, there will be an observable biological consequence. There are several precedents in the literature for this type of perturbation experiment. For example, injection of synthetic peptide ligands for the SH3 domain of Src into frog oocytes leads to an acceleration of the completion of meiosis [93], whereas electroporation of peptide ligands for the SH3 domain of Lyn blocks mast cell activation [94]. These results suggest that Src and Lyn play important regulatory roles in oocyte maturation and mast cell activation, respectively. Peptides with an LxxLL motif have been selected via phage display that bind estrogen receptor α or β when complexed with estradiol [95]. As these peptides appear to mimic an LxxLL sequence region in transcriptional coactivators, they block transcriptional activity of either receptor when overexpressed in mammalian tissue culture cells [96,97]. Also, affinity selections with phage-displayed libraries led to isolation of 6-mer peptides that can efficiently block binding of human CXCL8, a chemokine associated with many inflammatory diseases, to its cognate receptors CXCR1 and CXCR2. Prior incubation of hCXCL8 with synthetic peptides resulted in inhibition of hCXCR1/2-mediated chemotaxis in human neutrophils, thus showing the inhibitory potential of isolated peptides [98]. Finally, peptides, which bind to the active sites of enzymes and inhibit them, have been used in the same manner to evaluate the essentiality of the target proteins of bacteria [99,100]. Thus, peptide ligands have great utility in validating targets as being suitable for drug discovery.

Second, once a peptide ligand with antagonist activity has been identified, it can serve as a starting point in the design of a peptidomimetic inhibitor. The K_d values of peptides recovered by phage display, when chemically synthesized and tested in solution, range from 10 µM to 500 nM [101]. To enhance their affinity, it has been necessary to replace certain amino acids in the peptide with other natural and unnatural amino acids, as well as with different chemical entities. In this manner, the affinity of a peptide ligand for the Crk SH3 domain has been increased 20- [102] to 100-fold [103]. The design of an effective peptidomimetic can also be greatly aided by structure-based drug design, as demonstrated by the conversion of a phage-displayed peptide ligand for the major histocompatibility HLA-DR molecule [104] into a protease-resistant molecule that is 2000-fold more effective than the parent heptapeptide in inhibiting T-cell proliferation [105].

Third, the phage-displayed peptides can be used to establish a displacement assay in which collections of chemicals and natural products can be screened for antagonists. Since the peptides frequently bind at sites of protein–protein interactions of a target, any molecule that prevents binding of the peptide to the target presumably binds at the same site and would function as an inhibitor [106]. Therefore, synthetic forms of the peptides can be used to configure a competitive-binding assay for detecting chemical inhibitors. Many different types of such assays can be formatted in microtiter plate wells and monitored by measuring changes in well color, luminescence, fluorescence, fluorescence polarization, time-resolved fluorescence, or fluorescence resonance energy transfer [107,108]. As these assays are amenable to robotics and automation, large chemical and natural product libraries can be screened for chemical modifiers of target activity, some of which may be developed into drugs.

REFERENCES

1. Stumpf MPH, Thorne T, de Silva E et al. Estimating the size of the human interactome. *Proc Natl Acad Sci U S A* 2008; 105:6959–64.
2. Songyang Z. Recognition and regulation of primary-sequence motifs by signaling modular domains. *Prog Biophys Mol Biol* 1999; 71:359–72.
3. Davey NE, Van Roey K, Weatheritt RJ et al. Attributes of short linear motifs. *Mol BioSyst* 2012; 8:268–81.
4. Weatheritt RJ, Luck K, Petsalaki E, Davey NE, Gibson TJ. The identification of short linear motif-mediated interfaces within the human interactome. *Bioinformatics* 2012; 28:976–82.
5. Lam KS, Salmon SE, Hersh EM, Hruby VJ, Kazmierski WM, Knapp RJ. A new type of synthetic peptide library for identifying ligand-binding activity. *Nature* 1991; 354:82–4.
6. Liu R, Enstrom AM, Lam KS. Combinatorial peptide library methods for immunobiology research. *Exp Hematol* 2003; 31:11–30.

7. Scholle MD, Kehoe JW, Kay BK. Efficient construction of a large collection of phage-displayed combinatorial peptide libraries. *Comb Chem High Throughput Screen* 2005; 8:545–51.

8. Huang R, Fang P, Kay BK. Improvements to the Kunkel mutagenesis protocol for constructing primary and secondary phage-display libraries. *Methods* 2012; 58:10–7.

9. Sparks AB, Adey NB, Cwirla S, Kay BK. Screening phage-displayed random peptide libraries. In: McCafferty J, editor. *Phage Display of Peptides and Proteins: A Laboratory Manual*. San Diego, CA: Academic Press; 1996.

10. Kay BK, Kasanov J, Knight S, Kurakin A. Convergent evolution with combinatorial peptides. *FEBS Lett* 2000; 480:55–62.

11. Confalonieri S, Di Fiore PP. The Eps15 homology (EH) domain. *FEBS Lett* 2002; 513:24–9.

12. de Beer T, Carter RE, Lobel-Rice KE, Sorkin A, Overduin M. Solution structure and Asn-Pro-Phe binding pocket of the Eps15 homology domain. *Science* 1998; 28:1357–60.

13. Koshiba S, Kigawa T, Iwahara J, Kikuchi A, Yokoyama S. Solution structure of the Eps15 homology domain of a human POB1 (partner of RalBP1). *FEBS Lett* 1999; 442:138–42.

14. Rumpf J, Simon B, Jung N et al. Structure of the Eps15-stonin2 complex provides a molecular explanation for EH-domain ligand specificity. *EMBO J* 2008; 27:558–69.

15. Whitehead B, Tessari M, Carotenuto A, van Bergen en Henegouwen PM, Vuister GW. The EH1 domain of Eps15 is structurally classified as a member of the S100 subclass of EF-hand-containing proteins. *Biochemistry* 1999; 38:11271–7.

16. Salcini AE, Confalonieri S, Doria M et al. Binding specificity and in vivo targets of the EH domain, a novel protein–protein interaction module. *Genes Dev* 1997; 11:2239–49.

17. de Beer T, Hoofnagle AN, Enmon JL et al. Molecular mechanism of NPF recognition by EH domains. *Nat Struct Biol* 2000; 7:1018–22.

18. Yamabhai M, Hoffman NG, Hardison NL et al. Intersectin, a novel adaptor protein with two EH and five SH3 domains. *J Biol Chem* 1998; 273:31401–6.

19. Macias MJ, Wiesner S, Sudol M. WW and SH3 domains, two different scaffolds to recognize proline-rich ligands. *FEBS Lett* 2002; 513:30–7.

20. Karkkainen S, Hiipakka M, Wang JH et al. Identification of preferred protein interactions by phage-display of the human Src homology-3 proteome. *EMBO Rep* 2006; 7:186–91.

21. Kay BK. SH3 domains come of age. *FEBS Lett* 2012; 586:2606–8.

22. Sparks AB, Rider JE, Hoffman NG, Fowlkes DM, Quilliam LA, Kay BK. Distinct ligand preferences of SH3 domains from Src, Yes, Abl, cortactin, p53BP2, PLCg, Crk, and Grb2. *Proc Natl Acad Sci U S A* 1996; 93:1540–4.

23. Foster-Barber A, Bishop JM. Src interacts with dynamin and synapsin in neuronal cells. *Proc Natl Acad Sci U S A* 1998; 95:4673–7.

24. Holmes TC, Fadool DA, Ren R, Levitan IB. Association of Src tyrosine kinase with a human potassium channel mediated by SH3 domain. *Science* 1996; 274:2089–91.

25. Naisbitt S, Kim E, Tu JC et al. Shank, a novel family of postsynaptic density proteins that binds to the NMDA receptor/PSD-95/GKAP complex and cortactin. *Neuron* 1999; 23:569.

26. Ohoka Y, Takai Y. Isolation and characterization of cortactin isoforms and a novel cortactin-binding protein, CBP90. *Genes Cells* 1998; 3:603–12.

27. Cestra G, Castagnoli L, Dente L et al. The SH3 domains of endophilin and amphiphysin bind to the proline-rich region of synaptojanin 1 at distinct sites that display an unconventional binding specificity. *J Biol Chem* 1999; 274:32001–7.

28. Mongiov AM, Romano PR, Panni S et al. A novel peptide-SH3 interaction. *EMBO J* 1999; 18:5300–9.

29. Fazi B, Cope MJ, Douangamath A et al. Unusual binding properties of the SH3 domain of the yeast actin-binding protein Abp1: Structural and functional analysis. *J Biol Chem* 2002; 277:5290–8.

30. Tong AH, Drees B, Nardelli G et al. A combined experimental and computational strategy to define protein interaction networks for peptide recognition modules. *Science* 2002; 295:321–4.

31. Fields S, Song O. A novel genetic system to detect protein–protein interactions. *Nature* 1989; 340:245–6.

32. Phizicky EM, Fields S. Protein-protein interactions: Methods for detection and analysis. *Microbiol Rev* 1995; 59:94–123.

33. Naqvi SN, Zahn R, Mitchell DA, Stevenson BJ, Munn AL. The WASp homologue Las17p functions with the WIP homologue End5p/verprolin and is essential for endocytosis in yeast. *Curr Biol* 1998; 8:959–62.

34. Harris BZ, Lim WA. Mechanism and role of PDZ domains in signaling complex assembly. *J Cell Sci* 2001; 114:3219–31.

35. Ye F, Zhang M. Structures and target recognition modes of PDZ domains: Recurring themes and emerging pictures. *Biochem J* 2013; 455:1–14.

36. Nourry C, Grant SG, Borg JP. PDZ domain proteins: Plug and play! *Sci STKE* 2003; 2003:RE7.

37. Songyang Z, Fanning AS, Fu C et al. Recognition of unique carboxyl-terminal motifs by distinct PDZ domains. *Science* 1997; 275:73–7.

38. Cull MG, Miller JF, Schatz PJ. Screening for receptor ligands using large libraries of peptides linked to the C terminus of the *lac* repressor. *Proc Natl Acad Sci U S A* 1992; 89:1865–9.

39. Stricker NL, Christopherson KS, Yi BA et al. PDZ domain of neuronal nitric oxide synthase recognizes novel C-terminal peptide sequences. *Nat Biotechnol* 1997; 15:336–42.

40. Vaccaro P, Brannetti B, Montecchi-Palazzi L et al. Distinct binding specificity of the multiple PDZ domains of INADL, a human protein with homology to INAD from Drosophila melanogaster. *J Biol Chem* 2001; 276:42122–30.

41. Gee SH, Sekely SA, Lombardo C, Kurakin A, Froehner SC, Kay BK. Cyclic peptides as non-carboxyl-terminal ligands of syntrophin PDZ domains. *J Biol Chem* 1998; 273:21980–7.

42. Hillier BJ, Christopherson KS, Prehoda KE, Bredt DS, Lim WA. Unexpected modes of PDZ domain scaffolding revealed by structure of nNOS-syntrophin complex. *Science* 1999; 284:812–5.

43. Fuh G, Pisabarro MT, Li Y, Quan C, Lasky LA, Sidhu SS. Analysis of PDZ domain-ligand interactions using carboxyl-terminal phage display. *J Biol Chem* 2000; 275:21486–91.

44. Laura RP, Witt AS, Held HA et al. The Erbin PDZ domain binds with high affinity and specificity to the carboxyl termini of delta-catenin and ARVCF. *J Biol Chem* 2002; 277:12906–14.

45. Tonikian R, Zhang Y, Sazinsky SL et al. A specificity map for the PDZ domain family. *PLoS Biol* 2008; 6:e239.

46. Ilsley JL, Sudol M, Winder SJ. The WW domain: Linking cell signalling to the membrane cytoskeleton. *Cell Signal* 2002; 14:183–9.

47. Sudol M, McDonald CB, Farooq A. Molecular insights into the WW domain of the Golabi-Ito-Hall syndrome protein PQBP1. *FEBS Lett* 2012; 586:2795–9.

48. Chen HI, Sudol M. The WW domain of Yes-associated protein binds a proline-rich ligand that differs from the consensus established for Src homology 3-binding modules. *Proc Natl Acad Sci U S A* 1995; 92:7819–23.

49. Tamura H, Schild L, Enomoto N, Matsui N, Marumo F, Rossier BC. Liddle disease caused by a missense mutation of beta subunit of the epithelial sodium channel gene. *J Clin Invest* 1996; 97:1780–4.

50. Staub O, Abriel H, Plant P et al. Regulation of the epithelial Na$^+$ channel by Nedd4 and ubiquitination. *Kidney Int* 2000; 57:809–15.

51. Linn H, Ermekova KS, Rentschler S, Sparks AB, Kay BK, Sudol M. Using molecular repertoires to identify high-affinity peptide ligands of the WW domain of human and mouse YAP. *Biol Chem* 1997; 378:531–7.

52. Espanel X, Sudol M. Yes-associated protein and p53-binding protein-2 interact through their WW and SH3 domains. *J Biol Chem* 2001; 276:14514–23.

53. Gavva NR, Gavva R, Ermekova K, Sudol M, Shen CJ. Interaction of WW domains with hematopoietic transcription factor p45/NF-E2 and RNA polymerase II. *J Biol Chem* 1997; 272:24105–8.

54. Mosser EA, Kasanov JD, Forsberg EC, Kay BK, Ney PA, Bresnick EH. Physical and functional interactions between the transactivation domain of the hematopoietic transcription factor NF-E2 and WW domains. *Biochemistry* 1998; 37:13686–95.

55. Kasanov J, Pirozzi G, Uveges AJ, Kay BK. Characterizing Class I WW domains defines key specificity determinants and generates mutant domains with novel specificities. *Chem Biol* 2001; 8:231–41.

56. Tommasi di Vignano A, Di Zenzo G, Sudol M, Cesareni G, Dente L. Contribution of the different modules in the utrophin carboxy-terminal region to the formation and regulation of the DAP complex. *FEBS Lett* 2000; 471:229–34.

57. Rentschler S, Linn H, Deininger K, Bedford MT, Espanel X, Sudol M. The WW domain of dystrophin requires EF-hands region to interact with beta-dystroglycan. *Biol Chem* 1999; 380:431–42.

58. Huang X, Poy F, Zhang R, Joachimiak A, Sudol M, Eck MJ. Structure of a WW domain containing fragment of dystrophin in complex with beta-dystroglycan. *Nat Struct Biol* 2000; 7:634–8.

59. Verdecia MA, Bowman ME, Lu KP, Hunter T, Noel JP. Structural basis for phosphoserine-proline recognition by group IV WW domains. *Nat Struct Biol* 2000; 7:639–43.

60. Yan KS, Kuti M, Zhou MM. PTB or not PTB—That is the question. *FEBS Lett* 2002; 513:67–70.

61. Songyang Z, Shoelson SE, McGlade J et al. Specific motifs recognized by the SH2 domains of Csk, 3BP2, fps/fes, GRB-2, HCP, SHC, Syk, and Vav. *Mol Cell Biol* 1994; 14:2777–85.

62. Dente L, Vetriani C, Zucconi A et al. Modified phage peptide libraries as a tool to study specificity of phosphorylation and recognition of tyrosine containing peptides. *J Mol Biol* 1997; 269:694–703.

63. Songyang Z, Margolis B, Chaudhuri M, Shoelson SE, Cantley LC. The phosphotyrosine interaction domain of Shc recognizes tryosine-phosphorylated NPXY motif. *J Biol Chem* 1995; 270:14863–6.

64. Gram H, Schmitz R, Zuber JF, Baumann G. Identification of phosphopeptide ligands for the Src-homology 2 (SH2) domain of Grb2 by phage display. *Eur J Biochem* 1997; 246:633–7.

65. Cochrane D, Webster C, Masih G, McCafferty J. Identification of natural ligands for SH2 domains from a phage display cDNA library. *J Mol Biol* 2000; 297:89–97.

66. Li W. ORF phage display to identify cellular proteins with different functions. *Methods* 2012; 58:2–9.

67. Oligino L, Lung FD, Sastry L et al. Nonphosphorylated peptide ligands for the Grb2 Src homology 2 domain. *J Biol Chem* 1997; 272:29046–52.

68. Hart CP, Martin JE, Reed MA et al. Potent inhibitory ligands of the GRB2 SH2 domain from recombinant peptide libraries. *Cell Signal* 1999; 11:453–64.
69. King TR, Fang Y, Mahon ES, Anderson DH. Using a phage display library to identify basic residues in A-Raf required to mediate binding to the Src homology 2 domains of the p85 subunit of phosphatidylinositol 3′-kinase. *J Biol Chem* 2000; 275:36450–6.
70. Pero SC, Oligino L, Daly RJ et al. Identification of novel non-phosphorylated ligands, which bind selectively to the SH2 domain of Grb7. *J Biol Chem* 2002; 277:11918–26.
71. Smothers JF, Henikoff S. The HP1 chromo shadow domain binds a consensus peptide pentamer. *Curr Biol* 2000; 10:27–30.
72. Ulijasz AT, Kay BK, Weisblum B. Peptide analogues of the VanS catalytic center inhibit VanR binding to its cognate promoter. *Biochemistry* 2000; 39:11417–24.
73. Pillutla RC, Hsiao K, Brissette R et al. A surrogate-based approach for post-genomic partner identification. *BMC Biotechnol* 2001; 1:6.
74. Dintilhac A, Bernues J. HMGB1 interacts with many apparently unrelated proteins by recognizing short amino acid sequences. *J Biol Chem* 2002; 277:7021–8.
75. Huang J, Ru B, Dai P. Bioinformatics resources and tools for phage display. *Molecules* 2011; 16:694–709.
76. Huang J, Ru B, Zhu P et al. MimoDB 2.0: A mimotope database and beyond. *Nucleic Acids Res* 2012; 40:D271–7.
77. Yaffe MB, Leparc GG, Lai J, Obata T, Volinia S, Cantley LC. A motif-based profile scanning approach for genome-wide prediction of signaling pathways. *Nat Biotechnol* 2001; 19:348–53.
78. Brannetti B, Via A, Cestra G, Cesareni G, Citterich MH. SH3-SPOT: An algorithm to predict preferred ligands to different members of the SH3 gene family. *J Mol Biol* 2000; 298:313–28.
79. Zanzoni A, Montecchi-Palazzi L, Quondam M, Ausiello G, Helmer-Citterich M, Cesareni G. MINT: A Molecular INTeraction database. *FEBS Lett* 2002; 513:135–40.
80. Snyder K, Feldman H, Dumontier M, Salama J, Hogue C. Domain-based small molecule binding site annotation. *BMC Bioinformatics* 2006; 7:152.
81. Bader GD, Betel D, Hogue CWV. BIND: The Biomolecular Interaction Network Database. *Nucleic Acids Res* 2003; 31:248–50.
82. Gavin AC, Bosche M, Krause R et al. Functional organization of the yeast proteome by systematic analysis of protein complexes. *Nature* 2002; 415:141–7.
83. Kurakin A, Bredesen D. Target-assisted iterative screening reveals novel interactors for PSD95, Nedd4, Src, Abl and Crk proteins. *J Biomol Struct Dyn* 2002; 19:1015–30.
84. Deshaies RJ, Seol JH, McDonald WH et al. Charting the protein complexome in yeast by mass spectrometry. *Mol Cell Proteomics* 2002; 1:3–10.
85. Uetz P. Two-hybrid arrays. *Curr Opin Chem Biol* 2002; 6:57–62.
86. Peters EA, Schatz J, Johnson SS, Dower WJ. Membrane insertion defects caused by positive charges in the early mature region of protein pIII of filamentous phage fd can be corrected by prlA suppressors. *J Bacteriol* 1994; 176:4296–305.
87. Rodi DJ, Soares AS, Makowski L. Quantitative assesment of peptide sequence diversity in M13 combinatorial peptide phage display libraries. *J Mol Biol* 2002; 322:1039–52.
88. Kay BK, Adey NB, Yun-Sheng H, Manfredi JP, Mataragnon AH, Fowlkes DM. An M13 phage library displaying random 38-amino-acid peptides as a source of novel sequences with affinity to selected targets. *Gene* 1993; 128:59–65.
89. Luck K, Travé G. Phage display can select over-hydrophobic sequences that may impair prediction of natural domain-peptide interactions. *Bioinformatics* 2011; 27:899–902.
90. Kramer A, Reineke U, Dong L et al. Spot synthesis: Observations and optimizations. *J Pept Res* 1999; 54:319–27.

91. Ivarsson Y, Arnold R, McLaughlin M et al. Large-scale interaction profiling of PDZ domains through proteomic peptide-phage display using human and viral phage peptidomes. *Proc Natl Acad Sci U S A* 2014; 111:2542–7.

92. Lindgren M, Hallbrink M, Prochiantz A, Langel U. Cell-penetrating peptides. *Trends Pharmacol Sci* 2000; 21:99–103.

93. Sparks AB, Quilliam LA, Thorn JM, Der CJ, Kay BK. Identification and characterization of Src SH3 ligands from phage-displayed random peptide libraries. *J Biol Chem* 1994; 269:23853–6.

94. Stauffer TP, Martenson CH, Rider JE, Kay BK, Meyer T. Inhibition of Lyn function in mast cell activation by SH3 domain binding peptides. *Biochemistry* 1997; 36:9388–94.

95. Paige LA, Christensen DJ, Gron H et al. Estrogen receptor (ER) modulators each induce distinct conformational changes in ER alpha and ER beta. *Proc Natl Acad Sci U S A* 1999; 96:3999–4004.

96. Chang C, Norris JD, Gron H et al. Dissection of the LXXLL nuclear receptor–coactivator interaction motif using combinatorial peptide libraries: Discovery of peptide antagonists of estrogen receptors alpha and beta. *Mol Cell Biol* 1999; 19:8226–39.

97. Norris JD, Paige LA, Christensen DJ et al. Peptide antagonists of the human estrogen receptor. *Science* 1999; 285:744–6.

98. Houimel M, Mazzucchelli L. Identification of biologically active peptides that inhibit binding of human CXCL8 to its receptors from a random phage-epitope library. *J Leukoc Biol* 2009; 85:728–38.

99. Benson RE, Gottlin EB, Christensen DJ, Hamilton PT. Intracellular expression of peptide fusions for demonstration of protein essentiality in bacteria. *Antimicrob Agents Chemother* 2003; 47:2875–81.

100. Tao J, Wendler P, Connelly G et al. Drug target validation: Lethal infection blocked by inducible peptide. *Proc Natl Acad Sci U S A* 2000; 97:783–6.

101. Hyde-DeRuyscher R, Paige LA, Christensen DJ et al. Detection of small-molecule enzyme inhibitors with peptides isolated from phage-displayed combinatorial peptide libraries. *Chem Biol* 2000; 7:17–25.

102. Posern G, Zheng J, Knudsen BS et al. Development of highly selective SH3 binding peptides for Crk and CRKL which disrupt Crk-complexes with DOCK180, SoS and C3G. *Oncogene* 1998; 16:1903–12.

103. Nguyen JT, Turck CW, Cohen FE, Zuckermann RN, Lim WA. Exploiting the basis of proline recognition by SH3 and WW domains: Design of N-substituted inhibitors. *Science* 1998; 282:2088–92.

104. Hammer J, Valsasnini P, Tolba K et al. Promiscuous and allele-specific anchors in HLA-DR-binding peptides. *Cell* 1993; 74:197–203.

105. Bolin DR, Swain AL, Sarabu R et al. Peptide and peptide mimetic inhibitors of antigen presentation by HLA-DR class II MHC molecules. Design, structure–activity relationships, and X-ray crystal structures. *J Med Chem* 2000; 43:2135–48.

106. Kay BK, Kurakin A, Hyde-DeRuyscher R. From peptides to drugs via phage-display. *Drug Discov Today* 1998; 3:370–8.

107. Grøn H, Hyde-DeRuyscher R. Peptides as tools in drug discovery. *Curr Opin Drug Discov Dev* 2000; 3:636–45.

108. Houston JG, Banks M. The chemical–biological interface: Developments in automated and miniaturised screening technology. *Curr Opin Biotechnol* 1997; 8:734–40.

CHAPTER 10

High-Throughput and High-Content Screening Using Peptides

Robert O. Carlson, Robin Hyde-DeRuyscher, and Paul T. Hamilton

CONTENTS

10.1 INTRODUCTION

Phage-displayed peptide libraries can be used to isolate specific, high-affinity peptides to almost any protein target. Peptides have been identified for a wide range of protein targets including antibodies [1,2], enzymes [3,4], G proteins [5], nuclear receptors [6,7], cytokine receptors [8,9], cytokines [10], transcription factors [11], and protein interaction domains [12–14]. The peptides identified by affinity selection from phage-displayed peptide libraries bind to biologically relevant sites on these target proteins and therefore can serve as valuable reagents for dissecting the biological function of proteins as well as tools for drug discovery.

In this chapter, we will cover how peptides identified by phage display bind to functional sites on protein targets, such as enzyme active sites, protein–protein interaction sites, and DNA binding sites. These peptides can, therefore, be used as molecular probes of the protein target. For an enzyme, an active site-directed peptide can be used to screen for enzyme inhibitors. For proteins that undergo conformational changes based on activation by another protein or by ligand binding, the peptides can be used to monitor the conformational state of the target protein. In drug discovery research, these peptide characteristics can be exploited to develop rapid, in vitro assays for high-throughput compound screening and characterization.

10.2 PEPTIDES AS ENZYME INHIBITORS

10.2.1 Peptides Identified by Phage Display Are Directed to Biologically Relevant Sites

The evidence available from a number of laboratories doing peptide selections by phage display indicates that the peptides isolated with protein targets are directed to only one or a few sites on the protein, and in most cases, the sites the peptides bind to are functional sites. These peptides, therefore, often modulate the biological activity of the target protein and serve as "surrogate ligands" of the target.

A wide variety of enzyme classes have been used as targets for peptide selection by phage display (see review in Ref. [4]). Many of the peptide ligands identified by phage display for enzymes bind at the active site of the enzyme and inhibit the enzymatic activity with inhibition constant (K_i) or half-maximal inhibitory constant (IC_{50}) values ranging from 60 nM to 3 mM. In a study of seven diverse enzymes, Hyde-DeRuyscher et al. [3] were able to isolate peptides that bound to each of the seven enzyme targets. For six of the seven targets, they were able to isolate peptides that inhibited the enzymatic activity of the target enzyme. For example, hexokinase, which catalyzes the phosphorylation of glucose by ATP to yield glucose-6-phosphate, was used as a target for peptide selection, and seven peptides were identified. The two highest-affinity peptides were synthesized and used for kinetic analysis. Both peptides were competitive inhibitors of hexokinase with respect to glucose. The best inhibitor peptide had a K_i of 80 nM with respect to glucose and 100 nM with respect to ATP. In another example, the target enzyme was alcohol dehydrogenase, and again, peptides were identified that were competitive inhibitors of alcohol dehydrogenase with K_i values in the range of 100–800 nM. For hexokinase and alcohol dehydrogenase, the substrates for the enzymes are small-molecular-weight compounds with no peptidic or proteinaceous characteristics. Yet, peptide ligands were identified that are competitive inhibitors of the enzymes with K_i values in the hundred-nanomolar range.

In another study, Sperinde et al. [15] used phage display to identify a peptide that bound and inhibited Dnase II. When tested in activity assays, the peptide had a K_i of 2 μM but was sparingly soluble in aqueous solution. Extending the sequence of the peptide to increase solubility improved the inhibition of Dnase II to give a K_i of 0.4 μM. These researchers were able to use the Dnase inhibitor peptide to enhance the transfection of DNA into various cell types.

The finding that peptides identified by phage display bind to protein functional sites, and do not bind randomly or nonspecifically to a protein's surface, indicates that functional sites on proteins have a common set of physicochemical features that are recognized by the peptides. These features of a functional site are the basis of ligand binding, whether the ligand is an enzyme substrate, an interacting protein partner, or a phage-displayed peptide. A number of studies have looked at proteins to determine the characteristics of a functional site. Binding sites are usually grooves or depressions in the protein surface [16], often with exposed hydrophobic groups. Mattos and Ringe [17,18] have used small organic compounds as probes to define hydrophobic binding sites on the surface of a protein. They found that functional sites have, in addition to depressions and exposed hydrophobic groups, bound water molecules that make specific interactions with polar groups in the site. Ligand binding easily displaces the water molecules in the functional site. Another characteristic of functional sites appears to be flexibility of the residues that make up the site. DeLano et al. [19] examined a common binding site in the hinge region of an Fc (crystallizable fragment) of immunoglobulin G that interacts with four different natural proteins. In addition, they did phage-display selection experiments with peptide libraries and found high-affinity peptides that bound at this "consensus" binding site on the Fc. Crystallographic analysis of the "consensus" binding site with the peptide showed that the peptide adopted a structure that is different from the natural Fc binding proteins, despite the fact that the interactions of the peptide with the binding site mimicked the interactions of the other interacting proteins. In particular, the binding site on the Fc fragment underwent conformational changes in order to complement the surface residues of each binding partner.

The peptides, therefore, appear to be able to fit into a groove or depression on the protein surface and displace the bound water molecules. Hydrophobic residues on the peptide interact with the exposed hydrophobic residues in the functional site. These two features probably provide the majority of the energy for binding. Binding specificity may come from complementarity of charge and polar residues between the peptide and the binding site. In addition, the binding site is flexible and, therefore, can adapt its shape to accommodate the peptide ligand.

10.2.2 Peptides as Surrogate Ligands in High-Throughput Screening Assays

Since the peptides bind at a functional site of a target protein, a compound that prevents binding of the peptide to the target protein would also likely bind at the same site on the target and would therefore function as an inhibitor [3,20]. The peptides, therefore, can be used as surrogate ligands to configure simple, competitive-binding assays to detect inhibitors in high-throughput screening (HTS). Peptide-based, competitive-binding assays have been configured in a variety of detection formats including radioactivity-based scintillation proximity assay (SPA), luminescence, fluorescence polarization (FP), time-resolved fluorescence (TRF), and fluorescence resonance energy transfer (FRET).

These formats vary as to whether the peptide or the protein or both need to be labeled. In some formats, the protein is immobilized on a solid support; in other formats, the peptide is immobilized; and in still other formats, both protein and peptide are in solution. Regardless of the format, peptide-based, competitive-binding assays can detect inhibitors. For example, a peptide-based assay was developed for tyrosyl-tRNA synthetase using a peptide identified by phage display [3]. Using a variety of detection formats, a series of four known tyrosyl-tRNA inhibitors were tested in peptide-based assays [3,21]. In each assay format, the ability to detect the inhibitor and the potency of the compound were determined and compared to the activity of the compounds in a biochemical assay (Table 10.1). All four of the inhibitors were detected in each of the assay formats, and the potencies for the compounds in the peptide-based assays were comparable with the potencies of the compounds in the enzyme activity assay. The peptide-based assays were able to detect compounds with potencies ranging from nM to μM.

10.2.3 HTS Example: Deoxyxylulose-Phosphate Reductoisomerase

Isoprenoids are a group of compounds found in all living organisms. In bacteria, isoprenoids are synthesized from isopentenyl diphosphate (IPP) via a mevalonate-independent pathway. IPP is essential for bacterial growth; therefore, enzymes, such as deoxyxylulose-phosphate reductoisomerase (DXR), that are involved in the

Table 10.1   Comparison of Detection Methods for Peptide-Based Assays

Assay Format	Inhibitor NPC0101 IC_{50} (μM)	Inhibitor NPC0102 IC_{50} (μM)	Inhibitor NPC0103 IC_{50} (μM)	Inhibitor NPC0104 IC_{50} (μM)
Enzymatic activity	0.04	0.20	3.0	1.6
Target immobilized—SPA	0.03	0.19	7.0	18
Target immobilized—TRF	0.01	0.12	6.0	10
Peptide immobilized—TRF	0.02	0.24	3.7	10
Fluorescence polarization (FP)	0.02	0.22	2.1	3.6
Fluorescence resonance energy transfer (FRET)	0.03	0.45	6.8	12

Table 10.2   Compound Screening Results for DXR

32,000 compounds were screened
Screening statistics: coefficient of variation (CV) = 11%; Z' = 0.6
111 actives identified using DXR-specific peptide-based assay (0.35%) in a TRF assay format
30 of 89 compounds tested in a functional assay inhibit the enzymatic activity of DXR with potencies <20 μM
Potency of inhibitors ranged from 0.3 to 20 μM

biosynthesis of IPP are valid targets for antibacterial drug discovery [22]. For many targets in drug discovery, HTS assays are based on biochemical activity assays. For DXR, the activity assay involves organic separation of radioactive substrates and products, which is a process that is not very amenable to an HTS assay. We developed a peptide-based, HTS assay for DXR and used it to identify compounds that inhibited the enzymatic activity of DXR [23].

The *dxr* gene sequence was amplified by the polymerase chain reaction (PCR) from *Escherichia coli* genomic DNA, cloned into an expression vector with a biotinylation recognition sequence, expressed and biotinylated in *E. coli*, and purified. Purified, biotinylated DXR was immobilized on streptavidin-coated 96-well microtiter plates, and 20 different phage-displayed peptide libraries were used to select DXR-specific peptides by methods described previously [3]. A series of DXR-specific peptides were identified and used to build a TRF assay. The DXR-peptide-based assay was used to screen a collection of 32,000 structurally diverse compounds, from which 111 hits were obtained, for a 0.35% hit rate (Table 10.2). Each of these 111 compounds specifically blocked the binding of the DXR-specific peptide to DXR. From testing of the 89 most potent inhibitors in the DXR-peptide competition assay, 30 of the compounds (34%) were found to be DXR enzyme inhibitors with IC_{50} values <20 μM, making them candidates for further investigation for potential application as antibacterial agents.

The peptide-based, surrogate ligand assay for DXR proved to be a rapid, reliable method to identify enzyme inhibitors in a high-throughput screen of a compound collection. In addition to our internal efforts with DXR and other targets, other groups [24] have successfully used peptide-based, surrogate ligand assays to identify enzyme inhibitors. Since the peptides are directed to functional sites on the target protein, this method can be used to develop HTS assays for targets of unknown biochemical function. Surrogate ligand assays for targets of unknown function can identify compounds that can be tested directly for effects on the disease model. In addition, peptide-based assays can be used to develop HTS-compatible assays for targets where traditional biochemical assays are difficult or impossible to configure and run in a high-throughput format.

10.2.4 Peptides as Tools for Target Validation

The observation that the phage-display process selects peptides that bind at functional sites of target proteins rather than to random sites on the protein surface can be used to develop tools for target validation. The basic approach involves isolation of peptide ligands to a target protein using phage display followed by expression of the peptide inside the cell and monitoring for phenotypic effects. Peptides that bind at functional sites on the target will block that target function inside the cell and produce a change in phenotype. For targets that are essential for cell growth, the phenotype would be growth inhibition or cell death. This "protein knockout" technique has been applied successfully to several essential *E. coli* proteins [25,26]. Tao et al. [25] used phage display to select a peptide that bound to prolyl-tRNA synthetase, fused the peptide to glutathione S-transferases (GST), and demonstrated growth inhibition of the bacterial cells upon specific expression of the peptide–GST fusion. Similarly, Benson et al. [26] took a set of six targets involved in a wide range of basic bacterial cellular processes (DNA replication [DnaN], transcription [RpoD], DNA supercoiling [GyrA], lipid A biosynthesis [LpxA], protein secretion [SecA], and one target of unknown function [Era]) and identified target-specific peptides by phage display. When the target-specific peptides were expressed inside *E. coli* cells as GST fusions, the cells stopped growing. Overexpression of the target protein relieved the peptide-induced growth arrest, indicating that the peptides were inhibiting an essential function required for bacterial growth. For bacterial cells, this method validates the protein as a suitable target for antibacterial drug discovery, but it also validates the peptide as targeting an essential functional site. The peptide–protein pair can then be used to develop an HTS-compatible assay for antibacterial drug discovery as in Section 10.2.3 for DXR. This approach does not depend on prior knowledge of the biochemical activity or function of the target protein and therefore may be well suited for use with targets of unknown function that are identified by genomic and genetic techniques.

10.3 PEPTIDES AS CONFORMATIONAL PROBES

Another useful aspect of peptides identified using phage display is their ability to bind to targets in a conformationally sensitive fashion. This property can be used to develop specific probes useful for many drug discovery applications. We have used conformationally sensitive peptides to build assays for G-protein-coupled receptors (GPCRs) and nuclear receptors.

10.3.1 GPCRs

GPCRs are seven transmembrane receptor proteins that are expressed in almost all human tissues and play a fundamental role in signal transduction and physiology. GPCRs, therefore, are potential therapeutic targets for treating many diseases. Of the 483 marketed drugs in 1996, greater than 40% were directed at GPCR targets [27]. In 2000, 26 of the top 100 pharmaceutical products were therapeutics that modulate GPCR function [28]. This accounts for sales of more than $23.5 billion and represents approximately 9% of the total global pharmaceutical sales [29].

The human genome has been estimated to consist of 35,000 genes, of which approximately 750 are GPCRs. Almost half of these sequences are likely to encode sensory receptors. The remaining 400 or so receptors, however, represent potential targets for drug discovery, of which only about 30 are targets of existing therapeutics. In addition, an estimated 160 of these receptors are "orphan" receptors, for which there is no known ligand or function [29].

Most assays for GPCRs rely on monitoring levels of intracellular second messengers such as inositol trisphosphate or calcium. The problem with this approach is that many different cellular functions can effect the production of these second messengers, leading to a high rate of interference from compounds that have no effect on the activity of the GPCR in question. In addition, these assays are carried out in cells, requiring a large commitment of resources in cell culture to perform the assays. Our approach was to simplify the system by doing two things [30]. First, remove the assay from the cellular environment and utilize isolated membranes that contain all of the components necessary to monitor GPCR signaling. This includes the receptor itself and the heterotrimeric G protein, consisting of the alpha, beta, and gamma subunits. Second, we wanted to develop probes that would monitor GPCR activation by measuring the activation state of the G-alpha protein associated with the receptor. GPCRs are ligand-dependent guanine nucleotide exchange factors for G proteins. When a ligand binds to the receptor, it causes a signal to be transduced through the membrane from the GPCR to the associated G protein. This causes the G-alpha subunit to undergo a conformational shift whereby it releases guanosine diphosphate (GDP) and binds guanosine triphosphate (GTP). This results in a change in contact between the G-alpha and beta-gamma subunits of the heterotrimeric G protein, which then interact with other effector proteins, initiating the signal transduction cascade. Thus, the basic function of the receptor is to cause a ligand-dependent conformational change in the G-alpha subunit.

To identify peptides with the needed specificities, we produced the recombinant G-alpha subunit in *E. coli* and used this as a target for phage display. For this set of experiments, we presented the G-alpha protein in either the inactive GDP-bound state or the activated GTP-bound state. As shown in Figure 10.1, we were able to identify peptides that bound with very distinct nucleotide specificities. For example, one series of peptides, designated T-specific peptides, bound with high affinity to G-alpha protein with the GTP-bound state. Another class of peptides bound only to G protein complexed with GDP, designated D-specific peptides. Yet, a third class of peptides bound to the G-alpha protein in a nucleotide-independent fashion; that is, they bound to the G-alpha protein in both the GTP and the GDP states.

In addition, the peptides showed remarkable specificity for different subtypes of G proteins. For instance, peptides that bound to G-alpha I did not bind to G-alpha S, and similarly, peptides that bound G-alpha S did not bind G-alpha I. We have been able to identify peptides that are specific for not only the activation state of the G protein but also the subtype of G protein.

These peptides are very specific probes for the activation state of the G-alpha protein, providing a tool for the direct readout of the activation state of the receptor. As shown in Figure 10.2, these peptides can be used to format assays that monitor receptor activation. If T-specific peptides are linked to a reporter molecule, they can be used to format a simple binding assay or a homogeneous FRET assay. These

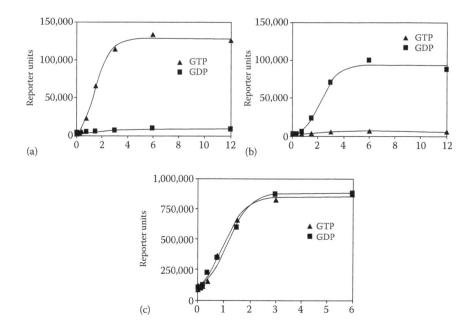

Figure 10.1 Binding specificities of the peptides identified using phage display. Synthetic peptides were conjugated with streptavidin-alkaline phosphatase and used in an ELISA format to probe increasing concentrations of Gα protein immobilized on a microtiter plate. (a) A peptide identified using Gα protein loaded with GTP binds specifically to GTP-loaded Gα. (b) A peptide identified using Gα protein loaded with GDP binds specifically to GDP loaded Gα. (c) A peptide identified using Gα protein loaded with GDP binds to Gα independent of nucleotide loading.

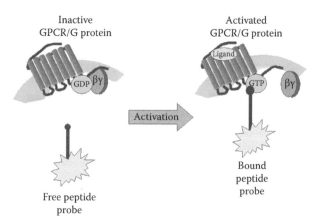

Figure 10.2 Schematic representation of detecting activation of a receptor using a peptide that binds to the activated form of Gα.

assays combine a number of features that are desirable in a HTS platform. They monitor only functional activation of the receptor, not merely a binding event, and so are an efficient way to find agonists for the receptor. The assay uses semipurified components, so cell culture is not required on a daily basis to set up the assay. The assays are routinely performed in volumes of 30 μL in 384-well plates; small amounts of materials are required for each assay, and the FRET-based assay does not require washing. All of these properties make this an ideal platform for automation.

As shown in Figure 10.3, an assay of this type was used to monitor the activity of the beta-2 adrenergic receptor. The agonist isoproterenol was titrated into the assay and showed a dose-dependent activity with a half maximal effective concentration (EC_{50}) value that is consistent with literature values. In addition, this activity could be antagonized with the inverse agonist ICI 118551 (Figure 10.3). This indicates that the assay monitors activation through the receptor and not nonspecific changes in this system. We also have formatted assays for G-alpha I-coupled receptors, such as the M2 muscarinic acetylcholine receptor. Figure 10.3 shows that the agonist carbachol also elicits a dose-dependent increase in activity when using this assay. The signal could also be reduced by the addition of the antagonist atropine.

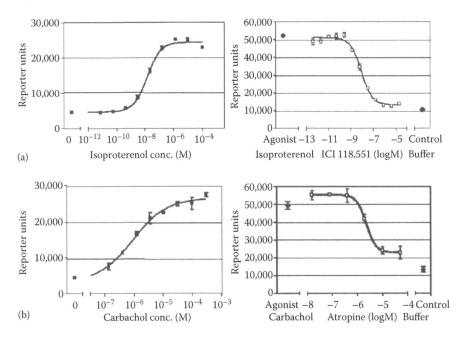

(a)

(b)

Figure 10.3 Assays of (a) beta-2 adrenergic receptor (Gαs coupled) and (b) muscarinic acetylcholine receptor M2 (Gαi coupled). Membranes from SF9 cells infected with baculovirus expressing the receptor, and the heterotrimeric G proteins were used with a peptide specific for activated Gα to monitor the activation of the receptor. Increasing concentrations of the agonist were added to the assay. Increasing signal from the bound probe is an indicator of receptor activation (graphs on left). Membranes were treated with agonist at 80% full activation, and increasing concentrations of antagonist were added (graphs on right).

Using the G-alpha I and the G-alpha S systems, we were able to demonstrate that there is specific coupling through each receptor. In addition, this system exhibits the expected activation profile for partial agonists. Several partial agonists as well as a full agonist were used for the beta-2 adrenergic receptor. In each case, the potency observed for each compound was consistent with published values, and their efficacy also mirrored the known activation profile for these compounds. Thus, the assay can also provide reasonable pharmacological data on the compounds in question.

The peptide probes used in these GPCR assays are specific for the G-alpha subunit. G proteins associate with many different GPCRs. This system, therefore, can be used with a variety of GPCRs by changing the receptor that is expressed in the cells. For example, we obtained the D-1 dopamine receptor and tested it using our G-alpha S system. To assess the activity of the receptor, we carried out a series of dose–response experiments with the agonist dopamine and evaluated the reproducibility over several experiments. We were able to obtain reproducible signal-to-background ratios for dopamine of 5:1 with no additional optimization. This highlights the modular nature of this GPCR assay system.

As a preliminary assessment of the ability of the system to be used as an effective tool to screen large collections of compounds, and to address the cross talk or interference by compounds, which would normally activate many pathways and cellular assays, we used a collection of 640 known pharmacologically active compounds. This set included 11 known agonists to the M2 acetylcholine receptor. All 11 agonists were detected by the assay, and only one "new" compound was determined to be an agonist. Upon further testing, we found that the activity of the "new" compound could be reversed by the inverse agonist for this receptor, indicating that it was a specific agonist. This "new" compound, 8-(N,N-diethylamino)octyl-3,4,5-trimethoxybenzoate-HCl (TMB8), has been described previously as an allosteric modulator of the M2 receptor but not an agonist [31]. Therefore, TMB8 may not have been previously identified as an M2 receptor agonist, since a direct binding assay would also not detect this compound, which further demonstrates the utility of this approach to find novel compounds that modulate GPCR function.

To summarize, for GPCRs, we have identified peptides that can be used to monitor the activation state of receptors in crude membrane preparations by assessing the nucleotide binding state of the G-alpha protein. This assay is functional and does not rely on live cells, and the data obtained reflect the pharmacological properties of individual receptors.

10.3.2 Nuclear Hormone Receptor

10.3.2.1 Complexity in Nuclear Receptor Signaling

Nuclear receptors are a group of transcription factors that regulate expression of target genes in response to ligand binding. The signal of ligand binding to nuclear hormone receptor is transduced through induced changes in receptor surface conformation that alter interactions with coregulatory proteins. These coregulatory proteins, in turn, regulate cellular functions through a variety of mechanisms, which include, but are not exclusively limited to, transcriptional regulation [32,33]. Different ligands induce distinct receptor surface conformations, leading to differences in the type and affinity of coregulatory protein interactions. The nature of these interactions thereby defines the biological activity of the ligand. Since the expression of coregulatory proteins can be cell-type dependent, such differences in expression may be a factor in the ability of nuclear receptor ligands to exert cell-type-specific effects, such as is observed for the selective estrogen receptor modulators (SERMs) [34].

There are a wide variety of possible coregulatory proteins. Most identified to date interact primarily with nuclear receptor activation function 2 (AF2), which is located in the ligand binding domain (LBD). These include p160 steroid coactivator family members, p300 and related integrator proteins, thyroid hormone receptor-associated protein (TRAP)/mediator complex, and various other coactivators [32]. The primary mechanism for ligand-mediated regulation of AF2 interaction with coactivators has been defined through mutagenesis and crystallography [35–38]. Using the estrogen receptor as an example, binding of the agonist 17-β-estradiol induces receptor helices 3, 4, 5, and 12 to create a hydrophobic "coactivator groove" that, in turn, binds LXXLL motifs present in coactivator proteins [39]. Estrogen receptor (ER) antagonists, such as 4-OH-tamoxifen, disrupt the position of helix 12, preventing coactivator binding through LXXLL motifs [40].

The identity of residues flanking and filling out the LXXLL motif has also been shown to be critical in directing affinity and specificity of binding to nuclear receptors. Using phage display with a focused LXXLL peptide library, Chang et al. [7] identified a multitude of LXXLL peptides that differentially bound ERα, ERβ, progesterone receptor (PR), glucocorticoid receptor (GR), and androgen receptor (AR) in the presence of their respective agonists. Recent studies of the binding of LXXLL containing sequences from p160/steroid receptor coactivator (SRC) family proteins to ERα and ERβ have also revealed significant differences in receptor and ligand specificity and affinity [41–43]. Studying the four variant LXXLL motifs of SRC-1, Parker and colleagues [35,38] demonstrated substantial differences in affinity for ERα, AR, and retinoid receptors (RARα and RXRα). They also defined a minimal core LXXLL motif of −2 to +6 in which a hydrophobic residue at −1 and a nonhydrophobic residue at +2 correlated with highest affinity binding. In the LXXLL motif of thyroid hormone receptor binding protein (TRBP), mutation of a single serine at the −3 position altered TRBP binding to ERα and ERβ and increased binding to thyroid hormone receptor (TR) and RXR [44]. Thus, not all LXXLL sequences are created equal with regard to ligand and/or nuclear receptor specificity and, hence, with regard to biological function.

LXXLL motifs are not the only sequences involved in ligand-dependent interaction of proteins with nuclear receptors. As demonstrated recently for the peroxisome proliferator-activated receptor gamma (PPARγ), the corepressor silencing mediator of retinoic acid and thyroid hormone receptor (SMRT) also binds the AF2 coactivator groove, but through an LXXIXXXL motif [45]. In contrast to coactivator binding, antagonist-mediated disruption of helix 12 leads to stabilization of binding for this corepressor motif. Ligand-dependent interactions of corepressors with nuclear receptors also occur outside of the AF2 domain [46]. AR has been shown to interact through its amino terminus with coregulators containing FXXLF or WXXLF motifs in an androgen-dependent fashion [47]. The coactivator nuclear receptor-interacting factor 3 (NRIF3) interacts specifically with TR or RXR through a domain that contains an LXXIL motif, but this motif alone does not determine this specificity [48]. Crystal structures of ERα with LXXLL containing peptide sequences from the p160 coactivator transcriptional intermediary factor 2 (TIF2) revealed that the box 3 sequence actually binds through an LXXYL rather the adjacent LXXLL motif present in the sequence [49].

There are also many proteins that do not contain either LXXLL or any of the other motifs described in the previous paragraph, which exhibit ligand-dependent interaction with nuclear receptors, based upon yeast two-hybrid, coimmunoprecipitation, or GST pull-down experiments. Table 10.3 contains a nonexhaustive list of these putative nuclear receptor interacting proteins that lack LXXLL or similar motifs. The proteins listed in Table 10.3 cover a diverse range of functions that are potentially regulated through ligand-dependent changes in nuclear receptor conformation. Some of these activities do not clearly have direct links to transcriptional regulation. Lacking LXXLL or related motifs, these proteins are likely to bind outside of the coactivator groove. Of course, based upon interaction studies alone, it is not clear that proteins bind directly to nuclear hormone receptors.

Thus, substantial evidence exists for complexity in ligand-mediated regulation of nuclear receptors with coregulatory proteins. In the context of novel nuclear receptor ligand discovery for therapeutic application, this complexity poses both difficulties

Table 10.3   Nuclear Receptor Interacting Proteins That Lack NR Box Motifs

Interacting Protein	Activity	Interacting Nuclear Receptor(s)	References
BAG-1	Apoptosis inhibitor	VDR, GR	[50]
C/EBP alpha	Enhancer binding protein	GR	[51]
CAPER	Coactivator	ER	[52]
Caveolin-1	Scaffold protein	AR	[53]
CDC37	Chaperone	AR	[54]
CIA	Coactivator	ER	[55]
DAP-3	Apoptosis mediator	GR	[56]
GMEB-1/2	Coactivator	GR	[57,58]
HBO1	Histone acetylase	AR	[59]
MDM2	Ubiquitin ligase	ER	[60]
Nucleolin	Undefined	GR	[61]
PAK6	Protein kinase	ER, AR	[62]
PBX	Homeodomain protein	TRα	[63]
PNRC2	Coactivator	SF1, ERR, other	[64]
REA	Corepressor	ER	[65]
Smad3	Signal transduction	AR	[66]
SP3	Transcription factor	ER	[67]
SRA	RNA coactivator	AR	[68]
Thioredoxin	Reducing enzyme	GR	[69]

and opportunities. The difficulties center on the daunting task of creating a novel ligand screening process sufficient to cover the potential nuclear receptor signaling complexity. This task, of course, is simplified in proportion to the level of understanding that exists for a given nuclear receptor mechanism of action within a specific biological response of interest for drug discovery. However, despite a rich database of potential nuclear receptor signaling mechanisms, the actual mechanisms involved in specific physiological events in most cases remains poorly defined. Therefore, the signaling complexity can create a burden in expanding the sphere of model systems that may need to be included in nuclear receptor drug discovery programs. On the other hand, the realization of this complexity is at the heart of a recent resurgence in interest in drug discovery for nuclear receptors. The understanding that ligands can direct changes in nuclear receptor signaling pathways through induced conformational changes, and that these ligand-directed events are highly cell-type dependent, has led to the appreciation that discovery of novel ligands with exquisitely selective activities is a distinct possibility for all nuclear hormone receptors.

What follows in this section is a description of a technology, termed Molecular Braille profiling, designed to detect ligand-induced changes in nuclear hormone receptor conformation. The basic principle involves use of molecular probes with differential selectivity and affinity for binding sites on the nuclear receptor surface that are altered in response to ligand binding. Work with an estrogen receptor is presented as an example of technology that can be applied to nuclear hormone receptors in general, and in principle, to any other protein that undergoes induced conformational changes.

10.3.2.2 Phage Display for Identifying Peptides to Probe Nuclear Receptor Surface Conformation

We currently use peptides exclusively as molecular probes in the Molecular Braille technology. Phage display provides a technique for identifying peptides that have altered binding affinity for nuclear receptors in response to ligand binding. Such peptides can be used as tools to probe the differences in conformation resulting from binding of different nuclear receptor ligands. We have had very good success in using phage display to find diverse peptide sequences that bind to nuclear receptors, including ERα, ERβ, GR, TR, and AR. A good example of the process is provided by a recent phage-display experiment at Karo Bio, which was designed to identify peptides that selectively bind the ERβ ligand binding domain, either unliganded (apo) or complexed with its natural agonist, 17-β-estradiol (E2) or the SERM raloxifene. The phage-display approach for nuclear receptors is essentially identical to the process described for enzymes [3,6]. After the second round of phage affinity selections, bound phage were eluted, and plated, from which about 1500 plaques in total were randomly picked for amplification from the three selection conditions. These amplified, clonal phage were then tested for specific binding using a phage ELISA technique, and about half were subsequently found to bind selectively to the original target, relative to irrelevant protein or uncoated plastic wells. Based upon signal intensity, a subset of over 100 of those selectively bound phage were then retested in the phage ELISA format for ligand selectivity using a panel of ER ligands bound to ERβ-LBD, including the SERMs

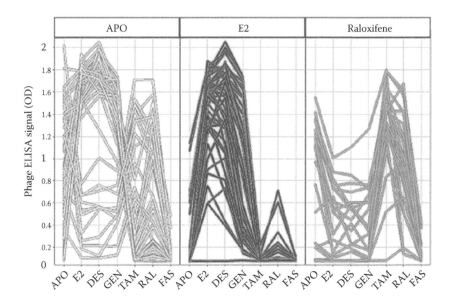

Figure 10.4   Ligand specificity of phage selected with ERβ. A set of 168 phage derived from affinity selections, using immobilized ERβ without a ligand, or saturated with estradiol or raloxifene were tested for ligand specificity using a phage ELISA protocol. Biotinylated ERβ-LBD was to bound to streptavidin-coated 384-well plates; ligands (E2, diethylstilbesterol [DES], genistein [GEN], 4-OH-tamoxifen [TAM], raloxifene [RAL], or faslodex [FAS]) were added to 1 µM final concentration, together with phage. After 1 h of incubation at room temperature, the plates were washed, and horseradish peroxidase (HRP)-conjugated anti-αM13 antibody was added, followed by another round of washing and then addition of HRP substrate 2,2'-azino-bis(3-ethylbenzothiazoline-6-sulphonic acid) (ABTS) to quantify bound antibody. The phage ELISA signal is based upon resulting absorbance at 405 nM. The resulting ligand specificity patterns are shown in relation to the original phage selection condition (ERβ unliganded [APO], or with E2 or raloxifene).

tamoxifen and raloxifene, the antagonist faslodex, and the agonists diethylstilbestrol (DES) and genistein. The results of this phage ELISA are depicted in Figure 10.4. In general, the ligand selectivity was consistent with the selection conditions. For example, phage selected against E2-bound ERβ-LBD subsequently recognize that complex, and the other agonist complexes (DES and genistein), but not the SERMs or antagonist. The converse was found for phage selected against raloxifene-bound ERβ. Curiously, phage selected against unliganded ERβ show higher binding to the apo receptor as would be expected, but also have either the E2 or raloxifene selection pattern. The latter may result from these peptides inducing a receptor conformation similar to that induced by E2 or raloxifene, or unliganded ER may spontaneously take on E2 or raloxifene conformations in the absence of the ligands, which is then recognized by peptides with affinity for these conformations.

In an effort to better understand the relative similarity of the isolated phage based upon this ligand selectivity phage ELISA, we submitted the ELISA data set to principal component analysis (PCA), using software from Spotfire. PCA provides a way of reducing multivariable data sets to fewer dimensions that aid visualization and can help in understanding which variables are changing the most. The seven component phage binding patterns were reduced through PCA to three dimensions, which are plotted in Figure 10.5 for the phage analyzed from all of the selection conditions. In that plot, each symbol represents one of the seven compound binding patterns of a single phage, and the closer the spheres are in space, the more similar their binding patterns. The phage isolated from using either E2- or raloxifene-bound ERβ-LBD clearly cluster separately on opposite sides of the plot, while the apo-ER selected phage spread throughout the plot. LXXLL motif-containing peptides were isolated using either apo- or E2-bound receptor, and as expected, all of those peptides show agonist conformation binding patterns. However, many non-LXXLL peptides also show agonist conformation binding patterns.

PCA was also applied to the same data set but for use in comparing the phage binding patterns relative to reference compound-bound (agonists DES, E2, and genistein; SERMs raloxifene and 4-OH-tamoxifen; antagonist faslodex) or unliganded receptors. In Figure 10.6, each sphere in the PCA plots represents the binding

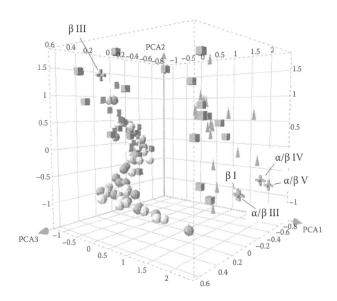

Figure 10.5 Discriminating the ligand specificity patterns of ERβ phage using PCA. The ligand specificity patterns based upon the phage ELISA data depicted in Figure 10.4 were submitted to PCA using statistical analysis software from Spotfire. Each symbol in the three-dimensional plot represents the complete ligand specificity pattern for each individual phage. The distance separating points in space is proportional to the similarity of the ligand specificity patterns. The shape of the symbol relates to the original selection condition: spheres = ERβ + E2; cubes = ERβ apo; pyramids = ERβ + raloxifene; crosses = phage from other ER selections. Amongst the spheres, white spheres represent nuclear receptor (NR) box motif-containing sequences, and gray spheres represent sequences without NR box motifs.

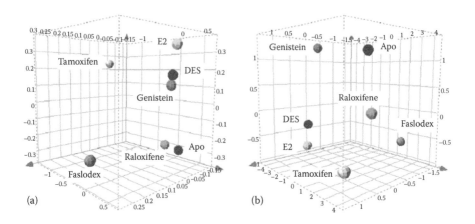

Figure 10.6   Discriminating the peptide-binding pattern for ligand-induced conformations of ERβ using PCA. The phage ELISA data depicted in Figures 10.4 and 10.5 was used to evaluate the extent to which phage could discriminate between ERβ conformations induced by different ligands. (a) The peptide-binding pattern is depicted for ERβ apo or bound with E2, genistein, DES, faslodex, raloxifene, or 4-OH-tamoxifen, based upon the phage-displayed peptides (β I, β III, α/β I, α/β III, α/β IV, and α/β V), which were previously used to discriminate differences in ligand-induced ER conformation. (b) The peptide-binding patterns for the same set of ligands are based upon the peptides in (a) plus 10 additional peptides from the ERβ phage affinity selections described in Figures 10.4 and 10.5. The distance separating each sphere in space is proportional to the similarity of the peptide-binding patterns.

pattern of multiple phage for each reference compound (or apo receptor). In Figure 10.6a, the phage binding pattern is based upon phage displaying the peptides β I, β III, α/β III, α/β IV, and α/β V, which we have previously used extensively as probes for ligand-induced conformational changes for ER [6,70]. As exhibited by the proximity of the spheres, this set of previously identified peptides can discriminate faslodex or tamoxifen-bound ERβ-LBD distinctly from the agonist or raloxifene-bound or the apo receptor. However, the agonists DES, E2, and genistein are poorly discriminated, and raloxifene-bound ERβ does not appear significantly different from the apo receptor. Based upon sequence and ligand-induced binding patterns from the recent E2, raloxifene, or apo phage affinity selections as described above, 10 new phages were chosen. These new phage were added to the six previously identified phage to evaluate the binding pattern with the same set of reference conditions. With the new phage added, the agonist-induced conformations appeared distinct, while the discrimination of the SERM, antagonist, and apo conformations was maintained (Figure 10.6b). Thus, the objective to find peptides for better ER agonist-induced conformation appears to have been achieved through the new selections. This use of PCA provides a good method for determining whether the isolated phage have the potential for discriminating the ligand-induced receptor conformations of interest.

Using PCA to visualize individual phage binding patterns and also phage binding patterns for specific receptor forms facilitates selection of a subset of phage for synthesis to be applied in assays for more precise detection of induced receptor conformational changes.

10.3.2.3 Methods for Use of Peptides to Detect Ligand-Induced Nuclear Receptor Conformational Changes

We utilize two general methods for detecting ligand-induced nuclear receptor conformational changes with peptides. One approach is entirely in vitro, using purified receptor and synthetic peptides. The precise format used to detect the amount of receptor complexed with peptide in response to ligand binding is quite flexible. We currently use TRF or FRET to measure the receptor–peptide complex. The TRF format is similar to an immunosorbent assay, with formats that involve binding either biotinylated synthetic peptide or biotinylated nuclear receptor to streptavidin-coated 384-well plates. Then, biotinylated receptor (for peptide coated plates) or peptide (for receptor-coated plates) is added with or without a ligand. Unbound receptor or peptide is washed away, and streptavidin–europium cryptate conjugate is added to detect bound receptor or peptide through TRF (Figure 10.7a). The FRET format does not require immobilization of any assay component. Streptavidin–europium cryptate conjugate is bound with biotinylated, synthetic peptide and then mixed with receptor that is complexed with allophycocyanin (APC), with or without a ligand. Excitation at 340 nm of the europium peptide tag leads to emission at 620 nm, or if the peptide is receptor bound, the excitation energy can directly transfer to the receptor APC tag, which emits at 665 nm. Fluorescence at 665 nm due to 340 nm excitation is therefore directly proportional to amount of peptide–receptor complex (Figure 10.7b).

Another method frequently used involves a mammalian two-hybrid protein–protein interaction assay, which we call Cellular Braille assay, in which peptide-Gal4 DNA binding domain fusion and nuclear receptor VP16-fusion are transiently expressed in cells along with a Gal4-driven reporter (Figure 10.7c) [71]. Gal4-peptide bound to VP16-receptor results in activation of the Gal4-driven reporter, which is therefore an indirect measure of peptide–nuclear complex concentration. Changes in the reporter readout in response to a ligand are thereby used to detect altered nuclear receptor surface conformation.

The results obtained from these alternate methods are typically equivalent. As shown in Figure 10.8, the effects of E2 or the SERM 4-OH-tamoxifen on interaction of ERβ-LBD with α/β I peptide show similar selectivity and affinity. The differences are primarily empirical or practical. Routinely, a smaller number of peptide sequences will be usable in the Cellular Braille assay. In particular, the Cellular Braille assay excludes peptides that require intact disulfide bonds for binding, because disulfide bonds are disrupted in the reducing environment of the cell. We also find that fewer peptides provide detectable ligand-induced signals in the FRET versus the TRF format, but the precise reason for this is unclear. In general, the in vitro formats allow for much higher throughput compared to the Cellular Braille assay. Perhaps the most important reason for using the Cellular Braille assay is that a source of purified, unliganded receptor is not required. For some of the nuclear receptors, obtaining purified protein in quantity is difficult, and purification in the absence of a ligand can be impossible due to receptor

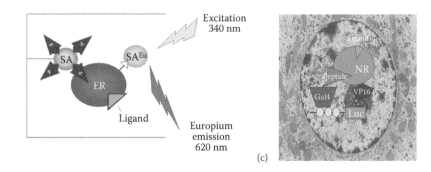

(a)

(c)

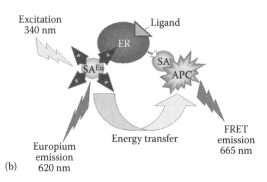

(b)

Figure 10.7 Examples of screening formats in which peptides are used as probes of ligand-induced nuclear receptor conformational changes. (a) *Immobilized*: peptide on plate (POP)-TRF Molecular Braille assay in which biotinylated peptide is bound to a streptavidin-coated plate. Test ligand and biotinylated receptor complexed to a streptavidin–europium cryptate conjugate are added, and quantity of receptor bound to peptide is measured through TRF from the europium. This assay may also be "turned over" into a target on plate (TOP)-TRF Molecular Braille assay in which biotinylated receptor is bound to a streptavidin-coated plate and the biotinylated peptide is linked to the streptavidin–europium cryptate conjugate. (b) *Homogeneous*: FRET-based Molecular Braille assay in which the peptide is complexed to streptavidin–europium and the receptor is complexed to streptavidin–allophycocyanin. The interaction between the peptide and receptor is measured by monitoring the signal at 665 nm. (c) *Cell based*: Cellular Braille assay in which GAL4-peptide fusion, nuclear receptor-VP16, and GAL4 promoter-driven luciferase reporter are transiently transfected into cells. Effects of ligand on luciferase transcription are directly proportional to the extent of ligand-induced receptor interaction with peptide.

instability in the absence of a ligand. For those receptors, the Cellular Braille assay is the method of choice. The Cellular Braille assay also allows surface conformation detection in the natural environment of the receptor, including the complexity of interactions with other intracellular proteins and with DNA. This ill-defined complexity, however, can also complicate the interpretation of results.

For ER, the Molecular Braille assay is currently our method of choice, based upon the relative robustness, throughput, and well-defined nature of the assay. Previous

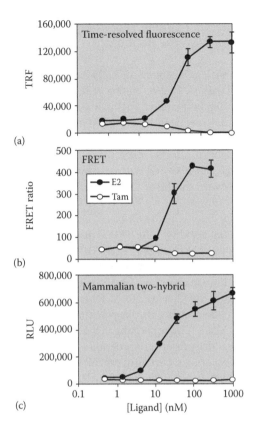

Figure 10.8 Comparison of methods for measuring ligand-induced peptide interaction with nuclear receptor. Comparison of E2 or 4-OH-tamoxifen-induced changes in interaction between ERβ and α/β I peptide using (a) POP-TRF Molecular Braille assay, (b) TOP-TRF Molecular Braille assay, or (c) Cellular Braille assay, as described in Figure 10.7.

publications described the use of the β I, β III, α/β I, α/β III, α/β IV, and α/β V peptides in the TRF format to discriminate ligand-induced conformational changes for ER [6,70]. That set of peptides is useful for discriminating SERM-induced conformations but not for agonist-induced conformations [6,70], as depicted in Figure 10.6a. When 10 peptides from the recent affinity selections using apo and E2 and raloxifene-bound ERβ-LBD were added to the previous set of six peptides for the Molecular Braille assay, discrimination among E2, DES, and genistein-induced conformations became apparent (Figure 10.6b). This is consistent with the characteristics of these peptides as displayed on phage (Figure 10.4). Thus, the phage ELISA data served as good bases for selecting peptides for synthesis, in that the ability to discriminate the agonist conformations apparent with phage was recapitulated with the synthetic peptides.

10.3.2.4 Role of Molecular Braille Technology in Drug Discovery for Nuclear Receptors

Drug discovery for nuclear receptors typically emphasizes optimization of ligand binding affinity and selectivity. Determination of relative agonistic and antagonistic activity follows, most often based upon a cellular assay involving some type of classical transactivation event mediated by a nuclear receptor response element. This approach is appropriate for finding novel agonists or antagonists that can mimic or block endogenous agonist effects, respectively, with high affinity and possibly nuclear receptor subtype selectivity, if desired. However, this approach is not adequate for finding SERMs or, similarly, tissue-specific modulators for other nuclear receptors.

To address tissue specificity, both cell-based and animal models are employed. For example, one potential liability of SERMs is agonistic activity in uterine epithelium, which may lead to uterine cancer [72]. Models used to screen for this activity include measuring agonism in uterine endometrial cancer cells in culture or tracking changes in uterine weight in mice or rats. Each screen has drawbacks but is accepted as an appropriate method for filtering out SERMs with potential for uterotrophic activity. Uterine cancer, however, is only one potential side effect arising from tissue-specific actions of SERMs, but incorporating cell-based or animal models for every tissue of interest within a drug discovery program is impractical.

Molecular Braille technology currently provides the most efficient and comprehensive method for evaluating the potential of nuclear receptor ligands to affect biological responses in a tissue-dependent manner. This is intrinsically based upon the fact that ligand-induced conformational changes, measured in the Molecular Braille assay, drive the coregulatory protein interactions that mediate tissue-specific signaling of nuclear receptors. Furthermore, the peptides utilized in the Molecular Braille assay may resemble actual sequences in coregulatory proteins that similarly bind nuclear receptors in response to a ligand. Therefore, binding of these peptides may mimic physiologically relevant events and, thus, could prove useful for correlating ligand-induced changes in peptide binding to ligand-induced biological responses. To the extent that this is true, Molecular Braille technology offers a surrogate, comprehensive method for assaying potential protein interactions nuclear receptors may encounter in the myriad of tissues in which the receptors are active. These potentially mimicked interactions may involve currently known and unknown coregulatory proteins. Consistent with the prominence of LXXLL motifs in ER agonist-induced protein interactions, our phage affinity selections using estradiol-bound ER primarily yield peptides containing LXXLL or related motifs. These LXXLL motif-containing peptides interact with ER with the same ligand specificity observed for interaction of coactivator proteins that contain LXXLL motifs. However, not all of the peptides found with estradiol-bound ER and none of the peptides found with ER bound with SERMs contain LXXLL motifs. Yet, these peptides can bind with the affinity and ligand specificity seen with LXXLL motif-containing peptides, and therefore, it stands to reason that these non-LXXLL peptides are also binding functionally relevant sites on ER. One example of this is the ability of the non-LXXLL peptide α/β III, but not the LXXLL peptide α/β I, to inhibit tamoxifen-induced, ER-dependent signaling through a C3 promoter [70]. The mechanism of ER transactivation through this C3 promoter remains undefined, yet it apparently can involve a tamoxifen-induced recruitment of a coregulatory protein that binds through a non-LXXLL binding site on ER [70]. This provides an example of how peptides discovered through phage display from our large Karo Bio combinatorial phage library may represent sequences in coregulatory proteins that have yet to be discovered but may be important for ER physiological actions of interest. As such, Molecular Braille technology may provide information that is unavailable from any other source.

Considering the central role ligand-induced conformational changes play in directing nuclear signaling, efforts to characterize this activity are essential for novel ligand discovery and should be included as early as possible in the discovery process. We envision use of the Molecular Braille assay immediately downstream from in vitro ligand affinity and selectivity characterization. As such, ligands subsequently tested in cell-based or animal models will have the desired affinity and selectivity and will also be associated with an induced conformation, as described by the Molecular Braille pattern or fingerprint, that can be correlated with desirable or undesirable in vivo effects.

10.4 SUMMARY

Peptides isolated from phage-display peptide libraries are powerful tools for drug discovery. These peptides are directed to functional sites on target proteins and can be used to develop assays for a wide range of different targets. For enzymes, the peptides target the active site and therefore function as inhibitors of the enzyme. These active site-directed peptides can be used as surrogate ligands for enzymes to develop HTS assays to screen chemical compound collections for small molecule inhibitors, such as was described in this chapter for the enzyme Dxr.

For targets involved in signal transduction, we have been able to isolate peptides that can specifically sense the conformation of the target protein allowing the development of unique assays for GPCRs and nuclear receptors. The assays for GPCRs are based on peptides that specifically bind to the active conformation of the G-alpha subunits of heterotrimeric G protein. The assay is formatted as a cell-free, homogeneous assay that detects functional activation of the GPCR and does not rely on any downstream reporter.

In addition, we have developed peptide-based assays for nuclear receptors, Molecular Braille technology, which probe the conformation of the nuclear receptors induced by ligand binding. Often, the peptides identified by phage-display affinity selections on nuclear receptors mimic proteins that interact with nuclear receptors. For example, we have identified peptides that contain the LXXLL motif found in nuclear receptor coactivators such as SRC-1. Using a panel of nuclear receptor-specific peptides, we can evaluate the receptor conformation induced by various ligands and classify the ligands based on peptide-binding profile. Biological activity of a given nuclear receptor ligand is determined by the induced receptor conformation; therefore, the peptide-binding profile should ultimately be able to predict the biological activity of a ligand.

REFERENCES

1. Scott JK, Smith GP. Searching for peptide ligands with an epitope library. *Science* 1990; 249:386–390.
2. Cortese R, Monaci P, Luzzago A, Santini C, Bartoli F, Cortese I, Fortugno P et al. Selection of biologically active peptides by phage display of random peptide libraries. *Curr Opin Biotechnol* 1996; 7:616–621.
3. Hyde-DeRuyscher R, Paige LA, Christensen DJ, Hyde-DeRuyscher N, Lim A, Fredericks ZL, Kranz J et al. Detection of small-molecule enzyme inhibitors with peptides isolated from phage-displayed combinatorial peptide libraries. *Chem Biol* 2000; 7:17–25.
4. Kay BK, Hamilton PT. Identification of enzyme inhibitors from phage-displayed combinatorial peptide libraries. *Comb Chem High Throughput Screen* 2001; 4:535–543.
5. Scott JK, Huang SF, Gangadhar BP, Samoriski GM, Clapp P, Gross RA, Taussig R et al. Evidence that a protein–protein interaction "hot spot" on heterotrimeric G protein betagamma subunits is used for recognition of a subclass of effectors. *EMBO J* 2001; 20:767–776.
6. Paige LA, Christensen DJ, Gron H, Norris JD, Gottlin EB, Padilla KM, Chang CY et al. Estrogen receptor (ER) modulators each induce distinct conformational changes in ER alpha and ER beta. *Proc Natl Acad Sci U S A* 1999; 96:3999–4004.
7. Chang C, Norris JD, Gron H, Paige LA, Hamilton PT, Kenan DJ, Fowlkes D et al. Dissection of the LXXLL nuclear receptor–coactivator interaction motif using combinatorial peptide libraries: Discovery of peptide antagonists of estrogen receptors alpha and beta. *Mol Cell Biol* 1999; 19:8226–8239.
8. Wrighton NC, Farrell FX, Chang R, Kashyap AK, Barbone FP, Mulcahy LS, Johnson DL et al. Small peptides as potent mimetics of the protein hormone erythropoietin. *Science* 1996; 273:458–464.
9. Cwirla SE, Balasubramanian P, Duffin DJ, Wagstrom CR, Gates CM, Singer SC, Davis AM et al. Peptide agonist of the thrombopoietin receptor as potent as the natural cytokine. *Science* 1997; 276:1696–1699.
10. Fairbrother WJ, Christinger HW, Cochran AG, Fuh G, Keenan CJ, Quan C, Shriver SK et al. Novel peptides selected to bind vascular endothelial growth factor target the receptor-binding site [in process citation]. *Biochemistry* 1998; 37:17754–17764.
11. Han Y, Kodadek T. Peptides selected to bind the Gal80 repressor are potent transcriptional activation domains in yeast. *J Biol Chem* 2000; 275:14979–14984.
12. Gee SH, Sekely SA, Lombardo C, Kurakin A, Froehner SC, Kay BK. Cyclic peptides as non-carboxyl-terminal ligands of syntrophin PDZ domains. *J Biol Chem* 1998; 273:21980–21987.
13. Fuh G, Pisabarro MT, Li Y, Quan C, Lasky LA, Sidhu SS. Analysis of PDZ domain-ligand interactions using carboxyl-terminal phage display. *J Biol Chem* 2000; 275:21486–21491.

14. Linn H, Ermekova KS, Rentschler S, Sparks AB, Kay BK, Sudol M. Using molecular repertoires to identify high-affinity peptide ligands of the WW domain of human and mouse YAP. *Biol Chem* 1997; 378:531–537.

15. Sperinde JJ, Choi SJ, Szoka FC Jr. Phage display selection of a peptide DNase II inhibitor that enhances gene delivery. *J Gene Med* 2001; 3:101–108.

16. Laskowski RA, Luscombe NM, Swindells MB, Thornton JM. Protein clefts in molecular recognition and function. *Protein Sci* 1996; 5:2438–2452.

17. Mattos C, Ringe D. Locating and characterizing binding sites on proteins. *Nat Biotech* 1996; 14:595–599.

18. Ringe D, Mattos C. Analysis of the binding surfaces of proteins. *Med Res Rev* 1999; 19:321–331.

19. DeLano WL, Ultsch MH, deVos AM, Wells JA. Convergent solutions to binding at a protein–protein interface. *Science* 2000; 287:1279–1283.

20. Gron H, Hyde-DeRuyscher R. Peptides as tools in drug discovery. *Curr Opin Drug Discov Devel* 2000; 3:636–645.

21. Christensen DJ, Gottlin EB, Benson RE, Hamilton PT. Phage display for target-based antibacterial drug discovery. *Drug Discov Today* 2001; 6:721–727.

22. Takahashi S, Kuzuyama T, Watanabe H, Seto H. A 1-deoxy-D-xylulose 5-phosphate reductoisomerase catalyzing the formation of 2-C-methyl-D-erythritol 4-phosphate in an alternative nonmevalonate pathway for terpenoid biosynthesis. *Proc Natl Acad Sci U S A* 1998; 95:9879–9884.

23. Gottlin EB, Benson RE, Conary S, Antonio B, Duke K, Payne ES, Ashraf SS et al. High throughput screen for inhibitors of 1-deoxy-D-xylulose 5-phosphate reductoisomerase by surrogate ligand competition. *J Biomol Screen* 2003; 8(3):332–339.

24. Jacobi A, Baum H, Nakano T, Annand RR, Anderson S, Breunig J, Bowes S et al. Isoprenoid biosynthesis as a novel antibacterial target. Abstracts of the 41st Interscience Conference on Antimicrobial Agents and Chemotherapy 247 [abstr 2124], 2001.

25. Tao J, Wendler P, Connelly G, Lim A, Zhang J, King M, Li T et al. Drug target validation: Lethal infection blocked by inducible peptide. *Proc Natl Acad Sci U S A* 2000; 97:783–786.

26. Benson RE, Gottlin EB, Christensen DJ, Hamilton PT. Intracellular expression of peptide fusions for demonstration of protein essentiality in bacteria. *Antimicrob Agents Chemother* 2003; 47(9):2875–2881.

27. Drews J. Drug discovery: A historical perspective. *Science* 2000; 287:1960–1964.

28. Klabunde T, Hessler G. Drug design strategies for targeting G-protein-coupled receptors. *ChemBioChem* 2002; 3:928–944.

29. Wise A, Gearing K, Rees S. Target validation of G-protein coupled receptors. *Drug Discov Today* 2002; 7:235–246.

30. Johnston CA, Willard FS, Jezyk MR, Fredericks Z, Bodor ET, Jones MB, Blaesius R et al. Structure of Galpha(i1) bound to a GDP-selective peptide provides insight into guanine nucleotide exchange. *Structure* 2005; 13(7):1069–1080.

31. Gnagey A, Ellis J. Allosteric regulation of the binding of [3H]acetylcholine to m2 muscarinic receptors. *Biochem Pharmacol* 1996; 52:1767–1775.

32. Rosenfeld MG, Glass CK. Coregulator codes of transcriptional regulation by nuclear receptors. *J Biol Chem* 2001; 276:36865–36868.

33. Coleman KM, Smith CL. Intracellular signaling pathways: Nongenomic actions of estrogens and ligand-independent activation of estrogen receptors. *Front Biosci* 2001; 6:D1379–D1391.

34. Jordan V. The secrets of selective estrogen receptor modulation: Cell-specific coregulation. *Cancer Cell* 2002; 1:215–217.

35. Heery DM, Kalkhoven E, Hoare S, Parker MG. A signature motif in transcriptional co-activators mediates binding to nuclear receptors [see comments]. *Nature* 1997; 387: 733–736.

36. McInerney EM, Rose DW, Flynn SE, Westin S, Mullen TM, Krones A, Inostroza J et al. Determinants of coactivator LXXLL motif specificity in nuclear receptor transcriptional activation. *Genes Dev* 1998; 12:3357–3368.

37. Feng W, Ribeiro RC, Wagner RL, Nguyen H, Apriletti JW, Fletterick RJ, Baxter JD et al. Hormone-dependent coactivator binding to a hydrophobic cleft on nuclear receptors. *Science* 1998; 280:1747–1749.

38. Mak H, Hoare S, Henttu P, Parker M. Molecular determinants of the estrogen receptor–coactivator interface. *Mol Cell Biol* 1999; 19:3895–3903.

39. Brzozowski AM, Pike AC, Dauter Z, Hubbard RE, Bonn T, Engstrom O, Ohman L et al. Molecular basis of agonism and antagonism in the oestrogen receptor. *Nature* 1997; 389:753–758.

40. Shiau AK, Barstad D, Loria PM, Cheng L, Kushner PJ, Agard DA, Greene GL. The structural basis of estrogen receptor/coactivator recognition and the antagonism of this interaction by tamoxifen. *Cell* 1998; 95:927–937.

41. Bramlett KS, Wu Y, Burris TP. Ligands specify coactivator nuclear receptor (NR) box affinity for estrogen receptor subtypes. *Mol Endocrinol* 2001; 15:909–922.

42. Wong CW, Komm B, Cheskis BJ. Structure–function evaluation of ER alpha and beta interplay with SRC family coactivators. ER selective ligands. *Biochemistry* 2001; 40: 6756–6765.

43. Kraichely DM, Sun J, Katzenellenbogen JA, Katzenellenbogen BS. Conformational changes and coactivator recruitment by novel ligands for estrogen receptor-alpha and estrogen receptor-beta: Correlations with biological character and distinct differences among SRC coactivator family members. *Endocrinology* 2000; 141:3534–3545.

44. Ko L, Cardona GR, Iwasaki T, Bramlett KS, Burris TP, Chin WW. Ser-884 adjacent to the LXXLL motif of coactivator TRBP defines selectivity for ERs and TRs. *Mol Endocrinol* 2002; 16:128–140.

45. Xu HE, Stanley TB, Montana VG, Lambert MH, Shearer BG, Cobb JE, McKee DD et al. Structural basis for antagonist-mediated recruitment of nuclear co-repressors by PPARalpha. *Nature* 2002; 415:813–817.

46. Jung D, Lee S, Lee J. Agonist-dependent repression mediated by mutant estrogen receptor alpha that lacks the activation function 2 core domain. *J Biol Chem* 2001; 276:37280–37283.

47. He B, Minges JT, Lee LW, Wilson EM. The FXXLF motif mediates androgen receptor-specific interactions with coregulators. *J Biol Chem* 2002; 277:10226–10235.

48. Li D, Wang F, Samuels H. Domain structure of the NRIF3 family of coregulators suggests potential dual roles in transcriptional regulation. *Mol Cell Biol* 2001; 21:8371–8384.

49. Warnmark A, Treuter E, Gustafsson J, Hubbard R, Brzozowski A, Pike A. Interaction of transcriptional intermediary factor 2 nuclear receptor box peptides with the coactivator binding site of estrogen receptor alpha. *J Biol Chem* 2002; 277:21862–21868.

50. Witcher M, Yang X, Pater A, Tang SC. BAG-1 p50 isoform interacts with the vitamin D receptor and its cellular overexpression inhibits the vitamin D pathway. *Exp Cell Res* 2001; 265:167–173.

51. Boruk M, Savory JG, Hache RJ. AF-2-dependent potentiation of CCAAT enhancer binding protein beta-mediated transcriptional activation by glucocorticoid receptor. *Mol Endocrinol* 1998; 12:1749–1763.

52. Jung DJ, Na SY, Na DS, Lee JW. Molecular cloning and characterization of CAPER, a novel coactivator of activating protein-1 and estrogen receptors. *J Biol Chem* 2002; 277:1229–1234.

53. Lu ML, Schneider MC, Zheng Y, Zhang X, Richie JP. Caveolin-1 interacts with androgen receptor. A positive modulator of androgen receptor mediated transactivation. *J Biol Chem* 2001; 276:13442–13451.

54. Rao J, Lee P, Benzeno S, Cardozo C, Albertus J, Robins DM, Caplan AJ. Functional interaction of human Cdc37 with the androgen receptor but not with the glucocorticoid receptor. *J Biol Chem* 2001; 276:5814–5820.

55. Sauve F, McBroom LD, Gallant J, Moraitis AN, Labrie F, Giguere V. CIA, a novel estrogen receptor coactivator with a bifunctional nuclear receptor interacting determinant. *Mol Cell Biol* 2001; 21:343–353.

56. Hulkko SM, Wakui H, Zilliacus J. The pro-apoptotic protein death-associated protein 3 (DAP3) interacts with the glucocorticoid receptor and affects the receptor function. *Biochem J* 2000; 349(Pt 3):885–893.

57. Chen J, Kaul S, Simons SS Jr. Structure/activity elements of the multifunctional protein, GMEB-1. Characterization of domains relevant for the modulation of glucocorticoid receptor transactivation properties. *J Biol Chem* 2002; 277:22053–22062.

58. Kaul S, Blackford JA Jr, Chen J, Ogryzko VV, Simons SS Jr. Properties of the glucocorticoid modulatory element binding proteins GMEB-1 and -2: Potential new modifiers of glucocorticoid receptor transactivation and members of the family of KDWK proteins. *Mol Endocrinol* 2000; 14:1010–1027.

59. Sharma M, Zarnegar M, Li X, Lim B, Sun Z. Androgen receptor interacts with a novel MYST protein, HBO1. *J Biol Chem* 2000; 275:35200–35208.

60. Saji S, Okumura N, Eguchi H, Nakashima S, Suzuki A, Toi M, Nozawa Y et al. MDM2 enhances the function of estrogen receptor alpha in human breast cancer cells. *Biochem Biophys Res Commun* 2001; 281:259–265.

61. Schulz M, Schneider S, Lottspeich F, Renkawitz R, Eggert M. Identification of nucleolin as a glucocorticoid receptor interacting protein. *Biochem Biophys Res Commun* 2001; 280:476–480.

62. Lee SR, Ramos SM, Ko A, Masiello D, Swanson KD, Lu ML, Balk SP. AR and ER interaction with a p21-activated kinase (PAK6). *Mol Endocrinol* 2002; 16:85–99.

63. Wang Y, Yin L, Hillgartner FB. The homeodomain proteins PBX and MEIS1 are accessory factors that enhance thyroid hormone regulation of the malic enzyme gene in hepatocytes. *J Biol Chem* 2001; 276:23838–23848.

64. Zhou D, Chen S. PNRC2 is a 16 kDa coactivator that interacts with nuclear receptors through an SH3-binding motif. *Nucleic Acids Res* 2001; 29:3939–3948.

65. Delage-Mourroux R, Martini PG, Choi I, Kraichely DM, Hoeksema J, Katzenellenbogen BS. Analysis of estrogen receptor interaction with a repressor of estrogen receptor activity (REA) and the regulation of estrogen receptor transcriptional activity by REA. *J Biol Chem* 2000; 275:35848–35856.

66. Chipuk JE, Cornelius SC, Pultz NJ, Jorgensen JS, Bonham MJ, Kim SJ, Danielpour D. The androgen receptor represses transforming growth factor-beta signaling through interaction with Smad3. *J Biol Chem* 2002; 277:1240–1248.

67. Stoner M, Wang F, Wormke M, Nguyen T, Samudio I, Vyhlidal C, Marme D et al. Inhibition of vascular endothelial growth factor expression in HEC1A endometrial cancer cells through interactions of estrogen receptor alpha and Sp3 proteins. *J Biol Chem* 2000; 275:22769–22779.

68. Watanabe M, Yanagisawa J, Kitagawa H, Takeyama K, Ogawa S, Arao Y, Suzawa M et al. A subfamily of RNA-binding DEAD-box proteins acts as an estrogen receptor alpha coactivator through the N-terminal activation domain (AF-1) with an RNA coactivator, SRA. *EMBO J* 2001; 20:1341–1352.

69. Makino Y, Yoshikawa N, Okamoto K, Hirota K, Yodoi J, Makino I, Tanaka H. Direct association with thioredoxin allows redox regulation of glucocorticoid receptor function. *J Biol Chem* 1999; 274:3182–3188.

70. Norris JD, Paige LA, Christensen DJ, Chang CY, Huacani MR, Fan D, Hamilton PT et al. Peptide antagonists of the human estrogen receptor. *Science* 1999; 285:744–746.

71. Finkel T, Duc J, Fearon E, Dang C, Tomaselli G. Detection and modulation in vivo of helix-loop-helix protein–protein interaction. *J Biol Chem* 1993; 268:5–8.

72. Hirsimaki P, Aaltonen A, Mantyla E. Toxicity of antiestrogens. *Breast J* 2002; 8:92–96.

CHAPTER 11

Engineering Protein Folding and Stability

Mihriban Tuna and Derek N. Woolfson

CONTENTS

11.1 PROTEIN REDESIGN AND DESIGN

Protein design tests our understanding of how protein sequence relates to structure and function, that is, our progress toward addressing the informational aspect of the protein-folding problem. In addition, successful protein design exercises provide tools, concepts, and rules for (1) engineering existing protein scaffolds (an area that we refer to as protein engineering or protein redesign) and (2) creating altogether new protein structures and functions (*de novo* protein design). This second endeavor is the ultimate quest for some protein designers, but it is difficult and, by and large, can only be achieved for a limited number of relatively straightforward protein-folding motifs—for instance, coiled coils and zinc-finger peptides—for which good sequence-to-structure relationship are available [1], although, very recently, Kuhlman and colleagues [2] describe the successful computer-aided design of a completely novel globular protein.

In general, however, *de novo* protein design attempts for such targets tend to deliver partly folded, molten-globule ensembles rather than specific, unique structures, though the secondary structure content and even chain topology may be correct. Thus, for globular structures, alternative approaches are required in design: one extreme is to choose an appropriate stably folded scaffold as a starting point and engineer changes in a rational and usually stepwise manner toward a target with altered and/or improved properties. At the other extreme, the starting protein, which may be folded or not, is subjected to combinatorial mutagenesis to create libraries of related protein sequences from which folded, stable, and perhaps even functional proteins are selected. The mutagenesis employed can be either random across part or all of the sequence or saturation mutagenesis targeted at specific residues chosen by the researcher on some rational basis.

Choosing where and how to mutate is not always clear, and herein lies a major problem in redesigning and designing globular proteins. In part, the problem is that, from an experimental point of view, for a given protein, it is not possible to create redundant libraries in which more than about eight residues are mutated to all possible amino acids. This problem has been solved to some extent by computational methods in which very large numbers of mutants can be assessed against a targeted structural model very rapidly [3–5]. A difficulty with this in silico approach, however, is that, compared with combinatorial experimental methods, only a relatively small number of the generated sequences can be made and tested experimentally. In addition, folded proteins are highly cooperative units; that is, their structures and stabilities are determined by many interdependent (i.e., cooperative) noncovalent interactions. In this respect, it is hard to fully disentangle the main contributors to protein stability. Thus, any multiple mutagenesis experiment carries the risk that as much damage can be done as good; in other words, many of the mutants in any given library will fail to fold altogether.

As said, it is widely accepted that the acquisition of a well-packed and complementary hydrophobic core is a general requirement for specifying the fold of a globular protein and for maintaining its stability [6]. Indeed, the lack of stability of many early protein designs, particularly those for four-helix-bundle proteins, has been put down to a lack of specificity in the prescribed hydrophobic cores [7–9]. Thus, tackling the protein design problem from the inside out may seem like a reasonable tack. Indeed, over the past decade, both computer-based and experimental protein design efforts have focused on developing methods for optimizing hydrophobic core packing [3–5]. This in silico work has been very successful. However, this chapter focuses on experimental efforts in this area. The tricks here are (1) to target a restricted (rationally chosen) set of sites for mutagenesis and (2) to design selection methods that allow the small fraction of competently folded and stable structures from the vast majority of mutants that, even with rational targeting, may fail to fold. We begin by summarizing the main experimental efforts that preceded phage-display work in this area. However, as we shall also describe, others have had considerable success in engineering protein stability by leaving the hydrophobic core well alone and optimizing surface residues; presumably, in these cases, a higher proportion of the mutant proteins do fold.

11.2 EARLY COMBINATORIAL STUDIES AIMED AT REPACKING THE CORES OF PROTEINS

11.2.1 Repacking the Cores of Natural Proteins

The studies of Lim et al. [10–12] on the N-terminal domain of λ-repressor provide a landmark in this area. A library of λ-repressor mutants is created by saturation mutagenesis to NN{C,G} codons (i.e., to give all of the possible 20 proteinogenic amino acids) at seven hydrophobic core positions. An in vivo functional assay is used to screen for competently folded mutants of the domain. The DNA for the library is transformed into *Escherichia coli*, which are subsequently challenged with a strain of phage λ lacking a functional λ-repressor gene. Thus, only those clones that express functional λ-repressor survive. The assertion is that in order to function, the mutant proteins must be folded correctly. The studies conclude that the core of this particular protein domain is tolerant to substitutions and, as such, is *plastic*. The main constraint on achieving functional mutants is the maintenance of hydrophobic residues within the core and, albeit to a lesser extent, the preservation of steric and core-volume constraints.

Axe et al. [13] repack the hydrophobic core of the ribonuclease barnase to investigate whether unique native-like packing is essential for maintenance of protein function. In this case, the library is made by mutagenesis of 12 of the 13 hydrophobic core sites to combinations of Phe, Ile, Leu, Met, and Val (FILMV), which are readily generated by the degenerate codon {A,G,T}T{C,G}. Again a functional screen is used to detect functional and presumably folded proteins. Barnase is extremely toxic to bacteria, and the assay developed is extremely sensitive, as mutants with as low as 0.2% of wild-type (WT) activity kill the host and test positive. Thus, a suppressible stop codon within the coding sequence for the barnase mutants is used in these studies, with active variants being identified by switching between nonsuppressor and suppressor strains of *E. coli*. As high as 23% of the hydrophobic core variants maintain enzymatic activity in vivo. Thus, the authors argue that, like the N-terminal domain of λ-repressor, the basic function of barnase is very tolerant of changes in the hydrophobic core. Hence, in terms of designing or maintaining crude function in proteins, hydrophobicity alone could be sufficient to generate a functional hydrophobic core; after this step, further mutagenesis and optimization of the packing could be employed to achieve a better (i.e., unique and more stable) scaffold [13].

One assumption in the studies by Lim and Sauer [10–12] and by Axe and colleagues is that correct folding is a prerequisite for correct function. Given the accepted link between protein sequence, structure, and function established by

Anfinsen, this seems reasonable. However, imagine a case where a protein is either unfolded or only partly folded in isolation but folds upon binding to its substrate. Such on-site assembly is well established in the area of protein–nucleic acid interaction. Furthermore, this idea of disorder-to-order transitions associated with protein function is gathering steam in the area of *natively unfolded proteins* [14]. In the approaches of Axe et al. [13] and Lim and Sauer, such transitions, if they occurred, would give false positives in the sense that certain proteins would not necessarily have to fold independently in order to elicit a functional response.

11.2.2 Creating Novel Proteins Using Binary Patterns of Hydrophobic and Polar Amino Acids

Hecht and coworkers [15–19] have championed a very different approach in which *focused* libraries for *de novo* designed proteins are created based on straightforward binary or *HP* patterns of hydrophobic (*H*) and polar (*P*) residues. Put simply, patterns of the type *HPHPHP* or *HPPHPPP* separated by turn-promoting sequences are used to guide folding to β-strand and α-helical elements of secondary structure, respectively. Limited amino-acid repertoires are also used, specifically *P* = K, H, Q, N, E, or D and *H* = F, I, L, M, or V. Hecht and his colleagues [17–19] have used this approach to generate four-helix-bundle motifs and, more recently, six-stranded β-sheet models. The approach differs considerably from those that form the main focus of this chapter because the libraries are "not subjected to high-throughput screens or directed evolution," although one imagines that some selection is at play within *E. coli*, the host into which the DNA libraries are transformed and expressed. Rather, clones are selected purely on the basis of expression of the mutant proteins. Several iterations of the four-helix-bundle designs have been performed and generate proteins that are water soluble and have α-helical structure, and some show native-like characteristics [15–19]. Nonetheless, in the earlier studies at least, most exhibited characteristics typical of molten-globule structures. However, this work has recently culminated with the determination of an nuclear magnetic resonance (NMR) solution structure for a four-helix-bundle protein retrieved from a binary library [20] and the observation that some of these proteins exhibit enzyme-like activity [21].

Interestingly, this approach does not appear to extend readily to the generation of globular β-structured scaffolds. Using libraries created to generate amphipathic β-strands separated by turn regions, the group retrieved proteins that formed amyloid-like fibrous assemblies [22]. Encouragingly, however, Wang and Hecht [23] have successfully managed to prevent amyloid-like fibrillogenesis in one of the selected β-structured mutants by incorporating rules proposed by Richardson and Richardson [24] for capping the edges of β-sheets.

To finish this section, we should like to mention the work of Silverman and colleagues [25], who describe the combinatorial dissection of the TIM-barrel structure, which is adopted by approximately 10% of all enzymes. This structure effectively has two hydrophobic cores, which Harbury and coworkers find to respond differently to mutation. The outer core between the α-helices and the β-barrel tolerates mutations, whereas the inner core, which is more conserved, is more sensitive to changes. This second finding is particularly important as it contrasts with most of the foregoing work in this area, which suggests that hydrophobic cores are generally plastic and permissive of (hydrophobic) mutations. This point, which might be termed "oil-drop versus jigsaw-puzzle models" for the packing of hydrophobic cores, is a theme that we will return in Section 11.4 when we describe studies on ubiquitin [26,27].

11.3 PHAGE DISPLAY IN ENGINEERING PROTEIN STABILITY

Phage display has been proved an effective tool in the armory of the protein engineer. In this method, a protein of interest is displayed on the surface of (usually) filamentous bacteriophage via fusion to one of the viral coat proteins (see Chapter 2). This carries a number of advantages. First, the protein is linked to the DNA that encodes it, which is encased within the virus particle. Thus, phenotype and genotype are physically linked, allowing the displayed proteins to be identified through DNA sequencing as necessary. Second, using standard mutagenesis techniques, the gene for the displayed protein can be mutated to create a library of variants. Each of these variants can then be displayed on distinct phage particles (to give what are termed "protein-phage") and are, hence, linked to their own encoding DNA. Third, using suitable selection methods, protein-phage variants with specifically sought properties can be retrieved from the pool. These can be amplified by infection into *E. coli*. At this point, the phage can be recovered for further rounds of selection and amplification, or clones can be identified by DNA sequencing of the phagemid DNA within the *E. coli*. All in all, the phage-display approach offers the opportunity of exploring a large number of sequences simultaneously. This method has most widely been applied to select protein variants, particularly those for antibody fragments, based on their ability to bind specific targets.

11.3.1 Nonprotease-Based Applications

Gu and colleagues [28] use binding to select stably folded proteins from a phage-displayed library of the IgG-binding domain of peptostreptococcal protein L. The selection rationale is that the correct folding of the domain is required for IgG binding. The group shows that, even without disruption or alteration to the binding interface, if the structure and folding of the domain are compromised, IgG binding is abolished. In this case, this clearly argues against the aforementioned potential caveat about cooperative folding and binding (i.e., on-site folding or disorder-to-order transitions). The group displays the domain via the major coat protein, g8p, and mutate 14 residues of the N-terminal β-hairpin of the protein. As noted by the authors, the mutagenesis and selections presented are too limited to allow conclusions about protein folding to be made. Thus, the experiments should be considered as proof of concept aimed at developing a method to study the sequence determinants of protein folding for structures that have a binding site that can be exploited in selection.

Chakravarty et al. [29] have taken a slightly different approach to achieve protein stabilization. They use fragment complementation in combination with phage display to improve the thermostability of RNase S. Specifically, the large proteolytic fragment S-protein (residues 21–124) is used to select tight binders from a random 15-mer peptide library presented on phage. One selected peptide complements S-protein to give a complex with melting temperature 10°C higher than the WT.

11.3.2 Combining Phage Display and Proteolysis in Protein Engineering and Design

In this chapter, we are primarily interested in selection methods in which the key selection step does not involve the ability of the displayed protein itself to bind some prescribed target. Therefore, the question is what alternative selection strategies can be conceived to retrieve proteins on the basis of folding and stability alone, rather than based on some *function* such as binding. Such a method would immediately remove any concern over false positives resulting from on-site assembly of the selected proteins; it would provide a route for assessing protein–sequence relationships in proteins and present new possibilities for designing new proteins from scratch free of constraints imposed by function. One option, which has considered independently by at least three groups [30–32], is that proteolysis could be combined somehow with phage-display selection to remove proteins that are either unfolded or only partly folded.

In many respects, the seminal publication from Matthews and Wells [33] provides the forerunner for such a strategy. These authors describe the concept of *substrate phage* in which a library of peptides is tethered between an affinity tag and the gene-3 minor coat protein (g3p). The resulting phage are bound to a surface and treated with protease. As phage are themselves protease insensitive, choosing a similarly resistant affinity tag means that this method reports directly on the protease sensitivity of the peptide linker: That is, phage released quickly from the surface harbored good protease substrates, whereas those that remain bound, harbored poor substrates. In this way, Matthews and Wells are able to identify pools of sensitive and resistant peptide sequences for a variant of subtilisin and factor $X(a)$. In this method and those that followed (below in this section), the binding properties of the affinity tag are of secondary importance, and the selection is predicated by the cleavage (or not) of the tag from the phage, which depends only on the susceptibility (or otherwise) of the peptide linker itself.

Initially, three independent groups reported successful studies utilizing phage display and protease selection for probing and/or increasing the stability of proteins [30–32]. In addition, these groups and a number of others have applied the approach in protein engineering and design [34,35]; these later studies will be described in Section 11.5 of this chapter. All of the selection methods are based purely on maintaining the structural integrity of the target proteins, and there is no requirement for any other assayable function. In each case, the selection rationale is that stable, fully folded proteins are more resistant to attack by proteases than partially folded or unfolded proteins [36].

In 1998, Kristensen and Winter [30] proposed that proteolysis could be used to select stably folded proteins from phage-displayed libraries. This method is a variant of selectively infective phage (SIP): a peptide or protein of interest is cloned between the D2 and D3 domains of g3p. D3 anchors g3p to the viral surface, while the D1 and D2 domains are required for infection of *E. coli*. Thus, cleavage of the D2–D3 linker renders the phage noninfective, which means that DNA carried by such phage is not propagated in the phage-display and selection experiment (Figure 11.1). The group demonstrates that protein folding and protease sensitivity are linked by monitoring and comparing phage infectivity of various peptide-phage and protein-phage fusions—specifically, a protease-sensitive peptide, and two variants of barnase with different stabilities and the villin headpiece—after treatment with trypsin over a range of temperatures. In addition, they show that this provides the basis for discriminating peptides and proteins of different stabilities (i.e., selection) by tracking phage retrieved after protease challenges on mixtures of (1) the two barnase variants and (2) the protease-sensitive peptide and the villin headpiece. This paper is also notable in a number of other respects: First, the authors demonstrate the ability of filamentous phage to endure a variety of insults, including incubations over a range of pH (pH 2.2–12, 37°C, 30 min) and in the presence of urea (60°C, 60 min) or

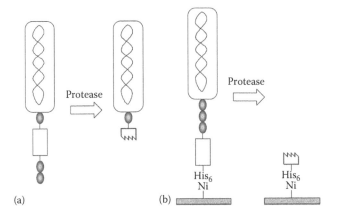

Figure 11.1   (a) Selectively infective phage takes advantage of the three-domain structure of the minor coat protein (g3p) of phage. The C-terminal domain anchors the protein in the viral coat, whereas the N-terminal domains are responsible for binding and infection into *E. coli*. Cloning a library into the flexible linker before the C-terminal domains allows protease-based selection because proteolysis of the insert removes the N-terminal domains and prevents infection into *E. coli*. This selects against unstable inserts [30,31]. (b) Alternatively, an uninterrupted g3p can be used as follows: His-tag–target–g3p–phage. This allows intact protein-phage fusions to be tethered to nickel-coated surfaces, which can be washed with protease to remove phage harboring unstable linkers [32]. In this case, selection can be monitored directly by surface plasmon resonance (SPR) in Biacore, which allows many conditions to be tested quickly and individual clones to be compared. Alternatively, nickel–nitrilotriacetic acid (NTA) agarose beads can be used for large-scale selections. (Reproduced from Woolfson DN, *Curr Opin Struct Biol*, 11:464–471, 2001.)

guanidine hydrochloride (37°C, 90 min); appreciable loss in phage infectivity is only reported for experiments using 5 M guanidine hydrochloride and higher. Similarly, resistance to a range of proteases is also reported. In this case, phage infectivity is only reduced after treatment with subtilisin, which is known to cleave g3p. In addition, to assist these experiments, the authors describe a protease-cleavable helper phage (KM13) and a phagemid vector system. These are created by introducing a protease-cleavable linker between the D2 and D3 domains of g3p. The KM13 helper phage is used to rescue target protein-phage, which means that after protease selection, only those phage with intact D1–D2–protein–D3 fusions will be infective and, hence, will be selected.

In the same year, Sieber et al. [31] describe a similar SIP-based approach for selecting stably folded proteins from phage-displayed libraries using protease selection. The group dubs their method *Proside* for "protein stability increased by directed evolution." Like Kristensen and Winter [30], the group describes a number of important and useful control and exploratory experiments. Specifically, they also demonstrate the stability of fd phage over a range of solution conditions, which may be of use in stability-based selections of protein-phage. They find that protease selection varies with different proteases—trypsin, chymotrypsin, pepsin, or proteinase K—with trypsin showing the lowest activity and proteinase K the highest. Also, they show that by fine-tuning the conditions of the selection process, variants with marginally different stabilities (differing by ≈1 kJ/mol) can be distinguished. Finally, the group demonstrates selection from a modest library of variants of RNase T1(4A) with mutations of surface residues. RNase T1(4A) is a destabilized variant of RNase T1 in which the four Cys residues are mutated to Ala to remove the two disulfide bonds. Positions Ser17, Asp29, and Tyr42 of the variant are targeted by saturation mutagenesis. After four rounds of protease selection, the preferred residues emerge at these sites: Ala, Leu, and Phe, respectively. Impressively, when the identified mutations are transferred to WT RNase T1, with its disulfide bonds intact, the resulting proteins are also stabilized significantly (by almost 10°C increases in their melting temperatures [T_Ms]). This is an important result, as it shows that surface engineering of proteins can be used to improve stability of proteins. Furthermore, it demonstrates that mutations identified in this way can be transferred between backgrounds, which is unlikely to be true of constellations of stabilizing residues discovered through core-directed design where residue–residue contacts are greater.

We published our own work in 1999, and we combined phage display and protease selection in a different way (Figures 11.1 and 11.2) [32,37]. In this case, the target protein is displayed on bacteriophage more traditionally, that is, as an N-terminal fusion to the minor coat protein g3p. In addition, a histidine tag (either hexa- or deca-His) is incorporated at the N-terminus of the target to provide an affinity tag for metal-based capture of the protein-phage onto surfaces. After the protein-phage are immobilized onto nickel-coated surfaces, they are challenged with a protease or protease mixtures. This results in cleavage of any unstable protein linkers and allows phage harboring such linkers to be washed away from the support. The remaining protein-phage are then eluted from the support using imidazole or EDTA washes, or with a change of pH of the buffer. The freed phage are then amplified for further

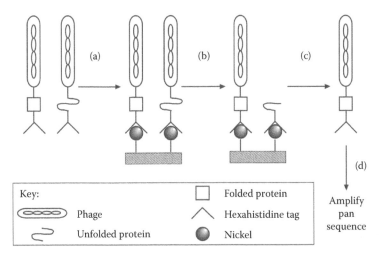

Figure 11.2   Details of the protease-selection method developed in the Woolfson laboratory [32]. (a) Protein-phage are captured onto nickel-derivatized support via N-terminal histidine tags. (b) Bound protein-phage are challenged with protease, phage-displaying proteins that are proteolyzed are washed away from the support. (c) Protein-phage that resist proteolysis are eluted from the nickel support and (d) amplified. (Reproduced from Tuna M, Finucane MD, Vlachakis NGM, Woolfson DN. Protease-based selection of stably folded proteins and protein domains from phage display libraries. In: Clackson T, Lowman HB, eds. *Phage Display A Practical Approach.* Oxford, UK, 2004.)

rounds of selection and/or identified by DNA sequencing. We find that a particular advantage of this approach is that the proteolysis reaction can be monitored in real time by following the release of phage by surface plasmon resonance (SPR) in Biacore. Note that this type of selection can also be done in solution with intact (protease-resistant protein-phage) being pulled down after proteolysis. As detailed in Section 11.4, we have applied this method in an approach that we refer to as *core-directed protein design* [1] to repack the hydrophobic core of ubiquitin [27]. In this case, we find that, even though we mutated one half of the protein's hydrophobic core simultaneously, all eight of the targeted residues showed a strong preference for WT or WT-like residues in the selected library.

11.4 WORKED EXAMPLE: REPACKING THE HYDROPHOBIC CORE OF UBIQUITIN

We have combined the ideas of core-directed protein design and protease-based phage-display selection to repack the hydrophobic core of ubiquitin [27]. To achieve this, the gene for mammalian ubiquitin was cloned between those for a (N-terminal) hexahistidine tag and gene *III* in phage-display vector, pCANTAB B2. This construct included two suppressible stop codons prior to gene *III* to allow expression of the hexahistidine-tagged protein both as a fusion to g3p and free of bacteriophage by using suppressor and nonsuppressor strains of *E. coli*, respectively. Residues corresponding to approximately half of the hydrophobic core of ubiquitin were mutated to combinations of hydrophobic side chains to produce a phage library. Specifically, residues 1, 3, 5, 15, 17, 26, and 30 were targeted for mutagenesis to combinations of FILMV using the aforementioned DTS degenerate codon ({A,G,T}T{C,G}). To avoid a bias of WT ubiquitin—which is exceptionally stable with respect to unfolding and proteolysis—a *nobbled* template was used for the combinatorial mutagenesis: Met 1 was replaced by Ala, and the remaining targeted sites were replaced by Leu to give an "AL_7" mutant. Theoretically at least, this combination of Ala and Leu has a similar volume to the WT targeted core residues. This mutant was destabilized with respect to proteolysis. The AL_7 plasmid was used as the template for sticky-feet mutagenesis [38] to create the library of ubiquitin hydrophobic core mutants. This library potentially contained 1.17×10^7 different DNAs and 390,625 protein variants including the WT sequence. Experimentally, DNA sequencing revealed that the naive library was 48% AL_7 parent and contained approximately 7×10^6, corresponding to approximately a 20-fold redundancy of the potential protein mutants.

Turning to the phage display and selection, two different nickel-coated surfaces were tested for immobilizing the protein-phage and performing protease selection on the library of ubiquitin hydrophobic core mutants: nickel–nitrilotriacetic acid (Ni-NTA) sensor chips for SPR (Biacore) and Ni-NTA agarose beads (QIAgen).

11.4.1 Following Protease Selection by Surface Plasmon Resonance

SPR technology has been developed specifically for monitoring biomolecular interactions [39]. The basic principle of this method is that changes in mass at a surface result in a change in refractive index of the solution above the surface. This gives rise to an SPR effect at the surface, which can be measured by monitoring changes in a laser beam reflected by the surface. The SPR response, measured in resonance units (RU), is proportional to the amount of material bound to the surface. In a typical SPR experiment, one of the interactants is immobilized to the surface of a sensor chip in a microflow cell into which the other interactant(s) is injected. Any interactions lead to a change in the effective mass at the surface and hence, a refractive index change. Typically, the response is followed with time to give a record, a sensorgram, of the interactions on the sensor chip. The method has been used to study interactions between biomolecules such as peptides, nucleic acids, carbohydrates, lipids, and small molecules. It is performed in real time and does not require labeling. It has mainly been used to evaluate binding, binding specificity, and kinetics [40], although other applications include ligand fishing, epitope mapping, molecular assembly, following purification, and small-molecule screening [41]. Various sensor chips have been designed for specific interactions with ligands. We used sensor chips with NTA immobilized on a dextran matrix to capture nickel and thence histidine-tagged protein-phage.

We used SPR to follow and to *visualize* binding and protease cleavage of both monoclonal and library protein-phage [32]. This established (1) specific binding of protein-phage to NTA sensor chips in Biacore; (2) that the binding capacity of such chips was limited but still useful; and (3) that the method could be used to follow protease cleavage.

Figure 11.3 shows a typical sensorgram for an SPR experiment performed using a Biacore experiment. Both the initial loading of nickel ions and the subsequent binding and release of protein-phage are detected in real time using this technique. In each case, binding is measured as net change in resonance response, in RU, before and after the addition of the nickel or the analyte.

Although we detected some background (nonspecific) binding, we only observed significant binding of protein-phage when both nickel had been loaded onto the chip and the protein-phage carried a histidine tag. On this point, we found that decahistidine tags performed better and more reliably than traditional hexahistidine tags.

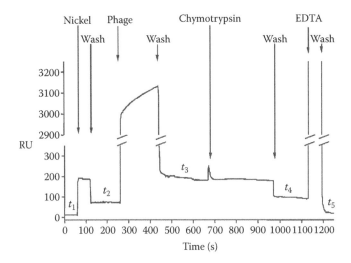

Figure 11.3 Sensorgram trace for an experiment of the type shown in Figure 11.2 and followed by SPR as carried out in a Biacore 2000 instrument. At time t_1, the resonance intensity (measured in RU) is at a minimum, and the intensity can be normalized to 0, as shown. While buffer containing additional components, such as nickel ions, phage, or enzyme, is passed over the chip, the resonance intensity will change because of the refractive index change, as shown by the immediate rise in intensity in the figure on adding the nickel solution. When the running buffer wash is resumed, the resonance intensity drops again, with the new equilibrium value (t_2) representing the material that bound in the previous step. Once the nickel has been bound, the phage are added, which then bind through their hexahistidine tags. After the wash is resumed, the intensity increases to t_3, indicating bound phage [bound phage $\approx (t_3 - t_2)$]. A further brief increase in resonance intensity indicates the change due to the chymotrypsin-containing solution, but this is followed by a decline as phage are stripped from the chip by proteolysis. After the resumption of washing, the new equilibrium value t_4 is lower than t_3, indicating the loss of phage from the surface (cleaved phage $\approx [t_3 - t_4]$). Finally, an EDTA solution strips all nickel ions from the surface and also any remaining phage and hexahistidine tags bound to the ions, restoring the resonance intensity (t_5) to the initial value (t_1). (Adapted from Finucane MD et al., *Biochemistry*, 38:11604–11612, 1999.)

The binding capacity of a flow cell on an NTA chip was determined by injecting varying amounts of protein-phage until saturation was reached—that is, until the SPR signal leveled off. The maximum response observed upon binding to one flow cell on the NTA chip was approximately 250–300 RU. According to the manufacturer (Biacore AB), 1000 RU corresponds to ≈ 1 ng/mm^2 bound material on the sensor chip surface, and the area used for binding of one flow cell on the chip is 1 mm^2. Thus, 250 RU corresponds to approximately 2.5×10^{-10} g of bound protein-phage, which, with the molecular weight of filamentous phage estimated at 1.5×10^6 g/mol, gives the maximum number of protein-phage binding within one flow cell as approximately 10^7. We also determined this value experimentally by recovering and quantifying the bound phage by infecting log-phase *E. coli* cells. This experiment gave an estimate of 9.6×10^6 bound protein-phage, in good agreement with the calculation.

We used two protein-phage constructs and chymotrypsin treatment to test the utility and specificity of our proposed approach to selection [32]. First, we chose a stable construct, His$_6$-WT-UBQ-phage, harboring WT ubiquitin; although the protein contains both Tyr and Phe, which are targeted by chymotrypsin, as well as Leu, which is cleaved to a lesser extent, it resists proteolysis for some time. The tyrosine residue was included in the mutagenesis to create the ubiquitin library described below. In addition, a negative-control construct, His$_6$-FLEXI-phage, was created, which harbored a Gly-Ser-based flexible linker peptide with a single Tyr cleavage site. Figure 11.4 shows a comparison of the response of these two constructs to chymotrypsin treatment.

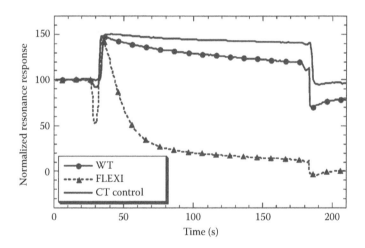

Figure 11.4   Sequence dependence of the protease cleavage profile as determined by SPR. The solid trace shows the cleavage of phage displaying WT ubiquitin, whereas the dashed line is for FLEXI-phage. The data have been normalized such that 100 on the *y*-axis is equivalent to 100% of the phage bound immediately before proteolysis. After 300 s of exposure to chymotrypsin, 83% of the WT-UBQ-phage remained bound to the surface of the Biacore chip; in contrast, only 23% of the FLEXI-phage remained. The plain solid line (CT) shows an underivatized control channel treated with chymotrypsin. (Adapted from Finucane MD et al., *Biochemistry*, 38:11604–11612, 1999.)

These data were collected simultaneously from two lanes on a Biacore sensor chip. The dramatic difference in loss of protein-phage from the surface between FLEXI and WT-UBQ confirmed that proteolysis was sensitive to the sequence of the displayed protein. Similarly, when His_6-AL_7-UBQ-phage were challenged with chymotrypsin the degree of proteolysis fell in between that for the FLEXI and WT-UBQ constructs. We confirmed and reproduced these results using both continuous and pulsed treatments with protease (Figures 11.4 and 11.5). Thus, the protease-selection method discriminated between folded proteins, destabilized protein, and flexible polypeptide chains.

Finally, we followed protease selection of His_6-LIBRARY-UBQ-phage by SPR in Biacore; LIBRARY-UBQ was the aforementioned library of ubiquitin hydrophobic core mutants. Interestingly, the rate of proteolysis of LIBRARY-UBQ-phage was initially fast but then slowed to a rate similar to that for His_6-WT-UBQ-phage. This suggests that the library contains some poorly folded and protease-susceptible mutants that are cleaved and removed rapidly, and some more resilient mutants that we cleaved slowly like WT ubiquitin.

In summary of this section, we were able to show that SPR could be used to follow the cleavage and release of protein-phage from the surfaces of Biacore sensor chips. In addition, we demonstrated that binding to the chips required N-terminally His-tagged protein-phage and that the rate and degree of proteolysis were related to the integrity of the folded structure of the displayed protein. While SPR provides a straightforward method for following protease treatment of monoclonal protein-phage, it is clearly limited for performing selection studies from phage-displayed libraries because the binding capacity of Biacore N-NTA sensor chips is limited to

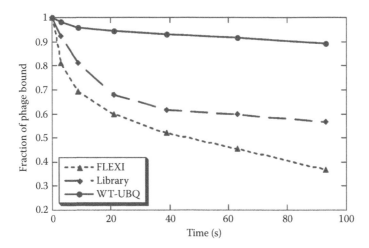

Figure 11.5   Pulsed proteolysis of various protein-phage bound to a Biacore sensor chip. In this experiment, changes in SPR signal were measured after pulses with 10 μm chymotrypsin. After each pulse, washing was resumed to reestablish equilibrium and determine the amount of phage remaining. The figure shows that cumulative loss of phage from the surface is plotted against the cumulative exposure time to protease for WT ubiquitin, library, and the control, FLEXI-phage. (Adapted from Finucane MD et al., *Biochemistry*, 38:11604–11612, 1999.)

approximately 10^7 protein-phage. Realistically, this limits library sizes to about 10^6 protein mutants. In addition, in our experience, it is not straightforward to reproducibly recover protein-phage from Biacore sensor chips. With these drawbacks in mind, we turned to the more traditional Ni-NTA agarose beads (QIAgen) as support for sequestering protein-phage prior to protease selection.

11.4.2 Preparative Protease Selection from Phage-Displayed Libraries

First, we tested that Ni-NTA agarose beads functioned as effectively as Ni-NTA Biacore sensor chips in protease selection using monoclonal protein-phage. For this experiment, we used a 1:1 mixture of His_6-FLEXI-phage and His_6-WT-UBQ-phage, which were the products of two different phagemids carrying different antibiotic resistance. The phage mixture was bound to the Ni-NTA agarose beads and treated with chymotrypsin. After various times, the beads were treated with the serine-protease inhibitor phenylmethylsulfonyl fluoride (PMSF) and washed, and the remaining phage were eluted with imidazole. The employment of two different antibiotic resistances allowed the two populations of protein-phage to be gauged through this competitive proteolysis experiment as follows: phage rescued from the beads were split and used to infect *E. coli*, which were then grown on separate agar plates containing the different antibiotics. The number of colonies resulting on each plate gave a measure of the relative fractions of His_6-WT-UBQ-phage and His_6-FLEXI-phage that remained intact and bound to the beads at each stage of the proteolysis experiment. This confirmed that UBQ-phage resisted proteolysis and that FELIX-phage was susceptible. It also demonstrated that a phage population could be enriched for a stable phenotype through protease selection.

We then turned to selection from the phage-displayed library, His_6-LIBRARY-UBQ-phage, using Ni-NTA agarose beads as a support [32]. Four rounds of selection were performed. Subsequent comparison of DNA sequences from the naive and selected libraries revealed a complete loss of the parental AL_7 mutant after selection, consistent with this being a heavily destabilized mutant. In addition and intriguingly, the selected library showed a clear preference for WT residues in seven of the eight positions targeted in the initial library. Several clones from the selected library were picked and expressed. The resulting proteins were characterized by NMR and circular dichroism (CD) spectroscopy. All gave spectra consistent with folding to native-like ubiquitin structures and showed cooperative thermal unfolding consistent with folding to unique structures. However, none of the selectants was more stable than WT ubiquitin. This result is intriguing as it demonstrates that although protease selection can be used to rescue competently folded and stable mutants from a library of hydrophobic core mutants, in the case of ubiquitin, no superstable variants emerge. Thus, while ubiquitin was a good model for testing the concept of protease selection applied to core-directed design [26,27,42] in some respects, in another, it was a poor choice; in hindsight, we might not have expected to stabilize ubiquitin. This is because the WT protein is extremely stable and also highly conserved in nature. Furthermore, to our knowledge, only one study has produced mutants that are more stable than the WT protein [42]. This led us to suggest that the WT ubiquitin sequences presents, or at least is close to, the optimal solution for packing its hydrophobic core [27].

Overall, using Ni-NTA agarose beads as an affinity matrix to capture hexahistidine-tagged protein-phage and in performing protease selection is successful. Well-folded ubiquitin variants were recovered from a library of hydrophobic core mutants, and the destabilized parental clone—that is, AL_7 with its heavily compromised hydrophobic core—was lost through four rounds of protease selection. However, we did encounter problems with using Ni-NTA agarose beads: the whole process was labor intensive and time consuming. In addition, extensive washes were required after binding and protease challenge due to the well-documented problems associated with nonspecific binding of protein-phage to agarose and plastics. The former problem can be reduced to some extent by using the beads and other nickel-coated surfaces to pull down intact phage after performing the selection in solution. In addition, the necessity to determine phage titers and extensive plating out for each step to follow selection were both laborious and time consuming. Hence, when applying our method, we recommend using SPR to examine monoclonal protein-phage on either side of protease selection and for establishing the best buffer conditions for selection, followed by preparative-scale selections of phage-displayed libraries using Ni-NTA agarose or other media.

11.5 STUDIES THAT BUILD ON THE ORIGINAL METHODS

Riechmann and Winter [43] describe an attempt to create novel protein domains by complementing a fragment of a known protein with randomly generated segments of polypeptide. Specifically, random fragments of *E. coli* genomic DNA are generated by randomly primed PCR and fused to a gene encoding the N-terminal 36 residues of the cold shock protein CspA, which correspond to the first three contiguous β-strands of the five-stranded β-barrel of CspA. The resulting library is then subjected to phage display and proteolysis to select stably folded chimeras. For this work, the group switches from the SIP-based method to using intact g3p and an N-terminal affinity tag as described by Finucane et al. [27,32]—except that the affinity tag is barnase—though they still employ KM13 helper phage to rescue phage prior to protease selection. Intriguingly, most of the selected chimeras contained genomic fragments in their orthodox reading frames. A number of the resulting proteins express and show characteristics of folded proteins: notably, weak but nonrandom coil CD spectra; dispersion in 1-D ^{1}H NMR spectra; evidence for cooperative thermal unfolding; and stability with respect to backbone-amide exchange. In a recent interesting development to the Riechmann and Winter experiment, Fischer et al. [44] combine two strands of an immunoglobulin domain with random fragments generated from human cDNA. In this case, one of the selected proteins, 2a6, expresses well in *E. coli*; gives a CD indicative of mixed α- and β-structure; and forms a cooperatively folded, thermostable dimer. The selection of multimers using phage display in this study and the foregoing work by Riechmann and Winter is interesting and is discussed to some extent in Ref. [44].

Martin et al. [45] also use Proside to select for stabilized variants of a cold shock protein (CspB) with optimized surfaces. They point out that 12 exposed surface residues differ between a mesophilic *Bs*-CspB from *Bacillus subtilis* and its thermophilic counterpart *Bc*-CspB from *Bacillus caldolyticus*. On this basis, they randomize six of these sites to create a library of *Bs*-CspB variants. Using several rounds of protease selection of varying stringency—(1) in the presence of 1.5 M guanidine hydrochloride at 25°C and (2) at 57.5°C and in a buffer of low ionic strength—mutants with higher stabilities than the WT *Bs*-CspB are identified, including several that are significantly more stable than the thermophilic *Bc*-CspB. Interestingly, the selected residues differ at all six positions from the thermophilic *Bc*-CspB and at five of the six positions from the homologous *Tm-Csp* from the hyperthermophilic *Thermotoga maritima*. The group has followed this workup with site-directed mutagenesis and thermodynamic studies to unravel the origins of the enhanced stabilities of the selected *Bs*-CspB mutants [46]. Through these studies, Martin and Schmid [47] argue very convincingly that protein surfaces offer good targets for optimizing protein stability. Finally, the group has used Proside to engineer a thermostable g3p to aid their selection studies.

In an interesting extension to his earlier studies with Winter, Kristensen (with colleagues) [48] describes protease-based selection on a library of barnase mutants. Library construction was guided by natural variations in the ribonuclease observed by comparing the sequences of barnase and binase. Bias toward specific amino-acid combinations was observed at four of the targeted sites, and at three of these, enrichments were with residues not present in either ribonuclease. In addition, for some of the selectants, improved proteolytic stability and selection were not mirrored by improved thermodynamic stability.

Chu and colleagues [35] have tested the approach of core-directed design further by attempting to convert a partially unfolded four-helix bundle, apocytochrome **b562**, into a stably folded protein. Three residues from the anticipated hydrophobic core were subjected to saturation mutagenesis. In addition, another core residue was changed to Trp to provide a fluorescence probe, and a proximal residue was substituted with Arg to provide a specific, Arg-c protease site. Interestingly, after selection, only one of the targeted sites showed any strong preference for hydrophobic residues. Nevertheless, the selected proteins were stable and amenable to high-resolution NMR and crystallographic studies, which confirmed four-helix-bundle structures. These results suggest that core hydrophobic interactions alone may not dictate protein stability and selection, although in this particular case, the location of the cutting site of the protease also appears to influence selection.

Phage display and proteolysis have also been used to select peptide sequences. For example, Dalby and coworkers [34] combine peptide binding and protease treatment to select new WW domain sequences from a phage-displayed library. This work identified novel peptide-binding WW motifs, some of which showed cooperative unfolding, while others did not. The approach has also been used in an attempt to select protease-resistant forms of a zinc-finger-based ββα-structure [49]. Although no monomeric and correctly folded selectants were identified, somewhat curiously, peptides that formed amyloid-like assemblies were retrieved.

11.6 SUMMARY

In conclusion, proteolysis can be used in phage display to select proteins based solely on their ability to fold competently to stable structures; that is, in the absence of any other more-traditional selectable function of the displayed protein such as binding. Subtly different approaches for applying protease selection in phage display have been presented independently by three groups [30–32]. In addition, all of these groups and, importantly, others have followed up the early work with adventurous applications of the method in the areas of protein engineering and design. For example, Finucane and Woolfson [27] and Chu and coworkers [35] have combined the method with core-directed design to engineer proteins from the inside out; Martin and colleagues [45] have succeeded in engineering proteins with improved stabilities by selection from libraries of surface mutants; Pedersen and coworkers [48] have retrieved variants of ribonucleases from a library based on combining sequence features of barnase and binase; and Winter and colleagues [43,44] have created novel protein chimeras by rescuing fragments of natural proteins with those encoded by randomly generated pieces of genomic DNA. Through these studies, the protease-selection method is now firmly established in the repertoire of phage-display experiments. It is hoped that it will find increased application in the fields of engineering protein folding and stability and of designing proteins *de novo*.

REFERENCES

1. Woolfson DN. Core-directed protein design. *Curr Opin Struct Biol* 2001; 11:464–471.
2. Kuhlman B, Dantas G, Ireton GC, Varani G, Stoddard BL, Baker D. Design of a novel globular protein fold with atomic-level accuracy. *Science* 2003; 302:1364–1368.
3. Lazar GA, Handel TM. Hydrophobic core packing and protein design. *Curr Opin Chem Biol* 1998; 2:675–679.
4. Street AG, Mayo SL. Computational protein design. *Struct Fold Des* 1999; 7:R105–R109.
5. Saven JG. Combinatorial protein design. *Curr Opin Struct Biol* 2002; 12:453–458.
6. Richards FM, Lim WA. An analysis of packing in the protein–folding problem. *Quart Rev Biophys* 1993; 26:423–498.
7. Betz SF, Raleigh DP, Degrado WF. De-novo protein design—From molten globules to native-like states. *Curr Opin Struct Biol* 1993; 3:601–610.
8. Betz SF, Bryson JW, Degrado WF. Native-like and structurally characterized designed alpha-helical bundles. *Curr Opin Struct Biol* 1995; 5:457–463.
9. Hill RB, Raleigh DP, Lombardi A, Degrado NF. De novo design of helical bundles as models for understanding protein folding and function. *Acc Chem Res* 2000; 33:745–754.
10. Lim WA, Sauer RT. Alternative packing arrangements in the hydrophobic core of lambda-repressor. *Nature* 1989; 339:31–36.
11. Lim WA, Sauer RT. The role of internal packing interactions in determining the structure and stability of a protein. *J Mol Biol* 1991; 219:359–376.
12. Lim WA, Farruggio DC, Sauer RT. Structural and energetic consequences of disruptive mutations in a protein core. *Biochemistry* 1992; 31:4324–4333.

13. Axe DD, Foster NW, Fersht AR. Active barnase variants with completely random hydrophobic cores. *Proc Natl Acad Sci U S A* 1996; 93:5590–5594.
14. Uversky VN. Natively unfolded proteins: A point where biology waits for physics. *Protein Sci* 2002; 11:739–756.
15. Kamtekar S, Schiffer JM, Xiong HY, Babik JM, Hecht MH. Protein design by binary patterning of polar and nonpolar amino-acids. *Science* 1993; 262:1680–1685.
16. Roy S, Helmer KJ, Hecht MH. Detecting native-like properties in combinatorial libraries of de novo proteins. *Fold Des* 1997; 2:89–92.
17. Roy S, Ratnaswamy G, Boice JA, Fairman R, McLendon G, Hecht MH. A protein designed by binary patterning of polar and nonpolar amino acids displays native-like properties. *J Am Chem Soc* 1997; 119:5302–5306.
18. Roy S, Hecht MH. Cooperative thermal denaturation of proteins designed by binary patterning of polar and nonpolar amino acids. *Biochemistry* 2000; 39:4603–4607.
19. Wei YN, Liu T, Sazinsky SL, Moffet DA, Pelczer I, Hecht MH. Stably folded de novo proteins from a designed combinatorial library. *Protein Sci* 2003; 12:92–102.
20. Wei YN, Kim S, Fela D, Baum J, Hecht MH. Solution structure of a de novo protein from a designed combinatorial library. *Proc Natl Acad Sci U S A* 2003; 100:13270–13273.
21. Wei Y, Hecht MH. Enzyme-like proteins from an unselected library of designed amino acid sequences. *Protein Eng Des Sel* 2004; 17:67–75.
22. West MW, Wang WX, Patterson J, Mancias JD, Beasley JR, Hecht MH. De novo amyloid proteins from designed combinatorial libraries. *Proc Natl Acad Sci U S A* 1999; 96:11211–11216.
23. Wang WX, Hecht MH. Rationally designed mutations convert de novo amyloid-like fibrils into monomeric beta-sheet proteins. *Proc Natl Acad Sci U S A* 2002; 99:2760–2765.
24. Richardson JS, Richardson DC. Natural beta-sheet proteins use negative design to avoid edge-to-edge aggregation. *Proc Natl Acad Sci U S A* 2002; 99:2754–2759.
25. Silverman JA, Balakrishnan R, Harbury PB. Reverse engineering the (beta/alpha) (8) barrel fold. *Proc Natl Acad Sci U S A* 2002; 98:3092–3097.
26. Lazar GA, Desjarlais JR, Handel TM. De novo design of the hydrophobic core of ubiquitin. *Protein Sci* 1997; 6:1167–1178.
27. Finucane MD, Woolfson DN. Core-directed protein design. II. Rescue of a multiply mutated and destabilized variant of ubiquitin. *Biochemistry* 1999; 38:11613–11623.
28. Gu HD, Yi QA, Bray ST, Riddle DS, Shiau AK, Baker D. A phage display system for studying the sequence determinants of protein-folding. *Protein Sci* 1995; 4:1108–1117.
29. Chakravarty S, Mitra N, Queitsch I, Surolia A, Varadarajan R, Dubel S. Protein stabilization through phage display. *FEBS Lett* 2000; 476:296–300.
30. Kristensen P, Winter G. Proteolytic selection for protein folding using filamentous bacteriophages. *Fold Des* 1998; 3:321–328.
31. Sieber V, Pluckthun A, Schmid FX. Selecting proteins with improved stability by a phage-based method. *Nat Biotechnol* 1998; 16:955–960.
32. Finucane MD, Tuna M, Lees JH, Woolfson DN. Core-directed protein design. I. An experimental method for selecting stable proteins from combinatorial libraries. *Biochemistry* 1999; 38:11604–11612.
33. Matthews DJ, Wells JA. Substrate phage—Selection of protease substrates by monovalent phage display. *Science* 1993; 260:1113–1117.
34. Dalby PA, Hoess RH, DeGrado WF. Evolution of binding affinity in a WW domain probed by phage display. *Protein Sci* 2000; 9:2366–2376.

35. Chu R, Takei J, Knowlton JR, Andrykovitch M, Pei WH, Kajava AV, Steinbach PJ et al. Redesign of a four-helix bundle protein by phage display coupled with proteolysis and structural characterization by nmr and x-ray crystallography. *J Mol Biol* 2002; 323:253–262.

36. Fontana A, deLaureto PP, DeFilippis V, Scaramella E, Zambonin M. Probing the partly folded states of proteins by limited proteolysis. *Fold Des* 1997; 2:R17–R26.

37. Tuna M, Finucane MD, Vlachakis NGM, Woolfson DN. Protease-based selection of stably folded proteins and protein domains from phage display libraries. In: Clackson T, Lowman HB, eds. *Phage Display A Practical Approach*. Oxford, UK, 2004.

38. Clackson T, Winter G. Sticky feet-directed mutagenesis and its application to swapping antibody domains. *Nucleic Acids Res* 1989; 17:10163–10170.

39. Szabo A, Stolz L, Granzow R. Surface-plasmon resonance and its use in biomolecular interaction analysis (bia). *Curr Opin Struct Biol* 1995; 5:699–705.

40. Lasonder E, Schellekens GA, Welling GW. A fast and sensitive method for the evaluation of binding of phage clones selected from a surface displayed library. *Nucleic Acids Res* 1994; 22:545–546.

41. Rich RL, Myszka DG. Advances in surface plasmon resonance biosensor analysis. *Curr Opin Biotechnol* 2000; 11:54–61.

42. Khorasanizadeh S, Peters ID, Roder H. Evidence for a three-state model of protein folding from kinetic analysis of ubiquitin variants with altered core residues. *Nat Struct Biol* 1996; 3:193–205.

43. Riechmann L, Winter G. Novel folded protein domains generated by combinatorial shuffling of polypeptide segments. *Proc Natl Acad Sci U S A* 2000; 97:10068–10073.

44. Fischer N, Riechmann L, Winter G. A native-like artificial protein from antisense DNA. *Protein Eng Des Sel* 2004; 17:13–20.

45. Martin A, Sieber V, Schmid FX. In-vitro selection of highly stabilized protein variants with optimized surface. *J Mol Biol* 2001; 309:717–726.

46. Martin A, Kather I, Schmid FX. Origins of the high stability of an in vitro-selected cold-shock protein. *J Mol Biol* 2002; 318:1341–1349.

47. Martin A, Schmid FX. Evolutionary stabilization of the gene-3-protein of phage fd reveals the principles that govern the thermodynamic stability of two-domain proteins. *J Mol Biol* 2003; 328:863–875.

48. Pedersen JS, Otzen DE, Kristensen P. Directed evolution of barnase stability using proteolytic selection. *J Mol Biol* 2002; 323:115–123.

49. Koscielska-Kasprzak K, Otlewski J. Amyloid-forming peptides selected proteolytically from phage display library. *Protein Sci* 2003; 12:1675–1685.

CHAPTER 12

Identification of Natural Protein–Protein Interactions with cDNA Libraries

Reto Crameri, Claudio Rhyner, Michael Weichel,
Sabine Flückiger, and Zoltan Konthur

CONTENTS

12.1 OVERVIEW

The genome projects provided us with a huge amount of information at the DNA level and lead to the identification of thousands of open reading frames. The demand for technologies allowing the functional analysis of gene products is, therefore, dramatically increased. Discovery and characterization of interacting gene products, molecular recognition, and molecular modeling became central to life sciences. Surface display technology based on two pivotal concepts—physical linkage between genotype and phenotype and rescue of individual clones from large libraries by affinity selection—has the potential to substantially contribute to functional genomics. The expansion of surface display technology in biosciences is facilitated by the adaptability of the systems to high-throughput screening formats for automated library handling. While recombinant DNA techniques allow construction of highly complex molecular libraries, high-throughput screening allows rapid exploration of molecular diversity using combinatorial methods. These technologies are becoming increasingly important as molecular tools for the understanding of protein–protein interactions and for the generation of lead compounds, which, hopefully, will attract the business community to make investments in this novel segment of biotechnology.

12.2 INTRODUCTION

All surface display technologies exploit the concept of linking the phenotype as a gene product displayed on a surface to its genetic information integrated into the host genome [1]. This concept is independent from the organism used and has been successfully applied to construct large molecular libraries in filamentous phage [2–4], phagemids [5–7], lytic bacteriophages [8,9], higher viruses [10,11] as well as prokaryotic [12–14] and eukaryotic [15,16] surface expression systems. When Smith [2] initially proposed the idea of phage display in 1985, he suggested that selection of genes from cDNA libraries could be one of the most significant applications of the technology. However, this potentially interesting area of research has lagged behind, despite the impressive progress of phage display technology achieved during the last few years. Among the over 2000 papers describing the use of phage display available to date from the literature, only a few deal with selection of cDNAs.

One of the reasons thereof is a direct consequence of the capsid structure. Most phage or phagemid cloning vectors take advantage of the ability to assemble phage decorated with hybrid versions of the receptor protein pIII or the major coat protein pVIII [17,18]. This strategy has proven to be useful for the N-terminal display of random peptide libraries [3,19–22], antibody fragments [23–26], and single proteins and protein domains [27–30], which are directly fused to the coat proteins [31] or to truncated forms thereof in phagemid vectors [32]. These approaches have been very successful because they allow direct fusions of the gene products to be displayed to the N-terminus of the capsid proteins. However, the integrity of the C-terminus of pIII and pVIII is essential for efficient phage assembly, and therefore, the original vectors can only tolerate insertion of foreign DNA at the N-terminus [5,9,33]. This represents the strongest limitation for the construction of cDNA display libraries. The cDNA inserts encoding the C-terminus of proteins as obtained after poly(A) priming and reverse transcription [34] always contain translation stop codons, which prevent the synthesis of hybrid coat proteins [5,33,35]. To overcome this limitation, several strategies, described in detail below, have been devised. Fewer efforts have been invested in the use of the remaining three capsid proteins, pVI [36–38], pVII, and pIX [39], but their adequacy as vectors to display cDNA libraries has not yet been tested extensively.

12.3 CLONING VECTORS

The basic idea of phage display technology consists in the synthesis of a recombinant protein as fusion with a phage coat protein, provided that the fusion does not interfere with phage infectivity or assembly. Historically, the first phage vectors allowing the fusion of polypeptides to pIII or pVIII contained the whole genome. According to the nomenclature proposed by Smith and Scott [40], these phage vectors can be described as type 3 and type 8 vectors (Figure 12.1). The strongest limitation of these types of vectors is related to the short length of the inserts tolerated by the phage. The major coat protein pVIII can only tolerate very small inserts of about six to eight amino acids between the N-terminal residues 3 and 4 [41,42]. The pIII allows larger peptides and small proteins to be presented as fusion between the export leader sequence and domain D1 without dramatically affecting its function [19,43]. Meanwhile, hybrid phage (types 33 and 88, Figure 12.1) has been developed, which contains both wild type and fusion coat proteins integrated into the genome [44]. The possible drawback of these vectors consists in recombination events between the homologous wild type and fusion DNA regions, resulting in the loss of the information required for the production of fusion proteins [45].

By combination of the best features of phage and plasmids, new types of vectors, termed phagemids, were created [17,46]. These vectors offer several advantages compared to filamentous phage, such as easy preparation of high yields of

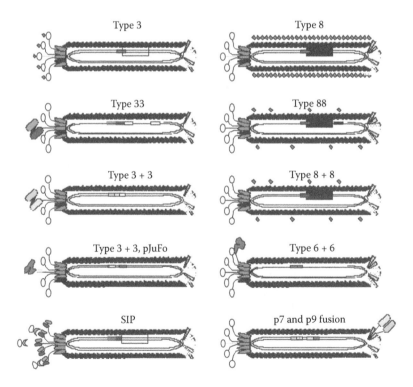

Figure 12.1 Different monovalent and multivalent M13-based phage surface display systems. Vectors of type 3, 8, 33, and 88 are modified wild-type phage. All other systems are phagemid vectors and require coinfection with wild-type phage for assembly of infective phagemid particles. See text for further explanations.

double-stranded DNA for cloning and sequencing, simple maintenance as replicative plasmid form in bacteria, and adaptability to robot-assisted high-throughput screening technologies [47–49]. Phagemids contain a bacterial and a phage origin of replication, the phage packaging signal, antibiotic resistance genes for selection of transformants, and the gene encoding a coat protein used to generate fusions to be displayed on the phage surface. As a consequence thereof, they replicate in the host as plasmids and are able to be packaged in a phagemid particle, or recombinant phage, upon infection with a helper phage that provides the genes for the production of the structural, packaging, and assembly proteins needed for phage morphogenesis. Sophisticated helper phage carry mutations in the origin of replication or packaging sequences. Therefore, during replication, the phagemid genome is packaged more efficiently than the helper phage genome. The big advantage of phagemid over phage vectors consists in the possibility of displaying not only small but also larger peptides [50]; large molecules, such as antibody fragments [51–53]; and many other proteins, including enzymes [54,55], enzyme inhibitors [56], and products of cDNA libraries [4,7,33,35,57–59]. This becomes possible because the helper phage carries the full complement of capsid-encoding genes. As a competition during phagemid assembly, a mixture of wild type and fusion coat protein can be incorporated into the phage coat (vectors of type 3 + 3 and 8 + 8, Figure 12.1). Moreover, the number of fusion protein copies incorporated into the recombinant phage particle (valency) can be influenced using inducible promoters inserted in front of the truncated coat protein gene on the phagemid genome [60].

More recently, other phage coat proteins have been exploited for the display of fusions, including pVI [36], used to display cDNA products, such as C-terminal fusions (type 6 + 6, Figure 12.1). The coat proteins pVII and pIX have been used as fusion partners for the display of heavy- and light-chain antibody fragments [39]; however, these approaches have, so far, been less commonly used. In addition to the mentioned "standard" phage display vectors, other pIII-based systems classified as "phage two-hybrid systems" have been reported. Formally, these vectors have been termed SAP for selection and amplification of phage [61] and SIP for selectively infective phage [62,63]. In these systems, the fusion proteins are expressed directly followed by the D2 and D3 domains of pIII, rendering all phage infection defective. Infectivity, for example, the ability of the phage to bind to the F-pilus and hence to infect *Escherichia coli* cells, is restored by the selection target fused to the pIII domain(s) D1 or D1 and D2. The SIP technology may represent a powerful tool for rapid selection of protein–protein interactions [64] in spite of the few applications reported so far. Possible display strategies and vectors have been reviewed recently [65] (see also Chapter 2) and will not be discussed here in further detail.

12.4 DISPLAY OF cDNA LIBRARIES ON PHAGE SURFACE

Highly diverse display libraries have been constructed by fusing either genomic [66,67] or cDNA fragments [5,7,35,36,57,59] to gene *III* or gene *VIII* of filamentous phage. In both cases, the display can be a challenge as the presence of stop codons can hamper the generation of N-terminal fusions to the coat proteins, a direct consequence of the capsid structure [33,47,49]. Since the integrity of the C-terminus of pIII and pVIII is considered essential for efficient phage assembly, insertions of foreign peptides can only be tolerated at the N-terminus. However, this problem has been alleviated using different strategies. Fusion of cDNAs to the C-terminus of the gene *VI* protein is compatible with phage propagation and packaging [36], as demonstrated in pilot experiments [37]. The feasibility of this approach has been clearly demonstrated by the isolation of peroxisomal proteins from human cDNA libraries [38,68] and of a collagen-binding protein from a *Necator americanus* cDNA library [69]. Moreover, Fuh and Sidhu [22] and Fuh et al. [70] have demonstrated that, in contrast to the common belief, polypeptides fused to the C-terminus of both the M13 pIII and pVIII coat proteins are functionally displayed on the phage surface. The C-terminal fusion approach, although not widely used until now, could be of considerable importance to phage display technology and would allow broad investigations of biological problems, which are not suited for N-terminal display. Main areas of interest in this field are the study of protein–protein interactions requiring free C-termini and functional screening of cDNA libraries.

More sophisticated approaches are based on the ability to separate the gene *III* product of filamentous phage into its functional domains: the N-terminal domains binding to the F-pilus and mediating infection, and the C-terminal domain morphologically involved in capping the trailing end of the filament according to the vectorial polymerization model [71,72]. Although cleavage of the gene *III* product into two separate functional entities is incompatible with phage propagation, infectivity can be restored by joining the segments through noncovalent protein–protein interactions [73,74]. The so-called SIP technology [63,64] can be efficiently used to screen cDNA libraries for selection of proteins that interact with a target molecule, as demonstrated in a few cases [75,76].

However, the most widely used systems for the construction and screening of cDNA libraries displayed on the phage surface involve an indirect fusion strategy where cDNA inserts fused to the 3′ end of the Fos leucine zipper are coexpressed with a truncated form of the gene *III* product decorated with the Jun leucine zipper [5]. The phagemid derived by modification of phagemid pComb 3 [77] and formally termed pJuFo (Figure 12.2) has been widely used for the isolation of IgE-binding molecules from complex allergenic sources as reviewed elsewhere [7,47,78–80]. Selective enrichment of IgE-binding molecules from cDNA libraries constructed using mRNA from *Aspergillus fumigatus* [81], *Malassezia furfur* [82], peanut [83,84], *Alternaria alternata* [85], *Cladosporium herbarum* [86], *Coprinus comatus* [87], storage mites [88], and wheat germ [89] yielded phage displaying hundreds of different IgE-binding proteins. Interestingly, some of these structures represent phylogenetically conserved proteins and share a high degree of sequence identity to their human counterparts. Human proteins, including manganese-dependent superoxide dismutase [90], acidic P_2 ribosomal protein [91], and cyclopilin [92], could also be directly selected from a human lung cDNA library displayed on the pJuFo surface using sera of patients sensitized to *A. fumigatus* as a ligand [93,94], thus demonstrating cross-reactivity to the environmental allergens [90,91].

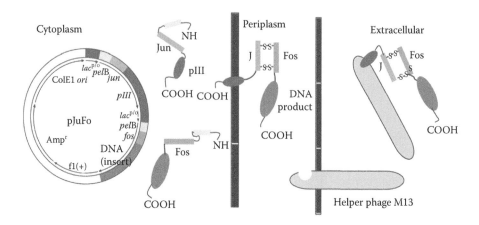

Figure 12.2   Genetic elements of the pJuFo phagemid and proposed pathway for the assembly of phage surface-displayed cDNA libraries. (Modified from Crameri R, Suter M, *Gene*, 137:69–75, 1993.)

Of course, filamentous phages and phagemids are not the vectors of choice for high-level expression of recombinant proteins. Therefore, all cDNAs isolated from phage surface display libraries need to be subcloned in high-level expression vectors and transformed to a suitable host if relevant amounts of protein are required, for example, for clinical studies [95]. In general, inserts subcloned from selected phagemids into high-level expression vectors are well expressed because, for display on the phage surface, the genetic information needs to be transcribed and translated by *E. coli*.

Filamentous phage display systems, like any other cloning system, are not universal as they are subjected to biological restrictions imposed by the host and by the codon usage of the cloned inserts. Possible serious biological limitations derive from the characteristics of the phage life cycle. Filamentous phage particles are released from the host cell without breaking the integrity of the cell membrane. The proteins, which assemble to form the capsid in the periplasmic space, must therefore cross the lipid bilayer of the inner membrane. Therefore, any fusion peptide or protein with biochemical characteristics preventing transmembrane transport will not be integrated into the capsid. To be recognized by the ligand used for selection, displayed proteins need to adapt a conformation able to interact with the ligand. The chemical characteristics of the periplasmic environment, which affect the folding and stability of the recombinant proteins displayed, may influence the ability of hybrid coat proteins to interact with the ligand used for selection. Since cDNA libraries encode very diverse protein domains with different biochemical properties, it is probable that a subset of these proteins or protein fragments will not be displayed and thus not be present in the surface display library. In addition, cDNA libraries, like any other molecular library displayed on the phage surface, suffer from host-specific biological limitations related to restriction in codon usage, refolding pathways, and potential toxicity of the expressed gene products for the heterologous host.

However, phage surface display of cDNAs allows for the survey of very large libraries using the discriminative power of affinity selection against homogenous and heterogeneous ligands. Although the most successful applications of the pJuFo-based cloning technology are related to IgE-binding molecules using serum IgE from allergic patients as a ligand, other successful applications have been reported. Examples are the mapping of protein–ligand interactions using whole genome phage display libraries [66], construction of vectors for stable immobilization of multimeric recombinant proteins [96], and detailed analysis of the C5a anaphylatoxin effector domain [97] followed by selection of a C5a receptor antagonist [98]. Pereboev et al. [99] have used pJuFo to display an adenovirus type 5 fiber knob as a tetrameric molecule able to bind to the coxsackie-virus-Ad receptor, demonstrating the versatile applicability of the cloning system. More recently, pJuFo has been used to select autoantigens from human cDNA libraries derived from patients suffering from vitiligo [100] and prostate cancer [101]. In the first study, purified IgG from serum of vitiligo patients was used to screen a melanocyte cDNA-phage surface display library, resulting in the discovery of the melanin-concentrating hormone receptor 1 (MCHR1) as a novel autoantigen related to this autoimmune disorder. Immunoreactivity against the receptor was demonstrated in sera of vitiligo patients using radiobinding assays. Among sera from healthy controls and from patients with other autoimmune diseases, no immunoreactivity to MCHR1 was found, indicating a high disease specificity of autoantibodies raised against the receptor. In the second study, a cDNA library constructed from mRNA isolated from a lymph node metastasis of a patient suffering from hormone refractory prostate cancer (HRPC) was screened with purified autologous and heterologous IgG of patients suffering from prostate cancer. Sequencing of single clones after four rounds of biopanning yielded different cDNAs depending on the amount of IgG used for screening, some of them corresponding to already known cancer-associated antigens. These results, together with the isolation of colorectal-tumor-associated antigens from a primary colorectal tumor cDNA library displayed on the surface as fusion to the gene *VI* protein [37], demonstrate the applicability of phage surface display for identification of cDNA expression products in such diseases as cancer and autoimmune disorders. The pJuFo vector was also used to clone proteins directly interacting with the cytoplasmic tail of the murine IgE-antigen receptor from a murine B-cell cDNA library displayed on the phage surface [102]. In contrast to the previous examples, which used heterogeneous ligands, a homogeneous synthetic 28-amino-acid-long peptide derived from the cytoplasmic tail of IgE was used as selection bait. Among the inserts from 30 randomly chosen clones sequenced after five rounds of biopanning, two carried cDNA fragments coding for the hematopoietic protein kinase 1 (HPK1). The BIACORE measurements showed that HPK1 interacts in vitro with the cytoplasmic tail of IgE as expected. The binding of HPK1 to the cytoplasmic domain of IgE indicates the existence of an isotype-specific signal transduction and may represent a missing link to upstream regulatory elements of HPK1 activation.

These examples clearly demonstrate that phage-displayed cDNA expression cloning can be a powerful tool for the isolation of unknown genes. The great advantage of cDNA surface display compared to conventional lambda phage-based methods is that in many cases, the functional activity of a protein structure can be used to select interaction partners together with the genetic information required for their production. Thus, sequencing of the DNA of the integrated section of the phage genome can readily elucidate the amino acid sequence of a displayed gene product.

12.5 PROBLEMS ASSOCIATED WITH THE DISPLAY OF cDNA LIBRARIES ON PHAGE SURFACE

A growing number of observations, published or not, indicate that filamentous phage display technology is subjected to several limitations, some of these have already been discussed. Obviously, the quality of any cDNA library depends directly on the quality of the cDNA ligated into the vector, which, in turn, is determined by the quality of the mRNA. Although the methods for the isolation of mRNA available are quite reliable, oligo(dT) priming might generate a high frequency of truncated cDNAs through internal poly(A) priming during reverse transcription [34]. Therefore, for genome-wide gene identification, reverse transcription should be carried out using anchored oligo(dT) primers, which diminish the generation of truncated cDNAs caused by internal poly(A) priming. During the construction of a cDNA-phage surface display library, every step should be optimized to create the highest frequency of potentially expressible full-length cDNA inserts. Transformation with empty phagemids or phagemids containing short inserts that have a growth advantage over large insert-containing phagemids may result in these undesirable clones becoming overrepresented during library amplification. Avoiding overgrowth by defective clones is especially important for large and highly heterogeneous cDNA expression libraries.

Translational problems related to the codon usage might be alleviated by the use of hosts harboring genes encoding limiting tRNA species like argU, ileY, and leuW to attain reasonable expression levels of proteins affected by rare codon usage [103]. Like other prokaryotic-based expression systems, phage display may not be suitable for selection of proteins that require posttranslational modifications (e.g., glycosylation, phosporylation, etc.) or heterodimeric assembly for functional activity. Moreover, due to the absence of mammalian chaperonins, conformationally dependent structures may not be efficiently expressed or refolded in these bacterial expression systems and therefore not be recognized by the ligand used for selection.

However, a successful screening depends not only on biological factors but also on the biopanning strategy used [1,18]. Selective enrichment of clones of interest becomes necessary since phage surface display libraries contain large numbers of cognate and uncognate phage. In a standard amplified library with a diversity of 10^8, each single clone is present in several thousand copies among a population of 10^{12}–10^{13} phage molecules. Therefore, the use of efficient selection and screening procedures is one of the key elements that determine the success of the combinatorial approach, as discussed in detail elsewhere [104]. Phage display techniques require immobilization of the target protein to a solid support during biopanning. The immobilization process must maintain the target in a native or native-like conformation for phage selection [105]. The commonly used method of protein immobilization through direct adsorption to plastic surfaces denatures many proteins, making them unsuitable targets for phage selection. Indirect immobilization of biotinylated ligands on streptavidin-coated surfaces or chemical cross-linking to bifunctional resins [106] has been reported to be more successful for the generation of native-like ligand surfaces and should be considered whenever possible.

In spite of these limitations, phage display technology has significant advantages over other screening methods. Compared to conventional bacterial or lambda phage-based expression systems, which are submitted to the same biological limitations, phage display technology enables rapid and selective enrichment of desired clones in small volumes using minimal amounts of ligand molecules, which can be, indeed, a limiting factor for selection.

12.6 ADAPTABILITY OF PHAGE DISPLAY TO HIGH-THROUGHPUT SCREENING TECHNOLOGY

The identification of the proteins produced in a given biological system started many years ago with the discovery and improvement of recombinant DNA technologies that allowed controlled expression of genes in many different hosts [107]. However, cDNA-cloning technology, including DNA sequencing, cannot be directly used to study protein–protein interactions; the challenge of functional genomics aimed to turn sequence information into function. Estimates of the total number of proteins resulting from transcription of the approximately 35,000 human genes vary from 300,000 to millions [108], thus allowing for a much greater number of potential protein–protein interactions. Although many phenotypes can already be pinpointed to their genetic origin through the sequence information of genome projects, many others remain unknown. Unfortunately, sequence information in itself is sufficient to provide significant knowledge neither of the underlying mechanisms of life nor of the biology of organisms. Rather, it provides a sound basis and framework for further investigations.

The rapid identification of complex networks of interacting molecules in cells and tissues requires technologies that provide logistic and/or physical links between proteins and the genes that encode them. The complex nature of molecular interactions and the large numbers of diverse molecules involved in biological processes require high-throughput technology, allowing a sufficient degree of parallelization [49]. New technologies for large-scale analysis of genes and proteins have been devised, such as differential display, RNA/DNA microarrays, and mass spectrometry [109] which, however, suffer from the lack of a physical link between sequence information and function. Phage display provides a physical link between genotype and phenotype [2,5,6] and allows the handling of large libraries based on the power of affinity selection [77]. This physical link can easily be combined with a logistic protein–DNA link provided by robot technology, enabling high-throughput picking and high-density arraying of single clones [49]. These high-density arrays have the advantage that each clone has a unique position defined by the coordinates on the microtiter plate and allow an unequivocal identification of each clone in later stages.

It has been shown that a human fetal brain cDNA expression library can be screened in parallel for either DNA hybridization or protein expression, and for antibody screening in a high-density array format on filter membranes [110,111]. This robot-based high-throughput screening technology has been successfully applied to phagemid libraries expressing complex allergen repertoires preselected with serum IgE of allergic individuals [47,78,80,94]. The potential of the combination of cDNA-phage surface display with selection for specific interaction by functional screening and robotic technology is illustrated by the isolation of more sequences potentially encoding IgE-binding proteins than postulated from Western blot analysis using extracts derived from raw material of complex allergenic sources [112]. Moreover, robot-based high-throughput screening technology has been applied to recombinant antibody arrays displayed on the phage surface to detect antibody–antigen interactions [48]. Therefore, high-throughput screening technology applied to complex surface-displayed cDNA libraries will play an important role in the postgenomic era by identifying potential ligands against large numbers of diverse molecules expressed by cell cultures, tissues, or organisms.

12.7 CONCLUSIONS

The major challenge in the postgenomic era is to turn sequence information into function. The key molecular players in cells and tissues, which are instrumental for the functioning of an organism, are the proteins. Built up from 20 different amino acids encoded by the DNA of a limited number of genes, proteins are produced through complex translational and posttranslational pathways generating a great deal of functional structures. This diversity enables complex networks of molecular interactions governing the functioning of an organism. The characterization of large numbers of genes, their expression patterns, and protein interactions demands the use of high-throughput technologies able to link information, deposited in the genome, to functions exerted by the proteins themselves.

Phage surface display of cDNAs as a biological approach linking genotype and phenotype, although subjected to intrinsic biological limitations, can be used for the efficient identification of gene products based on protein–protein interactions. Thus, the technology has the potential for substantially contributing to rapid developments in functional genomics. The basic knowledge accumulated from successful and unsuccessful applications of cDNA-phage surface display will improve our understanding of the biological limitations of the systems currently used and, thus, will help further improve the technology.

ACKNOWLEDGMENTS

We are grateful to Prof. K. Blaser and Prof. H. Lehrach for their continuous support and encouragement. This work was supported by the Swiss National Science Foundation grant 31.63382.00.

REFERENCES

1. Borrebaeck CA. Tapping the potential of molecular libraries in functional genomics. *Immunol Today* 1998; 19:524–527.
2. Smith GP. Filamentous fusion phage: Novel expression vectors that display cloned antigens on the virion surface. *Science* 1985; 228:1315–1318.
3. Scott JK, Smith GP. Searching for peptide ligands with an epitope library. *Science* 1990; 249:386–390.
4. Smith GP, Petrenko VA. Phage display. *Chem Rev* 1997; 97:391–410.
5. Crameri R, Suter M. Display of biologically active proteins on the surface of filamentous phages: A cDNA cloning system for selection of functional gene products linked to the genetic information responsible for their production. *Gene* 1993; 137:69–75.
6. Barbas CF III, Kang AS, Lerner RA, Benkovic SJ. Assembly of combinatorial antibody libraries on phage surfaces: The gene III site. *Proc Natl Acad Sci U S A* 1991; 88:7978–7982.
7. Rhyner C, Kodzius R, Crameri R. Direct selection of cDNAs from filamentous phage surface display libraries. Potential and limitation. *Curr Pharm Biotechnol* 2002; 3:13–21.
8. Mikawa YG, Maruyama IN, Brenner S. Surface display of proteins on bacteriophage lamda heads. *J Mol Biol* 1996; 262:21–30.
9. Castagnoli L, Zucconi A, Quondam M, Rossi M, Vaccaro P, Panni S, Paoluzi S et al. Alternative bacteriophage display systems. *Comb Chem High Throughput Screen* 2001; 4:121–133.
10. Grabherr R, Ernst W. The baculovirus expression system as a tool for generating diversity by viral surface display. *Comb Chem High Throughput Screen* 2001; 4:185–192.
11. Grabherr R, Ernst W, Oker-Blom C, Jones I. Developments in the use of baculoviruses for the surface display of complex eukaryotic proteins. *Trends Biotechnol* 2001; 19:231–236.
12. Hansson M, Samuelson P, Gunneriusson E, Ståhl S. Surface display on gram positive bacteria. *Comb Chem High Throughput Screen* 2001; 4:171–184.
13. Ståhl S, Uhlén M. Bacterial surface display: Trends and progress. *Trends Biotechnol* 1997; 15:185–192.
14. Georgiou G, Stathopoulos C, Daugherty PS, Nayak AR, Iverson BL, Curtiss R III. Display of heterologous proteins on the surface of microorganisms. From the screening of combinatorial libraries to live recombinant vaccines. *Nat Biotechnol* 1997; 15:29–34.
15. Boder ET, Wittrup KD. Yeast surface display for directed evolution of protein expression, affinity, and stability. *Methods Enzymol* 2000; 328:430–444.
16. Yeung YA, Wittrup KD. Quantitative screening of yeast surface-displayed polypeptide libraries by magnetic bead capture. *Biotechnol Prog* 2002; 18:212–220.
17. Mead DA, Kemper B. Chimeric single-stranded DNA phage-plasmid cloning vectors. *Biotechnology (N Y)* 1988; 10:85–102.
18. Kay BK, Winter J, McCafferty J. *Phage Display of Peptides and Proteins. A Laboratory Manual.* San Diego, CA: Academic Press Inc, 1996.
19. Devlin JJ, Panganiban LC, Devlin PE. Random peptide libraries: A source of specific protein-binding molecules. *Science* 1990; 249:404–406.
20. Barrett RW, Cwirla SE, Ackerman MS, Olson AM, Peters EA, Dower WJ. Selective enrichment and characterization of high affinity ligands from collections of random peptides on filamentous phage. *Anal Biochem* 1992; 204:357–364.
21. Cabilly S. The basic structure of filamentous phage and its use in the display of combinatorial peptide libraries. *Mol Biotechnol* 1999; 12:143–148.

22. Fuh G, Sidhu SS. Efficient phage display of polypeptides fused to the carboxy-terminus of the M13 gene-3 minor coat protein. *FEBS Lett* 2000; 480:231–234.

23. Burton DR, Barbas CF III, Persson MA, Koenig S, Chanock RM, Lerner RA. A large array of human monoclonal antibodies to type 1 human immunodeficiency virus from combinatorial libraries of asymptomatic seropositive individuals. *Proc Natl Acad Sci U S A* 1991; 88:10134–10137.

24. Fisch I, Kontermann RE, Finnern R, Hartley O, Soler-Gonzalez AS, Griffiths AD, Winter G. A strategy of exon shuffling for making large peptide repertoires displayed on filamentous bacteriophage. *Proc Natl Acad Sci U S A* 1996; 93:7761–7766.

25. Sblattero D, Bradbury A. Exploiting recombination in single bacteria to make large phage antibody libraries. *Nat Biotechnol* 2000; 18:75–80.

26. Liu B, Huang L, Sihlbom C, Burlingame A, Marks JD. Towards proteome-wide production of monoclonal antibody by phage display. *J Mol Biol* 2002; 315:1063–1073.

27. Zucconi A, Panni S, Paoluzi S, Castagnoli L, Dente L, Cesareni G. Domain repertoires as a tool to derivate protein recognition rules. *FEBS Lett* 2000; 480:49–54.

28. Schiffer C, Ultsch M, Aalsh S, Somers W, de Vos AM, Kossiakoff A. Structure of a phage display-derived variant of human growth hormone complexed to two copies of extracellular domain of its receptor: Evidence for strong structural coupling between receptor-binding sites. *J Mol Biol* 2002; 316:277–289.

29. Fazi B, Cope MJ, Douangamath A, Ferracuti S, Schiriwitz K, Zucconi A, Drubin DG et al. Unusual binding properties of the SH3 domain of the yeast actin-binding protein Abp1: Structural and functional analysis. *J Mol Biol* 2002; 277:5290–5298.

30. Reichmann L, Winter G. Novel folded protein domains generated by combinatorial shuffling of polypeptide segments. *Proc Natl Acad Sci U S A* 2000; 97:10068–10073.

31. Smith GP. Filamentous phages as cloning vectors. *Biotechnology (N Y)* 1988; 10:61–83.

32. Sidhu SS. Engineering M13 for phage display. *Biomol Eng* 2001; 18:57–63.

33. Crameri R, Hemmann S, Blaser K. pJuFo: A phagemid for display of cDNA libraries on phage surface suitable for selective isolation of clones expressing allergens. *Adv Med Exp Biol* 1996; 409:103–110.

34. Nam DK, Lee S, Zhou G, Cao X, Wang C, Clark T, Chen J et al. Oligo(dT) primer generates a high frequency of truncated cDNAs through internal poly(A) priming during reverse transcription. *Proc Natl Acad Sci U S A* 2002; 99:6152–6156.

35. Crameri R, Jaussi R, Menz G, Blaser K. Display of expression products of cDNA libraries on phage surfaces. A versatile screening system for selective isolation of genes by specific gene-product/ligand interaction. *Eur J Biochem* 1994; 226:53–58.

36. Jespers LS, Messens JH, De Keyser A, Eeckhout D, Van den Brande I, Gansemans YG, Lauwereys MJ et al. Surface expression and ligand-based selection of cDNAs fused to filamentous phage gene VI. *Biotechnology (N Y)* 1995; 13:378–382.

37. Hufton SE, Moerkerk PT, Meulemans EV, de Bruine A, Arends JW, Hoogenboom HR. Phage display of cDNA repertoires: The pVI display system and its applications for the selection of immunogenic ligands. *Immunol Methods* 1999; 231:39–51.

38. Amery L, Mannaerts GP, Subramani S, Van Veldhoven PP, Fransen M. Identification of a novel human peroxisomal 2,4-dienoyl-CoA reductase related protein using the M13 phage protein VI phage display technology. *Comb Chem High Throughput Screen* 2001; 4:545–552.

39. Gao C, Mao S, Lo CH, Wirsching P, Lerner RA, Janda KD. Making artificial antibodies. A format for phage display of combinatorial heterodimeric arrays. *Proc Natl Acad Sci U S A* 1999; 96:6025–6030.

40. Smith GP, Scott JK. Libraries of peptides and proteins displayed on filamentous phage. *Meth Enzymol* 1993; 217:228–257.
41. Greenwood J, Willis AE, Perham RN. Multiple display of foreign peptides on a filamentous bacteriophage. Peptides from *Plasmodium falciparum* circumsporozoite protein as antigens. *J Mol Biol* 1991; 220:821–827.
42. Petrenko VA, Smith GP, Gong X, Quinn T. A library of organic landscapes on filamentous phage. *Protein Eng* 1996; 9:797–801.
43. McCafferty J, Griffiths AD, Winter G, Chiswell DJ. Phage antibodies: Filamentous phage displaying antibody variable domains. *Nature* 1990; 348:552–554.
44. Haaparanta T, Huse WD. A combinatorial method for constructing libraries of long peptides displayed by filamentous phage. *Mol Divers* 1995; 1:39–52.
45. Bonnycastle LL, Mehroke JS, Rashed M, Gong X, Scott JK. Probing the basis of antibody reactivity with a panel of constrained peptide libraries displayed by filamentous phage. *J Mol Biol* 1996; 258:747–762.
46. Larocca D, Burg MA, Jensen-Pegakes K, Ravey EP, Gonzales AM, Baird A. Evolving phage vectors for cell targeted gene delivery. *Curr Pharm Biotechnol* 2002; 3:45–57.
47. Crameri R, Kodzius R. The powerful combination of phage surface display of cDNA libraries and high throughput screening. *Comb Chem High Throughput Screen* 2001; 4:145–155.
48. de Wildt RTM, Mundy CR, Gorick BD, Tomlinson IM. Antibody arrays for high-throughput screening of antibody–antigen interactions. *Nat Biotechnol* 2000; 18:989–994.
49. Walter G, Konthur Z, Lehrach H. High-throughput screening of surface displayed gene products. *Comb Chem High Throughput Screen* 2001; 4:193–205.
50. Wrighton NC, Farrell FX, Chang R, Kashyap AK, Barbone FP, Mucahy LS, Johnson DL et al. Small peptides as potent mimetics of the protein hormone erythropoietin. *Science* 1996; 273:458–464.
51. Rader C, Barbas CF III. Phage display of combinatorial antibody libraries. *Curr Opin Biotechnol* 1997; 8:503–508.
52. Winter G. Making antibody and peptide ligands by repertoire selection technologies. *J Mol Recognit* 1998; 11:126–127.
53. Sblattero D, Lou J, Marzari R, Bradbury A. In vivo recombination as a tool to generate molecular diversity in phage antibody libraries. *J Biotechnol* 2001; 74:303–315.
54. Soumillion P, Jespers L, Bouchet M, Marchand-Brynaert J, Sartiaux P, Fastrez J. Phage display of enzymes and in vitro selection for catalytic activity. *Appl Biochem Biotechnol* 1994; 47:175–189.
55. Forrer P, Jung S, Plückthun A. Beyond binding: Using phage display to select for structure, folding and enzymatic activity in proteins. *Curr Opin Struct Biol* 1999; 9:514–520.
56. Rottgen P, Collins J. A human pancreatic secretory trypsin inhibitor presenting a hypervariable highly constrained epitope via monovalent phagemid display. *Gene* 1995; 164:243–250.
57. Dunn IS. Phage display of proteins. *Curr Opin Biotechnol* 1996; 7:547–553.
58. Kemp EH, Waterman EA, Hawes BE, O'Neill K, Gottumukkala RV, Gawkrodger DJ, Weetman AP et al. The melanin-concentrating hormone receptor 1, a novel target of autoantibody responses in vitiligo. *J Clin Invest* 2002; 109:923–930.
59. Crameri R, Achatz G, Weichel M, Rhyner C. Direct selection of cDNAs by phage display. *Methods Mol Biol* 2002; 184:461–469.
60. Huang W, McKevitt M, Palzkill T. Use of the arabinose p(bad) promoter for tightly regulated display of proteins on bacteriophage. *Gene* 2000; 251:187–197.

61. Duenas M, Borrebaeck CA. Clonal selection and amplification of phage displayed anti-bodies by linking antigen recognition and phage replication. *Biotechnology (N Y)* 1994; 12:999–1002.

62. Spada S, Krebber C, Plückthun A. Selectively infective phage (SIP). *Biol Chem* 1997; 378:445–456.

63. Arndt KM, Jung S, Krebber C, Plückthun A. Selectively infective phage technology. *Meth Enzymol* 2000; 328:364–388.

64. Jung S, Arndt KM, Müller KM, Plückthun A. Selectively infective phage (SIP) technology: Scope and limitations. *J Immunol Meth* 1999; 231:93–104.

65. Irving MB, Pan O, Scott JK. Random-peptide libraries and antigen-fragment libraries for epitope mapping and the development of vaccines and diagnostics. *Curr Opin Chem Biol* 2001; 5:314–324.

66. Palzkill T, Huang W, Weinstock GM. Mapping protein–ligand interactions using whole genome phage display libraries. *Gene* 1998; 222:79–83.

67. Jacobsson K, Frykberg L. Shotgun phage display cloning. *Comb Chem High Throughput Screen* 2001; 4:138–143.

68. Fransen M, Van Veldhoven PP, Bubramani S. Identification of peroxisomal proteins by using M13 phage protein VI display: Molecular evidence that mammalian peroxisomes contain a 2,4-dienoyl-CoA reductase. *Biochem J* 1999; 340:561–568.

69. Viaene A, Carb A, Meiring M, Pritchard D, Deckmyn H. Identification of a collagen-binding protein from *Necator americanus* by using a cDNA-expression phage display library. *J Parasitol* 2001; 87:619–625.

70. Fuh G, Pisabarro MT, Li Y, Quan C, Lasky LA, Sidhu SS. Analysis of PDZ domain–ligand interactions using carboxyl-terminal phage display. *J Biol Chem* 2000; 275:21486–21491.

71. Chang CN, Model P, Blobel G. Membrane biogenesis: Cotranslational integration of the bacteriophage f1 coat protein into an *Escherichia coli* membrane fraction. *Proc Natl Acad Sci U S A* 1979; 76:1251–1255.

72. Stengele I, Bross P, Graces S, Giray I, Rasched I. Dissection of functional domains in phage fd adsorption protein. Discrimination between attachment and penetration sites. *J Mol Biol* 1990; 5:143–149.

73. Gramatikoff K, Georgiev O, Schaffner W. Direct interaction rescue, a novel filamentous phage technique to study protein–protein interactions. *Nucleic Acid Res* 1994; 22:5761–5762.

74. Krebber C, Spada S, Desplancq D, Krebber A, Ge L, Plückthun A. Selectively-infective phage (SIP): A mechanistic dissection of a novel in vivo selection for protein–ligand interactions. *J Mol Biol* 1997; 268:607–618.

75. Gramatikoff K, Schaffner W, Georgiev O. The leucine zipper of c-Jun binds to ribosomal protein L18a: A role in Jun protein regulation? *Biol Chem Hoppe Seyler* 1995; 376:321–325.

76. Hottiger M, Gramatikoff K, Georgiev O, Chaponnier C, Schaffner W, Hübscher U. The large subunit of HIV-1 reverse transcriptase interacts with beta-actin. *Nucleic Acids Res* 1995; 23:736–741.

77. Barbas CF III, Lerner RA. Combinatorial immunoglobulin libraries on the surface of phage (Phabs): Rapid selection of antigen-specific Fabs. *Methods: Companion Methods Enzymol* 1991; 2:119–124.

78. Crameri R, Walter G. Selective enrichment and high-throughput screening of phage surface-displayed cDNA libraries from complex allergenic systems. *Comb Chem High Throughput Screen* 1999; 2:63–72.

79. Appenzeller U, Blaser K, Crameri R. Phage display as a tool for rapid cloning of allergenic proteins. *Arch Immunol Ther Exper* 2001; 49:19–25.

80. Crameri R. High throughput screening: A rapid way to recombinant allergens. *Allergy* 2001; 54(suppl 67):30–34.

81. Crameri R. Molecular cloning of *Aspergillus fumigatus* allergens and their role in allergic bronchopulmonary aspergillosis. *Chem Immunol* 2002; 71:73–93.

82. Lindborg M, Magnusson CGM, Zagari A, Schmidt M, Scheyinus A, Crameri R, Withley P. Selective cloning of allergens from the skin colonizing yeast *Malassezia furfur* by phage surface display technology. *J Invest Dermatol* 1999; 113:156–161.

83. Kleber-Janke T, Crameri R, Appenzeller U, Schlaak M, Becker WM. Selective cloning of peanut allergens, including profilin and 2S albumin, by phage display technology. *Int Arch Allergy Immunol* 1999; 119:265–274.

84. Kleber-Janke T, Crameri R, Scheurer S, Vieths S, Becker WM. Patient-tailored cloning of allergens by phage display: Peanut (*Arachis hypogaea*) profilin, a food allergen derived from a rare mRNA. *J Chromatogr B Biomed Sci Appl* 2001; 756:295–305.

85. Weichel M, Schmid-Grendelmeier P, Flückiger S, Breitenbach M, Blaser K, Crameri R. Nuclear transport factor 2 represents a novel cross-reactive fungal allergen. *Allergy* 2003; 58:198–206.

86. Weichel M, Schmid-Grendelmeier P, Rhyner C, Achatz G, Blaser K, Crameri R. IgE-binding and skin test reactivity to hydrophobin HCh-1 from *Cladosporium herbarum*, the first allergenic cell wall component of fungi. *Clin Exp Allergy* 2003; 33:72–77.

87. Brander KA, Borbley P, Crameri R, Pichler WJ, Helbling A. IgE-binding proliferative responses and skin test reactivity to Cop c 1, the first recombinant allergen for the basidiomycete *Coprinus comatus. J Allergy Clin Immunol* 1999; 104:630–636.

88. Eriksson TLJ, Rasool O, Huecas S, Whitley P, Crameri R, Appenzeller U, Gafvelin G et al. Cloning of three new allergens from the dust mite *Lepidoglyphus destructor* using phage surface display technology. *Eur J Biochem* 2001; 266:287–294.

89. Rozynek P, Sander I, Appenzeller U, Crameri R, Baur X, Clarke T, Brünig B et al. BPIS—An IgE-binding wheat protein. *Allergy* 2002; 57:463.

90. Crameri R, Faith A, Hemmann S, Jaussi R, Ismail C, Menz G, Blaser K. Humoral and cell-mediated autoimmunity in allergy to *Aspergillus fumigatus. J Exp Med* 1996; 184:265–270.

91. Mayer C, Appenzeller U, Seelbach H, Achatz G, Oberkofler H, Breitenbach M, Blaser K et al. Humoral and cell-mediated autoimmune reactions to human ribosomal P_2 protein in individuals sensitized to *Aspergillus fumigatus* P_2 protein. *J Exp Med* 1999; 189:1507–1512.

92. Flückiger S, Fijten H, Whitley P, Blaser K, Crameri R. Cyclophilins, a new family of cross-reactive allergens. *Eur J Immunol* 2002; 32:10–17.

93. Appenzeller U, Mayer C, Menz G, Blaser K, Crameri R. IgE-mediated reactions to autoantigens in allergic diseases. *Int Arch Allergy Immunol* 1999; 118:193–196.

94. Crameri R, Kodzius R, Konthur Z, Lehrach H, Blaser K, Walter G. Tapping allergen repertoires by advanced cloning technologies. *Int Arch Allergy Immunol* 2001; 124:43–47.

95. Schmid-Grendelmeier P, Crameri R. Recombinant allergens for skin testing. *Int Arch Allergy Immunol* 2001; 125:96–111.

96. Grob P, Baumann S, Ackermann M, Suter M. A system for stable indirect immobilization of multimeric recombinant proteins. *Immunotechnology* 1988; 4:155–165.

97. Hennecke M, Kola A, Baensch M, Wrede A, Klos A, Bautsch W, Köhl J. A selection system to study C5a–C5a-receptor interactions: Phage display of a novel C5a anaphylatoxin, Fos-C5a^{Ala27}. *Gene* 1997; 184:263–272.

98. Heller T, Hennecke M, Baumann U, Gessner JE, zu Vilsendorf AM, Baensch M, Boulay F et al. Selection of a c5a antagonist from phage libraries attenuating the inflammatory response in immune complex disease and ischemia/reperfusion injury. *J Immunol* 1999; 163:985–994.

99. Pereboev A, Pereboeva L, Curiel DT. Phage display of adenovirus type 5 fiber knob as a tool for specific ligand selection and validation. *J Virol* 2001; 75:7107–7113.

100. Kemp EH, Waterman EA, Hawes BE, O'Neill K, Gottumukkala RV, Gawkrodger DJ, Weetman AP et al. The melanin-concentrating hormone receptor 1, a novel target of autoantibody responses in vitiligo. *J Clin Invest* 2002; 109:993–998.

101. Fossa A, Alsoe L, Crameri R, Funderud S, Gaudernack G, Smeland EB. Serological cloning of cancer/testis antigens expressed in prostate cancer using cDNA phage surface display. *Cancer Immunol Immunother* 2004; 53:431–438.

102. Geisberger R, Prlic M, Achatz-Straussberger G, Oberndorfer I, Luger E, Lamers M, Crameri R et al. Phage display based cloning of proteins interacting with the cytoplasmic tail of membrane immunoglobulins. *Dev Immunol* 2002; 9:127–134.

103. Kleber-Janke T, Becker WM. Use of modified BL21(DE3) *Escherichia coli* cells for high-level expression of recombinant peanut allergens affected by poor codon usage. *Protein Expr Purif* 2000; 19:419–424.

104. Levitan B. Stochastic modelling and optimization of phage display. *J Mol Biol* 1998; 277:895–916.

105. Suter M, Foti M, Ackermann M, Crameri R. In: Kay BK, Winter J, McCafferty J, eds. *Phage Display of Peptides and Proteins. A Laboratory Manual.* San Diego, CA: Academic Press Inc, 1996:195–214.

106. Chernukhin IV, Klenova EM. A method of immobilization on the solid support of complex and simple enzymes retaining their activity. *Anal Biochem* 2000; 280:178–181.

107. Baneyx F. Recombinant protein expression in *Escherichia coli. Curr Opin Biotechnol* 1999; 10:411–421.

108. Li M. Application of display technology in protein analysis. *Nat Biotechnol* 2000; 18:1251–1256.

109. MacCoss MJ, Yates JR III. Proteomics: Analytical tools and techniques. *Curr Opin Clin Nutr Metab Care* 2001; 4:369–375.

110. Büssow K, Nordhoff E, Lubbert C, Lehrach H, Walter G. A human cDNA library for high-throughput protein expression screening. *Genomics* 2000; 65:1–8.

111. Walter G, Büssow K, Lueking A, Glokler J. High-throughput protein arrays: Prospects for molecular diagnostics. *Trends Mol Med* 2002; 8:250–253.

112. Kodzius R, Rhyner C, Konthur Z, Buczek D, Lehrach H, Walter G, Crameri R. Rapid identification of allergen-encoding cDNA clones by phage display and high-density arrays. *Comb Chem High Throughput Screen* 2003; 6:147–153.

CHAPTER 13

Mapping Protein Functional Epitopes

Sara K. Avrantinis and Gregory A. Weiss

CONTENTS

13.1 INTRODUCTION

Essentially all events in biology require one molecule binding to another. With molecular recognition at the heart of biology, a number of phage-display-based techniques have been developed to dissect the details governing binding events. This chapter reviews phage-display applications that provide scalpels and microscopes to explore molecular recognition. These techniques capitalize upon the advantages of the phage-display format—rapid production of protein variants, simple protein purification, and the potential for powerful selections.

Modifying the primary structure of a protein can explore the intricate relationship between the primary sequence of amino acids and the properties of the resultant protein, including shape, stability, and activity. Mutagenesis of specific protein residues has proven invaluable in probing the contributions of individual amino acid side chains to such properties. For example, the seminal chemical syntheses of proteins with modified side chains by Hodges and Merrifield [1] provided a detailed understanding of the catalytic mechanism of RNase A. In another now classic experiment,

Fersht et al. [2] elegantly demonstrated the power of oligonucleotide-directed muta-genesis to measure hydrogen bond strengths within proteins. These experiments set the stage for the emergence of high-throughput techniques for protein modification and dissection using phage-displayed proteins.

The goal of the experiments described here is to identify the side chains and functional groups responsible for a particular protein function, often binding to a specific target. Quantitation of the energetic contributions made by specific side chains and functional groups is a welcome bonus made possible by some phage-display techniques. For the examination of receptor–ligand interactions, protein libraries can be generated by either random [3–5] or site-specific mutations. Several options are available to install mutations site-specifically. Saturation mutagenesis at specific sites replaces a particular side chain with all 20 naturally occurring amino acids [6,7]. Substitution with a few carefully chosen amino acids, on the other hand, can provide a thorough portrait of the contributions made by an individual side chain and potentially enable quantitative analysis of the data. In addition, limited mutagen-esis can allow for coverage of more protein residues in a single library.

Noncovalent binding interactions between receptors and ligands are mediated by directly contacting residues, which form a structural epitope. Biophysical tech-niques, such as x-ray diffraction and multidimensional nuclear magnetic resonance (NMR), can reveal such binding contacts. The structure and structural epitope alone, however, do not tell the whole story of how a protein works. Within the contact area, a subset of residues may contribute the majority of binding energy or protein func-tionality. Such residues can form key hydrogen bonds, salt bridges, dipole–dipole, and hydrophobic interactions within the binding interface. Residues with energeti-cally favorable contacts compose the functional epitope of a binding protein [8]. A tightly clustered functional epitope resembling a cross-section of protein structure (i.e., a hydrophobic center surrounded by hydrophilic groups) has been termed a "hot spot" of binding energy [9–11]. Identification of functional epitopes requires quan-tification of individual side chain contributions to protein function, a task eminently suited for phage-display techniques.

The functional epitope ultimately reveals how a protein works. However, under-standing protein function also requires identification of residues that position the functional epitope side chains ("second-sphere residues") [12] and even third-sphere residues, which may influence protein activity. In addition, thorough understand-ing of protein structure and dynamics includes identification of residues critical to protein folding and stability. Phage-display techniques can identify these important residues.

This chapter surveys experiments demonstrating the scope of functional epitope mapping by phage-display techniques. The examples cited are not meant to be com-prehensive, and the authors apologize in advance for not having the space to include all possible examples of functional epitope mapping by phage display. In general, this chapter focuses upon techniques with combinatorial, yet systematic, variation of spe-cific side chains. To introduce the topic, a few examples of single-point mutagenesis—both with and without phage display—will be described.

13.2 SINGLE-POINT ALANINE MUTAGENESIS

Alanine substitutions truncate amino acid side chains at the β-carbon. Thus, each alanine mutation explores the functional importance of side chain atoms past the β-carbon. Alanine is typically chosen because the substitution nullifies the amino acid side chain without introducing additional conformational flexibility into the protein backbone that might result from a glycine mutation. Alanine-scanning mutagenesis, systematic replacement of individual wild-type residues with alanine, is a particularly useful technique for the identification of functional epitopes and consequently for uncovering biological insight. For example, alanine scanning revealed that human growth hormone (hGH) binds to hGH-binding protein (hGHbp) with a remarkably compact patch of 8 out of 31 residues buried at the interface [8]. The hGHbp hot spot, also determined by alanine scanning, consists of a patch of residues with complementary size and functional groups [9]. In another example, researchers examined the first committed step in HIV infection (HIV gp120 binding to the cell surface receptor CD4). Alanine scanning of CD4 showed that the HIV gp120-binding site consists of several discontinuous segments that could be modeled onto a compact region of CD4 [13]. Also, through alanine scanning, Gibbs and Zoller [14] uncovered specific residues and regions of a protein kinase likely to be important in catalysis, binding MgATP, and binding a peptide substrate. Protein stability can also be probed by alanine scanning, as demonstrated by a series of alanine mutations in the helical region of phage T4 lysozyme, which identified three residues that have a substantial influence on protein stability [15]. These examples demonstrate the power of alanine scanning to correlate different aspects of protein function and stability with structure.

"Inverse alanine scanning" offers a labor-intensive alternative to conventional alanine scanning [16]. Each residue of a parent peptide consisting exclusively of alanines is separately and sequentially replaced by the 19 nonalanine amino acids. As usual, for scanning mutagenesis, each of these protein variants is separately synthesized and assayed. Inverse alanine scanning is perhaps suited only to short peptides. This differs from conventional alanine scanning, which works well for both large and small proteins.

Although alanine mutagenesis provides a detailed map of functional epitopes, the method can involve much effort. Each alanine-substituted protein must be separately constructed, expressed, refolded, characterized, and so forth. Only then can the effect of the truncated side chain on protein functionality be assessed by an in vitro assay of protein activity. In vivo assays can minimize effort spent on protein purification and other steps, but such assays are available for only a subset of interesting proteins. In addition, in vivo assays do not offer the control over receptor and ligand concentrations possible with in vitro assays, which can be necessary for rigorous assessment of binding thermodynamics. Combining alanine scanning with phage display can simplify some of the more tedious steps associated with traditional alanine scanning. For example, alanine mutants can be displayed on the phage surface by fusing the protein under study to a phage

Table 13.1   Receptor–Ligand Systems Representative of Functional Epitopes Studied by Alanine Scanning Phage-Displayed Proteins ("Turbo Alanine Scanning")

Receptor	Ligand	No. of Mutated Residues	References
Natriuretic peptide receptor-A	*Atrial natriuretic peptide*	25	[17]
ErbB3 and ErbB4 receptors	*Heregulinβ egf domain*	45	[18]
Microplasminogen	Staphylokinase	9	[19]
IGF binding protein 1 and 3	*Insulin-like growth factor (IGF)*	59	[20]

Note: Mutated proteins are shown in italics.

coat protein. The displayed proteins can then be amplified in an *Escherichia coli* host, isolated by phage precipitation, characterized by standard DNA sequencing, and examined by an in vitro phage-based assay. Alanine scanning of proteins displayed on the surface of phage, termed "turbo alanine scanning," has provided insight into the functional epitopes of many proteins, including those shown in Table 13.1.

While turbo alanine scanning expedites the process of functional epitope mapping, combinatorial libraries of phage-displayed alanine substitutions offer an alternative to scanning each position individually (Figure 13.1). However, to apply a combinatorial library technique, two basic issues must be addressed. First, synthesis of the library requires substitution of alanine and wild type in specific positions over large sections of the protein. Second, functional proteins must be selected from a library with diversities of up to 10^{11} alanine-substituted proteins. The use of phage-displayed protein libraries is appealing because the technique can address both issues inherent to combinatorial alanine scanning.

13.3 COMBINATORIAL SITE-SPECIFIC MUTAGENESIS

13.3.1 Binomial Mutagenesis

As an alternative to conventional alanine-scanning mutagenesis, several research groups have described methods to access multiple alanine substitutions. Oligonucleotide-directed mutagenesis is used for site-specific alanine incorporation into multiple positions. Binomial substitutions of either alanine or another amino acid are accessible by conventional oligonucleotide synthesis for seven amino acids (labeled with an asterisk in Table 13.2). For these seven amino acids, changing a single nucleotide encoding the wild-type amino acid can result in a codon for alanine. For example, the codon GAT encodes the amino acid aspartic acid; replacing A with C in the second position results in a codon for alanine (GCT). During oligonucleotide synthesis, addition of a 1:1 ratio of A and C phosphoramidites in the second position of the codon encodes a 1:1 ratio of aspartic acid and alanine in the translated protein

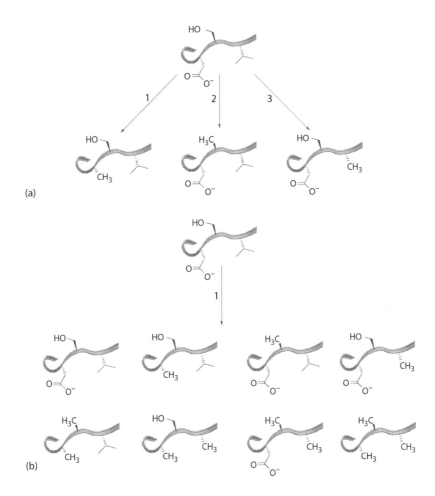

Figure 13.1   Alanine scanning using (a) single mutations where each amino acid is individually replaced with alanine (three separate mutagenesis reactions required) and
(b) combinatorial mutagenesis where many amino acids are simultaneously
mutated in a single step.

library. Libraries with alanine substitutions in multiple positions can be encoded by
degenerate oligonucleotides designed with mutations in multiple positions.

   Mutagenesis with the seven amino acids for which binomial mutagenesis is
accessible has been used to decipher functional epitopes of proteins. Gregoret and
Sauer [21] used the technique to analyze the effects of multiple alanine substitutions
on the function and stability of the DNA binding protein, λ repressor. Eleven positions of the λ repressor helix-turn-helix, a critical region for protein functionality,
were mutated to alanine or wild type. Approximately 25% of the alanine-substituted
proteins retained activity, which is indicative of the robust information encoding
protein folding and function. In general, the use of combinatorial libraries to study
functional epitopes relies upon the proper folding of alanine-substituted proteins.

Table 13.2 Degenerate Codons Used to Substitute Wild-Type and Another Amino Acid

Wild Type	Alanine Shotgun Scanning[a]				Homolog Scanning[b]		Proline Scanning[c]			
	Codon	Replacement aa			Codon	Homolog aa	Codon	Replacement aa		
Ala	GST	Gly			KCT	Ser	SCA	Pro		
Arg	SST	Ala	Gly	Pro	ARG	Lys	CST	Pro		
Asn	RMC	Ala	Asp	Thr	RAC	Asp	MMT	Pro	His	Thr
Asp*	GMT	Ala			GAM	Glu	SMT	Pro	Ala	His
Cys	KST	Ala	Gly	Ser	TSC	Ser	YST	Pro	Arg	Ser
Glu*	GMA	Ala			GAM	Asp	SMG	Pro	Ala	Gln
Gln	SMA	Ala	Glu	Pro	SAA	Glu	CMG	Pro		
Gly*	GST	Ala			GST	Ala	SST	Pro	Ala	Arg
His	SMT	Ala	Asp	Pro	MAC	Asn	CMT	Pro		
Ile	RYT	Ala	Thr	Val	RTT	Val	MYT	Pro	Leu	Thr
Leu	SYT	Ala	Pro	Val	MTC	Ile	CYG	Pro		
Lys	RMA	Ala	Glu	Thr	ARG	Arg	MMG	Pro	Gln	Thr
Met	RYG	Ala	Thr	Val	MTG	Leu	MYG	Pro	Leu	Thr
Phe	KYT	Ala	Ser	Val	TWC	Tyr	YYT	Pro	Leu	Ser
Pro*	SCA	Ala			SCA	Ala	YCT	Ser		
Ser*	KCC	Ala			KCC	Ala	YCT	Pro		
Thr*	RCT	Ala			ASC	Ser	MCT	Pro		
Trp	KSG	Ala	Gly	Ser	TKG	Leu	YSG	Pro	Arg	Ser
Tyr	KMT	Ala	Asp	Ser	TWC	Phe	YMT	Pro	His	Ser
Val*	GYT	Ala			RTT	Ile	SYA	Pro	Ala	Leu

Note: DNA degeneracies are represented in International Union of Biology (IUB) code (K = G/T, M = A/C, N = A/C/G/T, R = A/G, S = G/C, W = A/T, Y = C/T).

[a] Codons used for alanine shotgun scanning [22]. Binomial substitution (wild type or alanine) is accessible by exchanging a single nucleotide for seven amino acids (*). Combinatorial alanine scanning of other amino acids requires split-pool synthesis of degenerate oligonucleotides or the listed tetranomial substitutions.

[b] Codons used in homolog scanning to encode the wild-type residue and a homologous amino acid [23].

[c] Codons that could be used in proline scanning for introducing systematic breaks in protein secondary structure.

Studies of the λ repressor [21], T4 lysozyme [24], Arc repressor [25], and other systems (reviewed in Ref. [26]) have all demonstrated the resilience of protein folding despite extensive alanine substitutions.

To extend binomial mutagenesis beyond the seven amino acids for which exchange of a single codon nucleotide results in a codon encoding alanine, Chatellier and coworkers [27,28] reported the split-pool chemical synthesis of degenerate oligonucleotides for combinatorial alanine mutagenesis. Split-pool synthesis of oligonucleotides can be used to couple an alanine codon onto the nascent oligonucleotide in one reaction vessel and a wild-type codon in a different reaction vessel. After each triplet codon is added to the nascent oligonucleotide, the two fractions are pooled. Further splitting, coupling, and pooling creates combinatorial libraries of oligonucleotides with binomial substitution possible for all 20 amino acids. Split-pool oligonucleotide synthesis can be used to probe specific positions of any protein. Chatellier and coworkers [29] used combinatorial alanine scanning with mutagenesis by split-pool synthesized oligonucleotides to investigate the interface between heavy and light variable domains of an antibody. Combinatorial alanine scanning revealed a functional requirement for wild-type side chains bordering the antigen-binding site, which were ascribed to second-sphere effects. These results demonstrate the value of combinatorial alanine scanning for identification of residues contributing indirectly to protein function. Although the split-pool synthesis of oligonucleotides has not been used for phage display to date, the technique could be readily applicable to a phage-display system.

The combinatorial format of alanine-substituted libraries has also tested the following long debated question relevant to all functional epitope mapping studies. Are the energetic contributions of individual amino acids additive? Binomial mutagenesis of λ repressor, shotgun scanning (described in Section 13.3.2), and work by others [30] has demonstrated conclusively that, with a few exceptions, individual side chains contribute binding energy to the receptor–ligand interaction in an essentially additive fashion. In other words, contributions to the $\Delta\Delta G$ for a noncovalent binding interaction largely result from the sum of energetic contributions by individual side chains. This research set the stage for large-scale combinatorial alanine mutagenesis techniques, such as shotgun scanning. Shotgun scanning can apply combinatorial alanine mutagenesis to 20 or more positions, resulting in exceptionally diverse protein libraries, which are rapidly screened using phage-display techniques.

13.3.2 Shotgun Scanning

Shotgun scanning combines the concepts of alanine-scanning mutagenesis and binomial mutagenesis with phage-display technology (reviewed in Ref. [31]). Libraries of alanine-substituted proteins are displayed on the surfaces of filamentous phage particles for in vitro binding selections (Figure 13.2). By displaying libraries of alanine-substituted proteins on the surface of filamentous phage, successive rounds of selection for a specific binding activity can be used to enrich for side chains contributing binding energy to the receptor–ligand interaction (Figure 13.3).

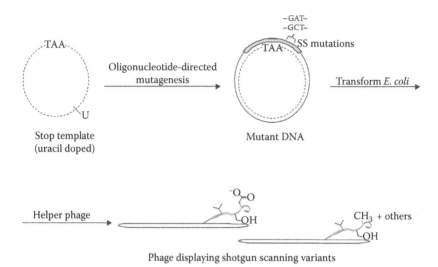

Figure 13.2 Flowchart indicating the steps needed to construct phage-display libraries of alanine-substituted proteins through oligonucleotide-directed mutagenesis.

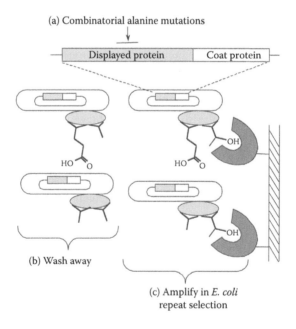

Figure 13.3 Screening combinatorial alanine libraries. (a) Libraries of proteins with alanine mutations are constructed using oligonucleotide-directed mutagenesis. (b) The phage library is applied to an immobilized receptor, and nonbinding phage are washed away. (c) Through successive rounds of binding selection and amplification in an *E. coli* host, wild-type side chains are enriched in positions with energetically favorable contacts.

Phage display simplifies the construction and screening of combinatorial libraries of alanine-substituted proteins in at least three important ways, which are discussed elsewhere in this volume. Briefly, large libraries of proteins (>10^{10} unique clones) are readily accessible, with each unique protein variant fused to the surface of a different phage particle. Second, after selection for displayed proteins that bind to a target molecule, phage can be amplified in an *E. coli* host. Additionally, the phage particles encapsulate DNA encoding the displayed protein; thus, DNA sequencing can be used to identify selected proteins. The final DNA sequencing step enables statistical analysis of the frequency of wild type and alanine in each mutated position.

Shotgun scanning applies degenerate oligonucleotides synthesized by conventional automated methods. Conventional DNA synthesis leads to tetranomial substitution for the 12 amino acids where single nucleotide exchange for alanine substitutions is unavailable (Table 13.2). A key simplifying assumption is made during statistical analysis of the selected and sequenced shotgun-scanning clones. Analysis of the energetic contribution by specific side chains to receptor–ligand binding focuses entirely on the distribution of alanine or wild type in each substituted position. With this simplification, combinatorial alanine mutagenesis becomes possible using standard oligonucleotide synthesis. However, secondary analysis of nonwild-type, nonalanine substitutions in each position is also possible. For example, statistical analysis to extract information about protein functionality from tetranomial mutagenesis has been described by Hu and coworkers [32]. Strong selection for nonalanine and nonwild-type amino acids could reveal unexpected interactions, such as mutations that confer improved affinity to the ligand. In addition, it could be possible to derive $\Delta\Delta G$ values for such nonalanine mutations.

In the first example of shotgun scanning, the 19 residues of hGH comprising a large part the high-affinity binding site for hGHbp were shotgun-scanned in a single library (Figure 13.4a) [22]. After multiple rounds of selection and amplification in an *E. coli* host, individual hGHbp binding phage were identified by a high-throughput phage-based ELISA and subjected to DNA sequencing. Focusing entirely upon the distribution of alanine and wild type in each scanned position revealed specific positions that were highly conserved as the wild-type amino acid and other positions with a roughly even distribution of alanine and wild type. By assuming that the shotgun-scanning selection occurred under equilibrium binding conditions, estimates of $\Delta\Delta G$ values were derived from the distribution of alanine and wild-type amino acids in each position following shotgun scanning. Analysis of hGH in this initial experiment allowed comparison of $\Delta\Delta G$ values measured by shotgun scanning with $\Delta\Delta G$ values from conventional alanine-scanning mutagenesis. Overall, data from hGH shotgun scanning compared favorably with measurements by alanine-scanning mutagenesis. The shotgun-scanning $\Delta\Delta G$ values confirmed that hGH binding to hGHbp is mediated by a compact hot spot of just 7 residues out of the 19 residues analyzed. Shotgun scanning offered the convenience of a single round of mutagenesis in multiple positions with the rapid purification and assay of phage-displayed alanine-substituted proteins.

Subsequently, several other proteins and receptor–ligand interactions have been examined by shotgun scanning. For example, the femtomolar interaction between

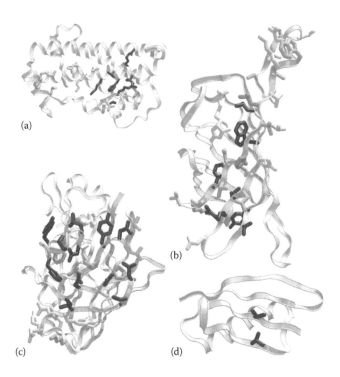

Figure 13.4 Proteins examined by shotgun scanning with stick models depicting mutated
side chains. (a) Shotgun scanning of hGH for binding to hGHbp identified seven
critical residues (black) out of the 19 residues in contact with the receptor. (From
Weiss GA et al., *Proc Natl Acad Sci U S A*, 97:8950–8954, 2000.) (b) Streptavidin
shotgun scanning revealed side chains essential for biotin binding (black), including some residues far from the biotin-binding site. (From Avrantinis SK et al.,
ChemBioChem, 3:1229–1234, 2002.) (c) Alanine scanning of the heavy chain of
anti-ErbB2 antibody revealed the side chains that contribute to antigen binding
(black). (From Vajdos FF et al., *J Mol Biol*, 320:415–428, 2002.) (d) Saturation
mutagenesis at positions 44 and 53 (black) of immunoglobulin G (IgG)-binding
protein (GB1) was used to study residue pairing in β-sheet stability. (From
Distefano MD et al., *J Mol Biol*, 322:179–188, 2002.)

the protein streptavidin and the small molecule biotin (molecular weight [MW] =
244 Da) has been dissected by shotgun scanning (Figure 13.4b) [33]. The results
demonstrated the importance of previously unreported hydrophobic residues contributing both direct and indirect contacts with biotin. These include residues that are
most likely responsible for forming the β-barrel structure of streptavidin, residues
that interact at the tetramer interface, and residues that form a network of extended
hydrophobic interactions buttressing residues in direct contact with biotin.

Vajdos and coworkers [23] used alanine shotgun scanning and a variation called
"homolog shotgun scanning" to map the antigen-binding site of an anti-ErbB2 antibody. Alanine shotgun scanning revealed the antibody side chains that contribute to
ErbB2 binding (Figure 13.4c). These included solvent-exposed residues that likely
comprise the functional-binding epitope and buried residues that maintain the

conformation of the residues required for the functional epitope. A separate homolog scan, in which library side chains were varied as wild-type or a similar amino acid residue, identified a subset of side chains that were intolerant to both alanine and homologous substitutions. The subset of side chains intolerant to homologous substitutions may be involved in precise contacts with the antigen. Homolog scanning demonstrates the potential utility of systematic mutagenesis of proteins by combinatorial libraries with substitutions different from alanine. For example, proline scanning could be used for introducing systematic breaks in protein secondary structure (Table 13.2).

In an example of using shotgun scanning to examine intrinsic protein stability, as opposed to receptor–ligand molecular recognition, Distefano and coworkers [34] used shotgun scanning to examine residue pairing across two strands of the GB1 β-sheet. The GB1 is a small, IgG-binding, model protein that has been used to examine individual residue contributions to β-sheet stability. GB1 shotgun-scanning libraries focused saturation mutagenesis upon two previously examined GB1 residues located at cross-strand positions in the β-sheet (Figure 13.4d). Properly folded variants were selected from the libraries through binding to human IgG$_1$ Fc. Selectants from the libraries displayed distinct preferences for specific amino acids at each position. However, the researchers were surprised to find that contributions to β-sheet stability from individual residues are more important than cross-strand contacts involving specific side chain–side chain interactions. In addition, these experiments demonstrated good agreement between shotgun-scanning methods with values for protein stability measured by extensive, conventional experiments. Shotgun scanning can, thus, be viewed as a method for rapidly identifying trends in binding activity and/or protein stability, which can be followed up by additional experiments.

13.4 OTHER APPROACHES TO PHAGE-DISPLAYED FUNCTIONAL EPITOPE MAPPING

Residues identified by alanine scanning can be used to guide further mutagenesis studies. To investigate the binding interface between tissue factor and factor VIIa, Lee and Kelley [35] created libraries in which two tissue factor residues shown by alanine scanning to be critical for high affinity, as well as surrounding residues, were randomized to include all amino acids. Unlike limited randomization throughout a protein, saturated randomization in specific positions allows all possible variants to be covered in a single phage-display library.

Another method to restrict diversity is through "biased" phage libraries, in which residues are mutated to a limited subset of amino acids. Dwyer and coworkers [36] used "biased" phage-display libraries with targeted residues of RNase S-peptide constrained as either polar or nonpolar substitutions. Such tailored libraries limited possible interactions present during a process of affinity maturation. High-affinity S-peptide variants retained a specific tryptophan residue, which provided a hot spot of binding energy. In general, limited mutagenesis, either by shotgun scanning or other forms of substitution bias, allows larger regions of the protein to be analyzed simultaneously.

13.5 CONCLUSION

With molecular recognition being key to essentially all biological events, methods are needed for rapid, yet detailed, analysis of functional epitopes of proteins. In the past, systematic mutagenesis of many residues in a protein required a tour de force effort, yet extensive mutagenesis may be required to map a complete functional epitope. Combinatorial alanine scanning connects the expedience of combinatorial libraries with the insight of site-directed scanning mutagenesis. Phage-display and combinatorial library techniques simplify these studies and make it possible to examine large libraries of protein variants. These large libraries can provide a more complete understanding of the principles of molecular recognition, protein folding, and the relationship between protein structure and function.

REFERENCES

1. Hodges RS, Merrifield RB. Synthetic study of the effect of tyrosine at position 120 of ribonuclease. *Int J Pept Protein Res* 1974; 6:397–405.
2. Fersht AR, Shi JP, Knill-Jones J, Lowe DM, Wilkinson AJ, Blow DM, Brick P et al. Hydrogen bonding and biological specificity analyzed by protein engineering. *Nature (London)* 1985; 314:235–238.
3. Jespers L, Jenne S, Lasters I, Collen D. Epitope mapping by negative selection of randomized antigen libraries displayed on filamentous phage. *J Mol Biol* 1997; 269:704–718.
4. Rossenu S, Dewitte D, Vandekerckhove J, Ampe C. A phage display technique for a fast, sensitive, and systematic investigation of protein–protein interactions. *J Protein Chem* 1997; 16:499–503.
5. Cain SA, Ratcliffe CF, Williams DM, Harris V, Monk PN. Analysis of receptor/ligand interactions using whole-molecule randomly mutated ligand libraries. *J Immunol Methods* 2000; 245:139–145.
6. Myers RM, Lerman LS, Maniatis T. A general method for saturation mutagenesis of cloned DNA fragments. *Science (Washington, DC)* 1985; 229:242–247.
7. Chen G, Dubrawsky I, Mendez P, Georgiou G, Iverson BL. In vitro scanning saturation mutagenesis of all the specificity determining residues in an antibody binding site. *Protein Eng* 1999; 12:349–356.
8. Cunningham BC, Wells JA. Comparison of a structural and a functional epitope. *J Mol Biol* 1993; 234:554–563.
9. Clackson T, Wells JA. A hot spot of binding energy in a hormone–receptor interface. *Science (Washington, DC)* 1995; 267:383–386.
10. Clackson T, Ultsch MH, Wells JA, de Vos AM. Structural and functional analysis of the 1:1 growth hormone–receptor complex reveals the molecular basis for receptor affinity. *J Mol Biol* 1998; 277:1111–1128.
11. Pons J, Rajpal A, Kirsch JF. Energetic analysis of an antigen/antibody interface: Alanine scanning mutagenesis and double mutant cycles on the HyHEL-10/lysozyme interaction. *Protein Sci* 1999; 8:958–968.
12. Arkin MR, Wells JA. Probing the importance of second sphere residues in an esterolytic antibody by phage display. *J Mol Biol* 1998; 284:1083–1094.

13. Ashkenazi A, Presta LG, Marsters SA, Camerato TR, Rosenthal KA, Fendly BM, Capon DJ. Mapping the CD4 binding site for human immunodeficiency virus by alanine-scanning mutagenesis. *Proc Natl Acad Sci U S A* 1990; 87:7150–7154.

14. Gibbs CS, Zoller MJ. Identification of electrostatic interactions that determine the phosphorylation site specificity of the cAMP-dependent protein kinase. *Biochemistry* 1991; 30:5329–5334.

15. Blaber M, Baase WA, Gassner N, Matthews BW. Alanine scanning mutagenesis of the α-helix 115–123 of phage T4 lysozyme: Effects on structure, stability and the binding of solvent. *J Mol Biol* 1995; 246:317–330.

16. Vetter SW, Keng Y-F, Lawrence DS, Zhang Z-Y. Assessment of protein-tyrosine phosphatase 1B substrate specificity using "inverse alanine scanning." *J Biol Chem* 2000; 275: 2265–2268.

17. Li B, Tom JYK, Oare D, Yen R, Fairbrother WJ, Wells JA, Cunningham BC. Minimization of a polypeptide hormone. *Science (Washington, DC)* 1995; 270:1657–1660.

18. Jones JT, Ballinger MD, Pisacane PI, Lofgren JA, Fitzpatrick VD, Fairbrother WJ, Wells JA et al. Binding interaction of the heregulin β *egf* domain with ErbB3 and ErbB4 receptors assessed by alanine scanning mutagenesis. *J Biol Chem* 1998; 273:11667–11674.

19. Jespers L, Van Herzeele N, Lijnen HR, Van Hoef B, De Maeyer M, Collen D, Lasters I. Arginine 719 in human plasminogen mediates formation of the staphylokinase: Plasmin activator complex. *Biochemistry* 1998; 37:6380–6386.

20. Dubaquie Y, Lowman HB. Total alanine-scanning mutagenesis of insulin-like growth factor I (IGF-I) identifies differential binding epitopes for IGFBP-1 and IGFBP-3. *Biochemistry* 1999; 38:6386–6396.

21. Gregoret LM, Sauer RT. Additivity of mutant effects assessed by binomial mutagenesis. *Proc Natl Acad Sci U S A* 1993; 90:4246–4250.

22. Weiss GA, Watanabe CK, Zhong A, Goddard A, Sidhu SS. Rapid mapping of protein functional epitopes by combinatorial alanine scanning. *Proc Natl Acad Sci U S A* 2000; 97:8950–8954.

23. Vajdos FF, Adams CW, Breece TN, Presta LG, de Vos AM, Sidhu SS. Comprehensive functional maps of the antigen-binding site of an anti-ErbB2 antibody obtained with shotgun scanning mutagenesis. *J Mol Biol* 2002; 320:415–428.

24. Heinz DW, Baase WA, Matthews BW. Folding and function of a T4 lysozyme containing 10 consecutive alanines illustrate the redundancy of information in an amino acid sequence. *Proc Natl Acad Sci U S A* 1992; 89:3751–3755.

25. Brown BM, Sauer RT. Tolerance of Arc repressor to multiple-alanine substitutions. *Proc Natl Acad Sci U S A* 1999; 96:1983–1988.

26. Bowie JU, Reidhaar-Olson JF, Lim WA, Sauer RT. Deciphering the message in protein sequences: Tolerance to amino acid substitutions. *Science (Washington, DC)* 1990; 247:1306–1310.

27. Chatellier J, Mazza A, Brousseau R, Vernet T. Codon-based combinatorial alanine scanning site-directed mutagenesis: Design, implementation, and polymerase chain reaction screening. *Anal Biochem* 1995; 229:282–290.

28. Chatellier J, Vernet T. Combinatorial scanning site-directed mutagenesis. *Curr Innovations Mol Biol* 1997; 4:117–132.

29. Chatellier J, Van Regenmortel MH, Vernet T, Altschuh D. Functional mapping of conserved residues located at the VL and VH domain interface of a Fab. *J Mol Biol* 1996; 264:1–6.

30. Wells JA. Additivity of mutational effects in proteins. *Biochemistry* 1990; 29:8509–8517.

31. Morrison KL, Weiss GA. Combinatorial alanine-scanning. *Curr Opin Chem Biol* 2001; 5:302–307.
32. Hu JC, Newell NE, Tidor B, Sauer RT. Probing the roles of residues at the e and g positions of the GCN4 leucine zipper by combinatorial mutagenesis. *Protein Sci* 1993; 2:1072–1084.
33. Avrantinis SK, Stafford RL, Tian X, Weiss GA. Dissecting the streptavidin–biotin interaction by phage-displayed shotgun scanning. *ChemBioChem* 2002; 3:1229–1234.
34. Distefano MD, Zhong A, Cochran AG. Quantifying b-sheet stability by phage display. *J Mol Biol* 2002; 322:179–188.
35. Lee GF, Kelley RF. A novel soluble tissue factor variant with an altered factor VIIa binding interface. *J Biol Chem* 1998; 273:4149–4154.
36. Dwyer JJ, Dwyer MA, Kossiakoff AA. High affinity RNase S-peptide variants obtained by phage display have a novel "hot spot" of binding energy. *Biochemistry* 2001; 40: 13491–13500.

CHAPTER 14

Selections for Enzymatic Catalysts

Julian Bertschinger, Christian Heinis, and Dario Neri

CONTENTS

14.1 INTRODUCTION

Enzymes are extremely powerful catalysts, and their proficiency and diversity typically exceed the performance of man-made chemical catalysts that are commonly used in industrial chemistry. An impressive example of the stunning efficiency of enzymes is provided by orotidine-5′-phosphate decarboxylase. The uncatalyzed reaction has a half-life of 78 million years, whereas the half-life of the catalyzed reaction is only 18 ms. This corresponds to a rate acceleration (k_{cat}/k_{uncat}) of 1.4 × 10^{17} [1].

The use of the catalytic power of enzymes for the synthesis of commodity products and fine chemicals could be an attractive, cost-effective, and environmentally friendly alternative to chemical catalysts. Enzymes may also find medical applications in areas such as prodrug activation and treatment of metabolic disease [2,3].

However, naturally occurring enzymes usually do not meet the requirements for industrial or medical applications. For the synthesis of chemicals, it is desirable to use high substrate concentrations, which may lead to inhibition of the biocatalyst [4]. In addition, the enzyme should be stable and accept a certain range of substrates while producing defined molecules without side products. For therapeutic applications, the enzyme should be nonimmunogenic, exhibit high specificity for the target molecule, and not interfere with other physiologic processes in the patient. Thus, the engineering of tailor-made enzymes, with improved properties, represents an important goal of modern protein engineering.

There are two general avenues to engineer enzymes. The first is rational design, which requires information about structure, catalytic mechanisms, and molecular modeling to design an enzyme de novo or to alter an existing one. The alternative to design is evolution: a process that involves several cycles of protein diversity generation, followed by selection or screening of mutants with desired characteristics. The procedure is analogous to a Darwinian process, where only the fittest survive. Applying Darwinian principles in the test tube to form desired phenotypes is called in vitro evolution. In vitro evolution has been successfully applied in the laboratory to routinely produce antibodies with high affinity to their target molecule [5], enzymes with novel activities [6,7], ribozymes [8], and even allosteric ribozymes [9]. This chapter will focus on the in vitro evolution of enzymes using phage display.

14.1.1 In Vitro Evolution of Enzymes

In vitro evolution of proteins consists of three basic steps: (1) sequence diversity must be generated; (2) the created mutant phenotypes (i.e., the function performed by the protein) have to be linked to the corresponding genotypes (i.e., the genetic information encoding the protein); and (3) the mutants with the desired biological activity must be selected from the ensemble of proteins. In the case of enzymes, the biological activity is their ability to accelerate chemical reactions. Thus, enzymes could be selected by virtue of their ability to form product molecules. This requires the linkage of the product molecules to the genotype, which is responsible for their formation.

A wide range of approaches have been developed to engineer proteins with catalytic activity. Various in vitro [10] and in vivo procedures, such as auxotrophic complementation [11] or, in the case of catalytic antibodies, immunization with transition-state analogues (TSAs) [12] and reactive immunization with a suicide inhibitor (SI) [13], demonstrated that catalysts with improved or novel activities could be generated. In vitro selection systems are potentially applicable in a much broader way compared to in vivo procedures, because they could allow the selection step to take place under nonphysiologic conditions, such as elevated temperatures, high or low salt concentrations, and extreme pH values, or even in organic media. In addition, due to their complexity, in vivo systems may find ways to pass the selection barrier of the experiment by a number of different strategies, which may lead to the evolution of undesired phenotypes.

14.1.2 Linking Genotype and Phenotype by Phage Display

There are two main strategies to link a genotype to the corresponding protein phenotype. The first possibility is to link the protein physically to the DNA encoding it. This way of linkage is fundamentally different from what nature does. The second possibility is to contain the protein and DNA encoding it in the same compartment. In nature, genes, the proteins they encode, and the products of their activity are held together in a cell. Therefore, the linkage between phenotype and genotype is achieved by compartmentalization. Most procedures applied in protein engineering use the physical linkage between genotype and phenotype. Phage display is one example of this. The protein to be engineered is displayed on the phage coat by fusing it to a phage coat protein (pIII, pVI, or pVIII). Several enzymes could already be displayed on phage [14–17].

Phage display does not take place entirely in vitro, as phage need to be assembled and amplified in living bacteria, but the selection step does. Phage are relatively resistant to a variety of environments, which makes them appropriate for use in enzyme engineering.

14.2 SELECTION METHODS

A variety of methods for the selection of catalysts using phage display have been developed. The different approaches can be divided in two groups: selection of catalysts by binding and direct selections for catalytic activity.

14.2.1 Selecting Catalysts by Binding

Whenever possible, selection of phage enzymes by means of their binding properties may be a practical methodology, as it may allow us to physically isolate phage enzymes with desired characteristics from the phage pool. For the selection of catalysts, affinity purification on TSAs or mechanism-based SIs have been considered.

14.2.1.1 Selections by Binding to Transition-State Analogues

The transition-state theory relates the rate of a reaction to the difference in Gibbs energy between the transition state and the ground state (Figure 14.1). The higher this difference is, the slower the reaction rate. By binding the transition state, enzymes can lower its energy level and accelerate the reaction.

The dissociation constant of the complex between a catalyst and a reaction's transition state is defined as K_{TS}. Efficient catalysts bind the transition state very tightly, and therefore, K_{TS} is very small. K_{TS} is an important parameter to characterize the

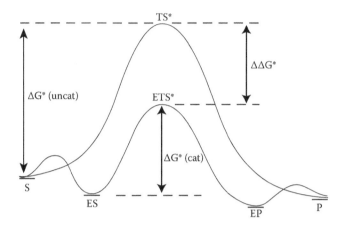

Figure 14.1　Free energy diagram for catalysis. The dissociation constant K_{TS} correlates quantitatively with $\Delta\Delta G^*$. S: substrate, P: product, ES and EP: enzyme–substrate/product complexes. TS* and ETS* denote the free and the enzyme-bound transition state, respectively.

proficiency of an enzyme. Using transition-state theory and the relations of the Gibbs free energy to equilibrium constants, K_{TS} can be described by the parameters k_{cat}, k_{uncat}, and K_S. k_{cat} denotes the rate constant for reaction of the enzyme–substrate complex, K_S its dissociation constant, and k_{uncat} the rate constant for the uncatalyzed reaction of substrate to product.

$$\frac{K_S}{k_{cat}/k_{uncat}} = K_{TS} \tag{14.1}$$

Equation 14.1 shows the relation between the three parameters K_S, k_{cat}, k_{uncat}, and the dissociation constant K_{TS} for the enzyme transition state complex. Enzymes can bind the transition state very tightly. Orotidine-5′-phosphate decarboxylase, which was mentioned at the beginning of this chapter, has a K_M of 7×10^{-7} M and a k_{cat}/k_{uncat} ratio of 1.4×10^{17} [1]. Using Equation 14.1 and assuming that $K_S \approx K_M$, it can be calculated that this remarkable enzyme binds the transition state with a dissociation constant of 5×10^{-24} M.

Many attempts have been made to create catalytic antibodies, which, by binding and stabilizing the transition state, accelerate a reaction. Molecules mimicking the transition state (TSAs) were synthesized. With an ideal TSA perfectly mimicking the actual transition state of the reaction, the rate acceleration of the selected catalysts would match the differential affinity to the TSA versus the substrate (Equation 14.1).

In most cases, catalytic antibodies were raised by immunization against a TSA. This has led to catalytic antibodies for a number of reactions. Rate accelerations are limited to about a factor of 10^2–10^4, occasionally 10^6 [18]. As a strategy to improve the activity of the antibodies, affinity selections against a TSA with antibody fragments displayed on phage were performed to optimize the affinity to TSAs [19]. In this study, the humanized antibody 17E8 with esterase activity was affinity-matured against a phosphonate TSA. The selections resulted in antibody variants with twofold to eightfold higher affinity to the phosphonate TSA. Surprisingly, none of the selected mutants showed improved catalytic activity compared to the parent antibody. In contrast, a weaker binding variant was identified to have a twofold higher catalytic activity. In a further study [20], residues remote from the active site of 17E8 were mutated. Panning of the library yielded a mutant with tenfold improved catalytic efficiency (k_{cat}/K_M). However, higher affinity to the TSA did not result in a parallel increase in catalytic activity. The increase of rate acceleration observed was due to a lower substrate affinity [20]. This is not surprising, because the selection pressure is for TSA binding and not for catalysis. Even if TSA binding were equivalent to catalytic activity, it would be very difficult to obtain binders with very low K_{TSA}. Rate enhancements by enzymes range from 10^6 to 10^{17}, with an average around 10^{10} [21]. When assuming that $K_M \approx$ 1 mM, K_{TSA} would have to be in the order of 10^{-13} M to obtain a catalytic antibody with activity comparable to enzymes. However, the selection of antibodies by phage display with affinities as high as 10^{-13} M remains a formidable challenge. Notably, one of the antibody–hapten complexes with the lowest dissociation constant (48 fM) was isolated by yeast display [5]. In addition, a TSA is never a perfect mimic of a true transition state and their synthesis can be very difficult. Therefore, to select catalytic antibodies against TSAs using phage display or any other method is of limited scope. It may be useful for finding catalysts without counterpart in nature. Better results may be achieved if, after selecting for binding, the library is screened [22] or selected for catalysis [23].

14.2.1.2 Selections by Binding to SIs

Mechanism-based SIs were used as an alternative to TSAs [24,25]. Enzymes tolerate an SI as a substrate, but the catalytic cycle does not proceed until its end. Instead, the reaction is trapped in an intermediate, where the SI is covalently bound to the active site of the enzyme. When affinity tags are coupled to the SI, enzymes labeled with the affinity tag through the SI can then be selected from repertoires of mutants using affinity purification.

Using SIs for selection by binding, the specificity of subtilisin 309 could be changed [24]. Subtilisins have broad substrate specificity but exhibit a preference for hydrophobic residues and very low reactivity toward charged residues in the P4 position. A wild-type subtilisin called Savinase from *Bacillus lentus* was displayed on phage fused to the pIII coat protein. The phage enzyme particles were successfully panned on streptavidin-coated beads after labeling with biotinylated SIs bearing the hydrophobic residue alanine in the P4 position.

In order to change the specificity of Savinase, libraries were created where residues 104 and 107 forming part of the S4 binding pocket were randomized. For the selections, biotinylated SIs were used with lysine instead of alanine in the P4 position. Savinase variants were isolated after three rounds of panning with a more than 100-fold enhanced activity for the substrate with lysine in the P4 position compared to the wild-type enzyme. The isolated clone did not have any detectable activity for the substrate with alanine in the P4 position.

Mechanism-based inhibitors may be superior to TSAs because the catalytic cycle is being selected more often. In some cases, the combined use of TSA and suicide inhibitors may yield biocatalysts of exceptional quality [26]. For example, the immunization of mice with a hapten combining an element mimicking the transition state of the reaction (a tetrahedral sulfone) with a diketone mechanism-based inhibitor has yielded antibodies with exceptional aldolase activities (rate accelerations $>10^8$ and efficient turnover). In principle, it should be possible to apply a similar strategy using phage display instead of immunization.

Similar to selections by binding to TSAs, selections by binding to SIs have their limitations. First of all, it is impossible to select for product release and turnover when using SIs. These steps can be rate limiting in catalysis [27]. Second, in reactions where the TSA or the trapped SI resembles the product, the selected catalysts may suffer from product inhibition and low turnover. In conclusion, selecting catalysts by binding (using phage display or any other method) is probably only effective where changing the substrate selectivity and not enhancement of rate acceleration or turnover is of main importance.

14.2.2 Selections for Catalytic Activity

In nature, enzymes are selected directly (but not solely) for their catalytic activity. When looking at the amazing proficiency of natural enzymes, direct selection for catalysis seems to be the most effective way of evolving catalysts. Proteins with catalytic activity must have several abilities.

- The catalyst must be able to bind the substrate efficiently at given concentrations. This property is characterized by the Michaelis–Menten constant K_M. K_M values of natural enzymes vary from 1 mM to 1 nM, and so do the effective, physiological concentrations of the corresponding substrates [28].
- Specificity is also important. The catalyst should be able to catalyze the conversion of one particular substrate in the presence of others.
- The catalyst must convert the substrate bound in the active site to a product with a higher rate than that of the free substrate in solution (rate acceleration, k_{cat}/k_{uncat}).
- The catalyst should be specific for the formation of one particular product (or few products). In solution, free substrate might turn into several, different products, whereas the enzymatic reaction only yields one product (e.g., one stereoisomer vs. a racemic mixture).
- And finally, the product should readily dissociate from the active site to yield a free enzyme, which is ready to bind new substrate (turnover, k_{cat}). Sometimes, the catalyst has to change its conformation before it is able to bind the substrate again. In order to select for turnover, the substrate should be in excess over the enzyme, and selection pressure should be directed to the conversion of all or most of the substrate molecules.

In order to obtain man-made catalysts of proficiency comparable to that of natural enzymes, selection pressure must be applied on all properties mentioned above simultaneously.

When phage display is used to engineer enzymes, either the reaction product must be immobilized on the enzyme–phage particle to allow the isolation of catalytic proteins using an antiproduct affinity reagent or the enzyme–phage particles must be selectively eluted from a solid support upon catalysis. Several procedures have been developed to select directly for catalytic activity. They can be divided in two main classes: intramolecular, single-turnover selections and intermolecular, multiple-turnover selections.

14.2.2.1 Intramolecular, Single-Turnover Selections

An approach in which the reaction substrate was covalently attached to the pIII minor coat protein of filamentous phage, displaying a nuclease, was first published in 1998 [29]. Intramolecular conversion of the substrate into product was meant to provide the basis for the selection of active catalysts from a library of mutants. Upon generation of the reaction product, the enzyme–phage particle would be released from a solid support (Figure 14.2).

In order to immobilize the substrate of the nuclease on the phage coat nearby the enzyme, a helper phage was generated displaying an acidic amphiphilic peptide at the N-terminus of the pIII coat protein. The acidic peptide and the enzyme were codisplayed on phage after infection with *Escherichia coli* (containing a phagemid) with the modified helper phage. The acidic peptide formed a heterodimeric coiled-coil complex with a basic amphiphilic peptide. By coupling the substrate of the reaction to the basic peptide, the substrate could be noncovalently attached to the phage adjacent to the enzyme. In order to achieve a covalent linkage between the two peptides, one cysteine was introduced at the C-terminus of each peptide, and therefore, upon removal of reducing agents, a covalent disulfide bridge between the two peptides was formed. In this way, the substrate could be coupled to the phage coat site specifically in the vicinity of the enzyme (Figure 14.3).

The nuclease used (staphylococcal nuclease [SNase]) preferentially hydrolyzes the phosphodiester bonds of single-stranded RNA, single-stranded DNA, and duplex DNA at A, U- or A, T-rich regions upon activation by Ca^{2+}. The substrate for the enzymatic

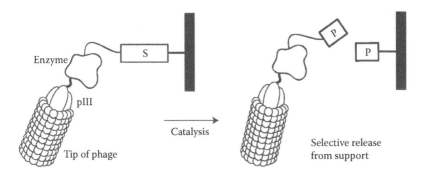

Figure 14.2 Upon intramolecular catalysis, the enzyme–phage particle is released from a solid support.

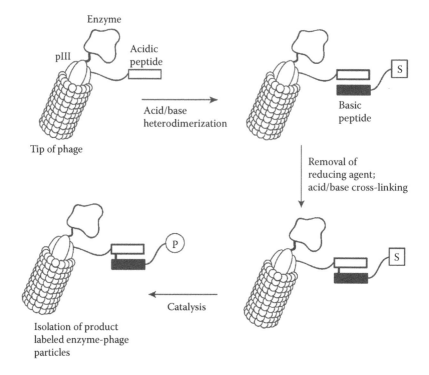

Figure 14.3   Immobilization of the substrate on phage with a disulfide-bridged, heterodimeric coiled-coil complex. The acidic peptide was codisplayed with the enzyme on phage.

reaction, a biotinylated oligodeoxynucleotide, was chemically coupled to the basic peptide. After addition of substrate basic peptide conjugate to the phage enzyme particles followed by removal of reducing agents, the enzyme–phage particles could be immobilized on streptavidin-coated beads via the biotin–streptavidin interaction. The cleavage reaction was initiated by addition of Ca^{2+}. By virtue of the catalytic activity of SNase, the enzyme–phage particles were released from the solid support. Phage displaying SNase were shown to be released 100 times more efficiently than a control antibody phage. It appeared that a small fraction of the phage leaked off the support during the assay.

The selection methodology described above, based on the display of enzymes on filamentous phage and focused on cleavage reactions, could in principle be applied to a variety of different reactions, involving substrate cleavage. However, the enzyme must be maintained in an inactive state during immobilization of the enzyme–substrate–phage particles on the solid support. Whenever reversible inactivation of the enzyme is not possible, capture of active enzymes must be performed with a product-specific reagent. In the case of bimolecular condensation reactions, in which bond formation results in phage immobilization on solid support, the regulation of enzymatic activity is not necessary.

A very similar approach as the one with SNase [29] was used for the selection of metalloenzymes by catalytic elution [30]. The method should be applicable to phage displaying enzymes whose activity depends on the presence of a cofactor, provided that the apoenzyme is still capable of binding to its substrate. In this method, a metalloenzyme, which required a metal ion cofactor, was displayed on phage. After complexation of the metal ion by a chelator, the inactivated enzyme–phage particles were immobilized on a solid support coated with substrate. After addition of the metal ion cofactor, active enzymes transformed the substrate into the product. Because the enzyme exhibited only low affinity to the product, the enzyme–phage particles were eluted from the solid support (Figure 14.4).

In this study [30], metallo-β-lactamase BCII (βLII) from *Bacillus cereus* was used as a model enzyme. The enzyme's structure contains two zinc ions. One zinc

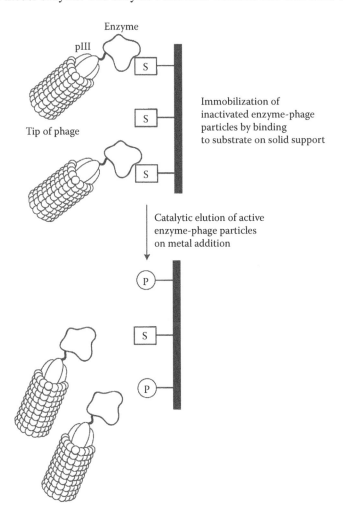

Figure 14.4 Selection of metalloenzymes by catalytic elution. Cofactor-depleted, inactivated enzyme–phage particles were adsorbed on a solid support. Upon catalysis, the enzyme dissociated from the product and was eluted.

ion, coordinated by three histidine residues, is essential for catalytic activity. Thus, phage-bound enzymes could be inactivated and reactivated by the incubation with EDTA followed by addition of zinc sulfate. In model experiments, phage displaying βLII were enriched over irrelevant phage displaying an inactive TEM-1 β-lactamase mutant [25] by a factor of up to 185. A library of 5×10^6 mutants was then generated by error-prone PCR. The mean activity of the library was 1.6% of the activity of the wild type. After two rounds of selection, 11 clones were randomly picked, mono-clonal enzyme–phage particles were prepared, and their activities were measured. The specificity constants, k_{cat}/K_M, ranged from 20% to 170% of the wild-type value. Further on, the thermostability of the selected clones was characterized. The selected variants had two to four mutations, all of which were distant from the active site. All the mutants were at least six times less stable than the wild-type enzyme on phage.

The method described above allowed the isolation of active mutants from a rep-ertoire of enzymes. However, the activity of the best mutant was not significantly higher than that of the wild-type enzyme. Better results may be achieved by choos-ing another strategy to generate the library (e.g., mutating defined regions of the enzyme) or by constructing a larger library.

In another selection method, the substrate of a reaction was fused to the enzyme itself [31], instead of being immobilized on a pIII coat protein neighboring the one displaying the enzyme [29]. Subtiligase, a double mutant of subtilisin BPN' [32] that catalyzes the ligation of peptides, was displayed on phage. To obtain an optimized subtiligase active site, 25 active site residues were randomly mutated in groups of four or five, yielding six different libraries with more than 10^9 individual members each. Mutants were selected according to their ability to ligate a biotinylated peptide onto their own extended N-terminus. The functional enzyme–phage particles were isolated by capture on immobilized neutravidin. Prior to selection experiments, it was demonstrated that ligation was occurring intramolecularly. By mixing active subtiligase phage lacking an N-terminal extension with a phage displaying catalyti-cally inactive subtiligase containing the N-terminal extension, it was shown that the biotinylated peptide was not ligated onto any phage.

After five to seven rounds of selection with each library, several clones were assayed for their catalytic activity, and the most active were subcloned into parent subtiligase that did not contain the N-terminal extension. Some of the selected mutants had increased ligase activity (about twofold). However, the most frequently isolated mutant showed a lower ligase activity than subtiligase, but display was improved by a factor of 10. Other mutations yielded mutants with better oxidative resistance. These results reflect that the selection method favored not only subtiligase variants with improved catalytic activity but also mutants with other properties improving functional enzyme display.

In a further experimental strategy, proximity coupling was used to retain the product of the enzymatic reaction linked to the enzyme–phage particle responsible for it [33]. The strategy could be classified as intramolecular, single-turnover, or intermolecular, multiple-turnover selection. It involves two chemically independent reactions. A catalytic reaction that leads to the product and a chemical cross-linking reaction in which substrate (and eventually the product) is linked to the phage. There are two avenues to a phage labeled with tagged product. First, the substrate reacts with the phage coat and then is converted into the product by the enzyme. Second, the substrate is converted into the product before the cross-linking reaction takes place. The first case would correspond to an intramolecular, single-turnover selection, whereas the latter would be consistent with an intermolecular, multiple-turnover selection. The order of the reactions will depend on the relative rates of the two reactions.

In this study, it was assumed that cis reactions would be favored over reactions in trans by proximity effects [33], thereby allowing the labeling of only those phage responsible for product formation. For the cross-linking reaction, maleimides were chosen because they are known to react with thiols and, in alkaline solutions, with amino groups. To test the approach, the Klenow and Stoffel fragments of DNA polymerase I from *E. coli* and *Thermus aquaticus* were displayed on phage as fusion to pIII. An oligonucleotide primer with a maleimidyl group at its 5′ end was used as substrate. The product was tagged by addition of biotinylated dUTP to the 3′ end of the primer by the catalytic action of the polymerase (Figure 14.5).

The display of the polymerases was difficult. It was estimated that only one in a thousand phage particles displayed a fusion protein [33]. In selection experiments, phage displaying Stoffel fragments could be enriched by a factor of 26 over Klenow phage when they were incubated at 60°C during the reaction. This reflects the higher thermostability of the Taq-polymerase-derived fragment compared to the Klenow fragment. Further selection experiments with Stoffel phage with different catalytic activities were performed. Enrichment factors of up to 123 in favor of the more active Stoffel fragment were achieved. Thus, proximity coupling could be used as a general strategy for the selection of catalytic activities from large repertoires. However, there are some points to be discussed.

As a first problem, products could diffuse to another phage that displays an inactive enzyme and react with the phage coat proteins. This would lead to the isolation of false positives and reduce the efficacy of the selection system. The higher the proficiency

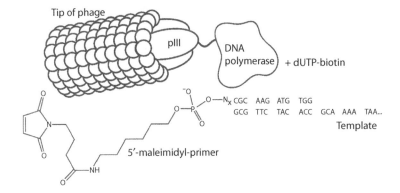

Figure 14.5   Selection of catalytically active DNA polymerase by phage display and proximity coupling. By the extension of the substrate primer with biotinylated dUTP in a template-dependent manner, the product of the reaction was tagged for selection. The maleimidyl group reacted with the phage coat or the enzyme, thereby linking the product of the phenotype to the corresponding genotype.

of the active phage enzymes, the more product molecules will probably be found on inactive enzyme–phage particles. In the work described above [33], only a small fraction of phage displayed polymerase. Phage exclusively carrying helper phage-derived pIII were treated with trypsin, so as to render them incapable of mediating infection [34]. Therefore, very few catalysts displayed on phage were surrounded by a large pool of noninfective phage, which possibly acted as a sink for the products diffusing away from the active enzyme–phage particles. The probability of a diffusing product being captured by an inactive phage enzyme would be low. In this way, the noninfective phage could prevent the isolation of infective phage displaying inactive enzymes. However, the recovery of the active enzyme–phage particles could be more efficient if all product molecules were retained on the phage responsible for their formation.

It was suggested for bimolecular condensation reactions to immobilize the reactive substrate before addition of the tagging compound [33]. But prelabeling of the enzyme–phage particles could possibly prevent most substrate molecules from reaching the active site of the enzyme displayed on pIII due to the linker length or sterical hindrance. A further possibility to prevent the labeling of inactive enzyme–phage particles with product could be the use of compartmentalization strategies [35].

Targeted immobilization of the substrate on phage, as described above for the peptides forming a heterodimeric coiled-coil complex [29], was also achieved by inserting calmodulin between pIII and the enzyme [36]. Calmodulin was functionally displayed on phage as pIII–calmodulin–enzyme fusion with different enzymes as fusion partners. Calmodulin binds very tightly to the peptide CAAARWKKAFIAVSAANRFKKIS in a calcium-dependent manner [37]. By fusing the substrate to the calmodulin-binding peptide, the substrate was kept in the vicinity of the enzyme displayed on phage. Catalytically active enzyme–phage particles were able to catalyze the formation of the product. Because of the linkage between product and calmodulin, it was possible to isolate phage displaying the active enzyme using an affinity reagent against the product. As the interaction between peptide and calmodulin is calcium dependent, captured phage were eluted by the addition of a calcium chelator (Figure 14.6).

Model selections were performed with three enzymes: submut (mutant of subtilisin from *Bacillus subtilis*) [36], biotin ligase (BirA) from *E. coli* [38], and the rat endopeptidase trypsin (His57Ala mutant) [38]. BirA catalyzes the biotinylation of a lysine residue in a 13-mer peptide. The H57A mutant of trypsin cleaves the sequence GGHR/DYKDE, whereas submut catalyzes the hydrolysis of AAHY/DYKDE. To adapt the reaction substrates to the selection scheme described above, the substrate peptides for the three reactions were fused to the calmodulin-binding peptide. Phage displaying submut were enriched over phage displaying an irrelevant enzyme by a factor of 54 [36]. In model selection experiments, phage displaying either BirA or trypsin H57A were isolated due to their catalytic activity. In the case of BirA, the phage titers recovered from the selection were 4–800 times lower when either the biotin–acceptor peptide or the biotin was omitted. When substrate peptides containing the wrong cleavage site were added to the calmodulin-tagged phage displaying the H57A mutant of trypsin, the phage titers recovered from the selection experiment were 15–2000 times lower compared to the one with the correct substrate peptide [38].

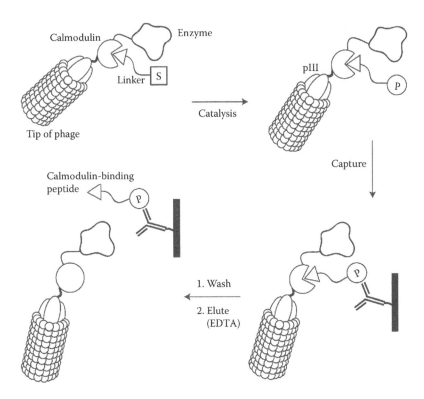

Figure 14.6 Calmodulin-tagged phage enzyme for the selection of enzymatic activity. The substrate was noncovalently tethered to the enzyme–phage particle. Upon catalysis, the reaction product was used as an affinity tag for the isolation of the corresponding phage. The calmodulin–peptide complex was dissociated by addition of a calcium chelator.

Since the results from the model selections were promising, a library with more than 5×10^5 trypsin mutants was constructed. The aim was to isolate a H57A trypsin variant with identical substrate specificity, but improved catalytic activity. After three to four rounds of panning, 30 trypsin mutants were screened for their endoproteolytic activity. None was found to show an endopeptidase activity similar to H57A trypsin. H57A trypsin was also present in the library but was not recovered. This result suggests that the selection conditions applied were probably not stringent enough. Because the substrate peptide is tethered adjacent to the enzyme, the effective substrate concentration was about 10^{-3} M. In order to cleave one peptide at this concentration during a time period that starts with the addition of substrate peptides and ends with the wash during affinity purification of the phage, only very moderate catalytic activity is required.

Intramolecular, single-turnover selections have two weaknesses. First, an enzyme–phage particle is only enriched over nonactive counterparts if the enzyme can accelerate product formation over the rate of the uncatalyzed reaction. The tight attachment of substrate near the enzyme leads to high effective substrate concentrations. If it is assumed that one substrate molecule is present in a sphere with the enzyme as its center and that the diameter of this sphere is 10 nm, the effective concentration of the substrate can then be calculated to be approximately 3 mM. Thus, it is difficult to select for lower K_M values. Irrespective of whether an enzyme mutant had a K_M of either 10^{-4} or 10^{-7} M, the enzyme would be saturated with substrate in both cases, and therefore, the formation of product would occur with comparable rates.

Second, selection pressure is for rate acceleration but not for product release and turnover. However, high turnover number is the prerequisite for an efficient catalyst, because one catalyst should be able to catalyze the formation of many product molecules. In intramolecular, single-turnover selection, it only is possible to select for higher turnover by shortending the reaction time. For a reaction with a k_{cat} of 1 s^{-1}, the half-life is 0.7 s. This value is reduced to 0.007 s for a k_{cat} of 100 s^{-1}. In practice, such differences in reaction time are difficult, if not impossible, to resolve.

On the other hand, the two disadvantages mentioned above are also the strength of intramolecular, single-turnover reactions. The high effective substrate concentration and the requirement for the formation of only one product molecule leads to the possibility that proteins with very modest catalytic activities could be selected. This could make the intramolecular, single-turnover approach especially useful when evolving catalysts de novo.

14.2.2.2 Intermolecular, Multiple-Turnover Selections

In contrast to the selection methods described so far, intermolecular, multiple-turnover selection strategies should allow the selection pressure for all parameters and steps of the catalytic cycle, which are important to be optimized to obtain an efficient enzyme. These parameters and steps are substrate–enzyme interaction (K_M), rate acceleration (k_{cat}/k_{uncat}), product release, and turnover (k_{cat}).

In order to select for product release and turnover, the enzyme to be selected should be forced to process a high number of substrate molecules within a given reaction time. For example, the selection can be based on the number of product molecules. The more product molecules are formed, the more likely it should be for an enzyme–phage particle to be selected. With a k_{cat} of 100 s^{-1}, an enzyme would yield 100 times more product molecules than another enzyme with a k_{cat} of only 1 s^{-1}, therefore increasing significantly the probability of the enzyme–phage particle to pass the selection barrier.

As already mentioned, the work of Jestin et al. [33] can be seen as either an intramolecular or an intermolecular approach, depending on which of the two reactions—cross-linking or primer elongation—is faster. To our knowledge, only one other study has been published so far that uses intermolecular, multiple-turnover selections combined with phage display [23].

In this study [23], catalytic antibodies were raised by immunization against a hapten mimicking the transition state of glycosidic bond cleavage. Approximately 100 clones that bound the TSA were generated by established hybridoma technology [39]. The heavy and light chains of these antibodies were shuffled to yield a library of about 10^4 individual phage-Fab fragments. A substrate molecule containing a difluoromethylphenol moiety was designed. Upon glycosidic action of the displayed Fab fragment on this substrate, the difluoromethylphenol moiety generated a reactive quinone methide species at or near the active site. Any neighboring nucleophile can be alkylated by this reactive intermediate, thereby potentially attaching the substrate molecule right after cleavage of the glycosidic bond covalently to the phage–Fab complex. By coupling the substrate to a solid support or to biotin prior to enzymatic treatment, phage immobilized or covalently labeled with the reaction product were selected.

In order to test this trapping strategy, substrate containing the difluoromethylphenol moiety next to the glycosidic bond to be cleaved was biotinylated. The substrate was incubated with β-galactosidase and was found to act as substrate-alkylating agent of this enzyme. Biotinylated β-galactosidase could be trapped on streptavidin-coated ELISA plates. Although the enzyme was biotinylated, it retained its catalytic activity. This nondestructive trapping of the enzyme is important because it may allow selecting clones with the ability to undergo multiple turnovers.

After four rounds of selection with the phage displaying the library of heavy and light chains, the best Fab fragment exhibited a rate acceleration (k_{cat}/k_{uncat}) of 7×10^4 for the hydrolysis of p-nitrophenyl-β-galactopyranoside ($k_{cat} = 0.0001$ s^{-1} and $K_M = 0.53$ mM) (Figure 14.7). This Fab fragment was compared with antibodies obtained from simple hybridoma screening. Twenty-two monoclonal antibodies

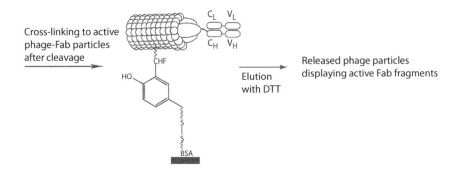

Figure 14.7 Procedure for the mechanism-based immobilization of antibody phage particles that catalyzed the hydrolysis of a galactopyranoside substrate. (Adapted from Janda, K.D. et al., *Science*, 275, 1997.)

were assayed for their catalytic activity, and the best hydrolyzed p-nitrophenyl-β-galactopyranoside with a rate acceleration of only 10^2. Therefore, by shuffling the heavy and light chains of only 100 antibodies, an increase of catalytic activity by 700-fold was achieved. Although mechanism-based inhibitors and suicide substrates have been described for a number of reactions, it will be difficult to generally convert this information into trapping compounds, which do not disturb the performance of the catalyst after their covalent linkage near the active site. The design of such sophisticated trapping substrates is even more complicated for reactions where only little is known about their mechanism.

The authors did not investigate whether the Fab displayed on phage were able to cleave several gylcosidic bonds. From the experiments performed with soluble β-galactosidase, it can be assumed that several turnovers should be feasible. However, it is possible that the active sites of the Fab fragments become more and more blocked with increasing number of cleaved glycosidic bonds. Thus, it remains to be seen whether this approach will be useful for the selection of enzymes exhibiting high turnover numbers.

14.3 DISCUSSION

14.3.1 Selection for Catalysis: Comparison of Phage Display with Other Methods

A number of systems to link a protein phenotype to the corresponding genotype have been reported. The linkage has been achieved by several methods such as phage display [14], peptides displayed as lac repressor fusions ("peptides on plasmids") [40], and cell-surface display systems [6]. All of these methods include a step of transformation of living cells, and therefore, the library sizes are limited to 10^9–10^{10} individual members. However, larger libraries are important to obtain access to a maximum of different mutants. Therefore, it would be desirable to perform directed evolution with libraries as large as possible. In order to increase library size, a number of in vitro selection systems based on cell-free transcription and translation have been developed. The in vitro translated polypeptide was coupled to its encoding messenger RNA (mRNA) in a ribosome complex [41] or through a puromycin derivative [42,43]. In addition, in vitro expressed proteins were fused to their encoding DNA through different strategies [44–46]. In vitro compartmentalization was used as a third class of genotype–phenotype linkage [35,47]. Of all these methods to link genotype and phenotype, only a few have been used to select for catalysis: cell-surface display [6], phage display [14], and in vitro compartmentalization [35,47].

Phage display has some advantages and disadvantages compared to other protein selection methods. An important strength of phage display in selections for catalysis is the relative robustness of the phage particles under a variety of conditions. It may be necessary to evolve a catalyst at low or high pH values, at high substrate concentrations (which are toxic for living cells), in high salt solutions, or at moderately elevated temperatures. In every one of these situations, cell-surface display would not be the system of choice, because the cells would most probably die during the selection procedure. Similarly, due to the sensitivity of RNA to degradation, RNA–peptide fusions or ribosome display cannot be an alternative to phage display. A compartmentalized system as proposed by Tawfik and Griffiths [35] is not well suited either, because the conditions for the enzymatic reaction must be compatible with the requirements for efficient in vitro transcription and translation. In order to evolve enzymes under nonphysiologic conditions, the expression of phenotype and coupling to its genotype must be separated temporally and spatially from the selection procedure. This is an important requirement for all in vitro systems. Phage display takes place partially in vivo and in vitro and therefore fulfills this criterion. An additional advantage of phage display is the ease with which enzyme–phage particles can be purified. This is an important practical aspect because it may be necessary to remove excess substrate or product molecules, which are not immobilized on the enzyme–phage particle before starting the capture of the phage tagged with product.

Covalent DNA–protein fusions are a very interesting alternative to phage display. After compartmentalized in vitro expression, they could be purified and then be used in the selection step under a large variety of conditions because of the robustness of DNA. In addition, through DNA–protein fusions produced in vitro, a much larger sequence space becomes accessible than with phage display because larger libraries can be created.

One of the biggest problems of phage is the inefficiency of protein display on the phage surface [33,36,38]. By the use of phage vectors instead of phagemids, display was improved. Better display correlated with increased efficiency of selection protocols [38]. Several attempts have been undertaken to improve display of proteins on phage. Mutants of pVIII coat proteins have been isolated that increased display on pVIII [17], a helper phage was developed that does not provide wild-type pIII after superinfection [48], and signal peptides were selected from a library to enhance display of a DNA polymerase fragment on phage [49].

14.3.2 Possible Improved Selection Systems Based on Phage Display

Intermolecular, multiple-turnover selections without using phage display have been reported. These selection systems take place fully [11] or partially in vivo [47], or in vitro [35]. The use of compartments to retain the product molecules in the vicinity of the genotype responsible for their formation is common to all three systems. The compartments used are cells [11], cells in combination with reversed micelles [47], or reversed micelles [35]. It might be possible to develop a selection system where enzyme–phage particles together with substrate are enclosed in a reversed micelle, which acts as a cell-like compartment. Depending on the substrate concentration, the activity of the enzyme variant in the compartment, the time given to the enzyme for the formation of product, and the size of the compartment (which influences the effective concentration of the enzyme in the reversed micelle), different amounts of substrate will be converted into product. If the product molecules could then be immobilized on the phage at a given time point before the water-in-oil emulsion is broken, enzyme–phage particles labeled with product molecules could be selected by affinity purification on a product-binding ligand. By applying stringent washes during affinity purification, phage labeled with many product molecules on their surface would be favored due to avidity effects. Such a selection scheme may allow high selection pressure for all the parameters that need to be optimized to obtain an efficient catalyst.

14.3.3 Implications for Biotechnology and Drug Discovery

Enzymes are becoming more and more important in biotechnology and thera-peutic applications. In chemical and pharmaceutical industries, enzymes and micro-organisms are used for the manufacture of a number of fine chemicals and drugs. In 1990, among the top 50 best-selling pharmaceuticals, 10 were produced from low-molecular-weight fermentation [50]. Industrially significant processes catalyzed by enzymes are starch hydrolysis and production of β-lactam antibiotics [50]. In medicine, enzymes are used in diverse fields including antibody-dependent prodrug therapy (ADEPT), therapy of metabolic diseases [2,3], and leukemia treatment [51].

However, as already mentioned at the beginning of this chapter, enzymes often do not meet all the requirements for the applications for which they are intended to be used. Therefore, technologies must be developed to engineer tailor-made biocatalysts. Several attempts have been made to achieve this goal, of which in vivo systems based on auxotrophic complementation [11] and reactive immunizations [26] were the most successful. But the former of these systems are limited to special cases where product formation is coupled to survival, and the latter to catalytic antibodies, which exhibit only moderate-rate accelerations. So far, in contrast to the engineering of antibodies, directed evolution using phage display or any other method has not led to proficient enzymes that are used in medicine or biotechnology. This may be due to the lack of powerful and general selection methods. Furthermore, the choice of the residues to be mutated is crucial for the success of the selection experiment. It is difficult to predict which of the systems described in this chapter will be important in the future, but we believe that an efficient selection procedure should be feasible in vitro and should make use of compartmentalization to retain reaction products close to the genotype encoding the enzyme (intermolecular, multiple-turnover selec-tions). Strategies must be found for the targeted immobilization of products on the genotype at any desirable time point. In addition, protocols must be devel-oped for the improved display of proteins.

We anticipate that the development of efficient in vitro selection strategies (based on phage display or not) will make it possible to isolate improved enzymes for a variety of applications. In medicine, for example, human nonimmunogenic proteases could be used as molecular knives to cleave cell-surface receptors or other proteins important for the survival of tumor cells. In industry, engineered enzymes may help to synthesize fine chemicals with less side products and lower energy input.

REFERENCES

1. Radzicka A, Wolfenden R. A proficient enzyme. *Science* 1995; 267:90–93.
2. Bhatia J, Sharma SK, Chester KA, Pedley RB, Boden RW, Read DA, Boxer GM et al. Catalytic activity of an in vivo tumor targeted anti-CEA scFv::carboxypeptidase G2 fusion protein. *Int J Cancer* 2000; 85(4):571–577.
3. Russell CS, Clarke LA. Recombinant proteins for genetic disease. *Clin Genet* 1999; 55(6):389–394.
4. Buchholz K, Kasche V. Reaktoren und prozesstechnik. In: Buchholz K, Kasche V, eds. *Biokatalysatoren und Enzymtechnologie*. Weinheim, New York, Basel, Cambridge, Tokyo: VCH Verlagsgesellschaft, 1997:273–279.
5. Border ET, Midelfort KS, Wittrup KD. Directed evolution of antibody fragments with monovalent femtomolar antigen-binding affinity. *Proc Natl Acad Sci U S A* 2000; 97:2029–2034.
6. Olsen MJ, Stephens D, Griffiths D, Daugherty P, Georgiou G, Iverson BL. Function-based isolation of novel enzymes from a large library. *Nat Biotechnol* 2000; 18:1071–1074.
7. Santoro SW, Schultz PG. Directed evolution of the site specificity of Cre recombinase. *Proc Natl Acad Sci U S A* 2002; 99:4185–4190.
8. Chapman KB, Szostak JW. In vitro selection of catalytic RNAs. *Curr Opin Struct Biol* 1994; 4:618–622.
9. Breaker RR. Engineered allosteric ribozymes as biosensor components. *Curr Opin Biotechnol* 2002; 13(1):31–39.
10. Griffiths AD, Tawfik DS. Man-made enzymes—From design to in vitro compartmentalization. *Curr Opin Biotechnol* 2000; 11:338–353.
11. Altamirano MM, Blackburn JM, Aguayo C, Fersht AR. Directed evolution of new catalytic activity using the alpha/beta-barrel scaffold. *Nature* 2000; 403:617–622.
12. Patten PA, Gray NS, Yang PL, Marks CB, Wedemayer GJ, Boniface JJ, Stevens RC et al. The immunological evolution of catalysis. *Science* 1996; 271:1086–1091.
13. Barbas CF, Heine A, Zhong G, Hoffmann T, Gramatikowa S, Bjornestedt R, List B et al. Immune versus natural selection: Antibody aldolases with enzymatic rates but broader scope. *Science* 1997; 278:2085–2092.
14. Smith GP. Filamentous fusion phage: Novel expression vectors that display cloned antigens on the virion surface. *Science* 1985; 228:1315–1317.
15. Hufton SE, Moerkerk PT, Meulemanns EV, de Bruine A, Arends JW, Hoogenboom HR. Phage display of cDNA repertoires: The pVI display system and its applications for the selection of immunogenic ligands. *J Immunol Methods* 1999; 231:39–51.
16. Malik P, Terry TD, Gowda LR, Langara A, Petukhov SA, Symmons MF, Welsh LC et al. Role of capsid structure and membrane protein processing in determining the size and copy number of peptides displayed on the major coat protein of filamentous bacteriophage. *J Mol Biol* 1996; 260:9–21.
17. Sidhu SS, Weiss GA, Wells JA. High copy display of large proteins on phage for functional selections. *J Mol Biol* 2000; 296:487–495.
18. Reymond JL. Catalytic antibodies for organic synthesis. In: Fessner WD, ed. *Topics in Current Chemistry*. Berlin, Heidelberg: Springer Verlag, 1999:59–93.
19. Baca M, Scanlan TS, Stephenson RC, Wells JA. Phage display of a catalytic antibody to optimize affinity for transition-state analog binding. *Proc Natl Acad Sci U S A* 1997; 94:10063–10068.

20. Arkin MR, Wells JA. Probing the importance of second sphere residues in an esterolytic antibody by phage display. *J Mol Biol* 1998; 284:1083–1094.
21. Fersht A. Chemical catalysis. In: Fersht A, ed. *Structure and Mechanism in Protein Science*. New York: W.H. Freeman and Company, 1999:60.
22. Tawfik DS, Green BS, Chap R, Sela M, Eshhar Z. catELISA: A facile general route to catalytic antibodies. *Proc Natl Acad Sci U S A* 1993; 90:373–377.
23. Janda KD, Lo LC, Lo CHL, Sim MM, Wang R, Wong CH, Lerner RA. Chemical selection for catalysis in combinatorial antibody libraries. *Science* 1997; 275:945–948.
24. Legendre D, Laraki N, Gräslund T, Bjornvard ME, Bouchet M, Nygren P, Borchert TV et al. Display of active subtilisin 309 on phage: Analysis of parameters influencing the selection of subtilisin variants with changed substrate specificity from libraries using phosphonylating inhibitors. *J Mol Biol* 2000; 296:87–102.
25. Soumillion P, Jespers L, Bouchet M, Marchand-Brynaert J, Winter G, Fastrez J. Selection of beta-lactamase on filamentous bacteriophage by catalysis activity. *J Mol Biol* 1994; 237:415–422.
26. Zhong GF, Lerner RA, Barbas CF. Broadening the aldolase catalytic antibody repertoire by combining reactive immunization and transtition state theory: New enantio- and diastereoselectivities. *Angew Chem Int Ed* 1999; 38:3738–3741.
27. Fersht A. Measurement and magnitude of individual rate constants. In: Fersht A, ed. *Structure and Mechanism in Protein Science*. New York: W.H. Freeman and Company, 1999:164–168.
28. Fersht A. Enzyme-substrate complementarity and the use of binding energy in catalysis. In: Fersht A, ed. *Structure and Mechanism in Protein Science*. New York: W.H. Freeman and Company, 1999:364–368.
29. Pedersen H, Hölder S, Sutherlin DP, Schwitter U, King DS, Schultz PG. A method for directed evolution and functional cloning of enzymes. *Proc Natl Acad Sci U S A* 1998; 95:10523–10528.
30. Ponsard I, Galleni M, Soumillion P, Fastrez J. Selection of metalloenzymes by catalytic activity using phage display and catalytic elution. *Chem Biochem* 2001; 2:253–259.
31. Atwell S, Wells JA. Selection for improved subtiligases by phage display. *Proc Natl Acad Sci U S A* 1999; 96:9497–9502.
32. Abrahmsen L, Tom J, Burnier J, Butcher KA, Kossiakoff A, Wells JA. Engineering subtilisin and its substrates for efficient ligation of peptide bonds in aequeous solution. *Biochemistry* 1991; 30:4151–4159.
33. Jestin JL, Kristensen P, Winter G. A method for the selection of catalytic activity using phage display and proximity coupling. *Angew Chem Ind Ed* 1999; 38:1124–1127.
34. Kristensen P, Winter G. Proteolytic selection for protein folding using filamentous bacteriophages. *Fold Des* 1998; 3:321–328.
35. Tawfik DS, Griffiths AD. Man-made cell-like compartments for molecular evolution. *Nat Biotechnol* 1998; 16:652–656.
36. Demartis S, Huber A, Viti F, Lozzi L, Giovannoni L, Neri P, Winter G et al. A strategy for the isolation of catalytic activities from repertoires of enzymes displayed on phage. *J Mol Biol* 1999; 286:617–633.
37. Montigiani S, Neri G, Neri P, Neri D. Alanine substitutions in calmodulin-binging peptides result in unexpected affinity enhancement. *J Mol Biol* 1996; 258:6–13.
38. Heinis C, Huber A, Demartis S, Bertschinger J, Melkko S, Lozzi L, Neri P et al. Selection of catalytically active biotin ligase and trypsin mutants by phage display. *Protein Eng* 2001; 14:1043–1052.

39. Kohler G, Milstein C. Continuous cultures of fused cells secreting antibody of pre-defined specificity. *Nature* 1975; 256:495–497.

40. Cull MG, Miller JF, Schatz PJ. Screening for receptor ligands using large libraries of peptides linked to the C terminus of the lac repressor. *Proc Natl Acad Sci U S A* 1992; 89:1865–1869.

41. Hanes J, Schaffitzel C, Knappik A, Pluckthun A. Picomolar affinity antibodies from a fully synthetic native library selected and evolved by ribosome display. *Nat Biotechnol* 2000; 18:1287–1292.

42. Nemoto N, Miyamato-Sato E, Husimi Y, Yanagawa H. In vitro virus: Bonding of mRNA bearing puromycin at the 3′-terminal end to the C-terminal end of its encoded protein on the ribosome in vitro. *FEBS Lett* 1997; 414:405–408.

43. Wilson DS, Keefe AD, Szostak JW. The use of mRNA display to select high-affinity protein-binding peptides. *Proc Natl Acad Sci U S A* 2001; 98:3750–3755.

44. Doi N, Yanagawa H. STABLE: Protein-DNA fusion system for screening of combinatorial protein libraries in vitro. *FEBS Lett* 1999; 457:227–230.

45. FitzGerald K. In vitro display technologies—New tools for drug discovery. *Drug Discov Today* 2000; 5:253–258.

46. Kurz M, Gu K, Al-Gawari A, Lohse PA. cDNA-protein fusions: Covalent protein-gene conjugates for the in vitro selection of peptides and proteins. *Chem Biochem* 2001; 2:666–672.

47. Ghadessy FJ, Ong JL, Holliger P. Directed evolution of polymerase function by compartmentalized self-replication. *Proc Natl Acad Sci U S A* 2001; 98:4552–4557.

48. Rondot S, Koch J, Breitling F, Dubel S. A helper phage to improve single-chain antibody presentation in phage display. *Nat Biotechnol* 2001; 19:75–78.

49. Jestin JL, Volioti G, Winter G. Improving the display of proteins on filamentous phage. *Res Microbiol* 2001; 152:187–191.

50. The application of biocatalysis to the manufacture of fine chemicals. In: Roberts SM, Turner NJ, Willetts AJ, Turner MK, eds. *Introduction to Biocatalysis Using Enzymes and Microorganisms.* Cambridge, UK: Cambridge University Press, 1995:140–187.

51. Ertel IJ, Nesbit ME, Hammond D, Weiner J, Sather H. Effective dose of L-asparaginase for induction of remission in previously treated children with acute lymphocytic leukemia: A report from Childrens Cancer Study Group. *Cancer Res* 1979; 39:3893–3896.

CHAPTER 15

Antibody Humanization and Affinity Maturation Using Phage Display

Jonathan S. Marvin and Henry B. Lowman

CONTENTS

15.1 INTRODUCTION

Three significant developments in molecular biology have been central in promoting the growing role of antibodies as tools for biochemical and biological research and as a significant class of molecules for drug development. First, following the establishment of reliable methods for the generation of high-affinity, high-specificity monoclonal antibodies from the natural immune response of mice and other species [1], recombinant DNA techniques have made it possible to isolate the corresponding complementary DNA (cDNA) encoding these antibodies, determine their sequences, and with the help of structural information, optimize their antigen-binding properties through manipulation of their genes. Second, combinatorial approaches to protein engineering such as phage display [2] have made it possible to explore many more variants of antibodies than would typically be accessible in a site-directed mutagenesis approach. Third, the elucidation of high-resolution molecular structures of antibodies has revealed some of the general features required for structural stability and antigen recognition [3,4]. These approaches provide not only a means for the rapid optimization of antigen-binding properties of existing antibodies as discussed in this chapter but also a means for the discovery of novel antibody specificities from immunized, nonimmunized (naïve), or synthetic sources of diversity. Humanization of nonhuman antibodies and affinity maturation of antibodies from hybridoma or diversity-library sources are two areas of antibody optimization that can often be most efficiently addressed using antibody–phage display.

In general, the display of antibodies on phage is similar to that described for other proteins. Polyvalent display may be useful for the identification of antigen-binding antibodies from naïve antibody–phage libraries [5]. However, monovalent display may offer advantages by reducing avidity effects during antigen-binding selections [6,7]. Two preferred formats have been used for the display of antibody variable domains on phage, often in the form of a phagemid vector [7–9]: (1) Fab-phage, in which the variable (VH) and first constant (CH1) domains of the heavy chain are fused to a portion of the gene *III* protein (gIIIp) of filamentous phage, while the variable (VL) and constant (CL) domains of the light chain are expressed as a soluble chain from the same vector, resulting in expression of a complete antigen-binding fragment (Fab), and (2) "single-chain" variable fragment (scFv)-phage, in which the VL and VH domains are linked and fused to a portion of the gIIIp as a single polypeptide, resulting in expression of an scFv. An advantage of the former method may be to preserve a monomeric form of the antibody for more efficient binding selections (Section 15.3.7), while an advantage of the latter can be the formation of diabodies [10], which can improve the avidity (apparent affinity) under appropriate conditions, leading to more efficient recovery of low-affinity binders [6,7].

In the first part of this chapter, we describe the application of phage display to humanization, often a key step in the process of converting monoclonal antibodies from nonhuman species to pharmaceutical drug candidates. The need for humanization was highlighted by early clinical experiences with murine and chimeric (i.e., containing murine variable and human constant domains) antibodies. Immunogenic responses in humans to antibodies originating in other species have the potential to render a therapeutic antibody ineffective after initial dosing and, worse, may give rise to potentially fatal hypersensitivity reactions (anaphylaxis). Transplanting the entire variable domains to an antibody scaffold of human constant domains, producing a chimeric antibody (Figure 15.1), does not remove the potential for antiframework antibodies. For example, when administered to humans, the murine antibody OKT3 gave rise to human antimurine antibodies, some of which were directed to the

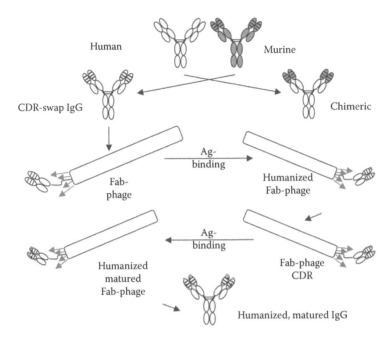

Figure 15.1 Strategies for using phage display in the humanization and affinity maturation of antibodies. Murine-derived antibody regions are shown as shaded domains; human antibody domains are shown as open domains. Murine residues selected from framework region (FR) libraries are depicted as gray dots, and optimized mutations selected from complementarity-determining region (CDR) libraries are depicted as black dots. The six CDRs are indicated as six lines within the variable domains (VH and VL) of the antibody.

variable region [11], which is also of murine origin in chimeric antibodies (Figure 15.1). A rat antihuman antibody directed to the CAMPATH-1 antigen also proved to be immunogenic in humans, and this was ultimately overcome through humanization [12,13]. Although transgenic mice producing human immunoglobulins are also a source of monoclonal antibodies for therapeutic development, humanization remains an important antibody engineering endeavor because of the remarkable and differing specificities of antibodies derived from various sources. Such differences may translate into dramatic differences in biological activity for molecules being considered for drug development. Indeed, antibodies derived by different means may often differ in their fine specificity for epitopes on a given antigen and hence in their biological effects, such as blocking an antigen from a binding receptor or cross-linking a cell-surface receptor to induce agonist activity; see, for example, Refs. [14] and [15]. Humanization by phage display represents a special case of in vitro affinity maturation in which only selected, often nonantigen-contacting residues are changed in order to optimize activity, while presenting the essential antigen-binding epitope (or paratope) in the context of the human framework and constant regions. Humanized antibodies derived from parental murine hybridomas such as anti-HER2

[16], anti-IgE [17], anti-CD11a [18], and anti-VEGF in both IgG form [19] and an affinity-matured Fab [20] are demonstrating success in clinical trials and as marketed therapeutics.

In the second part of this chapter, we discuss affinity maturation, which, together with humanization, can be seen as part of a unified process of antibody optimization (Figure 15.1). Antibodies derived from human, nonhuman, or synthetic sources may benefit from optimization for antigen binding through substitutions at positions within and outside the set of antigen-contacting positions. This is often termed "in vitro affinity maturation" by analogy to the process of natural affinity maturation, including recombination and somatic hypermutation, during maturation of the adaptive immune response in vertebrates. As applied to therapeutic antibodies, the motivations for this type of antibody engineering may include the need to improve potency and efficacy. However, relevant to therapeutic as well as diagnostic, reagent, or industrial uses of antibodies, motivations for affinity maturation may also include considerations of cost of goods, manufacturing capacity, assay sensitivity, or catalytic efficiency. In the second part of this chapter, we consider approaches to affinity maturation for human, humanized, or nonhuman antibodies.

15.2 HUMANIZATION USING PHAGE DISPLAY

The humanization of antibodies derived from mouse or other species is an optimization process by which the essential antigen-binding characteristics of a parental monoclonal antibody (often a murine IgG) are transferred to a human "scaffold" antibody—that is, an immunoglobulin with constant (CL, CH1–CH2–CH3) and variable (VL, VH) framework regions (FRs) derived from human genes (Figure 15.1). The process typically involves replacing the six complementarity-determining regions (CDRs) of the human antibody with those from the parental antibody. Despite the homology of human and nonhuman (e.g., mouse or rat) FRs, this direct "CDR swap" does not always result in a functional antibody with the same antigen-binding affinity observed for the parental [21]. Structural studies of antibodies suggested that a few key residues within the FRs of the antibody variable domains can greatly influence the conformation of the CDR loops in a given scaffold. Indeed, appropriate substitution of parental residues for human residues at these positions can restore antigen binding and activity [3,12,22–24]. The humanization of antibodies through structure-based, site-specific mutagenesis has been described and reviewed elsewhere [25]. These efforts often require the investigation of a relatively small number of point mutations; however, since the effects of combinations of mutations in the framework are not always additive, many more variants may need to be tested in order to produce a humanized antibody with similar activity to the parental one. Phage display provides a tool for the facile generation of diversity libraries easily large enough to encompass many of the framework positions that have been found key to achieving humanization. Humanized antibodies can then be selected for binding to antigens from libraries randomized to contain the parental (nonhuman) or human residue at these positions (Figure 15.1).

15.2.1 Choice of Human Scaffold

Several approaches to the selection of scaffolds for humanization have been described. In general, these can be grouped as follows: (1) selection of a common variable-region scaffold, for example, VH subgroup III, VL kappa subgroup I for CDR grafting, and subsequent FR mutations [26,27]; (2) selection of an antibody-specific, human variable FR region with close sequence similarity to the original nonhuman antibody, followed by specific FR mutations [28]; and (3) construction of variable-region libraries (representing the diversity of human germ line VH and VL families) with additional synthetic diversity at key framework positions [29].

Other approaches have abandoned the selection of a core FR in favor of simply "resurfacing" the parental antibody with residues found at the surfaces of human antibodies, the idea being that solvent-exposed residues are likely to dominate immunogenic reactions [30]. Another interesting approach involves purely a domain-shuffling strategy in which the VH domain of a parental antibody is first randomly recombined with a human VL library and selected for antigen binding; thereafter, the resulting human VL domain is similarly shuffled with a human VH domain library [31]. While clearly successful in some cases, this approach may not be generally efficient at preservation of the parental antibody epitope [29], a clear motivation for humanization as opposed to discovery of novel antibodies.

Here, we focus on the common-scaffold approach to humanization, with randomization at a small set of framework residues to confer the necessary structural determinants for presentation of the CDRs to provide antigen recognition. This approach offers several advantages in the development of therapeutic antibodies [16]. First, the heavy-chain VH subgroup II and light-chain VL kappa subgroup I families are abundantly represented in the human repertoire. To be sure, antihuman antibodies have been found in humans, and even antihuman-CDR antibodies are known; however, immunogenicity may be minimized by selecting the most abundantly represented germ line families as humanization scaffolds. Clinical experience with anti-HER2, anti-IgE, anti-CD11a, and anti-VEGF antibodies has proven that they are indeed low in immunogenicity, supporting the use of this scaffold. The use of a single scaffold also facilitates humanization of new antibodies because new CDRs can be modeled onto the same starting scaffold. Finally, protein expression and formulation and clinical experience with a given scaffold in the context of multiple indications provide a growing database for improving the development of therapeutic antibodies with similar chemical and physical properties.

15.2.2 Design of Humanization Libraries

Because humanization can typically be achieved by exchanging the parental framework residues for residues found in the human scaffold, the diversity of framework libraries designed for humanization can, in principle, be limited to twofold diversity at each randomized position (Table 15.1)—encoding the parental residue or the human residue [27]. On the other hand, slightly more diverse randomization can yield alternative FR residues that also restore antigen binding [27].

Table 15.1 Typical Set of Limited-Diversity Codons for Phage-Displayed Antibody Framework Region (FR) Libraries in Antibody Humanization

Chain	Consensus FR	Codon	Degeneracy	Residues[a]
VL	M4	MTG	ATG, CTG	M, L
VL	F71	TYC	TTC, TAC	F, Y
VH	A24	RYC	GCT, ACT, GTC, ATC	A, T, V, I
VH	V37	RTC	GTC, ATC	V, I
VH	F67	NYC	ATC, GTC, CTC, TTC, ACC, GCC, CCC, TCC	I, V, L, F, T, A, P, S
VH	I69	WTC	ATC, TTC	I, F
VH	R71	CKC	CGC, CTC	R, L
VH	D73	RMC	GAC, ACC, AAC, GCC	D, T, N, A
VH	K75	RMG	AAG, GCG, ACG, GAG	K, A, T, E
VH	N76	ARC	AAC, AGC	N, S
VH	L78	SYG	CTG, GCG, GTG, CCG	L, A, V, P
VH	A93	DYG	ATG, GTG, TTG, TCG, ACT, GCG	M, A, V, L, S, T
VH	R94	ARG	AAG, AGG	R, K

[a] The illustrated randomization has 1.6×10^6 diversity and is similar to that described in the humanization of murine A4.6.1; however, other schemes are possible [27]. The murine FR residues found in A4.6.1 are underlined.

Antibody humanization using phage display was reported for the murine anti-human VEGF antibody A4.6.1 using the common-scaffold approach [27], and the methodology described for this antibody illustrates the basic steps in humanization using phage display. As the starting point for phage-displayed libraries, the CDRs of the murine antibody were inserted in place of the CDRs of the human consensus framework to produce the variant hu2.0 (Figure 15.2). The Fab form of this antibody was displayed monovalently on phage using a phagemid construct in which the heavy-chain VH–CH1 domains were linked to the C-terminal domain of gIIIp, and the light-chain VL–CL domains were expressed as a soluble protein from the same phagemid. Compared to the murine–human chimeric form of A4.6.1, the CDR swap humanization variant was at least 4000-fold reduced in binding affinity to VEGF [27]. Following VEGF-binding selections with a library selectively randomized at 13 framework positions, a humanized variant (called hu2.10) was selected for binding to VEGF with affinity about sixfold weaker than that of the chimeric A4.6.1 antibody. Compared to the original CDR swap construct, hu2.10 contained selected substitutions (including two that were neither of human nor murine origin) at only eight framework positions: VL 71 and VH 37, 71, 73, 75, 76, 78, and 94; see Figure 15.2.

15.2.3 Further Optimization of Humanized Antibodies

While a framework library approach was successful in generating >400-fold improvement in antigen-binding affinity as compared with a simple CDR swap [27], higher-affinity variants of humanized anti-VEGF A4.6.1 were identified through

VL region

```
              1                    23            35            49
A461     DIQMTQTTSSLSASLGDRVIISC[SASQDISNYLN]WYQQKPDGTVKVLIY

human    DIQMTQSPSSLSASVGDRVTITC[RASQSISNYLA]WYQQKPGKAPKLLIY

hu2.0    DIQMTQSPSSLSASVGDRVTITC[SASQDISNYLN]WYQQKPGKAPKLLIY

hu2.10   DIQMTQSPSSLSASVGDRVTITC[SASQDISNYLN]WYQQKPGKAPKLLIY

Fab-12   DIQMTQSPSSLSASVGDRVTITC[SASQDISNYLN]WYQQKPGKAPKVLIY

Y0317    DIQLTQSPSSLSASVGDRVTITC[SASQDISNYLN]WYQQKPGKAPKVLIY

TKKT     DIQLTQSPSSLSASVGDRVTITC[SATKKITNYLN]WYQQKPGKAPKVLIY

              57                                 88
A461     [FTSSLHS]GVPSRFSGSGSGTDYSLTISNLEPEDIATYYC

human    [AASSLES]GVPSRFSGSGSGTDFTLTISSLQPEDFATYYC

hu2.0    [FTSSLHS]GVPSRFSGSGSGTDFTLTISSLQPEDFATYYC

hu2.10   [FTSSLHS]GVPSRFSGSGSGTDYTLTISSLQPEDFATYYC

Fab-12   [FTSSLHS]GVPSRFSGSGSGTDFTLTISSLQPEDFATYYC

Y0317    [FTSSLHS]GVPSRFSGSGSGTDFTLTISSLQPEDFATYYC

TKKT     [FTSSLHS]GVPSRFSGSGSGTDFTLTISSLQPEDFATYYC

              98
A461     [QQYSTVPWT]FGGGTKLEIKR

human    [QQYNSLPWT]FGQGTKVEIKR

hu2.0    [QQYSTVPWT]FGQGTKVEIKR

hu2.10   [QQYSTVPWT]FGQGTKVEIKR

Fab-12   [QQYSTVPWT]FGQGTKVEIKR

Y0317    [QQYSTVPWT]FGQGTKVEIKR

(a) TKKT  [QQYSTVPWT]FGQGTKVEIKR
```

Figure 15.2 Sequences of the variable regions of the human kappa (a) light-chain subgroup I consensus and four versions of A4.6.1, an antihuman VEGF antibody. The first sequence is that of the murine antibody A4.6.1. The second sequence corresponds to a human consensus sequence for VH subgroup III and VL subgroup I [32]. The third sequence, hu2.0, represents the CDR swap version used as the starting point for phage-displayed libraries. The fourth sequence, hu2.10, corresponds to the final phage-derived sequence. The fifth sequence, Fab-12 [19] corresponds to the final humanized antibody (known in IgG form as Avastin), including changes made after phage humanization. The sixth sequence corresponds to affinity-matured Fab-12 [20] and is also known as Lucentis™. The seventh sequence corresponds to an on-rate enhanced version [33]. The numbering system is that of Kabat et al. [32] and is shown to indicate the boundaries of the FRs. CDRs as defined by a combination of sequence hypervariability and structural data (see text for details) are shown in brackets. Mutations in the sequence of each humanized version as compared to the preceding sequence are underlined. Framework residues that have frequently been changed during humanization of other antibodies on this scaffold [16–19] are double-underlined in the human consensus sequence.

(Continued)

VH region

```
                  1                       25          36          49
A461        EIQLVQSGPELKQPGETVRISCKAS [GYTFTNYGMN] WVKQAPGKGLKWMG

human       EVQLVESGGGLVQPGGSLRLSCAAS [GFTFSSYAMS] WVRQAPGKGLEWVS

hu2.0       EVQLVESGGGLVQPGGSLRLSCAAS [GYTFTNYGMN] WVRQAPGKGLEWVG

hu2.10      EVQLVESGGGLVQPGGSLRLSCAAS [GYTFTNYGMN] WIRQAPGKGLEWVG

Fab-12      EVQLVESGGGLVQPGGSLRLSCAAS [GYTFTNYGMN] WVRQAPGKGLEWVG

Y0317       EVQLVESGGGLVQPGGSLRLSCAAS [GYDFTHYGMN] WVRQAPGKGLEWVG

TKKT        EVQLVESGGGLVQPGGSLRLSCAAS [GYDFTNYGMN] WVRQAPGKGLEWVG


                               66                          94
A461        [WINTYTGEPTYAADFKR] RFTFSLETSASTAYLQISNLKNDDTATYFCAK

human       [VISGDGGSTYYADSVKG] RFTISRDNSKNTLYLQMNSLRAEDTAVYYCAR

hu2.0       [WINTYTGEPTYAADFKR] RFTISRDNSKNTLYLQMNSLRAEDTAVYYCAR

hu2.10      [WINTYTGEPTYAADFKR] RFTISLDTSASTVYLQMNSLRAEDTAVYYCAK

Fab-12      [WINTYTGEPTYAADFKR] RFTFSLDTSKSTAYLQMNSLRAEDTAVYYCAK

Y0317       [WINTYTGEPTYAADFKR] RFTFSLDTSKSTAYLQMNSLRAEDTAVYYCAK

TKKT        [WINTYTGEPTYAADFKR] RFTFSLDTSKSTAYLQMNSLRAEDTAVYYCAK


                  103
A461        [YPHYYGSSHWYFDV] WGAGTTVTVSS

human       [G----------FDY] WGQGTLVTVSS

hu2.0       [YPHYYGSSHWYFDV] WGQGTLVTVSS

hu2.10      [YPHYYGSSHWYFDV] WGQGTLVTVSS

Fab-12      [YPHYYGSSHWYFDV] WGQGTLVTVSS

Y0317       [YPYYYGTSHWYFDV] WGQGTLVTVSS

(b)  TKKT   [YPHYYGRSHWYFDV] WGQGTLVTVSS
```

Figure 15.2 (Continued) Sequences of the variable regions of the human kappa (b) heavy-chain subgroup III consensus and four versions of A4.6.1, an anti-human VEGF antibody. The first sequence is that of the murine antibody A4.6.1. The second sequence corresponds to a human consensus sequence for VH subgroup III and VL subgroup I [32]. The third sequence, hu2.0, represents the CDR swap version used as the starting point for phage-displayed libraries. The fourth sequence, hu2.10, corresponds to the final phage-derived sequence. The fifth sequence, Fab-12 [19] corresponds to the final humanized antibody (known in IgG form as Avastin), including changes made after phage humanization. The sixth sequence corresponds to affinity-matured Fab-12 [20] and is also known as Lucentis™. The seventh sequence corresponds to an on-rate enhanced version [33]. The numbering system is that of Kabat et al. [32] and is shown to indicate the boundaries of the FRs. CDRs as defined by a combination of sequence hypervariability and structural data (see text for details) are shown in brackets. Mutations in the sequence of each humanized version as compared to the preceding sequence are underlined. Framework residues that have frequently been changed during humanization of other antibodies on this scaffold [16–19] are double-underlined in the human consensus sequence.

molecular modeling and additional point mutations [19]. In particular, the humanized antibody Fab-12 displayed antigen-binding affinity within twofold of that of the chimeric A4.6.1. This antibody, in the form of a full-length humanized IgG1, is known as Avastin™ (Figure 15.2) and is being studied for the treatment of solid tumors [34].

Ultimately, affinity maturation approaches as described in Section 15.3 can be employed to obtain the highest affinity for an antigen at a given epitope. As discussed in Section 15.3, these approaches, as well as structure-based design, have been applied to further optimization of the humanized A4.6.1 anti-VEGF antibody (Figure 15.2).

15.3 IN VITRO AFFINITY MATURATION OF ANTIBODIES

The affinity, or strength of binding, of any protein for its ligand is, by definition, the equilibrium ratio of the unbound antibody and ligand to that of the complex. The thermodynamic equilibrium constant can be defined by the relationship of the concentrations of the components of the binding reaction measured at equilibrium (see Refs. [35] and [36]) or by measuring association and dissociation rate constants prior to equilibrium using methods such as surface plasmon resonance (SPR) [37]. The association of the monomeric antigen-binding fragment (Ab) of the antibody with antigen (Ag) can be described by the following chemical equation Equation 15.1:

$$Ab + Ag \leftrightarrows Ab \cdot Ag \tag{15.1}$$

For the interaction of an antibody Fab fragment with its cognate antigen, we can calculate the equilibrium (dissociation) constant, K_d, as a function of the Fab, antigen, and Fab–antigen complex concentrations at equilibrium ([Ab], [Ag], and [Ab · Ag], respectively), according to Equation 15.2:

$$K_d = \frac{[Ab][Ag]}{[Ab \cdot Ag]} \tag{15.2}$$

When the component concentrations are expressed in molar units (nM, pM, etc.), it is apparent that K_d has units of concentration and that K_d decreases (approaching zero) as the antibody–antigen interaction becomes tighter (i.e., more complex and less free component at equilibrium).

The kinetic approach also permits calculation of the equilibrium dissociation constant via statistical mechanics (see Ref. [37]) using Equation 15.3:

$$K_d = k_d/k_a \tag{15.3}$$

Here, k_a represents the association rate constant for formation of the antigen–antibody complex (often expressed in units of $M^{-1} s^{-1}$ for a two-component reaction), and k_d represents the rate constant for dissociation of the complex (often expressed in units of s^{-1}). Hence, K_d again has units of concentration and decreases as the antibody–antigen interaction becomes tighter (i.e., faster association or slower dissociation).

Any measurement of thermodynamic binding affinity must also consider stoichiometry. The interaction of a single Fab domain with its cognate antigen generally represents a 1:1 or "monovalent" protein–protein interaction in which Equation 15.1 applies. However, for divalent proteins, such as full-length antibodies or Fab$'_2$ fragments with two antigen-combining sites, the observed affinity depends heavily on the conditions under which the binding constant is measured. In situations where the affinity is determined by measuring binding of the antibody to a surface coated with antigen, such as SPR, enzyme-linked immunosorbent assay (ELISA), fluorescence-activated cell sorting (FACS), or whole-cell binding assays, the observed affinity can be significantly enhanced by avidity effects, as simultaneous dissociation of both antigen-binding regions from surface-immobilized antigen occurs less frequently than single dissociation events [38,39]. This effect can be quite advantageous for antibodies that bind membrane-bound antigens. In fact, significant research has been focused on exploiting the avidity effect by generating polyvalent antibodies with four, six, or eight antigen-binding regions [40]. However, as avidity effects are necessarily highly dependent upon assay conditions [38,39], we will discuss affinity determinations here in the context of the 1:1 binding affinity of an antibody's antigen-combining site with a single site on its antigen. The measurement of this monovalent interaction may be made in the context of the bivalent IgG if care is taken to prevent bivalent interactions; see, for example, Ref. [37].

In most clinical applications of antibodies, the formation of a complex between the antibody and its cognate antigen is essential if not synonymous to its activity. The field of in vitro affinity maturation has developed to improve upon the potency and efficacy of many first-generation human, humanized, and nonhuman antibodies that may be limited by antigen-binding affinity. Enhanced binding affinity as measured in vitro has proved beneficial to potency in vivo [41].

During in vitro affinity maturation, an antibody is subjected to mutagenesis and selection with the hope that an "improved" antibody lies near the parent antibody in sequence space. Because it is impossible to sample all potential sequences, extensive effort has been focused on developing methods for constructing more efficient libraries and better methods for selecting high-affinity binders from them. Here, we review the literature on in vitro affinity maturation using phage display (Table 15.2), as well as other display techniques, including cell-surface display and ribosome display, and compare the strategies and methods employed.

In designing any in vitro affinity maturation experiment, the key considerations are how large the library can and should be; which amino acid positions should be randomized (specific positions, one particular CDR, all CDRs, just the heavy or light chain); by what technique genetic diversity will be generated (oligonucleotide-directed mutagenesis, error-prone polymerase chain reaction [PCR], DNA shuffling); and how much amino acid diversity will be allowed at each randomized position (all 20 amino acids, only hydrophilic residues, only amino acids resulting from single-point mutations). Additionally, one must consider the structural format for antibody display (Fab, scFv) and how the library will be selected (or "panned") for higher-affinity binders (solution binding and capture, immobilized antigen).

Table 15.2 Reports of in Vitro Affinity Maturation Surveyed in This Review

References	Antigen	Starting K_d (nM)	Ending K_d (nM)	Improved (fold)	Display Format	Mutagen Method	Mutant Targeting	Maximum Theoretical Diversity	Maximum Sampled Diversity
[42]	Fluorescein[a]	0.3	4.8×10^{-5}	6250	scFv	Mixed	Random	NA	2×10^7
[43]	HER2	16	0.013	1231	scFv	Oligo	Function	5.1×10^{11}	1×10^7
[44]	Gp120	6.3	0.015	420	Fab	Oligo	CDRs	6.4×10^7	6.7×10^7
[45]	IL-1β	0.5	0.050	10	Fab	Oligo	CDRs	1.2×10^2	1.6×10^5
[20]	VEGF	13	0.11	118	Fab	Oligo	Structure	3.2×10^6	3.2×10^6
[46]	gp120	2	0.2	10	scFv	Chain shuffling	CDRs	NA	3×10^6
[47]	Mesothelin	11	0.2	55	scFv	Oligo	CDRs[c]	8×10^3	8×10^3
[48]	Digoxin[b]	0.9	0.3	3	scFv	Oligo	Structure	1.6×10^5	1.6×10^5
[49]	$\alpha_v\beta_3$	28	0.3	92	Fab	Oligo	CDRs	2.6×10^3	2.6×10^3
[50]	EpCam	6	0.4	15	ScFv	Chain shuffling	Random	NA	6×10^7
[51]	phOx	320	1.1	291	ScFv	Chain shuffling	CDRs	NA	2.2×10^6
[52]	Fibronectin	110	1.1	100	ScFv	Oligo	CDRs	NA	NA
[53]	EGFR vIII	9	2	5	ScFv	Oligo	CDRs	8×10^3	8×10^3
[54]	NIP-caproic	42	9	5	ScFv	PCR[d]	Random	NA	4×10^4
[55]	Glycophorin	48,000	100	480	ScFv	Mutator strain	Random	NA	1×10^{13}
[56]	preS1 of HBV	8000	230	35	ScFv	Chain shuffling	Random	NA	3.2×10^8
[57]	Ars	3600	600	6	Fab	Oligo	Function	400	400
[58]	LewisY antigen	9900	700	14	Fab	Oligo	CDRs	NA	1×10^8
[59]	Progesterone	30,000	1000	30	Fab scFv	PCR[d]	Random	NA	5×10^6

a Yeast surface display.
b Bacteria surface display.
c Used DNA hot spots to target mutagenesis.
d Used error-prone PCR for mutagenesis.

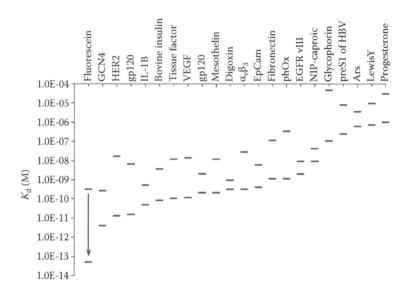

Figure 15.3 Results of antibody affinity maturation vary widely, in both overall affinity (48 fM for fluorescein–biotin, 1 μM for progesterone) and magnitude of improvement (6250-fold for fluorescein–biotin, 3-fold for digoxin). (Compiled from Boder ET et al., *Proc Natl Acad Sci U S A*, 97:10701–10705, 2000; Daugherty PS et al., *Protein Eng*, 11:825–832, 1998; Gram H et al., *Proc Natl Acad Sci U S A*, 89:3576–3580, 1992.)

As seen in Figure 15.3, successes in the in vitro affinity maturation of antibodies reported span a broad range of absolute affinities (from 1 μM to 48 fM) and relative improvements in affinity (from 2- to 6250-fold improvement). The starting affinities of these antibodies for their antigens have ranged from subnanomolar to nearly 50 μM, with examples of hundreds-fold improvement in each affinity range (Table 15.2). With more than 20 reports of antibody affinity maturation studies, we can begin to compare the approaches and experimental methods for their effectiveness in identifying tighter binding antibodies.

15.3.1 Library Size

The central problem in affinity maturation using combinatorial diversity is to efficiently select from among the astronomical number of possible amino acid sequences at least one sequence that is optimal for any given application. It is inferred that the likelihood of discovering a better binder will increase with the number of sequences sampled [60]. Although there is a clear advantage to sampling more sequence space, a larger library does not guarantee that the highest-affinity binders will be isolated. For instance, in one case [49], a 92-fold-improved, subnanomolar antibody was obtained from a library of just 2592 sequences, while in another [50], a library of 6×10^7 sequences improved binding only 15-fold.

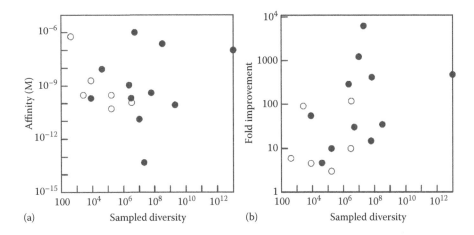

Figure 15.4 Relationship between library size and (a) absolute affinity or (b) affinity improvement relative to the parental antibody fragment. For oversampled libraries (empty circles), the sampled diversity is the maximum theoretical diversity for mutagenic oligonucleotides used. For undersampled libraries (filled circles), the sampled diversity is the constructed library size. The library with 10^{13} members was constructed by passage through a mutator strain. (From Irving RA et al., *Immunotechnology (Amsterdam)*, 2:127–143, 1996.)

To evaluate whether larger libraries more generally generate tighter binders, we have plotted both these values (Figure 15.4a and b) as a function of the maximum number of sequences sampled for each reported study. Figure 15.4b shows that there may be some general trends toward increasing improvements in affinity with increasing library size. However, there is no obvious correlation between absolute affinity and library size (Figure 15.4a).

Also revealed by Figure 15.4 is the fact that there is no clear distinction between the successes of libraries that oversample or undersample the theoretical maximum diversity of the library. Oversampling has the advantage of providing valuable information in the event that a better binder is not retrieved from the library—one is confident of the exact region of sequence space around the parental sequence that has been searched and that higher-affinity antibodies are unlikely to arise from searching that space again. Unfortunately, this requires the spectrum of diversity to be predetermined and thus is limited by the biases inherent to the design of the experiment. Undersampling has the advantage of accessing an extremely broad and unbiased spectrum of diversity but has no guarantee that repeating the selection will yield the same (or even similar) results. Thus, little is learned by obtaining a negative result from poorly sampled libraries.

15.3.2 Targeting Positions for Random Mutagenesis

Another critical facet of the library design process is determining which amino acid positions should be randomized (Figure 15.5; Ref. [61]). Intuitively, one would

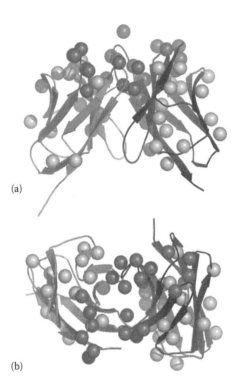

(a)

(b)

Figure 15.5   Location of mutations identified by in vitro affinity maturation in the variable domain of various antibodies. (a) "Side view." (b) "Bird's eye view" of binding interface. Beneficial mutations identified in the reports surveyed here are mapped onto the structure of a humanized Fv (1FCV). The light-chain (V_L, gray) and heavy-chain (V_H, black) components of the variable domain are shown in ribbon form. Mutations obtained through oligonucleotide-directed mutagenesis, primarily within the CDRs, are shown as dark gray spheres. Mutations obtained through undirected, random mutagenesis are shown as light gray spheres. Image constructed with PyMol. (From Delano WL. *The PyMOL Molecular Graphics System.* DeLano Scientific, Palo Alto, CA, 2002.)

expect that targeting positions that make direct contact with the antigen would be more likely to result in significant improvements in affinity than would targeting positions that are distant from the antigen-binding site. Early affinity maturation studies indicated that both residues intimately involved in intermolecular contacts as well as those at the periphery of the interface could yield affinity improvements [62]. If a structure of the antibody–antigen complex is available, identifying these contact residues is straightforward, as in the case for an anti-VEGF antibody, which was affinity-matured by a factor of >20-fold [20]. In this case, a total of four CDR changes, identified by targeting contact and noncontact residues, resulted in the improved variant Y0317 (Figure 15.2). This variant, also known as Lucentis™ in Fab form, is being investigated for the treatment of macular degeneration [63].

Without a structure, the library design can be guided by selecting surface-exposed CDR residues in general or by identifying residues important for binding

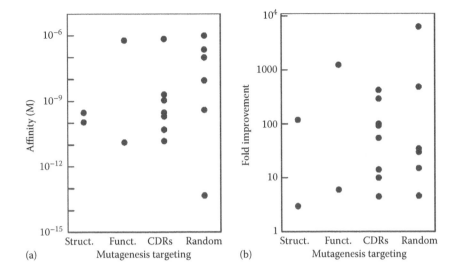

Figure 15.6   Relationship between (a) absolute affinity or (b) affinity improvement relative to the parental antibody fragment and selection of target residues for mutagenesis. Two studies used structural information (Struct.) to guide mutagenesis (Refs. [20] and [48]), and two studies used functional analysis (Funct.) (Refs. [43] and [57]). The remaining studies either targeted CDR residues in general (CDRs) or introduced random mutations (Random) throughout the molecule via PCR or mutator strain.

(in another application of antibody–phage display) via site-directed [64] or shotgun [65] alanine-scanning mutagenesis of the CDRs. These approaches have produced targeted libraries that resulted in a number of successful affinity optimizations [57]. It is also possible to improve affinity by "walking though" the CDRs [44], limiting mutations to one CDR at a time, and combining the best mutations from each.

Alternatively, significant increases in affinity can be achieved by constructing a library, in the heavy chain, the light chain, or the entire antibody gene, without any specific targeting. In fact, to our knowledge, the highest-affinity antibody reported (and the one with the greatest improvement in affinity) was isolated by an untargeted mutagenesis scheme [42], with mutations important for increasing affinity located as far as 25 Å away in a "fourth shell" surrounding the direct contact residues. All of these methods show varying degrees of success, as shown in Figure 15.6, with no single method clearly surpassing others. With respect to therapeutic antibodies, however, limiting the sites of mutagenesis to normally hypervariable residues [32] has the presumed advantage of reducing the risk of immunogenicity in selected variants without the necessity of testing the effects of mutations at conserved sites.

15.3.3 Method of Mutagenesis

One must also consider whether the constructed library will oversample or under-sample the theoretical maximum diversity of the library. For a statistical analysis of the degree to which the actual library size must exceed the theoretical maximum library size, see Ref. [66]. Oversampled libraries can be obtained by oligonucleotide-directed mutagenesis (either template or cassette based) at a restricted number of sites. Oligonucleotide-directed mutagenesis also allows the experimenter to control exactly where mutations are introduced, thus alleviating concern about introducing potentially immunogenic mutations to framework residues. Any nonsystematic mutagenesis method that introduces mutations at completely random positions (e.g., via passage through a mutator strain, error-prone PCR, DNA "shuffling") requires no information on the importance of each residue to binding and permits the maturation of antibodies without the imposed biases of the experimenter. However, such methods also increase the theoretical maximum diversity to the point that it cannot be effectively sampled. Although some of these methods can limit mutagenesis to specific regions of the amino acid sequence (e.g., by using error-prone PCR to randomly mutate just one or two CDRs), the technical expertise required can offset the simplicity and ease that make random mutagenesis so attractive.

The technique used to introduce mutations is highly coupled to the position-targeting scheme. However, whether oligonucleotide-based, PCR, or other methods

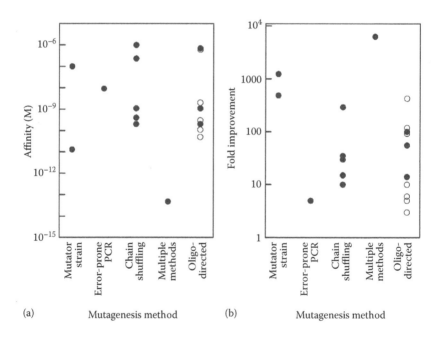

Figure 15.7 Relationship between mutagenesis method and (a) absolute affinity or (b) affinity improvement relative to the parental antibody fragment. Methods are described in the text. "Multiple methods" refers to a combination of mutator strains, chain shuffling, and PCR. For oversampled libraries (empty circles), the sampled diversity is the maximum theoretical diversity for mutagenic oligonucleotides used. For undersampled libraries (filled circles), the sampled diversity is the size of the library actually constructed. (From Boder ET et al., *Proc Natl Acad Sci U S A*, 97:10701–10705, 2000.)

are used, the mutagenesis method alone is not correlated with the success of the antibody selection (Figure 15.7).

15.3.4 Diversity and Codon Degeneracy

In vivo affinity maturation relies on somatic mutation and recombination of the CDRs, yet all natural antibodies originate from a relatively small set of germ line variable-domain genes, which themselves have arisen from evolutionary gene duplication. There are two schools of thought on how closely the in vitro affinity maturation process should parallel the in vivo one. One perspective is that if antibodies with high affinity are derived in vivo by somatic mutation, which introduces primarily single-nucleotide point mutations (and not triple-nucleotide codon replacements), then high-affinity antibodies should be selected in vitro from libraries themselves constructed by methods that introduce primarily single-nucleotide point mutations. Another perspective is that there is no intrinsic thermodynamic reason for the genetic diversity of an in vitro library to be biased by the evolutionary history of the genetic code, and thus all 20 amino acids should comprise the library. Although both perspectives are based on compelling hypotheses, a survey of reported improvements in affinity shows that significant improvements in affinity can be obtained through either approach (Figure 15.8).

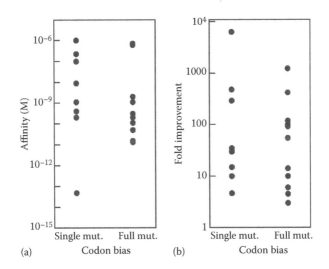

Figure 15.8 Relationship between (a) absolute affinity or (b) affinity improvement relative to the parental antibody fragment and any codon bias that may result from the mutagenesis method. In nonoligonucleotide-directed mutagenesis (error-prone PCR or shuffling), it is assumed that most amino acids changes arise from single-nucleotide point mutations (Single mut.) and thus will be biased toward amino acids that are more closely related in the genetic code to the wild-type amino acid at any given position. Oligonucleotide-directed mutagenesis incorporates amino acids with only the bias of the degeneracy of the genetic code (Full mut.).

15.3.5 Antigen-Binding Selection Methods

Once the library is designed, constructed, and transformed into *Escherichia coli*, it must be subjected to antigen-binding selections for antibodies with higher affinity than the parent. Depending on how the library was designed, it is possible that a significant number of the variants may be very similar to the wild type, making it essential that the selections be performed under conditions that can finely discriminate between antibodies with similar affinities. The most common technique, panning for binders with surface-immobilized antigen [2], has provided very impressive results (Figure 15.9). In this technique, a 96-well plate (usually modified polystyrene or polycarbonate) is coated with antigen at 1–10 µg/mL and blocked with bovine serum albumin (BSA), casein, or another blocking reagent to prevent adsorption of phage. The phage library (~10^{10}–10^{12} phage/mL) is allowed to bind for a period of minutes to hours and then nonbinding and weaker binding phage are washed off. To discriminate among many phages with high affinities and isolate phages with the slowest off-rates, washing can take place over a period of hours (or even days) with soluble antigen in the wash buffer to prevent rebinding. Although this technique is simple to implement, it takes experience to know how to adjust the washing parameters to get the best binders.

Another selection method is solution sorting [67]. In this method, the phage-displayed library is incubated with soluble antigen that has been tagged with biotin. Bound phages are captured on streptavidin-coated wells or beads, washed, eluted, and propagated. By controlling the concentration of antigen, one can control the

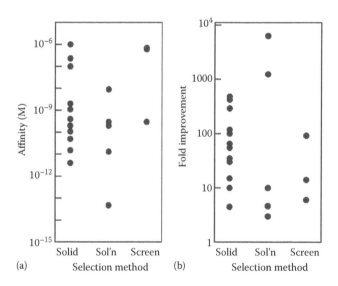

Figure 15.9 Relationship between (a) absolute affinity or (b) affinity improvement relative to the parental antibody fragment and the selection method used. Most selections were performed on a solid support (Solid, indicating intact cells or antigen immobilized on a plastic surface), while some used a solution binding and capture technique (Sol'n) or relied on screening individual variants (Screen).

stringency of the selection. For example, selecting with 1 nM antigen should favor recovery of phages with a K_d of 1 nM or lower.

It is also possible to identify affinity-improved variants by simply screening each clone. This method precludes the technical intricacies inherent to display and ensures that rare, high-affinity clones will not be lost or outgrown during the propagation step used in phage display. However, it requires either a very small library or high-throughput screening machinery. The most "successful" implementation of this technique is the affinity maturation of an anti-$\alpha_v\beta_3$ antibody 92-fold [49]. By screening all possible single amino acid mutations at each CDR position (2336 mutants), Wu et al. [49] were able to identify individual mutations that increased affinity 2- to 13-fold and then combinatorially combine those (256 mutants) and identify multiple clones with even higher affinity. Similarly, but not as dramatically, Casson and Manser [57] improved the affinity of an antiarsenate antibody by sixfold by screening all 400 possible mutations at two amino acid positions.

15.3.6 Merging Results from Separate Libraries

In a number of cases using targeted mutagenesis schemes, in vitro affinity maturation experiments have used multiple libraries to sample mutations in each CDR. A key question that must then be addressed is how to combine the mutations derived from each library. There are three basic approaches one can take (Figure 15.10). One

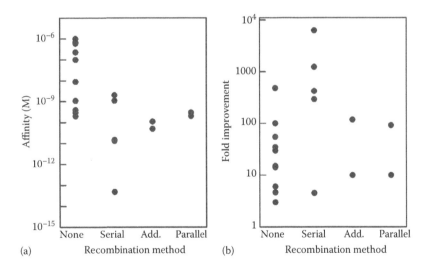

Figure 15.10 Relationship between (a) absolute affinity or (b) affinity improvement relative to the parental antibody fragment and the means by which mutations from separate selections or pannings were recombined to yield yet higher-affinity antibodies. The majority of studies performed only one panning experiment (None). Others used the winners from one panning as the framework for a second panning (Serial). Two cases of simple additivity were reported (Add.), as were two cases of making an additional library containing the best mutations from previous libraries followed by additional selections (Parallel).

option is to construct a number of separate libraries, identify critical mutations in each, and assume that their effects on affinity will be roughly additive. Sometimes, this approach is successful [20,45], but it is naïve to think it always will be, as the intricacies of protein–protein interactions are not yet completely understood. Another option is to proceed serially, sometimes called "CDR walking" [44], making a library of one CDR, selecting the best binder, and using that improved variant as the framework for the next library with different amino acid positions randomized. This approach has the obvious advantage that binders from each successive library are guaranteed to work within the context of the previously identified beneficial mutations. However, synergistic mutations bridging multiple CDRs will not be identified by this technique. A third option, which addresses the inadequacies of the assumptions of additivity in the first described approach, is more parallel in nature, with separate libraries being constructed and sorted for each CDR (or other targeted mutagenic region), and then recombined combinatorially to identify synergistically interacting residues [49].

15.3.7 Antibody Format

Another factor to consider in an affinity maturation experiment is the antibody format. Both antigen-binding fragments (Fabs), usually with the heavy-chain VH–CH1 domains fused to the phage protein and VL–CL coexpressed as a soluble second chain, and scFvs, with VH and VL domains connected via a peptide linker, have been used for phage display. The single-chain format may have expression advantages in general; however, the formation of diabodies can produce misleading results in binding and activity assays. Both formats continue to be widely used, and neither

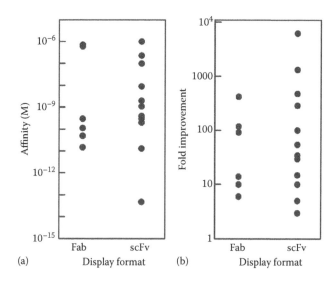

Figure 15.11 Relationship between display format (Fab or scFv) and (a) absolute affinity or (b) affinity improvement relative to the parental antibody fragment.

has demonstrated a clear advantage over the other with respect to in vitro affinity maturation (Figure 15.11).

15.4 EMERGING APPROACHES

Recently, two additional display technologies have been used for affinity maturation: FACS and ribosome display. FACS separates populations of cells based on their fluorescent properties. This technology has been used to affinity-mature an antifluorescein–biotin antibody by display on the surface of yeast [42], resulting in the greatest improvement in affinity (6250-fold) and the tightest overall antibody (48 fM). Fluorescence-activated cell sorting has also been used to separate bacterial cells, displaying scFv via fusion with the OmpA bacterial surface protein, though resulting in only minor affinity improvements [48]. One obvious limitation of this technology is that it depends on the use of a soluble fluorescent target.

Ribosomal display is a completely ex vivo approach in which the protein of interest is tethered to the actual ribosome (and messenger RNA) that encodes it [68,69]. Although this technology has been used not to affinity-mature an antibody per se, but rather, to isolate antibodies with picomolar affinity [70,71], it is likely that this technology will be adapted soon for affinity maturation.

In addition to these new methods for sorting/selecting improved binders from large libraries of variants, progress has been made in the use of computational methods for improving affinity. In the case of anti-VEGF antibodies, variants with improved affinity were designed by specifically targeting residues within and outside the set of antigen-contacting residues [33]. In this case, although extensive phage selection strategies had led to improved affinity through off-rate selections, no significant improvements in on-rate had been found. The computational approach led to a new set of mutations that provide enhanced affinity, primarily through enhanced on-rate in the "34-TKKT" variant (Figure 15.2).

15.5 CONCLUSIONS

With many examples of humanization and affinity maturation of antibody fragments in the literature, one can begin to compare the methods employed to address the variables of display-based affinity maturation. Since a variety of mutagenesis, residue-targeting, and selection methods have been used, comparisons between specific methodologies can only be made with a limited number of examples of each. However, the picture that emerges at this point is that no "magic bullet" (to borrow a long-used reference to therapeutic antibodies) has been found for systematically improving the affinity of antibodies, but rather, each method has its own benefits and limitations. The success of a variety of approaches likely arises from the fact that multiple high-affinity solutions often exist for optimized protein–protein interfaces and the fact that single amino acid changes (or the combination of a few changes) can often have dramatic effects on binding affinity.

Ultimately, the combination of structure-based design, structure-based diversity design, and stringent selection strategies is likely to yield the most useful antibodies for reagent and therapeutic use.

REFERENCES

1. Kohler G, Milstein C. Continuous cultures of fused cells secreting antibody of pre-defined specificity. *Nature* 1975; 256:495–497.
2. Smith GP. Filamentous fusion phage: Novel expression vectors that display cloned anti-gens on the virion surface. *Science* 1985; 228:1315–1317.
3. Chothia C, Lesk AM. Canonical structures for the hypervariable regions of immuno-globulins. *J Mol Biol* 1987; 196:901–917.
4. Wilson IA, Stanfield RL. Antibody–antigen interactions: New structures and new con-formational changes. *Curr Opin Struct Biol* 1994; 4:857–867.
5. Kang AS, Barbas CF III, Janda KD, Benkovic SJ, Lerner RA. Linkage of recognition and replication functions by assembling combinatorial antibody Fab libraries along phage surfaces. *Proc Natl Acad Sci U S A* 1991; 88:4363–4366.
6. Cwirla SE, Peters EA, Barrett RW, Dower WJ. Peptides on phage: A vast library of peptides for identifying ligands. *Proc Natl Acad Sci U S A* 1990; 87:6378–6382.
7. Barbas CF III, Kang AS, Lerner RA, Benkovic SJ. Assembly of combinatorial anti-body libraries on phage surfaces: The gene III site. *Proc Natl Acad Sci U S A* 1991; 88:7978–7982.
8. Hoogenboom HR, Griffiths AD, Johnson KS, Chiswell DJ, Hudson P, Winter G. Multi-subunit proteins on the surface of filamentous phage: Methodologies for displaying anti-body (Fab) heavy and light chains. *Nucleic Acids Res* 1991; 19:4133–4137.
9. Marks JD, Hoogenboom HR, Bonnert TP, McCafferty J, Griffiths AD, Winter G. By-passing immunization. Human antibodies from V-gene libraries displayed on phage. *J Mol Biol* 1991; 222:581–597.
10. Perisic O, Webb PA, Holliger P, Winter G, Williams RL. Crystal structure of a diabody, a bivalent antibody fragment. *Structure* 1994; 2:1217–1226.
11. Jaffers GJ, Fuller TC, Cosimi AB, Russell PS, Winn HJ, Colvin RB. Monoclonal anti-body therapy. Anti-idiotypic and non-anti-idiotypic antibodies to OKT3 arising despite intense immunosuppression. *Transplantation* 1986; 41:572–578.
12. Riechmann L, Clark M, Waldmann H, Winter G. Reshaping human antibodies for ther-apy. *Nature* 1988; 332:323–327.
13. Hale G, Dyer MJ, Clark MR, Phillips JM, Marcus R, Riechmann L, Winter G et al. Remission induction in non-Hodgkin lymphoma with reshaped human monoclonal anti-body CAMPATH-1H. *Lancet* 1988; 2:1394–1399.
14. Sugo T, Mizuguchi J, Kamikubo Y, Matsuda M. Anti-human factor IX monoclonal antibodies specific for calcium ion-induced conformations. *Thromb Res* 1990; 58: 603–614.
15. Suggett S, Kirchhofer D, Hass P, Lipari T, Moran P, Nagel M, Judice K et al. Use of phage display for the generation of human antibodies that neutralize factor IXa function. *Blood Coagul Fibrinolysis* 2000; 11:27–42.
16. Carter P, Presta L, Gorman CM, Ridgway JB, Henner D, Wong WL, Rowland AM et al. Humanization of an anti-p185HER2 antibody for human cancer therapy. *Proc Natl Acad Sci U S A* 1992; 89:4285–4289.

17. Presta LG, Lahr SJ, Shields RL, Porter JP, Gorman CM, Fendly BM, Jardieu PM. Humanization of an antibody directed against IgE. *J Immunol* 1993; 151:2623–2632.

18. Werther WA, Gonzalez TN, O'Connor SJ, McCabe S, Chan B, Hotaling T, Champe M et al. Humanization of an anti-lymphocyte function-associated antigen (LFA)-1 monoclonal antibody and reengineering of the humanized antibody for binding to rhesus LFA-1. *J Immunol* 1996; 157:4986–4995.

19. Presta LG, Chen H, O'Connor SJ, Chisholm V, Meng YG, Krummen L, Winkler M et al. Humanization of an anti-vascular endothelial growth factor monoclonal antibody for the therapy of solid tumors and other disorders. *Cancer Res* 1997; 57:4593–4599.

20. Chen Y, Wiesmann C, Fuh G, Li B, Christinger HW, McKay P, de Vos AM et al. Selection and analysis of an optimized anti-VEGF antibody: Crystal structure of an affinity-matured Fab in complex with antigen. *J Mol Biol* 1999; 293:865–881.

21. Foote J, Winter G. Antibody framework residues affecting the conformation of the hypervariable loops. *J Mol Biol* 1992; 224:487–499.

22. Jones PT, Dear PH, Foote J, Neuberger MS, Winter G. Replacing the complementarity-determining regions in a human antibody with those from a mouse. *Nature* 1986; 321: 522–525.

23. Verhoeyen M, Milstein C, Winter G. Reshaping human antibodies: Grafting an antilysozyme activity. *Science* 1988; 239:1534–1536.

24. Chothia C, Lesk AM, Tramontano A, Levitt M, Smith-Gill SJ, Air G, Sheriff S et al. Conformations of immunoglobulin hypervariable regions. *Nature* 1989; 342:877–883.

25. Gussow D, Seemann G. Humanization of monoclonal antibodies. *Methods Enzymol* 1991; 203:99–121.

26. Carter P, Abrahmsen L, Wells JA. Probing the mechanism and improving the rate of substrate-assisted catalysis in subtilisin BPN. *Biochemistry* 1991; 30:6142–6148.

27. Baca M, Presta LG, O'Connor SJ, Wells JA. Antibody humanization using monovalent phage display. *J Biol Chem* 1997; 272:10678–10684.

28. Queen C, Schneider WP, Selick HE, Payne PW, Landolfi NF, Duncan JF, Avdalovic NM et al. A humanized antibody that binds to the interleukin 2 receptor. *Proc Natl Acad Sci U S A* 1989; 86:10029–10033.

29. Rosok MJ, Yelton DE, Harris LJ, Bajorath J, Hellstrom KE, Hellstrom I, Cruz GA et al. A combinatorial library strategy for the rapid humanization of anticarcinoma BR96 Fab. *J Biol Chem* 1996; 271:22611–22618.

30. Roguska MA, Pedersen JT, Keddy CA, Henry AH, Searle SJ, Lambert JM, Goldmacher VS et al. Humanization of murine monoclonal antibodies through variable domain resurfacing. *Proc Natl Acad Sci U S A* 1994; 91:969–973.

31. Jespers LS, Roberts A, Mahler SM, Winter G, Hoogenboom HR. Guiding the selection of human antibodies from phage display repertoires to a single epitope of an antigen. *Biotechnology (N Y)* 1994; 12:899–903.

32. Kabat EA, Wu TT, Perry HM, Gottesman KS, Foeller C. *Sequences of Proteins of Immunological Interest*. Bethesda, MD: National Institutes of Health, 1991.

33. Marvin JS, Lowman HB. Redesigning an antibody fragment for faster association with its antigen. *Biochemistry* 2003; 42:7077–7083.

34. Ferrara N. Role of vascular endothelial growth factor in physiologic and pathologic angiogenesis: Therapeutic implications. *Semin Oncol* 2002; 29:10–14.

35. Klotz IM. Ligand–receptor interactions: Facts and fantasies. *Q Rev Biophys* 1985; 18:227–259.

36. Neri D, Montigiani S, Kirkham PM. Biophysical methods for the determination of antibody–antigen affinities. *Trends Biotechnol* 1996; 14:465–470.

37. Karlsson R, Michaelsson A, Mattsson L. Kinetic analysis of monoclonal antibody–antigen interactions with a new biosensor based analytical system. *J Immunol Methods* 1991; 145:229–240.

38. Crothers DM, Metzger H. The influence of polyvalency on the binding properties of antibodies. *Immunochemistry* 1972; 9:341–357.

39. Muller KM, Arndt KM, Pluckthun A. Model and simulation of multivalent binding to fixed ligands. *Anal Biochem* 1998; 261:149–158.

40. Miller K, Meng G, Liu J, Hurst A, Hsei V, Wong WL, Ekert R et al. Design, construction, and in vitro analyses of multivalent antibodies. *J Immunol* 2003; 170:4854–4861.

41. Schlom J, Eggensperger D, Colcher D, Molinolo A, Houchens D, Miller LS, Hinkle G et al. Therapeutic advantage of high-affinity anticarcinoma radioimmunoconjugates. *Cancer Res* 1992; 52:1067–1072.

42. Boder ET, Midelfort KS, Wittrup KD. Directed evolution of antibody fragments with monovalent femtomolar antigen-binding affinity. *Proc Natl Acad Sci U S A* 2000; 97: 10701–10705.

43. Schier R, Marks JD. Efficient in vitro affinity maturation of phage antibodies using BIAcore guided selections. *Hum Antibodies Hybridomas* 1996; 7:97–105.

44. Yang WP, Green K, Pinz-Sweeney S, Briones AT, Burton DR, Barbas CF III. CDR walking mutagenesis for the affinity maturation of a potent human anti-HIV-1 antibody into the picomolar range. *J Mol Biol* 1995; 254:392–403.

45. Jackson JR, Sathe G, Rosenberg M, Sweet R. In vitro antibody maturation. Improvement of a high affinity, neutralizing antibody against IL-1 beta. *J Immunol* 1995; 154:3310–3319.

46. Thompson J, Pope T, Tung JS, Chan C, Hollis G, Mark G, Johnson KS. Affinity maturation of a high-affinity human monoclonal antibody against the third hypervariable loop of human immunodeficiency virus: Use of phage display to improve affinity and broaden strain reactivity. *J Mol Biol* 1996; 256:77–88.

47. Chowdhury PS, Pastan I. Improving antibody affinity by mimicking somatic hypermutation in vitro. *Nat Biotechnol* 1999; 17:568–572.

48. Daugherty PS, Chen G, Olsen MJ, Iverson BL, Georgiou G. Antibody affinity maturation using bacterial surface display. *Protein Eng* 1998; 11:825–832.

49. Wu H, Beuerlein G, Nie Y, Smith H, Lee BA, Hensler M, Huse WD et al. Stepwise in vitro affinity maturation of Vitaxin, an $\alpha_v\beta_3$-specific humanized mAb. *Proc Natl Acad Sci U S A* 1998; 95:6037–6042.

50. Huls G, Gestel D, van der Linden J, Moret E, Logtenberg T. Tumor cell killing by in vitro affinity-matured recombinant human monoclonal antibodies. *Cancer Immunol Immunother* 2001; 50:163–171.

51. Marks JD, Griffiths AD, Malmqvist M, Clackson TP, Bye JM, Winter G. By-passing immunization: Building high affinity human antibodies by chain shuffling. *Biotechnology (N Y)* 1992; 10:779–783.

52. Neri D, Carnemolla B, Nissim A, Leprini A, Querze G, Balza E, Pini A et al. Targeting by affinity-matured recombinant antibody fragments of an angiogenesis associated fibronectin isoform. *Nat Biotechnol* 1997; 15:1271–1275.

53. Beers R, Chowdhury P, Bigner D, Pastan I. Immunotoxins with increased activity against epidermal growth factor receptor vIII-expressing cells produced by antibody phage display. *Clin Cancer Res* 2000; 6:2835–2843.

54. Hawkins RE, Russell SJ, Winter G. Selection of phage antibodies by binding affinity. Mimicking affinity maturation. *J Mol Biol* 1992; 226:889–896.

55. Irving RA, Kortt AA, Hudson PJ. Affinity maturation of recombinant antibodies using E. coli mutator cells. *Immunotechnology (Amsterdam)* 1996; 2:127–143.
56. Park SG, Lee JS, Je EY, Kim IJ, Chung JH, Choi IH. Affinity maturation of natural antibody using a chain shuffling technique and the expression of recombinant antibodies in *Escherichia coli. Biochem Biophys Res Commun* 2000; 275:553–557.
57. Casson LP, Manser T. Random mutagenesis of two complementarity determining region amino acids yields an unexpectedly high frequency of antibodies with increased affinity for both cognate antigen and autoantigen. *J Exp Med* 1995; 182:743–750.
58. Yelton DE, Rosok MJ, Cruz G, Cosand WL, Bajorath J, Hellstrom I, Hellstrom KE et al. Affinity maturation of the BR96 anti-carcinoma antibody by codon-based mutagenesis. *J Immunol* 1995; 155:1994–2004.
59. Gram H, Marconi LA, Barbas CF III, Collet TA, Lerner RA, Kang AS. In vitro selection and affinity maturation of antibodies from a naive combinatorial immunoglobulin library. *Proc Natl Acad Sci U S A* 1992; 89:3576–3580.
60. Perelson AS. Immune network theory. *Immunol Rev* 1989; 110:5–36.
61. Delano WL. *The PyMOL Molecular Graphics System*. DeLano Scientific, Palo Alto, CA, 2002.
62. Lowman HB, Wells JA. Affinity maturation of human growth hormone by monovalent phage display. *J Mol Biol* 1993; 234:564–578.
63. Ferrara N. VEGF and the quest for tumour angiogenesis factors. *Nat Rev Cancer* 2002; 2:795–803.
64. Muller YA, Chen Y, Christinger HW, Li B, Cunningham BC, Lowman HB, de Vos AM. VEGF and the Fab fragment of a humanized neutralizing antibody: Crystal structure of the complex at 2.4 Å resolution and mutational analysis of the interface. *Structure* 1998; 6:1153–1167.
65. Vajdos FF, Adams CW, Breece TN, Presta LG, de Vos AM, Sidhu SS. Comprehensive functional maps of the antigen-binding site of an anti-ErbB2 antibody obtained with shotgun scanning mutagenesis. *J Mol Biol* 2002; 320:415–428.
66. Lowman HB, Wells JA. Monovalent phage display: A method for selecting variant proteins from random libraries. *Methods: Companion Methods Enzymol* 1991; 3:205–216.
67. Parmley SF, Smith GP. Antibody-selectable filamentous fd phage vectors: Affinity purification of target genes. *Gene* 1988; 73:305–318.
68. Mattheakis LC, Bhatt RR, Dower WJ. An in vitro polysome display system for identifying ligands from very large peptide libraries. *Proc Natl Acad Sci U S A* 1994; 91: 9022–9026.
69. Hanes J, Pluckthun A. In vitro selection and evolution of functional proteins by using ribosome display. *Proc Natl Acad Sci U S A* 1997; 94:4937–4942.
70. Hanes J, Jermutus L, Weber-Bornhauser S, Bosshard HR, Pluckthun A. Ribosome display efficiently selects and evolves high-affinity antibodies in vitro from immune libraries. *Proc Natl Acad Sci U S A* 1998; 95:14130–14135.
71. Hanes J, Schaffitzel C, Knappik A, Pluckthun A. Picomolar affinity antibodies from a fully synthetic naive library selected and evolved by ribosome display. *Nat Biotechnol* 2000; 18:1287–1292.

CHAPTER 16

Antibody Libraries from Immunized Repertoires

Jody D. Berry and Mikhail Popkov

CONTENTS

16.1 INTRODUCTION

Polyclonal antibodies have been used from diverse species such as rabbit, goat, sheep, donkey, chicken, pigs, cats, dogs, minks, and cattle for almost a century as specific and high-affinity probes for a variety of immunological assays in research and clinical laboratories. While polyclonal-pooled immunoglobulin (Ig) is still the gold standard for the treatment of some human diseases, the high purity and specificity of monoclonal antibody will eventually oust these serum preparations from this position; monoclonal antibodies (MAbs) promise to be safer and more specific in therapy. While there are a few examples where MAbs have been developed for all of the above species [1,2], MAb development in these species has been very slow mostly because of a lack of reliable myeloma fusion partners resulting in unstable clones and complex back fusions to produce heterohybridomas [1]. Repertoire cloning by phage display has offered a new method for obtaining MAbs from immunized repertoires from all of these species, and for the immune systems of all jawed vertebrates in the future.

Antibody libraries represent a major means by which MAbs are produced today. Monoclonal antibodies can be made easily and reproducibly in large quantities without lot variation seen in polyclonal antibody preparations; therefore, MAbs allow many experiments to be performed, which were not previously possible or practical. The use of MAbs avoids undesirable cross-reactivity and high-affinity MAbs can be achieved easily. The purity of MAbs produced *in vitro* is extremely high. The uniform nature of pure MAbs has led to widespread use in biotechnology and biopharmaceutical research. Monoclonal antibodies produced *in vitro* also have less risk of carrying unknown passenger viruses and thus are safer compared to pooled human gamma globulin preparations of MAbs from humans for use in therapy. Other applications include research tools for cell, antigen, or pathogen identification; ligands for column chromatography and molecule purification; as diagnostic reagents; therapeutic antibody preparations; and vaccine development. This is in addition to the many procedures in which MAbs are used in the basic science lab. Clearly, MAbs have indisputable value in modern biological laboratories. Antibody libraries, to date, have had their greatest application in the development of therapeutic MAbs in humans. In many ways, the lure of therapeutic MAbs has driven the field of immune libraries, and it is only more recently that experiments on novel animal species have been initiated.

The antibody library approach provides a recombinant means to derive MAbs from various sources *in vitro*. The libraries themselves allow scientists to easily

Table 16.1 Antibody Libraries

Library	V-Gene Source	Diversity
Immune	Activated B cells (IgG)	Matured *in vivo*
Naive	Resting B cells (IgM)	Immature
Synthetic	Cloned V genes	Synthetic

obtain the genetic information that encodes a given MAb. This has been useful, in itself, for elucidating fundamental immunological principles and has contributed to the knowledge of antibody genetics. The three common formats of antibody library available include immune, naive, and synthetic libraries [3,4]. Immune libraries are made from the antigen sensitized B-cell IgG mRNA. While the antibody fragments selected from immune libraries are generally of high quality and high affinity, a library has to essentially be set up (animals immunized ahead of time) for each antigen. Large naive libraries offer the advantage of being able to produce many antibodies to an assortment of antigens but have, in general, lower affinity. When additional diversification is introduced into naive libraries in the complementarity determining region (CDR) regions, they are called semisynthetic [5]. The libraries are, in general, made from the IgM mRNA but can be made from IgG by using alternative primer sets in the library construction stage (Table 16.1). Synthetic libraries are made using modular consensus scaffolds (frameworks) with the CDR antigen contact domains encoded by random synthetic oligonucleotides [6].

We have divided this chapter into four sections. In Section 16.1, we review the development of antibody libraries. We also provide a general classification scheme to encompass all immune libraries derived from sensitized B lymphocytes for exposed, immunized, and infected hosts. In Section 16.2, we review the general aspects common to all immune libraries. This section will also emphasize aspects of the strategy to help scientists determine if they should make immune libraries or alternatively make informed decisions about using another method to produce a particular MAb. There are far too many examples of immune libraries for us to review them all within the scope of this chapter. However, in Section 16.3, we review some recent key examples of the use of immune libraries to derive MAbs from various animal species as well as humans. We highlight points upon which our experience with immune antibody libraries gives practical information. These studies have expanded the utility of antibody libraries and show that libraries can be derived from the immune repertoire of virtually any species for the selection of MAbs. In Section 16.4, we predict that the use of immune libraries will continue to expand greatly for other animal species especially with the development of xenogenic mice. Overall, this review summarizes progress in antibody libraries generated from immune repertoires.

16.1.1 History of Phage Display Development

The originator of filamentous phage display was Smith [7]. While developing (f-phage) vectors as a biological "way station" for cloned DNA, Smith made an important leap. He displayed a solvent exposed peptide upon the surface of an (f-phage) by creating a fusion protein with a native phage pIII protein. This linkage of genotype and phenotype has, nearly 20 years later, developed into an entire new field of phage-based ligand selection systems. Smith went on to display large poly-peptides [8] but found that they significantly reduced the phage viability, and the phage vector was biologically inefficient compared to today's phagemid vectors. The first selectable phage display libraries were published in 1990 using peptide ligands expressed upon the surface of f-phage [9–11].

The ability of prokaryotes to express functional antibody binding domains of eukaryotic origin was critical to the development of antibody phage display. The simultaneous publication of prokaryotic expressed variable fragments (Fv) (no linker) [12] and significantly more stable, and larger, antigen-binding fragments (Fab) [13] demonstrated that *Escherichia coli* was capable of expressing functional antibody-binding domains, and, moreover, that some functionality was still present in these molecules. The next major development was a procedure to clone, en masse, the B-cell pool of expressed *V* genes. This came about from the efforts of several inde-pendent groups. Studies in the late 1980s and early 1990s by James Larrick's lab demonstrated the power of using polymerase chain reaction (PCR) for cloning unknown *V* genes from hybridomas [14–16]. Using oligonucleotide primers with designed degeneracy in the leader regions, or consensus primers, in combination with isotype-specific back primers to the constant domains, the specific I_g cDNA could be amplified and cloned. While the oligonucleotide PCR-based approaches work well with hybridoma-derived RNA encoding MAbs, it was not clear at the time if it would work for the amplification of a library of polyclonal antibody responses produced in an immune response. What followed next was the extension of this tech-nique to pools of I_g RNA. These efforts were driven largely by increasing confidence in PCR that was progressing rapidly at the time. It was realized that a representative polyclonal *V* gene cDNA pool from an immunized host will inherently reflect the increased representation of antigen-specific *V* genes due to clonal expansion and plasmablast formation. Thus, regardless of the inherent bias in the PCR itself, a large pool of *V* gene cDNA is representative of the immune repertoire of an immune host and is rich in antigen-specific B-cell RNA.

The first antibody libraries were from immune sources and were cloned into lambda phage vectors. Lambda phage vectors were used to set up combinatorial anti-body libraries, which required screening via exhaustive plaque lift screening assays. Initially, immune libraries from mice and humans were assembled in this fashion. The expressed *V* gene cDNA pool encoding the heavy- and light-chain antibody-binding domains was assembled and cloned separately as distinct PCR amplified cDNA "cassettes" encoding the V_H and V_L regions as a Fab fragment [17–19]. This marked the beginning of a series of very rapid advances in antibody display libraries.

The next major advance was the ability to select binding clones from antibody libraries. The development of antibody libraries originated mainly from the com-peting interests of two labs. The initial work in immune antibody libraries done in the United States was by the Scripps Research Institute in La Jolla, California, by Dr. Burton, Dr. Barbas, and Dr. Lerner. The other group, in the United Kingdom, was led by Dr. Winter and Dr. Milstein at Cambridge. These two groups developed original different approaches for obtaining libraries of antibody-binding domains, scFvs, or Fabs displayed on phage. Those important new advances to the display systems include scFv display on pIII [20], the first immune mouse scFv library [21], Fab display on pVIII [22], Fab display on pIII [23], and the first immune human Fab library [24]. Soon thereafter, naive human scFv [25] and fully synthetic human Fab libraries [26] were published.

Immunoglobulin mRNA from a mouse immunized with a hapten was used to produce a lambda phage antibody library and was screened (not selected) for expression of Fabs produced by the random combinatorial assortment of heavy- and light-chain fragments [17]. The development of recombinant antibody phage display technology and combinatorial immune libraries soon followed [27,28]. Display libraries brought about selection of binding clones, whereby specific MAbs are selected *in vitro* upon antigen. Similarly, in the first monovalent combinatorial antibody display vectors, the V_L and V_H cDNAs were amplified with primer sets bearing four different restriction sites corresponding to unique restriction sites in the phagemid vector, with two unique cutters each for a total of four enzymes [24]. The ability to link genotype and phenotype lays in the genetic fusion of the heavy-chain region to the gene encoding the C-terminal domain of the minor phage coat protein pIII. This in-frame fusion protein, in the *E. coli* coinfected with helper phage, is coloaded onto the surface of phagemid particles containing the plasmids with the f-phage packaging signals. This ushered in the era of phagemid biopanning.

Phenotypic mixing in an *E. coli* host coinfected with helper phage resulted in the assembly of fully infectious phage particles. The particles carry the combinatorial libraries of displayed Fab fragments linked to the *V* genes, which encode them. The *V* gene cDNAs used in these early libraries were cloned sequentially with two independent rounds of ligation into the phagemid vectors. Generally, the V_L cDNAs were cloned in first, and then the V_H cDNAs second. The oligonucleotide primers were designed to have restriction sites to match either the heavy- or light-chain "site" in the vector using two unique restriction sites each to achieve directional cloning [29].

Today immune phage antibody libraries can select MAbs from virtually any species with enough information known about the *V* genes. This is because the methods are founded upon an understanding of the molecular genetics of the species under study, rather than the availability of stable myeloma cell lines or methods of cell immortalization. The phage antibody technique has been used to generate MAbs from a wide spectrum of species. In this review, we confine ourselves to phage-based display systems, although several other systems are available, including yeast [30], ribosome [31], and bacterial display [32].

16.1.2 Classification of Immune Libraries

Immune antibody libraries are created from the antigen-sensitized I_g repertoire of the host's available B lymphocytes. A sensitized host is required to produce an immune library. The technique is dependent upon the ability to recover the expressed I_g repertoire from recoverable B-lymphocyte mRNA, and the construction of a representative antibody library displaying functional antibody fragments for the selection of individual clones against an antigen source. We define "immune libraries" as antibody cDNA expression libraries from the available B-cell lympho-cyte pool, wherever its origin, of any host, which has been immunized, infected, or exposed to an antigen (endo or exo origin). This definition includes experimentally immunized animals; any species having experienced an infectious disease; expo-sure to a pathogen, toxin, or venom; neoantigen exposure through the development of cancer; breakdown in tolerance in autoimmune responses; or any other example where an antibody response has occurred. Clearly immune libraries exist as a spec-trum of immune states with differing antigen reactivity depending upon the B-cell source and multiple factors including properties of both the host and the immunogen.

Immune libraries usually represent the IgG subclass of B cells of an immune animal [33]. All immune antibody libraries have several important characteristics, which include the following: (1) immune libraries are enriched in antigen-specific antibodies; and (2) immune libraries include affinity-matured-binding domains when derived from species that undergo this antigen-dependent process. Clonal selection leads to the enrichment of antigen-specific B cells, and somatic mechanisms such as hypermutation and receptor editing followed by selection lead to affinity maturation. These processes are discussed further in the following.

Immune libraries are inherently biased for binding clones against the immuno-gen. The *in vivo* immune system, which has evolved over millions of years, is still the most efficient means to produce high-affinity antibody to a foreign antigen simply by injecting it into a host and allowing a normal immune response to proceed with its natural molecular and physiological mechanisms. Additionally, it is of great interest and importance biologically to clone the heavy- and light-chain domains of antibody responses naturally produced *in vivo* to given antigens and pathogens.

We have implemented a taxonomical classification of immune antibody librar-ies. Whereas in the past antibody libraries were classified based upon display frag-ment formats or antigen selection strategies, the rapidly expanding field of immune antibody libraries no longer permits this. We have classified immune libraries based upon the species of origin (Figure 16.1). Figure 16.1 depicts all of the species in which antibody libraries have been published to date as well as the potential libraries that may be constructed from immunized fish species.

Monoclonal antibodies have been selected from antibody libraries produced from immune repertoires of laboratory animals (mice, rabbits, and chickens); large animals (sheep, cattle, and camelids); nonhuman primates (macaque and chimpan-zee); and humans. This species-specific classification is broken down further for humans, and it is based upon the type of disease under investigation. Several other species, including axotol [34], rat [35], horse [36], dogs [37,38], cats [39], mink [40], and swine [41], have enough molecular information known to develop reasonable antibody libraries, but MAb selection from these types of libraries remains to be demonstrated.

Libraries from chimeric systems are considered to be derived from the species, which donated the I_g genes. The use of chimeric animals presents a unique immune repertoire and allows experimentation that otherwise could not be performed. For example, human antibody libraries have been derived from SCID mice engrafted

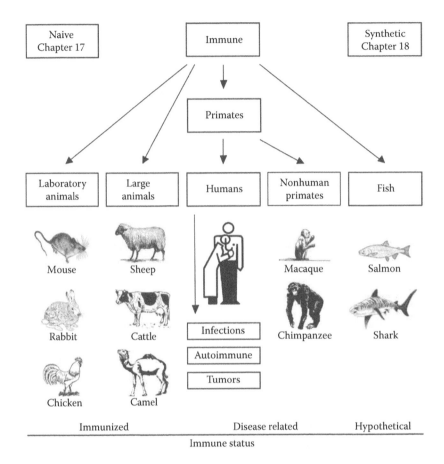

Figure 16.1   Taxonomical classification of antibody libraries. This figure depicts the main animal species from which immune antibody libraries have been used to select monoclonal antibodies. Laboratory animals and humans have been used more frequently due to earlier efforts to understand the antibody genetics. Libraries from large animals (including cattle, sheep, camelids, horses, pigs, and dogs) are not far off. Human immune libraries can be broken down into the types of disease. In many cases, a specific antibody is sought for protective capacity, to prove or imply role of a protein in pathogenesis of disease, or to target a tumor or other self-proteins. Nonhuman primates are also used because of their close relationship to humans. MAbs from primates should prove to be interchangeable for human therapy and are also of great interest. Primates have been used to perform "mixed" immunization experiments with hepatitis and have revealed that multiple antigens can be used to make immune libraries for selection of MAbs to multiple antigens. Fish represent a hypothetical immune antibody library system where the genes are known and scaffold has been made, but no publications on immunized fish antibody libraries to derive fish MAbs have yet appeared in the literature.

with human peripheral blood lymphocyte (human-PBL-SCID) [42,43]. In this case, the B cells are fully human and carry the repertoire from the individual who it was derived from. However, the system allows boosting of the B-cell compartment to a specific antigen. This is thought by some investigators to be capable of enriching the human B-cell compartment in these huPBL-SCID mice with antigen-specific B cells. Similarly, mice-made transgenic for human I_g gene loci is another form of chimera [44]. In this case, the I_g genes, which are human, are transferred rather than B cells. Thus, in V gene transgenic mice, I_g gene rearrangement, antibody expression, and B-cell tolerance are all carried out in the context of the mouse system. In the future, when human antibody libraries are derived from these transgenic "XenoMice," they will also be considered to be of human origin and the primers will need to match the species of the donor of the V genes.

16.2 IMMUNE ANTIBODY LIBRARY CONSTRUCTION

The immune system evolved to produce a diverse spectrum of high-affinity antibodies to foreign antigens. It does this while accommodating the individuality of each host. Thus, each MAb discovery method has inherent biases, which affect the repertoire of MAbs, which can be obtained from immune animals. Biases in phage display include those inherent to PCR amplification, restriction digestion, bacterial expression, and folding/toxicity, which can affect binding to the original antigen as a recombinant form. Obviously, it may be very difficult to select anti-*E. coli* lipopolysaccharide MAbs using f-phage display.

The development of phage antibody libraries from the B lymphocytes of antigen-sensitized hosts represents a major advance in recombinant antibody technology. The ability to select phage-borne ligands with a linked genotype adds a powerful tool for antibody discovery. In all areas of biotechnology and biopharmaceutical development, there is a high demand for more efficient generation of MAbs. This section considers some general points regarding antibody libraries. In particular, we emphasize the appropriate choice of immune or other libraries, host immune status, species, and tissue source.

16.2.1 Monoclonal Antibody Technology

Antibody libraries, at best, reflect the immune status of the host B-cell repertoire from which they are derived. The immune status of the host directly impacts upon the success of retrieving monoclonal antibodies from the antibody repertoire. In any case, the RT-PCR–based cloning procedure captures diversity from the antigen-sensitized available B-cell repertoire. For example, the repertoire in the newborn mouse is restricted by the absence of N-region diversification and the preferential recombination of certain V gene elements [45]. In contrast, adult mice have developed long-term and highly reactive memory B cells within the B-cell repertoire. These somatic changes are intimately linked to I_g gene assembly and diversification processes. These assembly and diversification mechanisms are remarkably conserved among most jawed vertebrate species. Subtle differences in germline immune repertoires likely arose as a result of evolutionary selection pressure over time [46].

While the actual number and chromosomal location of I_g genes differ from species to species, the elements involved in the construction of I_g repertoires are highly similar. Thus, not only is the available repertoire shaped in a different background of self-proteins between species, but also the germline armament itself is different from species to species and even within a species.

Each method used to develop MAbs from immune animals captures a slightly different "snapshot" of the immune repertoire, and the biases inherent in each method prevent any one method from representing the entire available repertoire. However, all methods of MAb development from immune repertoires require immunization or exposure of an animal to an antigen. This sensitization results in a skewing of the antigen specificity of the host's B-cell pool toward the foreign antigens through the development of immune responses.

16.2.1.1 Methodological Options for Monoclonal Development

Monoclonal antibodies have been and continue to be routinely produced by commercial companies using the hybridoma technique for research and diagnostic tools [47]. This is because the hybridoma procedure is quite robust for rodents and is traditionally the most efficient means of producing monoclonal antibodies to date. However, despite recent advances in the establishment of new myeloma partners for various species [1], hybridomas are not as reliable for producing nonrodent MAbs. The use of immune libraries to select MAbs is a specialized use of phage libraries, which require knowledge of immunology and the molecular genetics of antibody gene expression and f-phage display cloning, which collectively has contributed to a less widespread use of immune phage libraries compared to hybridoma production. The latter is simpler, requiring only knowledge of immunology and the availability of mammalian tissue culture facilities.

The method used to produce MAbs is important and can affect the type of MAbs discovered. In what has historically been a controversial issue, it is now clear that the identical monoclonal antibody can be isolated to the same antigen by using either the main technique or its adaptation such as ribosome display. However, this may be a rare find as each system captures a distinct representative cross section of the B-cell response [18,48–50]. We would like to point out that these techniques are highly complementary and that neither is likely to provide an exhaustive sampling of the immune response [51,52].

Hybridomas themselves represent another "immune pool" to exploit using repertoire cloning. Many laboratories have valuable hybridomas to important targets, but no ability to create recombinant versions of the MAbs or to even clone and sequence the I_g V genes. The PCR cloning of V genes from hybridomas is how many scientists become interested in recombinant antibody technology. Many excellent papers have been produced on the optimization and PCR cloning of V genes from human and murine hybridomas using PCR and are a good source for primer sequences [14,15,53–56]. More recently, improved methods for the PCR cloning and *in vitro* recombinant expression of murine scFv [52] and Fabs [23,57] or methods for both [58] have been developed and provide an excellent resource for cloning *V* genes from hybridomas as well as for setting up immune libraries from murine, chicken, human, and primate B cells. A good example of the complementary nature of libraries and hybridomas was the production of a Fab library from antigen-specific hybridomas of a mouse immunized with a chemical hapten [59]. They developed a method that allowed for the direct chemical selection for catalysis from antibody libraries, wherein the positive aspects of hybridoma technology were preserved and were simply reformatted in the f-phage system to allow direct selection of catalysis *in vitro*. Through this chain shuffling approach, they identified novel catalytic reactivates that would likely not be available through the direct screening approach. This two-step procedure based upon an initial hybridoma screen, followed by catalytic selection with phage antibody, is clearly cumbersome and requires expertise in both phage display and hybridoma development. However, it is an incremental advance toward new approaches, which may in the future combine the initial selection and screen for catalytic antibodies in a single step.

V gene cloning preserves the specificity of a hybridoma clone. The cloning and sequencing of the *V* genes from a hybridoma are increasingly important as it further characterizes the binding domain for proprietary purposes as each binding domain has unique identity inherent in the *V* genes themselves. Laboratories with programs in both recombinant antibody technology and hybridoma development are at a distinct advantage as these technologies support each other very well for patent protein.

The choice of method used to produce MAbs depends upon a large number of factors. This even includes the downstream purpose for making the MAb. Figure 16.2 outlines the general flow of producing MAbs from immune libraries compared to hybridoma production followed by recombinant cloning. The important thing to keep in mind is that the type of tests used to screen/select for a given MAb should be as close as possible to the intended use for the MAb. For example, if you need to develop MAbs for a diagnostic competition ELISA (C-ELISA), then either the first or second level of screening should be in the format of the C-ELISA. This is provided that antigen is optimized and used to identify binders in primary screen in a viable ELISA. ELISA conditions must be within the range of MAS sys 50–400 mg/mL.

Several characteristics of repertoire cloning and immune libraries will continue to make phage display an advantageous technique relative to hybridoma fusion methods. These include the following:

1. The ability to select isotype-specific libraries (including IgA [60] or IgE [61]) or to rapidly convert selected binding fragments to full-length antibodies of any human or other animal's isotype *in vitro* [62].
2. The ability to affinity-select binding clones from a library and not merely screen.
3. The ability to simply freeze down libraries as cDNA and screen again in the future against other antigens (there is only one shot at screening with hybridomas).
4. The ability to negatively subtract background phage allows antibodies to complexes of antigens to be selected.

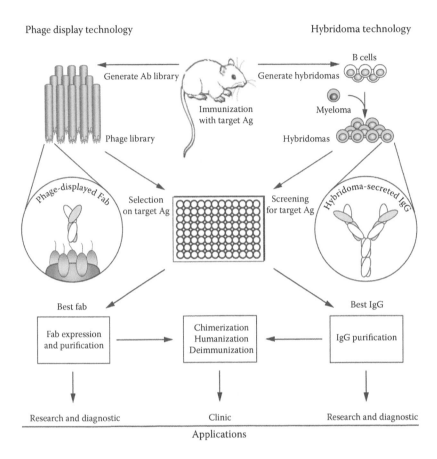

Figure 16.2 MAb generation exemplified by phage display and hybridoma technologies. These two technologies are used to produce MAbs for research, diagnostics, and clinical development. Depicted on the left is the procedure for obtaining MAbs from antibody libraries using phage display. Depicted is Fab display, but third-generation vectors like pComb3X are capable of displaying scFv as well. Binding phage is enriched through successive rounds of selection on antigen. The best MAb fragments are identified through screening of the eluted phage. The right side shows the procedure for generating MAbs from hybridoma technology. Immune splenocytes are immortalized through fusion with myelomas and drug selection. Clones producing binding MAbs are identified through screening for binding antigen. While an IgG is depicted, other isotype such as IgA or IgM could also be identified. The pathway is identical, and only secondary reagents need to be adjusted for genetically engineered chimeric mice or normal mice. The middle part shows the procedures for taking MAbs produced in either pathway into the clinic. MAbs produced from nonprimates in either pathway need to be genetically engineered to be more human in sequence. These modified MAbs must be tested for safety, modified *in vitro* for reduced immunogenicity, and reconstructed for optimal full-length Ig expression. Any time an *in vitro* modification is performed, the new version of the MAb molecule must be rechecked for binding and safety characteristics. The current industrial practice, regardless of the technology used, is to clone the final candidate MAb into mammalian expression vectors and to establish stable cell lines for high-level expression through gene amplification or other procedures. Ag, antigen.

5. Phage display lends itself to all molecular approaches as you can immediately sequence the clones based on vector-specific primers and pursue any *in vitro* modifications promptly.
6. The potential for high-throughput screening of antibody libraries [63]. The randomization of V_H and V_L pairs, which occurs inherently in phage display, may be considered as a drawback of the technique but lends itself to sample an even more diverse repertoire not shaped by *in vivo* tolerance.
7. The ability to select nonrodent MAbs.

The only prerequisite for immune antibody libraries is that the I_g genes of the species in question are known in sufficient detail to allow design of oligonucleotide primers for PCR cloning of antibody-binding domains from immune lymphocytes.

Decisions concerning which approach to use must be based largely upon the expertise of the lab, as well as the end use for the MAbs. For example, when human IL-5 was used to produce MAbs from both hybridoma and the immune library methods, completely different representative MAbs were discovered [51]. However, this is not surprising given the vast redundancy of the immune system, the differences between the methods, and the fact that the antigen is conserved. Also, the number of animals used in the study on IL-5 is small, and the library sizes are large so it is really not useful to conclude that one method is better than the other, as both work.

If a purely diagnostic or research MAb is needed to a given antigen, which has been cloned and is available in pure and large amounts, then the development of MAbs via the classical murine hybridoma is the simplest and most recommended approach for the majority of cases. In keeping with the complementary nature of the technologies, hybridoma-derived MAbs can always be converted to recombinant forms as the hybridoma line provides an unlimited source of specific I_g mRNA [64]. This is particularly useful for previously identified hybridoma-derived Mabs, which have demonstrable potent biological properties. Furthermore, as Ames et al. [51] point out, by back-shuffling hybridoma-derived neutralizing V_H genes into the light-chain immune libraries, novel V_H/V_L-neutralizing Fabs were isolated, showing the powerful combination of these two methods of MAbs discovery. It is clear that for generating nonrodent-derived MAbs, the immune library is the most reliable method of choice.

16.2.1.2 *Factors Influencing Choice of Category of Antibody Libraries*

There are many factors that impact upon the choice of library format. We recommend the best choice of library according to the application of the desired MAbs in Table 16.2. Laboratory animals are perhaps the best choice for producing MAbs for research tools. They are small, easily handled and housed, and inexpensive. For veterinary diagnostics, large animals of the homologous species, as the disease, are optimal for the highest accuracy in validated tests. For human therapeutic antibodies, fully human or some primate MAbs are a recommended source of lymphocytes as they require no humanization and will have much lower odds of containing cryptic T-cell epitopes.

Table 16.2 Factors Influencing the Library Choice

| Antigen Restriction | Library | Antibody Application | | |
		Research	Diagnostic	Clinic
Available immunogenic nontoxic	Immune			
	Laboratory animals	Best choice	Recommended	Recommended
	Large animals	Recommended	Best choice	Worst
	Primates (human)	Worst	Worst	Best choice
Unavailable nonimmunogenic toxic	Naive synthetic	Chapter 17 Chapter 18		

For therapeutic purposes, every MAb produced regardless of the technology must be tested for pathological contraindications. These would likely be due to potential T-cell epitopes (in synthetic libraries, in mice-bearing human B cells, and in V gene transgenic mice), alternative glycosylation patterns (all murine-derived hybridomas), or even serendipitous autoreactivity possibly generated through somatic or *in vitro* modifications. Indeed, it is incorrect to imply that fully synthetic human MAbs have therapeutic potential without advising that they will need as rigorous testing as nonhuman antibody. Humanization (see Chapter 14) or deimmunization [65] offers strategies to avoid, mask, or redirect this human immune surveillance of these molecules. These strategies include the genetic sewing of murine variable regions to human constant regions as "chimeric" antibody molecules, the deimmunization by removal of potential T-cell epitopes, and the humanization of nonhuman antibodies by grafting nonhuman (rabbit, mouse) binding residues onto a human antibody framework [66]. These procedures are in themselves very technical and extremely time-consuming and should be avoided if possible as they can also lead to the appearance of modified fine specificity or unwanted cross-reactivity on their own [3], whereas the natural immune system inherently, and rapidly, disallows these potentially deleterious reactivities. Conversely, humans are a poor choice for the production of antibodies for diagnostic or research purposes, as you cannot hyperimmunize them and the amounts of B cells harvested must be within ethical limits. Finally, if an antigen is toxic to all forms of life, we would recommend either the inactivation (toxoid) of the molecule if possible or to use premade very large synthetic and/or naive libraries.

Immune libraries can be custom-designed to select specific classes of antibody. Scientists mainly seek IgG MAbs as they have excellent neutralizing properties and have usually undergone affinity maturation and thus have the I_g V genes, which encode MAbs, which have been optimized for antigen recognition. In general, it takes around 2 weeks for B cells specific for conventional antigens to class-switch and undergo affinity maturation [67]. This must be taken into consideration when trying to derive IgG MAbs regardless of the downstream method used to produce the MAbs. For example, laboratory animals are generally hyperimmunized in order to produce high-titer IgG antibody responses, which are largely expressed from affinity-matured B lymphocytes. Antibody of non-IgG classes can also be selected in order to study particular diseases or immune compartments. For example, human IgA [60] and IgE [61] have been selected from immune libraries to study respiratory syncytial virus (RSV) and allergy, respectively.

Immune libraries usually provide specific MAbs against one antigen (immunogen), although there are a few cases where libraries have been used to select MAbs to multiple antigens or infectious agents; for example, when animals were initially coimmunized with multiple immunogens simultaneously and thus had enriched the B-cell pool in antigen-specific B lymphocytes to multiple antigens [68], or in humans where the patients were known to be convalescent to multiple infectious diseases [69].

It is possible to reduce the time needed to elicit an IgG antibody response for some antigens. The serum IgG antibody response, which normally takes about 2 weeks to develop, has been elicited in a matter of days to peptides, proteins, and whole virus under specific conditions (Refs. [70,71], respectively). In the case of peptides and proteins, the antigens were delivered *in vivo* to CD11c$^+$-dendritic cells by linking the antigens to hamster antimouse CD11c in an immunotargeting approach [72]. Similarly, extremely rapid B-cell kinetics is characteristic of the high-affinity neutralizing antibody response produced in mice to vesicular stomatitis virus (VSV). This fast response does not incur further affinity maturation and shows only a subtle shift in the *V* gene usage as the response develops [73,74]. This would serve to reduce the costs of MAb production by reducing the amount of antigen needed to produce an immune response and by reducing the time needed to house animals.

It is difficult to produce antibody responses against endogenous self-antigens. This is especially true for self-antigens with high conservation between species. Rodents are phylogenetically related to humans and have high identity in many of the important regulatory molecules and ligands of the immune system. This can pose problems in terms of immunogenicity and immune recognition of highly conserved antigens for which classic tolerance may be observed. In this case, immune libraries from other more distantly related animals such as rabbits or birds may be useful as they are more likely to recognize these antigens as foreign.

Immune libraries are not particularly useful for fishing out MAbs to targets that the host immune systems were not sensitized against. In principle, large prefabricated nonimmune (naive) or synthetic antibody libraries are slowly becoming a source of recombinant MAbs with increasingly useful affinities. These kinds of libraries offer the possibility of selection of high-affinity antibodies to any desired antigen without the need for involving live animals or immunization, or for the investigator to be involved in the library construction [3,4]. Theoretically, a single library made in one laboratory could be used by many others to select antibodies to diverse targets.

In general, the size of the library is proportional to the chances for selecting a good clone. For antibody libraries created from immune repertoires, only 10^6 individual bacterial clones are likely an adequate size to ensure selection of a high-affinity clone [75]. This is in contrast to naive or synthetic antibody libraries, which require enormous diversity on the order of at least 10^{10} clones to ensure selection of rare high-affinity clones [6,33,76–79]. One advantage to these types of libraries from large synthetic repertoires is that a single library can be used for the selection of antibodies against any antigen in theory [4] although the affinity of these antibodies tends to be lower [80].

The availability of synthetic or nonimmune libraries offers a potential solution to the problem of identifying antigen binders in cases where it is impossible to immunize (lack of antigen, antigen is toxic or not immunogenic). However, we and others [81] feel that the naive or synthetic libraries may be inferior to immune libraries. The current success rate in utilizing premade naive and synthetic libraries to isolate binders, as inferred from the literature, is not widespread, although we understand that most of these data have been obtained by companies and are not published. Only the antibody recovered from human lymphocytes or chimpanzees, which are essentially genomically identical to humans, and which is expressed in a human cell line, should be most suitable for use in humans. All others (naive, synthetic, chimerized, humanized, transgenic mice, etc.) could fail in clinical trials because they are targeted by the host, containing new T-cell epitopes, improper glycoslyation, and improper folding. They could also be dangerously cross-reactive with epitope on human tissues. Moreover, since the MAbs obtained from these nonimmune or synthetic libraries are not matured against the antigen of interest, we expect that the affinities will be low so that extensive *in vitro* affinity maturation steps are required to improve on their binding capacity [81]. This is as much work as the initial library construction and screening for an entire second library. Therefore, in the end, it is likely much faster to use the potential of the immune system to enrich the antigen-binding B cells via clonal expansion and to perform the affinity maturation, and then to generate a relatively small immune library.

16.2.1.3 Sources of Blast Cells

There are several important aspects to consider about the source of the B cells before creating an immune library. In terms of both sources, the species of origin and the tissues are the most important. For example, if human therapeutic antibodies are required, then human libraries may seem to be the most logical source of B cells. However, B cells may not be readily available from convalescent, immune, or exposed humans. There must be a bona fide B-cell response to the antigen of interest. In other cases, it may not be ethical to acquire the best tissue sample from the best source. For example, it may not be ethical to extract bone marrow from human immunodeficiency virus (HIV-1) resistant or even long-term nonprogressor individuals living in Africa who are already at high risk for acquiring HIV or other debilitating diseases. As an alternative, closely related nonhuman primates such as chimpanzees or macaques have been used as a source of B cells for immune libraries to hepatitis E and simiar immunodeficiency virus (SIV) [82,83].

The source of tissue is critical for immune library generation. The number of antibody reactivities is influenced by properties of the antigen, the number of antigen exposures, or boosts as well as inherent properties of the host immune system.

However, success depends upon adequate sampling of the sensitized B cells. Clearly, these issues raise another question regarding the stage at which a B cell is most useful for immune library generation. While comprehensive studies have not yet been carried out, the plasmablast stage is thought to be the most useful for immune library generation as they contain about 100–300 times more specific I_g mRNA than a resting B cell [84,85]. Once the B cell becomes terminally differentiated, it becomes greatly enlarged. This is in order to accommodate the increased size of the endoplasmic reticulum, increased cellular volume, and high volume transcription, which is occurring in the nucleus (Figure 16.3). Fully differentiated, antigen-activated blast cells are rapidly present in presensitized animals following a recall response as the specific-IgG serum titer is elevated by day 7. In immune mice, it has been estimated that somewhere between 1000 and 10,000 antigen-specific B cells are generated in response to a very complex antigen-like whole virus [86].

Tissues rich in antigen-specific plasmablasts, such as bone marrow [87] and spleen [88], are ideal for generating antibody libraries provided that they are harvested at appropriate times. Table 16.3 outlines most of the B-lymphocyte sources, which have been used for laboratory and large animal as well as for human/primate library production. Clearly, some tissue sources are not easily accessed for some species, such as the bone marrow in mice. Peripheral blood contains a limited number of B cells secreting antibody (few blasts) and thus is not an optimum source of lymphocytes for the generation of representative immune antibody libraries [89], but it is easily sampled and does not require cadaver or surgical procedures. The spleen of adult mice contains about 54% B cells, which make it ideal for immune library

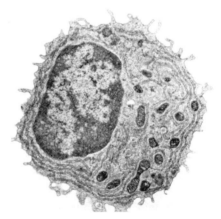

Figure 16.3   B-cell blast. B cells become activated and differentiate into memory or blast cells. The blast cell becomes greatly enlarged to accommodate the increased production of antibody. Note the large Golgi and endoplasmic reticulum and the engorged nucleus, which likely occurs to handle higher transcriptional activity. Blasts are enriched 100–300-fold in the specific Ig mRNA. Each B-cell blast produces a different antibody according to its own rearrangement and somatic modifications. (Reproduced from Kincade PW, Gimble J. B lymphocytes. In: Paul WE, ed. *Fundamental Immunology*. With permission from Lippincott Williams and Wilkins.)

Table 16.3 Possible Sources of Blast Cells

Library	IgG mRNA Source[a]				
	PBL	Spleen	Bone Marrow	Lymph Node	Tissue Biopsies[b]
Laboratory animals	+	+	+	−	−
Large animals	+	+	−	+	+
Primates (human)	+	−	−	−	+

[a] For references, please see corresponding species sections.
[b] Live samples of any origins.

generation or hybridoma production [90]. Of note, the lymph nodes of large animals and even the cervical lymphocytes collected from swabs of humans have served as successful B-cell sources for immune antibody libraries [69,91].

For repertoire cloning, one must try to maximize the antigen-specific I_g messages in the mRNA population. While this is done proactively by using primer sets that amplify the mRNA of isotypes of heavy and light chains found to bind the antigen of choice, the timing of harvest can be critical as well. Lymphocytes harvested from different sites, or different antigens, have different properties, and we would recommend that investigators anticipate that some optimization of the boost step and the actual day of harvest is required to increase the likelihood of successful MAb discovery. For example, there is a delayed response to some proteins expressed *in vivo* from viral or naked DNA vectors, and thus, the harvest may have to be extended to 8 or 9 days to ensure maximal B-cell stimulation on the day of tissue collection and recovery of specific mRNA.

In the case of immune libraries derived from laboratory animals, the immunization strategy must try to target tissue harvest to coincide with the time of maximal B-cell blast formation. Hyperstimulation of B cells is routinely used for the production of diagnostic antiserum from larger animals and birds. However, overstimulation of the murine B-cell pool with too many or too frequent repeated injections was found to produce negative effects upon the yield of antigen-specific hybridomas [92,93]. This is likely due to both clonal exhaustion and deletion. Prolonged rest following hyperimmunizations, to allow for the recovery of the B cell, has contrasting effects upon recovery of this pool. Thus, immune responses should be allowed to wane between boosters to maximize the effects of affinity maturation *in vivo* and the peak antibody titer determined. Thus, in general, B-cell harvesting should be done in lab animals about 3–5 days or 7–8 days postboost for hybridomas and immune repertoires, respectively.

16.2.2 Cloning Strategies

Once an immune source is decided upon, the expressed antibody gene pool needs to be cloned and expressed in a selectable form. Immune libraries focus upon the amplification of cDNA produced from the B cells of an antigen-sensitized host. The use of the highly specific PCR approach removes the need to purify B lymphocytes from other immune cells. Clean preparations of total RNA are sufficient and all that is required to build good libraries. The expressed I_g mRNA pool, the genetic material encoding the antigen-binding regions of Igs, is reverse-transcribed into cDNA, amplified by specific primer pairs in PCRs, and ligated en masse into phagemid display vectors. There is no need to purify mRNA away from total RNA. The emphasis is on the quality of the RNA. The quality is greatly affected by the contaminants, which are produced during harvest and the length of time the RNA is exposed to the contaminants before being used to generate cDNA. It is advisable not to delay the library construction at this point and to generate the cDNA as soon as the RNA is made.

The cloning strategy is usually dictated by the choice of display system. The choice of vector, the primers, and all necessary supplies must be on hand and working by the time the cDNA is prepared. However, before considering the various methods used to assemble antibody libraries from cDNA, one must consider the development of the cDNA amplification process. When RNA is isolated from pools of B cells, there is inherent scrambling of the natural heavy- and light-chain pairs. This is due to the fact that, upon harvest and lysis of the lymphocytes, all of the mRNA is released into a general pool. These pools are PCR-amplified en masse by heavy- or light-chain specific primers and randomly repaired by the assembly processes. It is likely that natural pairs are formed in antibody libraries and have contributed to the success of this strategy, as clonal expansion will clearly increase the chances of the library being able to recapitulate these pairings [94].

The genomes of most phagemid vectors are relatively small and easily purified in the replicative form as a double-stranded DNA using standard plasmid purification procedures. In general, the PCR-amplified V_L cDNA is gel purified and pooled. The light-chain cDNA is cut with the "light-chain" cloning enzymes, purified again and ligated to the precut phagemid vectors [24]. This "light-chain library" is transformed into *E. coli*, amplified by bacterial growth, purified as a plasmid, and cut again with the "heavy-chain" enzymes. Following this, the V_H cDNA are ligated into the vector, which was again transformed into *E. coli*, amplified, and double-stranded phagemid DNA once again purified. This is the purified immune library, which is used to transform *E. coli* for packaging into phage particles upon coinfection with helper phage. The order of cloning the V_L first then the V_H was believed to be important to avoid the potential loss of V_H genes by deleterious internal cutting by the light-chain enzymes as the V_H genes tend to be more important in determining antigen specificity. This meant that the representative cDNA of the cloned antibody pool was subjected to the effects of cutting by four restriction endonucleases as well as the inherent biases of the amplification of the light-chain repertoire in the prokaryotic systems and left room for improved next-generation vectors.

Detailed protocols for cloning most species that have been used successfully for the selection of MAbs from immune repertoires are described in detail elsewhere [29,81,95,96].

16.2.2.1 *Antibody Diversity, Clonal Expansion, and Affinity Maturation*

In many animal species, the molecular genetics of I_g genes is now sufficiently understood to apply the technology of immune antibody library cloning for the isolation of species-specific MAbs. Multiple divergent *V* gene families make up the primary antibody repertoire of mice and humans [97]. In contrast, other mammals such as the pig or chicken may have a *V* gene pool composed of a single main *V* gene family. Some species tend to utilize one light-chain class or the other in immune responses for reasons that are not entirely clear. What is clear, however, is that each species represents a unique pool of antibody diversity (Table 16.4).

Somatic modifications create huge levels of diversity. In humans, despite estimates, which place total possible diversity at 10^{12}, at any one given time, there are only about 10^6–10^7 different B-cell specificities comprising the available repertoire. Each method used to develop antibody libraries from immune animals captures a slightly different "snapshot" of the immune repertoire, and the biases inherent in each method prevent any one method from representing the entire available repertoire. However, all methods of MAb development from immune repertoires require immunization or exposure of an animal to an antigen, which results in a skewing of the antigen specificity of the host B-cell pool toward the foreign antigens through the development of immune responses.

Knowledge of the host antibody repertoire is essential for cloning I_g genes and for the generation of immune libraries. In the immune system of most vertebrates, there are several levels of diversity in the I_g (*V* genes) usage [98]. The first level of diversity is the fundamental germline diversity. This is the actual number of *V* gene cassettes and the individual sequence of the I_g gene cassettes. The next level of diversity is junctional diversity, which includes the combinatorial diversity and the way minicassettes

Table 16.4 Species-Specific Immunoglobulin Gene Diversity and Primer Requirements

Species	Total Genes[a] V_H	V_L	Families[b] V_H	V_κ	V_λ	Serum LC (%) κ	λ	Nb Primer Sets[c] ($V_H/V_\kappa/V_\lambda$)	Diversity Mechanism
Mouse	145	> 97	> 15	4	3	95	5	22 (15/4/3)	CJ/SH
Rabbit	> 200	> 4	4?	3	1	90	10	8 (4/3/1)	GC/SH
Chicken	100	25	1	0	1	0	100	2 (1/0/1)	GC/SH
Sheep	10	?	> 1?	3	6	5	95	10 (1/3/6)	GC/SH
Cattle	15	20	> 1?	?	3	2	98	5 (1/1/3)	GC/SH
Camel[d]	40	n/a	1	n/a	n/a	n/a	n/a	1 (1/0/0)	GC/SH
Primate (human)	44	82	7	7	10	60	40	24 (7/7/10)	CJ/SH

Source: Based on the data available at http://www.medicine.uiowa.edu; http://www.ncbi.nlm.nih .gov/; Muyldermans S, *Rev Mol Biotechnol*, 74:277–302, 2001; O'Brien P, Aitken R, *Methods Mol Biol*, 178:73–86, 2002; Flajnik MF, *Nat Rev Immunol*, 2:688–698, 2002; Berek C, Milstein C, *Immunol Rev*, 96:23, 1987.

Note: CJ, combinatorial joining; GC, gene conversion; LC, light chain; SH, somatic hypermutation.

[a] Total number of genes in genome including pseudogenes.
[b] Number of Ig chain families based upon expressed proteins.
[c] Minimum number for an IgG library to represent every family; for precise numbers of primers used, refer to the corresponding section.
[d] For camel single-heavy chains antibodies (HCAbs).

physically recombine. The variability of the N-terminal domain of V gene was subsequently shown to be due to the fact that the N-terminal domains are assembled from modular genes (VDJ). This includes N (nongermline nucleotides added to the coding minicassette joints mediated by TdT) and P addition (addition of nucleotides at the end of a minicassette that form a palindrome) [97,99]. Occasionally, multiple D_H regions are inserted or a minicassette is deleted, which can result in further nongermline changes [100]. Insertions and deletions in the hypervariable loops of antibody heavy chains also contribute to molecular diversity [101]. Recently, receptor editing of Ig chains, or the conversion of an expressed V gene for another upstream V gene, has been shown to be another source of antibody diversification [102]. Thus, *in vivo* immune responses serve to generate higher affinity antigen-specific antibody responses and a more responsive memory B-cell population in the event that the antigen returns to the host.

Despite the general similarities of the immune repertoire of most vertebrates, divergent evolution and shuffling of the receptors are still ongoing today [103,104]. Differences in the mechanisms used to somatically diversify I_g genes also further differentiate the species. Indeed, the repertoire of I_gs is heterogeneous between all species and can even vary between members of the same species [103]. Indeed, germline V gene polymorphisms in humans and mice can be attributed to differences in nucleotide sequences among allelic V genes [105–107], D region sequences, and J_H region elements [108,109], as well as differences in the absolute numbers of V genes [110–112]. The fact that individuals exist with unique antibody genes supports the notion that the species has more diversity than any one individual and that the germline antibody repertoire continues to evolve [113].

Immunoglobulin (B-cell receptor) diversity is generated in different ways and in alternative primary lymphoid tissues in different vertebrate taxa. In general, the cumulative use of mechanisms for antigen-dependent and -independent somatic diversification of antigen-binding domains correlates inversely with the amount of germline combinatorial V gene joining. There are two general systems under which most vertebrates fall for the generation of preimmune B-cell diversity [114]. These are the classical human/mouse diversification systems, where B cells are generated throughout life in the bone marrow, and the gut-associated lymphoid tissue (GALT) system, where a single wave of progenitor B cells populates secondary tissues early in life and after which they are not renewed. In mice and humans, a large preimmune diversity is generated by the combinatorial joining of the numerous VDJ element for V_H and V_J for V_k/L [98]. In contrast, most other mammals (including the rabbit, sheep, and perhaps all birds) generate the preimmune repertoire by diversifying a single or few rearranged V gene segments through somatic hypermutaion and/or gene conversion in GALT. In birds and sheep, this occurs in what is essentially an antigen-independent process, which in birds takes place in the bursa of fabricus [115,116] and for sheep in the gut ileal peyer's patches [116–118]. However in rabbits, the B cells of the GALT require coincident antigenic exposure to the normal gut microflora during development for optimal preimmune diversity [119].

Affinity maturation is the selection of B cells bearing high-affinity B-cell receptor (BCR) in the germinal center reaction, which leads to a selective clonal expansion, entry into the memory compartment, and the emergence of higher affinity soluble antibody. Somatic hypermutation [120,121] is the substitution, addition, or deletion of untemplated nucleotides to a rearranged VDJ or VJ segment that occurs in the mature B cells of all jawed vertebrates. This is an antigen- and helper T cell-dependent response in mature B cells but also occurs in immature B cells at the lambda locus of sheep and the rabbit heavy chain in an antigen-independent manner to generate those species preimmune repertoire [114]. Error-prone gene conversion is defined as a homologous recombination event where upstream *V* genes are recombined into the recombined *V* gene expression site [122]. In the antigen-dependent response, each species tends to use somatic hypermution and/or gene conversion (which may turn out to be homologous to receptor editing in mice) to further diversify the immune repertoire. When the high-affinity memory B cells are restimulated upon booster, they terminally differentiate into RNA-rich blast cells.

The reasons for these vastly different approaches to antibody diversification are not clear. It is clear, however, that somatic hypermutation [123,124], class switch [123], and gene conversion [125] (three B cell-specific DNA modification pathways that are not present in the B cells of all jawed vertebrates) all depend upon the same enzymes to be carried out. Clearly, this shows that other differences exist to produce alternative diversification processes and that much remains to be learned from studying the *V* gene diversification processes of various species [114]. The diversification mechanisms used by each respective species are included in Table 16.4. Rabbits, chicken, sheep, and cattle all appear to have a GALT system and use somatic gene conversion to mutate their preimmune repertoire and somatic hypermutation to make high-affinity variants during immune responses. Regardless of the mechanisms of affinity maturation, many of which differ somewhat from species to species, it is the affinity-matured, antigen-selected pool of blasts that dominate immune libraries.

The ability to generate MAbs from a variety of species will be important not only for veterinary medicine and comparative immunology but also for human therapeutics. Antibodies from different species conceivably target a different array of epitopes on a given antigen, which is likely going to be very important for human antigens, which are highly conserved [4]. Epitopes, which are immunodominant to the B cells of one species, may not be immunogenic in a different species. While the ability to target functional-binding sites with MAbs may be of no importance to diagnostic applications, it is critical for therapeutic development and biomedical research. Thus, each species produces a novel range of antibody structure to select potential agonists and/or antagonists. The repertoire of epitopes targeted on a given antigen is not going to be identical between species because the antibody repertoire is different. Antibodies from nonrodent species have increasing value in this regard as murine antibodies still dominate the majority of human disease models and inherent tolerance mechanisms make it difficult for murine B cells to mount antibody responses to antigens conserved between humans and mice.

16.2.2.2 *Primer Design*

Repertoire cloning and library selections of MAbs are not based upon the random immortalization and cloning of the antibody-producing lymphocytes themselves.

Instead, one needs to know the nucleotide sequences of the expressed I_g genes and to design oligonucleotide primers for RT-PCR-based cloning. The more the diversity in the expressed variable gene pool, the more the primers needed to adequately clone the immune repertoire. However, the cloned I_g pool may reflect a predominantly expressed allele, even if the genetic diversity of a species is quite large. The oligonucleotide primers were all in general designed from databanks, personal or public, of expressed antibody *V* gene sequences from hybridoma clones or simply cloned and sequenced from B cells of the given species or a related species. For example, murine primers have been used to clone rat immune repertoires as they are related rodent species, and the *V* gene sequences of the rat were not as well characterized as those of the mouse [126]. Many excellent databases exist today on the World Wide Web to assist scientists in the alignment and identification of new *V* genes, *V* gene variants, and immunogenetics of variable regions in general (Immunogenetics database; NCBI I_g Blast).

The highly variable nature of the I_g domains argued against the possibility of en masse cloning of immune antibody repertoires with any kind of representative cloned B-cell pool. However, the original analysis of V-domain variability by Wu and Kabat clearly shows that the framework regions (FRs) of antibody variable domains are less variable than the antigen-contacting complementarity determining regions (CDRs) [127]. Thus, PCR primers used for immune library construction are mainly targeted to the relatively conserved FRs 1 and 4, which also tend to incur fewer somatic mutations during affinity maturation. However, highly mutated *V* genes, which do incur mutations within the FRs, may not be amplified but would be unknowingly and inherently lost from the immune library at the outset. The framework homology can be grouped between similar genes, which have been grouped into families according to the degree of relatedness for a given host. For example, *V* genes with greater than 80% identity were designated to be in the same family, and genes with no less than 70% are also included [128]. This information is very important when trying to clone immune libraries from hosts with diverse germline repertoires, for example, in mice and men. Mice have a lot of germline sequence variation in the N-terminal regions of their *V* genes, which poses a disadvantage as it necessitates the use of many PCR primers to capture a good representation of the expressed repertoire. It is expected that the mouse genome project will reveal important information regarding the de facto germline repertoire of the mouse, and this will allow more accurate design of murine immune libraries in the future.

The more complex the expressed repertoire of a species is, in general, the more the primers are needed to clone the repertoire *in vitro*. The primers are generally designed to roughly correspond to a group or family of related *V* genes but can cross-prime members of other families due to homology either of germline or somatically induced. Thus, the number of *V* gene families, and perhaps more accurately the number of functional germline V_H and V_L genes, impacts upon the number of oligonucleotide primers, which need to be designed and used to clone an immune repertoire [129]. Some animals, such as rabbits and chickens, use very few *V* gene segments to encode their I_gs, which greatly reduces the number of primers needed to clone the immune repertoire (Table 16.4). For example, the primer sets need only to focus upon single expressed V genes to clone the expressed immune repertoire of other species like the heavy-chain V gene cDNA of chickens. Usually, for most species, a number of primers corresponding to framework 1 are used in combination with framework 4 or isotype-specific back primers in order to amplify a library of antibody cDNA corresponding to a single class (for example, IgG).

Species-specific primer sets are generally needed to clone immune repertoires. Although the I_g gene families between species are highly related, they are not exactly the same. For this reason, primers must be designed for each respective species. The PCR amplification and library construction may still work as it has for highly related species such as mice and rats [126] and humans and nonhuman primates [130]. However, the representation of the V genes within the expressed pool is not necessarily the same between species, and the primers would have to be evaluated in a case-by-case setting. Thus, the number of consensus primers needed to accurately amplify and clone the representative immune repertoire of a given species roughly reflects the diversity of the germline antibody repertoire for that species.

In Table 16.4, we also suggest a minimum number of oligonucleotide primers needed in order to generate a representative species-specific IgG library based upon the known or deduced numbers of V gene families. The actual number of primers used could be higher or lower and is discussed in more detail in Section 16.3.

16.2.2.3 Display Fragments

There are two main fragments used to display the binding domains of MAbs. There is the Fab, or the recombinant version of the fragment responsible for antigen binding, and the scFv (single-chain Fv) (Figure 16.4). The Fab contains the heavy-chain variable region in-frame with the heavy-chain first constant domain to form what is known as the Fd portion ($V_H + C_{H1}$ = Fd) and whole light chain. In most cases, the C-terminus of the C_{H1} region is genetically fused in frame to the C-terminal domain of the gene encoding the f-phage protein pIII, which loads the Fab onto the extruding phage.

The Fab display systems have several unique features. In Fab libraries, the light chain is encoded as a separate gene and must cofold around the pIII-fused Fd fragment as it is extruded into the bacterial periplasm. It is possible to coscreen positive clones produced from *in vitro* affinity selection using light-chain detection ELISA in the downstream screening of phagemid colonies, which acts as a suitable control for expression. This is done simultaneously by simply coating a duplicate plate with unconjugated antilight-chain polyclonal antibody, and detecting with antiheavy-chain-specific enzyme-labeled antibody (either anti-Fc-isotype or anti-HA tag in the CH–pIII junction of the pComb3X vector). This shows if the vector system is actually working and if bacterial supernatants contain any Fab molecules.

There are several configurations of scFv MAbs. The first examples of recombinant Fv linked together the two variable domains using a polypeptide linker [131].

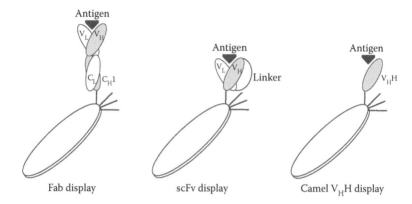

Figure 16.4   Representative antibody fragments expressed on the surface of M13-like phagemid particles. Monoclonal Fab (fragment, antigen binding; Fd + Lc), scFv (single-chain Fv; light-chain variable region linked to heavy-chain variable region), and V_HH (camel heavy-chain Fv; no light chain and no CH1 region) are represented as pIII fusion proteins expressed on the surface of M13-like filamentous phagemid virus particles. These phage-borne proteins link the I_g genotype to the respective phenotype (physical binding properties) in a selectable unit. The ability to select phage antibodies highlights the key advantage antibody libraries have over classical hybridoma screening procedures. Full I_g molecules cannot be selected upon the surface of phage and if desired must be built from these binding domains.

Because the linker is unnatural, many arbitrary configurations have been designed for scFvs. In some cases, the light-chain variable region is N-terminal to the heavy chain, while in others, the heavy chain is N-terminal to the light chain. In general, the linkers have been based upon flexible glycine residues. Perhaps, the most comprehensive study of the scFv and linker configurations was by Whitlow et al. [132] in which the effects of linker length on binding affinity and aggregations were evaluated for an scFv MAb 4-4-20. Linkers may be suitably modified to incorporate epitope tags and cleavage sites, to remove proteolytic sites, or for optimal binding to an antigen. It is likely that the optimal linker will vary depending upon the MAb and the purpose, and could be improved for each clone. Generally, a single linker is used for immune antibody generation for a given vector system [29] and is encoded as an overlap in the oligonucleotide primers used to PCR amplification, V_H, and V_L cDNAs. The use of a unique linker sequence allows forward-and-backward sequencing of V_H and V_L inserts at the junction of the two and obviates the need to sequence the entire scFv from the vector sites at the ends of V_H and V_L. In this way, it is easy to create variations of the linker once a potent clone is found.

The scFv libraries offer several distinct advantages. The scFv library is easier to assemble, as it requires one overlap PCR to join the V_L and V_H instead of two for Fab libraries (one for C_H/C_L fusion; then for Fd/light chain fusion), which reduce the number of steps and time needed to construct the library. Other properties of *in vitro*-modified affinity-selected scFvs and Fabs are discussed elsewhere in Ref. [129].

There are natural examples of single-chain antibodies. SHARKS [114] and camelids [81] naturally express true single-chain antibodies as a part of their humoral response. Immune libraries from camelid species have enabled the selection of *in vivo* affinity-matured V_HH (no C_{H1}) MAbs. This is outlined in detail in Section 16.3.2. The point is that antibody fragments have a demonstrable role in natural protective immunity, and thus, recombinant antibody fragments likewise have potential value in this area (Figure 16.4).

Reliable antibody libraries for the selection of intact full-length antibody molecules have not been demonstrated. Thus, if whole recombinant MAbs are needed, they must be rebuilt from the fragments (Fab or scFv) selected from immune libraries. In some cases, the properties of the fragments may make them useful for therapy. For example, the small size of recombinant scFvs enables penetration of MAbs into the back of the eye by simply adding the protein in droplets [133]. Topically applied scFv was found to penetrate into the anterior chamber fluid of rabbit eyes *in vivo*. The engineered fragments were stable and resistant to ocular proteases. In some cases, better tumor penetration may be achieved with the smaller scFvs, and this may warrant their use in anticancer treatments [134]. Collectively, these studies show that there may be a need to produce the fragments themselves, for example, as a therapeutic modality, and that there are diverse uses for recombinant antibody fragments.

16.2.2.4 Display Vectors

The limitations and plasticity of the much larger phage genomes naturally led to the use of phagemid systems for the display of libraries of protein domains. The development of a helper phage with defective packaging signals enabled scientists to focus upon cloning into much smaller plasmids, which encoded an antibody-binding domain as pIII or pVIII fusion proteins. The early discovery of the two functional domains of pIII by Smith revealed that the C-terminal portion of the pIII protein [7] was all that was necessary to have proteins loaded onto f-phage particles. More recently, other proteins have been revisited for the display of proteins (see Chapter 2).

The combinatorial approach of cloning antibody domains was founded upon the successful cloning of the repertoire into the prokaryotic system. The original experiments, which demonstrated that antibody-binding domains could be expressed and functionally assembled in *E. coli*, were successful because the antibody fragments were secreted from the cytoplasm into the oxidizing environment of the periplasmic space under the guidance of bacterial leader sequences [27]. It is believed that the oxidizing environment and the secretory event collectively possibly contributed to the correct formation of disulfide bonds and proper folding of antibody domains. Analogies are obvious between the recombinant expression of antibody domains in prokaryotic cells and the natural production of antibody in eukaryotic cells, where the heavy and light chains are probably translated as they are extruded into the lumen of the endoplasmic reticulum. A major caveat is the absence of the full-length I_g molecules and the eukaryotic posttranslational modification machinery in *E. coli.* This is another bias inherent to the immune library system in that not all antibody-binding domains will be correctly assembled in the prokaryotic periplasm. However, the selection inherently reveals those that are functional.

Many different phage and phagemid systems have been developed for the display of peptides and polypeptides upon the surfaces of f-phage particles. There are two main display systems for f-phage. The first one for peptide display consists of modi-fied phage genomes with cloning sites in frame with the major coat protein pVIII for which there are about 2500 copies per particle [135]. The pVIII protein is very small and fusions result in multivalent display, which can affect the avidity of antibody domains selected. Moreover, larger polypeptides are not well tolerated as fusions to pVIII in phage systems and thus phagemid systems were developed to accommodate expression of two genes for pVIII and the recombinant proteins are displayed via phenotypic mixing [136,137]. The multivalent display offered by the major coat pro-tein pVIII may be useful for selecting MAbs from non-IgG libraries [22].

In some cases, the use of phage instead of phagemids can allow more efficient selection. The naive human scFv repertoire was subcloned from the phagemid vector pHEN1 into the phage vector fdTET to create an antibody library of 5×10^8 phage clones, which was selected on several recombinant proteins [80]. In this case, multi-valent phage display, compared to monovalent scFv display, resulted in the selection of a larger panel of clones with improved efficiency of display and expression. The average affinity of the clones from phage libraries was relatively low and also lower than the clones derived from the monovalent phagemid display. While this may be overcome by utilizing further *in vitro* affinity maturation steps, this is not simple and clearly shows that for the selection of the highest affinity clones, monovalent display is still the best solution. However, in many cases, affinity is not directly related to biological effect, and the increased efficiency of multivalent phage libraries to iden-tify large panels of binders after a single round of panning may facilitate automation.

Phagemids are hybrids of phage and plasmid vectors. They are bacterial plasmids with f-phage packaging signals, the phage origin of replication, appropriate multiple cloning sites designed to genetically fuse the antibody fragment at the N-terminus of the pIII C-terminal domain as a fusion protein, and a suitable selection marker (usually ampicillin resistance). There are about four or five copies of the pIII pro-tein at one terminus of the particle. The pIII is added last to the phage particle as it comes out of the periplasm and plays a critical role in terminating phage particle length [138]. Phagemids lack all other structural and nonstructural gene products required for generating a complete phage. Phagemids can be grown as plasmids or alternatively packaged as recombinant M13 phage with the use of a helper phage that contains a slightly defective origin of replication (such as VCSM13), which supplies, in trans, all the structural proteins required for generating a complete phage [33].

In general, monovalent display of antibody fragments has been adopted for use as a phagemid system based upon the pIII system. For the most part, phage vectors are limited in the size of the genetic material, which can be carried in the phage chromosome as their genomes are highly plastic. The use of helper phage again helps circumvent this drawback. Unlike lytic phage, f-phages and phagemids are replicated and extruded from the bacterial periplasm leaving the cell unharmed. As the phage/phagemid DNA is extruded, it is coated with the phage proteins and recombinant versions are coloaded. Fully infective helper phage with defective packaging signals is used to provide all the structural and nonstructural phage proteins to package the phagemid DNA. The creation of an N-terminal fusion of a peptide or antibody domain with the C-terminal domain of pIII allows loading of the recombinbant molecule upon the surface of phage particles [139]. The recombinant pIII-antibody fragments encoded upon the phagemid DNA are expressed and are coloaded onto these particles through phenotypic mixing in the same *E. coli* host cell. The phagemid particle displaying antibody molecules with the desired specificity and highest affinity can be selected and enriched in a process known as biopanning.

Antibody libraries generally use phagemid display systems to circumvent potential problems of random deletion of the antibody gene inserts from the much larger phage genome. Most phagemid antibody display vectors are about 3.5–5 kb in size (Table 16.5). The large majority of these systems use the low copy number pIII [24] system for selection of higher affinity antibodies. The original phagemid vector, pComb3, developed by Lerner's group was designed for Fab display on pIII

Table 16.5 Vectors Most Commonly Used for Immune Library Construction

Principal Vector and Derivatives	Display Fragment	Display Type	References
pCom3			[24]
pCom3H			[140]
pCom3X	Fab/scFv	Phagemid pIII	[58]
pComBov			[91]
pHEN1	scFv		[23]
pHEN4	V_HH	Phagemid pIII	[141]
pUR4536	V_HH		[142]
pSEX81	scFv	Phagemid pIII	
pCANTAB5E	scFv	Phagemid pIII	Unpublished
pCANTAB6			[143]
pMOC1	Fab	Phagemid pIII	[144]
pAK100	scFv	Phagemid pIII	[52]
pSD3	scFv	Phagemid pIII	[145]
pSD3a			[146]
pDM1	scFv	Phagemid pIII	[147]
pComb8	Fab	Phagemid VIII	[24]
fdDOG1	scFv	Phage pIII	[21]
fUSE5	scFv	Phage pIII	[8]

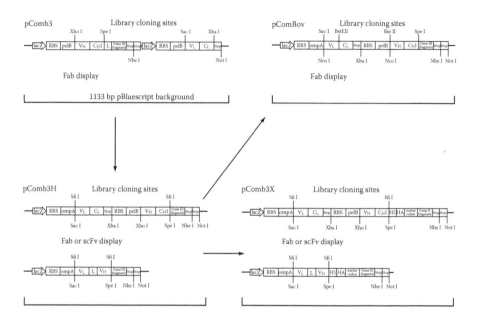

Figure 16.5 A series of pComb3 family vectors for phage display. The pComb3 series of
phagemid vectors [58] were designed to express recombinant antibody fragments on the surface of filamentous phage as fusions with the pIII-loading
domain or to express them as soluble proteins. In the original pComb3 vector,
the Fd fragment of the heavy chain and the light chain was cloned separately
into two different directional cloning sites. Separate lacZ promoters with two
ribosome-binding sites (RBSs) gave rise to separate polypeptides that are
directed by the same signal peptide to the periplasm, where they are assembled and displayed on the surface of the phage particle as a Fab fragment.
Expression of soluble protein required the excision of *gene* III by restriction
digest with SpeI and NheI, followed by re-ligation of the remaining vector. This
feature was maintained in all the subsequent pComb3 vector variants. The
pComb3H phagemid vector has a different order of the genes (reversed) and
a single lacZ promoter, which gives rise to a dicistronic message containing
the light- and heavy-chain Fd. However, the presence of two RBS sequences
still gives rise to separate polypeptide chains. The light- and heavy-chain pIII
fragments are directed to the periplasm by two different signal peptides. Two
asymmetric Sfi I restriction enzyme sites allow single-step directional cloning
of Fab or scFv. The pComb3X vector has three additional sequence insertions.
The amber codon has been inserted between the 3′ Sfi I restriction site and the
5′ end of *gene* III. This allows for soluble protein expression in nonsuppressor
strains of bacteria without excising the gene III fragment. The 6× histidine (H6)
tag has been inserted carboxy-terminal to the Fd fragment for universal protein purification. The hemagglutinin (HA) decapeptide tag has been inserted
at the 3′ end of the H6 tag for universal detection using an anti-HA antibody.
This vector also allows the expression of both Fab and scFv. The pComBov
phagemid vector is another modified version of the pComb3H designed specifically for bovine library construction. The bovine CH1 is produced via PCR
amplications in the usual Fd fragment. However, the bovine lambda constant
region (the predominant bovine isotype) has been preinserted into the vector
in order to silence an internal Sac I site [148]. Therefore, bovine light-chain
variable regions need to be amplified and cloned without including the lambda
constant domain in the cDNA.

[24]. Similarly, Winter's group developed the phagemid pHEN systems for Fab and phage fdDOG1 systems for scFv display [21,23]. Many second- and third-generation phagemid display vectors have been developed with improved on/off systems and specific constant domains for modular library assembly and cloning (see the references in Table 16.5).

Most commonly used vectors for immune antibody library construction have evolved from the pCom3 phagemid display vector (Figure 16.5). The third-generation phagemid system, pComb3X, assembles Fabs or scFvs by PCR with modular insertion that is cloned in a single step [56,130]. This improved vector also incoporates epitope and histidine purification tags for improved detection and purification, respectively. In some cases, Fab phagemid vectors have been remodeled in order to express the constant regions of a particular species. For example, in order to facilitate expression of bovine Fabs with the correct structural format, the bovine C_{H1} and C_k domains were swapped into the pComb3 vector in order to create pCom-Bov [91]. There are many other published and unpublished vectors with similar small but important modifications in public and private research laboratories. See Chapter 2 for more information about phage display vectors.

16.3 IMMUNE ANTIBODY LIBRARY SELECTION

Advances in the understanding and molecular analysis of expressed I_g genes have been made for many animal species. While being of no consequence to hybridoma production, this knowledge directly translates into basis used to generate immune antibody libraries from mice and humans. Why would one need to use novel species to generate MAbs? In addition to the fundamental knowledge gained from each species unique immune system, perhaps the most important reason to use other species for the development of antibody libraries is that they provide a new source of antibody diversity to tap into.

Genetic and somatic differences exist within the antibody repertoires of each species. These can be subtle differences or even completely different mechanisms of diversification. Thus, each species has a different germline set of V genes, which have evolved to contact antigens over evolutionary time. Many species use different combinations of mechanisms of somatic diversification and affinity maturation. Some species predominantly use a particular V gene or V gene class. These pools are shaped by the unique developmental patterns, which occur during ontogeny and the generation of self-tolerance in the context of the speciated antigen milieu. Differences such as unique MHC molecules, alternative posttranslational modifications, and genetic drift ensure a collectively vast natural antibody diversity for the planet. Some examples of these differences include the absence of affinity maturation in the antibody response in the axolotl [34], the localized V gene diversification in the peyer's patches of sheep [149], the use of mutated versions of a single V_H gene in the chicken B-lymphocyte pool [150], and an extra Cys bridge in rabbit kappa light chain [151]. Thus, each species provides very different opportunities for the *in vitro* selection of MAbs to the same target. This is especially true when you consider that the recognition of foreign antigens by each species is not necessarily the same. That is, the epitopes "seen" as immunodominant to the immune system of a mouse may not be dominant to the human immune system. That is also another reason why vaccines must progress through clinical trials in humans for efficacy, and they sometimes may not translate seamlessly between species. Thus, novel repertoires offer the possibility of targeting alternative epitopes, perhaps never targeted by rodent or human B-cell repertoires against the same antigen. Potent MAbs derived from any of these species can, in turn, be modified *in vitro*, for example, for higher affinity or humanization.

Monoclonal antibodies have been derived from antibody libraries using a variety of selection strategies. These include selection for binding using purified antigen, selection for binding on unpurified antigens (cell panning) with and without negative subtraction, selections for function, selection linked to phage infectivity, altered elution conditions, competition, bait capture, and high-throughput selection and screening. These methods are described in more detail in Chapter 4 and reviewed by Hoogenboom and Chames [3]. In phage display, each round of selection should show some kind of enrichment of antigen-specific phages over background. This is usually determined by whole phage ELISA and by enriched numbers of eluted phages in each successive round of selection. For detailed protocols, the readers are directed to an excellent laboratory manual on phage display [29].

Once binding clones have been identified, it is useful to optimize bacterial expression even further. This is usually done most efficiently by using a panel of different *E. coli* nonsuppressor hosts to identify the best strain for a given recombinant antibody [152]. This is done by transforming the plasmid clone into each strain and then testing a small number of individual supernatants containing the expressed binding domains in ELISA or flow cytometry. This will allow identification of the optimal strain for expression of a given antibody.

16.3.1 Laboratory Animals

Laboratory animals have been used extensively for the production of immune libraries. This is because the parameters involved in generating potent antibody responses are well known, and the animals are smaller and easier to handle in close confines. Monoclonal antibodies have been produced in both scFv and Fab formats utilizing vectors originally designed for murine/human I_g expression systems.

Mice, rabbits, and birds have received perhaps the most attention historically for use in MAb production. Mice have been very popular due to the production of hybridoma-derived MAbs. However, there have been many papers published on primer design and the generation of immune antibody libraries from mice. The *in vitro* modifications required to humanize a recombinant nonhuman MAb, or to alter affinity, or to modify specificity of selected nonimmune MAbs are not trivial. Success is protein-dependent and each MAb's original properties must be reconfirmed in the end product to ensure that desired properties are maintained. The difficulty is inherent in the fact that nearly each humanized MAb had to be completed by modifying the humanization selection or screening strategy.

16.3.1.1 Mice

We emphasize that for the routine generation of murine MAbs for diagnostic or reagent MAbs especially when an antigen is in good supply, one should strongly consider using the hybridoma system. This will simplify the screening and production enormously, and give you all the MAb you need through simple scale-up. Phage display of murine immune repertoires is mainly used to supplement the hybridoma technology in cases where humanization and affinity maturation are anticipated as downstream modifications for rodent MAbs with therapeutic potential in humans [64,153]. In these cases, generation of clones with an immune library inherently facilitates further *in vitro* modifications.

Murine immune systems have very complex genetics. The murine I_g loci for V_H, V_κ, and V_λ are located upon chromosomes 12, 6, and 22, respectively. While mice and humans have similar diversification mechanisms and high homology between some I_g gene families, the V genes are unique in both sequence and physical organization on the chromosome. The murine V genes are more clustered and less interdigitated than the human V genes. The heavy-chain locus is likely the most diverse among species as there are predicted to be more than 100 V_H genes [154], 14 D_H [155], and 5 J_H [156] genes found on chromosome 12. It has been estimated that there are about 140 V_κ genes located on chromosome 6 that can be organized into about 20 families [157]. The kappa genes, however, are overall less diverse than the V_H genes among themselves [97]. These genes have been loosely organized into families based upon sequence relatedness [128]. Information about the mouse genome is growing daily and can be found at Jackson Labs (http://www.informatics.jax.org).

To fully express the murine antibody repertoire, the comprehensive sets of PCR primers were designed by Orum et al. [57] based on a collection of 90 heavy- and 70 light-chain murine IgG sequences from the Kabat and genebank databases. Two equimolar mixtures of 25 individually synthesized primers (one set for the V_H and another for V_L) were designed to represent all known V_H and V_L subgroups as defined by Kabat et al. [158]. These primer sets, however, do not cover the lambda light-chain repertoire, which in mice represents only about 5% of expressed light chains found in serum IgG. This diminished abundance of lambda light chains is associated with a diminished molecular diversity as there are fewer germline V_λ alleles [97].

There are numerous other studies on primer optimization for cloning murine immune repertoires. For example 10 "forward" and 10 "back" primers were designed for cloning V_H and V_κ sequences with additional primer pair to cover the V_λ light-chain amplification [159]. This demonstrates that as many as 81 separate first-round PCR reactions are required in order to clone murine immune repertoires into an scFv repertoire for phage display library construction. Collectively, this highlights two major characteristics of murine antibody libraries produced from immune repertoires. First, given the inherently large number of PCR reactions, the scFv format, which requires fewer PCR manipulations by removing the need to amplify C_H1 and C_L domains, is heavily preferred over Fabs for murine repertoires. Second, in most protocols, only a representative set of heavy- and light-chain variable region primers are used [21].

The first example of phage display of antibody binding domains was with a murine heavy-chain Fd fragment [20]. Filamentous phage carrying V genes that encode binding activities was selected directly with antigen, albeit only the heavy chain and not the light chain was present. Indeed, this showed that antigen-specific phage-borne Fds bound specifically to antigen and that rare phage (one in a million) could be isolated after affinity chromatography.

Winter's group went on to develop f-phage libraries displaying both heavy and light chains of mice [21]. Using a random combinatorial library of the rearranged heavy and kappa light chains from mice immunized with the hapten2-phenyloxazol-5-one (phOx), they selected scFv expressing specific phage after a single pass over a hapten affinity column. Indeed, fd phages with a range of phOx binding activities were detected, and at least one exhibited high affinity (dissociation constant, $K_d = 10^{-8}$ M). No phOx-binding clones were selected after two rounds of panning from unimmunized mice, suggesting that immunization seems to be necessary to ensure strong biasing of the B-cell pool toward blasts making antigen-specific antibody for the library construction of the relatively small size (~10^6 members).

In the 10 years since those initial publications, over 100 articles have appeared describing the successful selection of MAbs from murine antibody libraries from immune repertoires by phage display. However, only the examples where the use of phage display rather than hybridoma technology was justified in our opinion are discussed in the following.

Immune display libraries offer the advantage of selection rather than screening, which is especially valuable for certain rare or poorly immunogenic antigens. For example, a phage mid-display library was successfully used to obtain high-affinity MAbs against Lex (lewis antigen) [160]. All of the anti-Lex MAbs of hybridoma origin that they previously produced were of the IgM isotype and have low affinity for antigen [161]. The two MAbs obtained from GM1 ganglioside-immunized mice using phage display have affinities for Lex antigen of approximately 1000-fold higher than MAb PM81, which is currently used to purge autologous bone marrow of leukemic cells before bone marrow reconstitution [162]. The higher affinity of the phage-derived IgG antibody fragments may improve their use in immunodiagnostic and immunotherapy.

Murine immune libraries but not hybridomas allowed for the successful selection of prion protein (PrP)-specific MAbs. Antibody libraries were generated from PrP-deficient mice, which had been immunized with mouse PrP [163,164]. In normal mice, normal regulatory mechanisms of self-tolerance to the murine PrP restrict the anti-PrP antibody response to only nonmurine epitopes. Thus, the MAbs derived from normal mice would not be useful for the development of an animal model of prion disease in that they would not bind the mouse PrP proteins. Furthermore, in this case, the hybridomas produced were extremely unstable for reasons that are not entirely clear. All attempts at producing MAbs via hybridoma production from PrP-deficient mice yielded hybridoma cells that failed to secrete anti-PrP antibodies beyond a period of 48 h. However, the use of phage display as an alternative approach resulted in isolation of the novel antimouse PrP antibody Fab28[DLPC]. Clearly, the technique of phage display has, in this case, circumvented the block in the generation of anti-PrP MAbs demonstrating that immune libraries provide a complementary approach for the production of MAbs. In a similar approach, an immunized fibroblast-activation protein (FAP)-deficient mouse was used for the generation of scFvs with cross-reactivity for both human and mouse FAP [165]. A phagemid display strategy was also used to successfully isolate scFvs against mesothelin after several attempts to produce antimesothelin hybridomas from spleen cells of immunized mice were unsuccessful [166].

Reiter and Pastan [167] found that many scFvs made from hybridoma-derived MAbs could be inherently unstable. This instability is likely due to the fact that these proteins were not subjected to the biases of prokaryotic expression and display, and this may limit their antitumor activity in animal models [167]. Phage display technology has been used to bypass the hybridoma step in order to isolate scFv MAbs to a mutant EGFR-vIII from immunized mouse spleens [168]. The immunotoxin MR1 was created from one of these scFvs and had a higher binding affinity than all of the recombinant immunotoxins made previously ($K_d = 11$ nM). Therefore, it seems likely that standard phage display inherently selects for stable scFvs.

Mice have been used to directly generate partially human scFvs. Recently, Rojas et al. [169] offer a fast and simple way to produce half-human scFv fragments, while keeping the advantage of being able to immunize animals for high-affinity antibodies. Their strategy was to simplify scFv library construction by using a library made up of a single human light-chain variable region gene as a general partner for mouse immune heavy chains. This light-chain gene was constructed from genomic cassettes of cloned germline *A27* gene to the $J_k 1$ minigene segment, both of which are prominently involved in human antibody responses. This simplified protocol thus involved only a single cloning step of V_H regions from RNA of the mouse immunized with human prostate-specific antigen (PSA) into the vector already carrying the invariant human V_L region. This work clearly demonstrated that it is possible to select phage-displayed scFvs with high affinity ($K_d = 3.5$ nM for 10A clone) and desired specificity from a relatively small (1.5×10^5 members) hybrid immune library.

Several examples of antibodies selected from mouse libraries are given in Table 16.6. In summary, the examples discussed illustrate several important advantages of using antibody libraries from immune repertoires for the generation of murine MAbs. First, murine antibodies can be readily selected in some cases where conventional hybridoma technology was unsuccessful [163,164,166,168]. Second, MAbs with higher affinity against poorly immunogenic antigens can be isolated by phage display [160]. Third, phage display proved to be the right choice for generation of scFvs with cross-reactivity for human and mouse antigen [165], which is important for the development of useful animal models. Finally, chimeric mouse/human MAbs can be directly derived from immunized mice by phage display technology [169].

In the future, with the advent of the human *V* gene transgenic mice, it will be conceivable to use immune antibody libraries from XenoMice as a tool for the library-based humanization of already discovered and potent murine MAbs to infectious agents. This is especially the case for the use of genetically manipulated mice. We predict that transgenic mice with representative immune systems from other animals

Table 16.6   Mouse

Antigen	Library (Size)	K_d (nM)	References
phOx	scFv (2×10^5)	10	[21]
EGFRvIII	scFv (8×10^6)	22	[168]
MUC-1	scFv ($>10^7$)	nd	[170]
RSV	scFv (3.8×108)	3.6	[171]
CD13	scFv (10^6)	3.3	[172]
Amp	scFv (10^7)	1500	[173]
CD30	scFv (1.6×10^6)	76	[174]
FAP	scFv (10^8)	9	[165]
PSA	scFv (1.5×10^5)	3.5	[169]

Note: Amp, ampicillin; CD13, human CD13, aminopeptidase N; CD30, human CD30 antigen; EGFRVIII, epidermal growth factor mutant receptor vIII; FAP, fibroblast activation protein; MUC-1, MUC-1 mucin; phOx, 2-phenyloxazol-5-one; PSA, prostate specific antigen; RSV, respiratory syncytical virus.

will be developed over time. These would be valuable and much easier to manipulate and care for than the larger more sentient higher animals. Mice with gene knockouts, gene knock-ins, and immune deficiencies and even the human antibody transgenic *V* gene mice all have potential value for the generation of immune antibody libraries.

16.3.1.2 Rabbits

Rabbits as a species have been widely used for the production of polyclonal antibody for diagnostics and research. The widespread historical use of rabbit antiserum in many laboratory and diagnostic tests makes this animal a natural choice for the generation of MAbs. While MAbs are routinely generated in mice and rats via hybridoma technology, it has not been widely used for rabbits. Recently, a plasmacytoma cell line has been described as a fusion partner for rabbit B cells to establish stable hybridomas as a source of antigen-specific rabbit MAbs [175]. However, despite this report, the generation of rabbit MAbs via hybridoma fusion is not a standard technique as compared to murine hybridoma technology. This is mainly because of the low stability of rabbit heterohybridoma clones. The development of phage display of antibody fragments derived from immune rabbits [176,177] has made rabbits an interesting alternative source of MAbs.

Compared with the other existing sources of MAbs, immune rabbits have several advantages for use in the production of immune libraries. Rabbit I_g genes and MAbs have been well studied primarily through the work of the Mage and Knight laboratories. First, rabbits are known as excellent producers of high-affinity polyclonal antibodies against many antigens that are weak immunogens in mice. Second, rabbit antibodies are usually of relatively high affinity. Third, because most MAbs are generated in mice and rats, there are relatively few MAbs available that react against mouse or rat antigens. This is particularly important for the development of therapeutic human antibodies to self-components, comparative immunology, and other areas of biopharmaceutical development. These antibodies must be evaluated in rodent models and are required to recognize and mediate biological effects through both the human antigen and its corresponding mouse homologue. Finally, the limited V_H and V_L gene usage in the recombination events by rabbit B cells (reviewed in Ref. [178]) facilitates the generation of rabbit MAbs by requiring fewer oligonucleotide primers for repertoire cloning. Thus, the rabbit has become recognized as a particularly appropriate animal for the production antibody fragments by phage display [179–181].

The rabbit I_g gene repertoire is well characterized [182]. Rabbits are unusual among mammals, although they posses more than 200 functional V_H genes; most of the time, only the V_H1 gene, the most proximal to the D segment, is used in VDJ recombination in immature B cells [183]. Indeed, more than 80% of rabbit B cells express the V_H1 heavy chain of a1, a2, or a3 allotype specificity [184]. This contrasts sharply with the situation in murine or human B cells. Indeed, murine and human B cells express antibody recombined from diverse germline genes, which are grouped into 10 and 6 V_H gene families, respectively [154,185,186].

Rabbits use multiple light-chain genes in the primary repertoire in contrast to what occurs at the V_H loci [187]. Similar to mice, most of the rabbit light chains are encoded by a C_κ gene. However, in contrast to other mammals including mice, the rabbit has two C_κ genes, $C_\kappa 1$ and $C_\kappa 2$ [188–190]. In normal rabbits, approximately 90% of all light chains are derived from $C_\kappa 1$, termed the K1 isotype [178]. The $C_\kappa 2$ gene encodes the K2 isotype of light chains, which is only expressed in wild-type rabbits, Basilea mutant rabbits, and allotype-suppressed rabbits [188,189,191]. The lambda light chains comprise only 5%–10% of total serum I_gs in the rabbit. However, the organization of V_λ and C_λ genes appears to be similar to that of other mammals [192]. Like the chicken, rabbit V genes are further modified in secondary lymphoid tissues by somatic hypermutation/gene conversion events during immune responses [193]. More detailed information concerning rabbit immunogenetics is available in several recent reviews [151,178].

The practical application of the genetic information developed on rabbits has allowed for the generation of combinatorial rabbit antibody libraries displayed upon phage. The construction of antibody libraries derived from rabbit immune repertoires has resulted in the selection of rabbit monoclonal antibodies [176,177,179]. The first published report of the use of antibody libraries from rabbit immune repertoires described the construction of a phage-displayed rabbit scFv library and the selection of high-affinity monoclonal scFvs against the recombinant human leukemia inhibitory factor (rhLIF) [176]. One of the selected clones, scFv3-3, exhibited a K_d of 2.8 × 10^{-8} M as measured by surface plasmon reasonance. This finding opened up the use of rabbit immune repertoires as a new source of antibody diversity.

Rabbit Fabs have been produced against human type-1 plasminogen activator (PAI-1). In this second report on the use of rabbit immune repertoires, a diverse Fab library was constructed from the spleen and bone marrow of a rabbit immunized with purified human platelet a-granules [177]. Several Fabs against PAI-1 were isolated after three rounds of affinity selection. This study demonstrates the power of phage selection, as PAI-1 is present only in low amounts within the platelet a-granule.

The a-granule itself represents only about 0.01% of total platelet protein [194]. Clone R012 exhibited a K_d of 1×10^{-9} M and did not recognize several other members of the serine protease inhibitor superfamily. The ability to sort the rabbit Fab library against individual a-granule proteins (e.g., PAI-1) suggested that the strategy could be readily adapted for the identification of panels of rabbit Fabs.

Rabbit PBLs have been used to produce Fabs against model antigen. In the report by Foti et al. [179], the authors show the feasibility of using immune PBL rather than spleen cells and/or bone marrow for antibody library construction. In this paper, they showed that they could still select Fabs with relatively good affinity. A small library (2×10^6 clones) was used to select anti-KLH Fab C1.3IV3, which exibited a K_d of 6.4 nM. The authors conclude that peripheral B cells from rabbits appear to be mostly CD5 positive recirculating B cells, which represent the primary immune repertoire in adult rabbits and, therefore, are equally suited as the source of mRNA for library construction as spleen cells. Furthermore, because PBL and plasma can be prepared simultaneously, this enabled the authors to follow the immune response of individual rabbits by analyzing plasma for antibody reactivity. This allowed them to selectively choose, by extrapolation, what they concluded were the best sources of sensitized B cells from the corresponding stored PBLs for mRNA extraction for library construction. A similar approach was used by Chi et al. [195] to successfully isolate rabbit scFvs, rather than Fabs, against human Reg Ia protein from a PBL-derived library after only one round of panning.

Rabbits are large enough to easily provide sensitized B lymphocytes from tissue sources other than spleen. Hawlish et al. [196] generated rabbit scFv specific for the guinea pig complement protein C3 by repertoire cloning from three different rabbit immune libraries. The libraries were obtained from spleen, bone marrow, or PBLs of the same animals. The DNA sequence analysis of C3-specific MAb clones revealed that the blood-derived (PBL) clone BA8 comprises a $V_H \times 32$ heavy chain. This indicates that V_H a-allotypes [197] are not only of theoretical interest for a high degree of diversity within the libraries, but they may also be of functional relevance in the VDJ recombination events leading to antigen-specific antibodies.

Rabbit scFvs have been selected against multiple antigens from multiply-immunized rabbits. Single-chain Fvs with subnanomolar affinities against the group of haptens, including mecoprop, atrazine, simazine, and isoproturon, were isolated from a phage display library (8.7×10^8 clones) derived from the rabbit [68]. In this work, the authors demonstrated that antihapten scFvs with affinities in the subnanomolar range could be readily isolated from an immune phage display library derived from a single rabbit immunized with multiple antigens. This avoids the expensive necessity of constructing separate libraries de novo for each antigen. The highest affinity MAb exhibited an extremely low K_d value of 6.75×10^{-10} M. The general utility of the multiple immunization strategy challenges previously suggested limitations of using antibody libraries from immune repertoires to obtain high-affinity antibodies [198].

Rabbit monoclonal antibody can be used directly as a source of chimeric human Fabs. To facilitate rabbit MAb humanization, a chimeric rabbit/human Fab library was constructed (2×10^7) clones where rabbit V_L and V_H sequences were combined with human light chain and C_{H1} sequences [180]. A rabbit was immunized with human A33 antigen, which is a target for immunotherapy of colon cancer. Selected high-affinity Fabs exhibited K_d values as low as 390 pM. In contrast, rabbit antibodies selected by scFv display tend to exhibit lower affinities [176,179]. This is likely based on the Fab being displayed more or less monovalently at the phage surface and thus being selected for on the basis of affinity and expression. The selection of scFvs is also influenced by avidity due to their tendency to dimerize or form higher-order aggregates depending upon the length of the linker region [199,200]. For the humanization, the authors used a selection strategy that combines CDR3 grafting with framework fine-tuning and found that the resulting humanized antibodies retained both high specificity and affinity for human A33 antigen. In the follow-up article by Steinberger et al. [201], Barbas' group isolated a human CCR5-specific antibody, ST6, from a phage-displayed antibody fragment library (5×10^7 clones) generated from an immune rabbit. The presence of human constant domains, rather than rabbit, at the C-terminus of the variable regions means that antihuman antibody reagents can be used as another means of detection of binding or for purification.

Recently, Barbas' group demonstrated that rabbits with mutant bas and wild-type parental b9 allotypes are excellent sources for therapeutic monoclonal antibodies [181]. Featured among the selected clones with b9 allotype is a rabbit/human Fab that binds with a dissociation constant of 1 nM to both human and mouse Tie-2, which could facilitate its evaluation in mouse models of human cancer. That study also revealed that rabbits exhibit an HCDR3 length distribution more closely related to human antibodies than mouse antibodies [181].

Rabbit monoclonal antibody may be useful for intracellular therapy of humans. Recently, Goncalves et al. [202] developed a chimeric rabbit/human Fab library of approximately 9×10^7 independent clones from a rabbit immunized with HIV-1-encoded Vif protein, which is important for viral replication and infectivity. A selected Vif-specific Fab was converted into an scFv, which was expressed intracellularly in the cytoplasm. Folding of the rabbit scFv in the reducing environment of the cytoplasm can form functional binding sites as was demonstrated by coimmunoprecipitation. Furthermore, the toxicity of the scFv was low as assessed by cell viability in intrabody-expressing human T lymphocytes. These results suggest that gene therapy approaches, which deliver Vif intrabody, may represent a new therapeutic strategy for inhibiting HIV reverse transcription.

More recently, rabbit scFvs were used for the design and engineering of a bispecific, tetravalent endoplasmic reticulum-targeted intradiabody for simultaneous surface depletion of two endothelial transmembrane receptors, Tie-2 and vascular endothelial growth factor receptor 2 [203]. The findings suggest that simultaneous interference with the VEGF and the Tie-2 receptor pathways results in at least additive antiangiogenic effects, which may have implications for future drug developments.

The antibodies selected from rabbit libraries are summarized in Table 16.7. We have summarized this section with the following points. First, it was demonstrated that antibodies could be successfully selected from relatively small (1.5×10^6 clones) rabbit immune libraries [176]. Second, the same immune library can be used to select antibodies against several antigens if the animal was coimmunized with those antigens [177]. Third, subnanomolar affinity Fabs can be obtained from

Table 16.7   Rabbit MAbs from Immune Libraries

Antigen	Library (Size)	K_d (nM)	References
rhLIF	scFv (1.5×10^6)	28	[176]
PAI-1	Fab (2×10^7)	1	[177]
KLH	scFv (2×10^6)	6.37	[179]
C3	scFv (5×10^7)	nd	[196]
Simazine	scFv (8.7×10^8)	0.67	[68]
A33	Fab (2×10^7)	0.39	[180]
CCR5	Fab (5×10^7)	2.7	[201]
Vif	Fab (9×10^7)	nd	[202]
Tie-2	Fab (10^9)	1	[181]

Note: A33, human colon cancer A33 antigen; C3, guinea pig complement protein C3; PAI-1, type-1 plasminogen activator; rhLIF, recombinant human leukemia inhibitory factor; Vif, HIV-1-encoded Vif protein.

rabbit immune libraries [180]. Fourth, rabbit scFvs can be successfully expressed as cytoplasmic [202] or endoplasmic reticulum-targeted [203] intrabodies. With the increasing availability of transgenic rabbits, the use of animals expressing foreign proteins as endogenous molecules from other animals will help generate MAbs to complexed antigens. This will greatly facilitate MAb selection from immune libraries, which specifically bind to complexed antigens and to cryptic epitopes, which are only exposed when bound to ligands.

16.3.1.3 Chicken

Birds represent another source of MAbs now becoming recognized as a new source of antibody diversity, especially for the production of MAbs against mammalian antigens whose structure is highly conserved and therefore renders only a limited immune response in mice and rabbits due to immunological tolerance. The chicken, therefore, may be the most useful small host for the development of new antibody specificities since it is located on a different evolutionary branch from mammals on the phylogenetic tree [204]. The ability to harvest large amounts of chicken IgG (often called IgY) from eggs makes chicken a desirable target host for the generation of large amounts of antibody. For example, the possibility to produce human xeno-chickens expressing human Ig-gene loci could result in large-scale antibody production against enteric pathogens for food additives.

There are a few reports of hybridoma-derived monoclonal antibodies from immunized chickens [205–207], which correspond with the availability of some myelomas for this species [208,209]. However, the simplicity of the avian expressed antibody gene repertoire makes it very suitable to create recombinant antibody libraries.

The naive B-cell repertoire of birds is derived somatically through seemingly random and antigen-independent events. B-cell diversity is a result of gene conversion events from upstream pseudo V-region genes and takes place in the unique microenvironment of the bursa of fabricus of young chickens [210–212]. Immature

B cells are diversified by intrachromosomal gene conversions of a single rearranged functional $V_H 1$. There are about 100 upstream V_H genes [213,214] that diversify the rearranged *VDJ* genes until they migrate to secondary lymphoid tissues. A similar mechanism is considered to operate in the light chain, which is composed of one functional V_L and J_L with pseudo V_L-genes clustering upstream of the functional gene [210]. The repeated gene conversions of even similar *V* genes result in changes to each of the CDRs and can alter the length of the CDRs via codon additions and deletions. B-cell development is thus developmentally regulated in birds and occurs for a limited time in contrast to the continuous development in mice and humans. Furthermore, in birds, B cells with rearranged I_g genes migrate in a single "wave" from the bone marrow to the bursa where they then undergo gene conversion [114].

Chickens also use antigen-dependent somatic hypermutation in combination with gene conversion to further diversify mature B cells during an immune response. These processes combined with antigen selection leads to affinity maturation of the chicken B-cell response [215]. Somatic changes, in general, make it very difficult to judge whether the origin of some somatic changes is germline templated or novel as the sequences can be very mutated or very small [216,217]. Avian B-cell development is covered in more detail in Ref. [150].

Chicken antibody libraries are simple to construct. Despite the fact that extensive random gene conversion events occur during bursal B-cell development, they tend not to effect the termini of the *V* genes as much as the internal stretches [218]. Thus, the 5' and 3' ends of rearranged *V* genes are highly conserved, which means a single set of primers for V_H and for V_L is sufficient to amplify the expressed *V* gene repertoire from immune chickens. The feasibility of producing specific phage antibody from chickens was first shown by Davies et al. [219]. They generated scFv libraries from isolated bursal lymphocytes of an 8-week-old naive chicken providing the foundation for antibody library development from chickens. In the same year, the RNA from chicken hybridoma clones was used as a substrate for constructing a recombinant antibody library for the selection of antigen-specific recombinant chicken MAbs [220]. In this case, the authors initially screened avian hybridomas from immune chickens injected with a human protein involved in the pathogenesis of cystic fibrosis. Next, they cloned the *V* genes from the antigen-specific hybridomas into a whole IgG expression system in order to produce chimeric chicken/human IgG. This work demonstrated the advantages to having the chicken as an alternative system to generate MAbs.

Antibody libraries have been derived from avian immune repertoires. The first demonstration of an antibody library from an immune chicken repertoire was in 1996 and produced MAbs to a model antigen. Yamanaka et al. [221] used the spleen cells of outbred White Leghorn hyperimmunized with mouse serum albumin (MSA) to construct an immune scFv library of 1.4×10^7 members. All five selected MAbs were highly specific for MSA and demonstrated different degrees of cross-reactivity to rat serum albumin, but not to human and bovine serum albumin. Thus, the chicken immune repertoire can provide simple to construct antibody libraries for the efficient selection of MAbs against murine proteins or other proteins whose structure is highly conserved in mammalian species.

Modular-format vectors have been used to select chicken scFv or Fab MAbs from immune birds. It was not until 4 years later that the value of using animals located at phylogenetic distance became a hot topic. Simultaneously, two more articles were published from Barbas and Silverman's collaborating groups at the Scripps Research Institute and the University of California at San Diego, respectively. These articles described the generation of avian MAb fragments by phage display [58,222]. In the first article, Andris-Widhopf et al. optimized methods for constructing chicken I_g phage display libraries in the modular pComb3 system and generated combinatorial antibody libraries from spleen and bone marrow of Red/Black Cornish Cross chickens immunized with fluorescein-BSA [58]. In the same study, they went on to develop methods to construct scFv, diabody, and Fab format libraries all from the same immune chicken [58]. This indicates the versatility of phage display in obtaining different types of chicken MAb fragments from the same animal. The same V_H and/or V_L regions were selected from different types of libraries indicating that the format of the MAb molecule (scFv vs. Fab) had little or no impact with regard to avidity vs. affinity in this case [58]. The chicken Fabs library was constructed using human constant regions, which facilitated detection with readily available antihuman secondary reagents [58]. Moreover, this vector system is expected to greatly facilitate possible downstream *in vitro* modification processes.

The evolutionary distance of the chicken immune system from humans was directly demonstrated in the study by Silverman's group [222]. In this case, they produced chicken MAbs against the evolutionarily conserved antibody domains themselves. They immunized Leghorn chickens with the human clan III I_g (V3-23/clan III Vk2) proteins and selected MAb fragments to these highly conserved mammalian antigens. They needed these reagents in order to directly investigate the expression of highly related antibody clan-defined sets, and thus, they needed to exploit the chicken immune system and the selection power of phage display in order to derive diagnostic MAbs for clan III Ig. Using a specially tailored immunization and selection strategy, they selected recombinant avian scFvs specific for the clan III products, including those from the human V_{H3} family and the analogous murine families. Reactivity with the representative LJ-26 scFv was completely restricted to clan III I_g and had complete nonreactivity with other clan I and clan II Ig. They went on to demonstrate the utility of a novel recombinant serologic reagent for studying the composition of the B-cell compartment and also the consequences of B-cell superantigen exposure. This clearly shows how I_g gene diversity of nonmammalian antibodies can be useful in the development of new reagents for biomedical research.

The antibodies selected to date from immune chicken libraries are summarized in Table 16.8. In conclusion, the vast variety of domestic avian species and the simple immunogenetics of birds make them an attractive and wonderful new species for creating immune libraries for MAb development. The studies with chicken libraries also provided the evidence that, in the suitable host, the highly conserved antigen surfaces can be recognized by the immune system. This also allows speculation about the relatedness of proteins to which the immune species reacts.

Table 16.8   Chicken

Antigen	Library (Size)	K_d (nM)	References
MSA	scFv (1.4×10^7)	nd	[221]
FITC	scFv (9.6×10^7)	nd	[58]
FITC	Fab (3.8×10^7)	nd	[58]
Clan III Ig	scFv (8.5×10^8)	nd	[222]

16.3.2 Large Farm Animals

Large animals of agricultural importance represent another area where the development of immune antibody libraries can be of great advantage. In many species, the understanding of the basis of I_g formation is now sufficient for the application of antibody phage display, and this technology has been successfully applied to animal species including sheep [146], cattle [91], camel [141,223], and llamas [224].

There is a general trend to develop validated diagnostic tests for infectious diseases in veterinary medicine that can be done in the level 2 laboratory without the need to use bovine MAbs, for example, to pathogens such as foot and mouth disease virus, in order to standardize and verify serological responses and for the development of validated C-ELISAs. The ability to produce recombinant MAbs to infectious pathogens, which affect our livestock, whether from immune libraries or derived from hybridoma clones, is expected to benefit species of veterinary and economic importance.

16.3.2.1 Sheep

Sheep polyclonal antibodies are routinely used as laboratory and veterinary reagents. Sheep also represent a promising new system for deriving high-affinity MAbs. In contrast to sheep polyclonal antibody usage, sheep MAb technology is still in its infancy. The attempts to generate sheep MAb by fusing sheep lymphocytes to a mouse myeloma line were too inefficient to meet the needs in using these molecules for biomedical and agricultural purposes [2,225].

Sheep I_g genetics are relatively well studied mainly by the work of the Reynaud and Weill laboratory. Sheep create a diverse preimmune repertoire through antigen-independent somatic mechanisms. Genetic studies have shown that sheep have about 10 germline V_H genes, which contribute to the generation of preimmune antibody diversity of the sheep [147,226]. There is strong evidence to indicate that sheep B cells are diversified in ileal peyer's patches [117,227]. Unlike chickens and similar to rabbits, in sheep these additional germline V_H genes are all functional (not pseudogenes) for both heavy- and light-chain rearrangement and expression [114,212,228]. Sheep have around 60–90 V_λ members that account for 75%–80% of sheep Ig light chains [229]. Studies on the lambda light chain locus, which tend to preferentially rearrange only a few of these genes, clearly showed that the rearranged V_L genes become modified by somatic hypermutation to generate a diverse preimmune repertoire [117]. Indeed, sheep mainly generate antibody diversity in their immunoglobulin repertoire by hypermutating the mature rearranged V genes in postrearrangement diversification [212,230]. This is an antigen-independent process, which also occurs in sheep B cells *in vitro* under appropriate conditions. Recent findings, however, demonstrate that combinatorial rearrangement plays a much larger contribution to the sheep I_g diversity generation than is currently acknowledged [231].

There is evidence for affinity maturation in sheep B cells during immune responses. The substantial number of expressed sheep heavy- and light-chain genes indicates that gene conversion may exist as a possible mechanism of generating antibody diversity [118,147], in addition to somatic hypermutation [114]. These mutations seem to be targeted specifically to the CDRs, which support the notion that active immune responses provide a beneficial bias in the enrichment of antigen-specific B cells [116].

The design of primers for the generation of sheep MAb fragments from immune repertoires opens up the use of caprine V genes for the generation of MAbs. Until recently, sheep were thought to have a single V_H gene family with homology to the human V_{H4} family, composed of just 10 germline genes. Thus, only three primers were needed to clone the majority of V_H genes [226].

Antibody libraries were used to study antibody diversity in sheep [147]. During this study, eight new V_H gene families were identified with homology to human V_H families, which had not been previously reported in sheep. These findings indicate a greater level of germline diversity than anticipated, although this did not necessarily imply by itself a larger functional gene pool. In total, the sheep V_H loci are composed of at least nine V_H gene families, and there is additional evidence for the usage of J_H pseudogenes to encode diverse CDR3s [147]. The light chains of sheep are more diverse than the heavy chains, and more primers have been designed to clone sheep V_L genes [212,229]. This requirement is similar to that for rabbit antibody library construction, yet is still less than that required for cloning of human or mouse V_L genes [176].

Sheep scFv MAbs have been produced against model antigens. In 2000, Li et al. used total RNA isolated from the spleens of sheep immunized with the model antigens human serum albumin (HSA) and chicken egg conalbumin (CONA) to generate an scFv phage display library [146]. The I_g V gene repertoires were PCR-amplified and used to construct an scFv library in a modified phagemid vector [232]. A total of 14 different scFvs were isolated and chosen for further characterization. The sequences of the MAbs revealed typical ovine characteristics, and diverse I_g genes were selected from the immune library revealing the distinct clones, which were grouped into three V_λ and one V_H family, respectively. The sequence analysis indicates that VDJ recombination can contribute significantly to V gene diversity in sheep immune responses. Due to very low expression levels (<0.2 mg/L), only one HSA binder H3 was affinity purified. The affinity of the H3 scFv produced from the sheep immune library was in the low nanomolar range ($K_d = 1.83$ nM). A total of only 1.5 g of immune spleen tissue was used to generate the sheep immune antibody library. Clearly, this shows that it is unnecessary to process the whole spleen of such a large animal in order to generate an immune library capable of generating antigen-specific MAbs, and this observation may save time and considerable effort. Furthermore, this shows that the entire cellular repertoire does not have to be recapitulated *in vitro* and raises the possibility of live (nonmortal) retrieval of spleens from large immune animals via splenectomy for MAb generation.

High-affinity sheep scFv MAbs have been produced against an herbicide using immune libraries. Splenic mRNA was prepared from a 10-year-old Welsh breed/ Suffolk sheep, which had been hyperimmunized against the hapten atrazine, conjugated to bovine thyroglobulin. The PCR amplification of sheep V_H and V_L cDNAs was achieved using a diverse set of primers representative of the immune repertoire of sheep [147]. While the sheep immune library was composed of 1.1×10^9 members, the library used for selection was reportedly made up of 8.5×10^8 independent clones [233], which is still a massive immune library. Affinity selection was performed using several variations of the immunotube techniques on atrazine-BSA conjugates as antigen. Eluted phage antibodies were screened first in a phage-ELISA, and later on, soluble scFvs were evaluated in a C-ELISA to determine specificity and to gauge affinity. This produced a panel of sheep antiatrazine scFv MAbs with high affinity. These high-affinity antibodies were highly specific for the atrazine molecule and showed low cross-reactivity with related molecules in ELISA tests. Two of the selected clones (4D8 and 6C8) exhibited K_d values of 0.13 and 0.2 nM, respectively. These monoclonal antibodies will improve current diagnostic tests, which utilize sheep polyclonal antibody by lowering background reactivity. Moreover, these scFvs are extremely stable under nonphysiological conditions. Indeed, two of the sheep scFv antiatrazine MAbs have a limit of detection of 1–2 parts per trillion and are well within the required EC-legislated limit of 100 parts per trillion.

The antibodies selected from sheep libraries are summarized in Table 16.9. In conclusion, sheep MAbs have been generated by antibody (phage display) libraries from immune repertoires and can provide specific high-affinity probes for biorecognition.

Table 16.9   Large Animals MAbs from Immune Libraries

Antigen	Library (Size)	K_d (nM)	References
	Sheep		
HAS	scFv (4.2×10^6)	1.83	[146]
Atrazine	scFv (8.5×10^8)	0.13	[233]
	Cattle		
GST	Fab (2×10^7)	nd	[91]
	Camelids		
Lysozyme	V_HH (10^7)	5	[141]
AM	V_HH (5×10^6)	3.5	[142]
RR6	V_HH (nd)	18	[234]

Note: AM, porcine pancreatic amylase; GST, glutation S-transferase; HSA, human serum albumin; RR6, azodye RR6.

16.3.2.2 Cattle

Immune cattle libraries are another important large animal system in which progress has been made. Although hybridomas have been derived from cattle [145], the unstable fusion partners and the simple *V* gene expression pattern in cattle make repertoire cloning an attractive alternative. In cattle, B cells are diversified in both GALT and spleen [235]. Therefore, it is expected that cattle engage in multiple I_g diversification mechanisms.

The preferential expression of lambda light chains is a feature common to many domesticated species including cattle [212], and the cattle light-chain pool is dominated by a single V_λ family [236,237]. Like the sheep system, the available V_H-gene repertoire of cattle is nearly entirely derived from a single gene family comprising around 15 nearly identical genes. The bovine I_g repertoire is diversified by both somatic hypermutations and gene conversion events. Indeed, cloned adult cattle mu-V_H-gene transcripts revealed evidence of extensive somatic hypermutation and very long CDR3s. The length of CDR3 from V(D)J rearrangements averaged 21 amino acids, which is larger than other mammalian CDR3s [238].

Cattle antibody repertoires arise by somatic hypermutation and/or gene conversion, which indicates a GALT-mediated diversification process [239]. Robust somatic diversification mechanisms and stringent cellular selection must be taking place in these animals as antibodies derived from them in most cases have higher affinities ($K_d = 10^{-10}$ to 10^{-16} M) compared to the average affinity of murine MAbs ($K_d \sim 10^{-9}$ M [1]). The relative simplicity of cattle I_g genetics also makes cloning of immune repertoires from these animals relatively simple as fewer primers are needed. Interested readers are invited to review these references for more information on cattle I_g genetics [95,240].

The first and only immune library from cattle to date [91] demonstrates that enough information about the bovine I_g locus is at hand to develop immune libraries *in vitro*. O'Brien et al. performed three rounds of affinity selection of Fab library derived from a Simmental calf immunized with a GST-fusion protein and selected 23 anti-GST clones (Table 16.9). However, the expression levels of the recombinant Fabs produced *in vitro* by *E. coli* could not be detected by SDS-PAGE or immunoblotting, suggesting that a very low level of antibody was being expressed. O'Brien et al. went on in 2002 to optimize the expression of bovine Fabs G77 and L250 using the third-generation pCom Bov expression vector with cloned bovine C_H1 and C_k1 [148]. This time, high levels of Fab protein were found in the culture growth medium, which implies that a reliable method for the generation of bovine MAbs from immune antibody libraries is now available.

16.3.2.3 Camelids

The serum of camels and llamas (camelids) contains a unique type of antibody devoid of light chains in addition to conventional antibodies [241]. The heavy chains of these single-heavy chain antibodies (HCAb) have a lower molecular weight due to the absence of the first constant domain, the C_H1. Camel serum contains about 75% HCAbs, while llamas contain no more than 45% HCAbs [242]. The variable domain of the heavy chain is referred to as $V_H H$ to distinguish it from classic V_H [243]. The cloning of $V_H H$ in phage display vectors offers an attractive alternative to obtain the smallest antigen-binding fragment (MW ~ 15 kDa) compared to scFv with both V_L and V_H (MW ~ 30 kDa). The $V_H H$s selected from immunized camels or llamas have a number of advantages compared to the Fabs and scFvs derived from other animals because only one domain has to be cloned and expressed.

The camel and llama $V_H H$ sequences belong to a single gene family, namely, family three, which shows a high degree of homology with human $V_H 3$ [243]. It is unlikely that camels possess other $V_H H$ gene families, as a representative database of germline $V_H H$ sequences revealed the presence of about 40 different $V_H H$ genes that all evolved within the $V_H 3$ subgroup [171].

Cloning the repertoire of antigen-binding $V_H H$ fragments from an immunized camel is straightforward. The single-domain nature of $V_H H$ simplifies the effort considerably as only one set of PCR primers is required to amplify the entire *in vivo* matured $V_H H$ repertoire of an immunized animal [141]. Also, no scrambling of V_L and V_H pairs occurs because the V_L cloning is not required. However, two distinct methods were designed to avoid cloning of V_H fragments into a $V_H H$ pool. The use of PCR primers that anneal selectively onto the hinge of the HCAb and thus will only amplify the $V_H H$ genes [242]. Another method is to use pan-annealing primers to amplify all IgG isotypes followed by separation on an agarose gel and recovery of the desired shorter fragment [141]. More detailed information concerning the current status of single-domain camel antibodies is available in a recent excellent review by Muyldermans [81].

High-affinity camel MAbs can be selected from immune camel $V_H H$ repertoires. In the first report [141], two $V_H H$ libraries were constructed from the PBLs of a camel that had been immunized with two model antigens, tetanus toxoid (TT) and lysozyme. Several $V_H H$ isolated from the immunized libraries are extremely stable, highly soluble, and highly specific for the target antigens. The affinity determination of one of the selected antilysozyme clones (cAb-Lys3) yielded a K_d value of 5 nM. Later analysis of this camel $V_H H$ clone showed that residues at the tip of the CDR3 loop mimicked the carbohydrate substrate of lysozyme, thus making it a true enzyme inhibitor [244]. This report showed that immune camel $V_H H$ can be viewed as a route to obtain a new class of high-affinity antibodies capable of targeting epitopes rarely accessed by conventional V_H/V_L-based antibodies.

Camel $V_H H$ heavy-chain antibodies have the ability to inhibit enzyme activity. In 1998, Lauwereys et al. [234] demonstrated that functional HCAbs from camels behave quite differently in comparison with conventional antibodies. In this study, they immunized one adult male dromedary with two enzymes, porcine pancreatic α-amylase (AM) and bovine erythrocyte carbonic anhydrase (CA), and constructed a $V_H H$ library of 5×10^6 members from camel PBLs. Four different inhibitory $V_H H$ fragments were selected (two for each enzyme) after three rounds of panning. The AM-specific clone AM-D9 binds the target with a K_d of 3.5 nM and inhibited enzymatic activity of AM with an IC_{50} of 10 nM, whereas the CA-specific clone CA-06 binds the target with a K_d of 20 nM and inhibited enzymatic activity of CA with an IC_{50} of 1.5 μM. In summary, these findings suggest that the selection of $V_H H$

antibody fragments from immunized camels is a powerful strategy for the selection of a new type of potent and specific enzyme inhibitor.

Llama $V_H H$ MAbs can also be selected from immune libraries. The first example of the isolation of llama antihapten-specific $V_H H$ fragments was published by Frenken et al. [142]. Three young male llamas were immunized with three azo-dyes (RR6, RR120, and human pregnancy hormone chorionic gonadotropin). From the constructed immune libraries, the authors were able to select antigen-specific $V_H H$ fragment-producing clones by simple colony screening. For all antigens, the percentage of antigen-specific clones was over 5%. This high percentage of binders in the immune pool abrogates the need to use phagemid or other selection/display systems, and simple screening will reveal binders. The anti-RR6 fragments were further investigated, and affinities determined for one of the anti-RR6 fragments (R5) were found to be in the low nanomolar range ($K_d \sim$ 18 nM). This work demonstrated that the llama could be an even more practical source of antigen-specific $V_H H$ than the camel. The last two examples show once again the ability of immune libraries to be used to generate MAbs to multiple antigens.

The antibodies selected from camel and llama libraries are summarized in Table 16.9. In conclusion, the $V_H Hs$ selected from immune camel and llama repertoires have strikingly high-affinity constants for their target proteins that are comparable to those of Fabs and scFvs [141,142,234]. More importantly, the small size of the $V_H H$ fragment and extended CDR3 loops allow access to antigenic sites not generally accessed by conventional antibodies (e.g., enzyme active sites) [141,234]. The unique features of HCAbs may facilitate other specific therapeutic applications such as drug delivery vehicles and as intrabodies where $V_H H$ fragments should perform better than other antibody formats.

16.3.3 Humans

Antibodies and their derivatives constitute the largest class of biotechnology-derived molecules in clinical trials. Collectively, MAbs represent about 25% of therapeutics currently in development [245] and 30% of biopharmaceuticals in clinical trial [65], and have a prospective market value of several billion dollars [78]. Antibodies have a long and proven track record for therapy including prevention of hemolytic diseases of the newborn with antirhesus D preparations [246], of chronic hepatitis B in high-risk infants [247], and of Argentine hemorrhagic fever [248]. These and other examples are discussed in more detail elsewhere by Burton and Barbas [27].

All of the 12 MAbs currently approved by the US Food and Drug Administration (FDA) contain rodent protein sequences (Table 16.10). These antibodies, which all bear murine sequences, have the potential to elicit pathological complications when used in humans. For example, patient trials, which compared the half-life of murine vs. chimeric mouse/human MAbs, revealed that the human immune system specifically responds to the protein sequences of murine origin, even in the chimera, and the proteins are treated as foreign proteins by the human immune system [249]. This sensitizes the human immune system against future repeated therapy, thus reducing the efficacy of a MAb treatment. Glycosylation patterns on the MAbs themselves are altered when produced in other species (including mice, insect cells, and plants) and can affect the MAb function as well [250].

Table 16.10 FDA-Approved MAbs[a]

Product Name	Target	IgG Format	Disease Indication	Company
Oncology				
Rituxan[b]	CD20	Chimeric	NHL	Genentech/IDEC
Herceptin	erbB2	Humanized	Breast cancer	Genentech
Mylotarg	CD33	Humanized	AML	Wyeth/Celltech
Avastin	VEGF	Humanized	Colorectal cancer	Genentech
Campath	CD52	Humanized	CLL	Millenium/Ilex
Transplantology				
Orthoclone	CD3	Murine	Transplant rejection	OrthoBiothech
Zepanax	CD25	Humanized	Transplant rejection	Centocor
Simulect	CD25	Chimeric	Transplant rejection	Novartis
Autoimmune				
Remicade	TNFα	Chimeric	RA, Crohn's disease	Centocor
Xolair	IgE	Humanized	Allergy	Genentech/Novartis
Infections				
Synagis	F-protein	Humanized	RSV	MedImmune
Cardiology				
ReoPro	$\alpha_{IIb}\beta_3$/IIIa	Chimeric	Cardiovascular	Centocor/EliLilly

Note: AML, acute myeloid leukemia; CLL, chronic lymphocytic leukemia; NHL, non-Hodgkin's lymphoma; RA, rheumatoid arthritis; RSV, respiratory syncytial viral disease.

[a] Latest FDA approvals can be found at http://www.fda.gov.

[b] Also available as 90Y mAb conjugate (Zevalin).

Antibody libraries and hybridoma technology are now bringing fully human therapeutic antibodies to the clinic. Most of the more than 100 new MAbs in clinical development are derived from hybridoma technology from normal mice, which results in either a fully murine, chimeric human–mouse or a humanized antibody molecule. The growing trend is to develop therapeutic MAbs of fully human origin, which to date can be derived from either phage display of human or primate libraries or from hybridoma screening from transgenic animals [245,251]. Significantly, around 15% of these new MAbs in the clinic are fully human and were derived from hybridoma fusions from transgenic mice or from a human antibody library. Antibody libraries thus remain an important source of human MAbs and have produced about 30% of all the fully human antibodies in the clinic [252]. Monoclonal antibody represents the future and final improvement of the current gold standard in human antibody therapy, which is the use of pooled human Igs.

Molecular approaches for the generation of human MAbs offer several advantages over traditional methods such as hybridoma technology or Epstein–Barr virus (EBV) immortalization. These traditional methods often can result in a bias toward certain B-cell populations and the creation of cell lines that produce only low levels of antibodies or are unstable [253]. Since the phage display method removes any technical limitations to the production of fully human MAbs, there is a huge effort to produce fully human MAbs for therapeutic use in humans for the treatment of infectious diseases, autoimmune disorders, transplant rejections, and cancers. Several key examples of the use of immune repertoires for the selection of human MAbs to these diseases are highlighted in the following.

Humans have a diverse germline I_g gene repertoire. Large-scale sequencing has revealed the entire V_H locus of an individual human being. This has shown that humans have a total of 123 V_H genes with 39 functional V_H gene segments [254]. This also revealed that the total human combinatorial diversity of the V_H locus is much smaller than first anticipated at about 6000 possible combinations. Humans have a total of about 82 V_L genes making up their light-chain germline repertoire. This is composed of 46 V kappa and 36 lambda genes, which are located on chromosomes 2 and 22, respectively (NCBI database). The human preimmune repertoire is mainly made up from combinatorial joining (junctional) processes.

The preimmune repertoire is highly diversified during an immune response. The repertoire of antibodies expressed in human memory responses is highly selected by antigen. Somatic hypermutation contributes significantly to the shaping of the immune repertoire in humans and leads to a shift in the repertoire of V_L genes expressed in naive vs. memory B cells [255]. The sequencing of the human V_H gene locus has made it clear that primers can be designed to specifically amplify the expressed repertoire, based upon the functional germline genome. However, it is likely that more than this minimal number of primers will continue to be used to clone human antibody libraries as the nonfunctional V_H genes potentially can become functional by gene conversion and/or receptor editing mechanism. Moreover, there is also a lot of polymorphism in the V gene loci [112,256,257], which can alter the repertoire in individuals, in particular of certain ethnic backgrounds [258]. It is expected that individuals will have a slightly different repertoire, which shows the total I_g repertoire in humans as a species remains unknown and continues to evolve [259].

There are many publications on the design and use of oligonucleotide primers for the amplification of human Ig variable region genes. The minimum number of primers needed to clone the whole human I*g* repertoire, which consists of the 44 V_H and 82 V_L genes (Table 16.5), would be enormous and total about 130 primer sets. However, in general, only a representative sublibrary of the immune repertoire is cloned based upon the phenotype (heavy- and light-chain class) of the desired MAbs [24,260]. In the case of the most popular format IgG1$_{\kappa\lambda}$, a minimum of about 25 primers is enough to clone a representative immune library to select binding clones from an immune human repertoire. In some cases, rather than amplify IgG1 subclass alone, all four IgG subclasses should be amplified based on the patient serum containing IgG2 and IgG4 autoantibodies [261]. The use of the popular IgG1$_{\kappa\lambda}$ format for immune library construction should be broadened to include other isotypes for the study of the *in vivo* repertoire, because the use of the IgG1$_{\kappa\lambda}$ format alone imposes severe limitations in the diversity of Igs. For an up-to-date comprehensive review of primer design, we recommend the readers consult one of several authoritative reviews [130,262].

Although phage display has been used to generate human MAbs without immunization, there remains considerable interest in cloning antibodies from immune individuals for prophylaxis, therapy, and study of the human humoral response. Contrary to experimentally immunized animals, the majority of immune human libraries are from naturally exposed patients. This includes, but is not limited to, exogenous antigens such as bacterial or viral infections, rare cases of vaccination in humans, and exposure to endogenous self-antigens in cancer or autoimmune disorders. All of these situations fall under our initial definition of immune library in Section 16.1.

Mice can be used to directly derive human MAbs in special cases. Usually antibodies are prepared from a recently boosted animal such that the B-cell pool reflects ongoing immune responses [263]. In humans, this is restricted as ethical constrains generally prevent antigen boosting. However, the use of transgenic mice expressing fully human antibodies [264,265] or of severe combined immune deficiency mice populated with hu-PBL-SCID allows for the restimulation of B-cell antibody responses without these constraints [266]. The earliest example was the use of immune antibody libraries derived from sensitized human lymphocytes via the hu-PBL SCID mouse [42]. The hu-PBL mouse was boosted *in vivo* with TT to which the human donor had been immunized over 17 years earlier. The splenic mRNA was used to generate immune human Fab library. TT-specific human Fab fragments were isolated through three rounds of panning with apparent binding affinities in the nanomolar range. Following this, more rapid techniques of using SCID mice to derive human MAbs were developed. RSV neutralizing human MAbs, with therapeutic potential, was isolated following a single round of stringent biopanning [267]. This was done by combining the hu-PBL-SCID mouse model with an scFv phage display library technique. The authors were not only able to bypass the meticulous hybridoma route but also avoided the inherent skewing of human antibody responses toward the dominant, nonneutralizing epitopes of RSV. This was noted previously as a confounder for the generation of potent MAbs to RSV when attempting to clone virus-neutralizing MAbs directly from human donors [267].

The use of human I_g transgenic chimeras will also allow for the simple generation of human MAbs via the antibody library approach. The hybridoma technology has already led to the development of fully human MAbs against *Neisseria meningitidis* [268], the shiga-toxin [269], the human HIV-1 [270], cancer [271,272], and autoimmune/inflammatory diseases [273,274]. The use of antibody libraries to derive fully human MAbs from transgenic mice has yet to be described in a scientific publication.

There is a spectrum of immune states from which immune libraries have been made from human B-cell repertoires. This spectrum depends upon the clinical pathogenesis of the disease and the level of immune exposure of the host (Figure 16.1). It is advantageous to be able to collect an enriched pool of immune B cells. These tend to be found in the marrow of convalescent patients. The optimal time to collect would be following recovery of a patient from an acute infection. In most cases, the patient will have a vigorous antibody response and then will go on to clear the infection. Immune libraries generated from such an individual with a high serum antibody titer would be expected to provide a large number of pathogen-specific MAbs by selection upon the inactivated agent or predominant antigens from the agent *in vitro*. The next best case for successful selection of MAbs from immune libraries is perhaps to use the B cells of long-term nonprogressor, patients chronically infected with viruses, like the HIV-1. In this case, while the infection is not cleared, the B cells are driven to produce very strong immune responses, which typically result in the production of high-titer neutralizing antibody to the homologous infecting viral strain.

Phage display was successfully used to isolate MAbs from individuals with demonstrable serum antibody responses to a variety of antigens, including infectious agents such as HIV-1 [275], self-antigens in autoimmune diseases [276], and mutated protein in malignancy [260]. Several examples of the use of immune repertoires for the selection of human MAbs to these diseases are discussed in the following.

16.3.3.1 Infections

There are numerous examples of successful MAb selections from human immune libraries against microbes [277,278]. We present several examples of human antibody selection from immune libraries, which represent the selection of MAbs against some of the worst and most feared infectious scourges on earth.

The first immune repertoire used to select anti-HIV-1 MAbs from humans was prepared from a 31-year-old long-term nonprogressor who had been HIV-1 positive for 6 years [275]. The individual had high-titer serum IgG to HIV-1 gp120 envelope protein of strain LAI. A Fab library of 1×10^7 primary clones was made from the RNA isolated from bone marrow samples from this person. Selection on gp120 *in vitro* led to the discovery of a panel of related clones. This led to the discovery of one of the most potent HIV-1-neutralizing MAb, b12. They went on to improve the affinity of these neutralizing MAb, which resulted in the most potent human anti-HIV-1-neutralizing MAb to date [279]. The importance of this antibody cannot be overstated for the development of an active vaccine. This renewed hope for the quest for an HIV-1 vaccine and that antibodies may actually play a role in limiting infection. Indeed, the holy grail of HIV-1 vaccine design would be to use b12-like MAbs to rationally develop immunogens capable of engendering similar protective antibody responses in humans. It is known that b12 and human MAbs of similar potency are capable of preventing mucosal infection *in vivo* in the SHIV-macaque model [280–282]. Along with MAbs to HIV-1 envelope proteins, scientists at the Scripps Research Institute have made human MAbs to many other important viruses (including RSV, Cytomegalo virus [CMV], herpes simplex virus [HSV-1], and Varicella zoster virsus [VZV] from the lymphocytes of infected or exposed individuals under informed consent). For an excellent review on much of this work, the readers are referred to Ref. [27].

The production of human MAbs to highly pathogenic organisms can sometimes be facilitated through the use of surrogate organisms. Similarly, the recombinant production of antigen in some cases reduces the need to use live infectious organisms for immunization. However, a live organism is still required at some stage in order to obtain enough nucleic acids for cloning of the important antigens. What happens when not only the organism is lethal, but also the protective antigens are not clearly defined?

The creation of neutralizing human monoclonal antibodies to Pox viruses is an excellent illustration of the difficulties sometimes encountered in producing MAbs [283]. Pox viruses are among the largest of viruses in terms of genetic complexity. Vaccinia virus has a genome of about 192 kb, which encodes more than 100 polypeptides [284]. In this study, a panel of vaccinia-specific Fabs were selected from a combinatorial phage display library made up of IgG and light chains from a library prepared from about 2×10^7 PBLs of a vaccinia virus immune donor. Plaque reduction–neutralization tests revealed that six of the Fabs were able to neutralize vaccinia virus infectivity *in vitro*. This is the strongest evidence to date to suggest that antibodies to vaccinia virus may be responsible for neutralization and protection to smallpox (closely related to vaccinia virus). Furthermore, ELISA studies revealed that 15 of 22 Fabs recovered from the library were cross-reactive with the monkeypox virus, a highly virulent zoonotic relative of the vaccinia and smallpox viruses. However, clone 14, which had the best vaccinia virus neutralizing activity, failed to bind to the monkeypox virus, again revealing the antigenic complexity of these viruses.

There are currently no vaccines or effective treatments for filovirus infections. Viruses such as Ebola and Marburg cause a severe hemorrhagic fever with high mortality in humans [285]. A small percentage of humans infected with these viruses do survive and may be the important clue to deciphering the role of antibodies in protection from these viruses [286]. While there are no published reports of immunity to Ebola virus infection after a primary infection, transfusion of convalescent phase whole blood to infected patients in the 1995 Kikwit outbreak was described to confer increased resistance in treated patients [287]. A panel of human MAbs to Ebola virus Zaire was generated from immune libraries constructed from the bone marrow lymphocytes of two donors who recovered from infection with the Kikwit Ebola virus in 1995. Several Fabs were selected from two independent immune libraries of 6×10^6 and 2.2×10^6 clones from the marrow of two individuals, and another with diversity of 5×10^6 from the PBLs of 10 donors. Binding clones were selected using gamma-irradiated whole virus or infected cells as the selective antigen, and immunoprecipitation to determine reactivity to either the nucleoprotein or the envelope glycoprotein. The specificity of the binding Fabs to Ebola virus was confirmed using immunofluorescence to detect binding to live and to fixed Ebola virus-infected cells. One of the Fabs to the envelope glycoprotein, KZ52, neutralized Ebola virus in both the recombinant Fab and recombinant whole human IgG form. Of note, one of the nucleoprotein-specific Fabs was inhibited from binding by 10 of 10 seropositive convalescent donor serum. This suggests that this Fab may be useful as a serological diagnostic assay in a C-ELISA for Ebola in humans and possibly other species [288]. The inability of equine immune serum, produced against whole inactivated Ebola virus, to protect infected macaques [289] reveals the present limitations of polyclonal antibody preparations and, furthermore, supports continued studies on the development of therapeutic MAbs with high specific activity against filoviruses.

Bacterial exotoxins provide an exquisite example of stringent selection and of the immediate protection afforded by passive antibody therapy. In the past, toxoids have been used as both active vaccines and to generate equine immune serum for some of the more common bacterial toxins. Today, there are no technological limitations holding back the replacement of equine immune serums with human monoclonal antibodies. Collectively, bacterial exotoxins, scorpions, spiders, bee venoms, stonefish, box jelly-fish, and snake toxins cause a lot of human morbidity and mortality worldwide each year. However, the diversity of these toxins makes it unlikely or impractical to produce active vaccines for each, and again, therapeutic antibodies offer the best new hope for protection. There are therapeutic antivenoms produced for many of these toxins.

Clostridial neurotoxons are the most toxic substances known to man [290], with murine LD_{50} values ranging from 0.1 to 1 ng/kg of body weight. Human immune serum produced against botulinum toxin (where the toxin moiety is denatured and thus not lethal to administer) neutralizes the toxin *in vitro* compared to nonimmune serum, which does not [287]. In cases of food poisoning, equine immune serum is still used today as a passive vaccine and protects humans from lethal intoxication. However, there are some adverse reactions to the horse antiserum, and cleaner and more homo-geneous human MAb preparations will one day replace the equine source.

There are seven different known serotypes (A–G), and there is evidence that recombinant combinatorial forms of the toxin may exist [291]. The diverse genetic locations of the botulinal neurotoxins and recent genome sequence data on *Clostridium botulinum* species collectively support the notion that they are encoded within transposons or other highly mobile genetic elements (Dr. S. Hayes, University of Saskatchewan, personal communication). Neutralizing antibodies are thus needed to be able to neutralize all forms of this toxin either by targeting conserved sites or by producing pools of MAbs containing type-specific neutralizing antibody.

Recently, neutralizing human scFvs were selected from immune repertoires of volunteers immunized with pentavalent botulinum toxoid [292]. The immune library was prepared from PBLs with a measurable protective titer in the mouse serum neu-tralization bioassay. The scFvs were selected from immune and nonimmune libraries of 7.7×10^5 and 6.7×10^7 clones, respectively, upon botulinum neurotoxin (BoNt)/serotype A. Of note, while binding clones were identified from both libraries, neu-tralizing scFv was derived only from the immune library, but not from the nonimmune human library. Moreover, scFvs specific to each of the five toxins used in the penta-valent vaccine were derived only from the immune library. Neutralization of the toxin correlated with affinity and competition with holotoxin for binding sites on the heavy chain of BoNt. Moreover, neutralization was synergistic among some of the scFvs revealing that multiple epitopes were targeted by the scFvs. The neutral-izing scFvs selected from the immune library exhibited K_d values of 36.9 and 7.8 nM, which are comparable to values reported from hybridomas [293]. Nonimmune scFvs had lower affinities with K_d values ranging from 460 to 26 nM. This clearly demon-strates some of the advantages that can be had by using immune libraries. While it is likely that large nonimmune libraries can have binding clones to the solvent acces-sible areas of a given toxin, it is less likely that these clones bind well to toxins like BoNt, which contain a limited number of antigenically variable protective epitopes. Immunization and selective expansion direct the recognition of toxins to a limited number of immunodominant epitopes by clones, which produce protective antibod-ies. It is logical to rationalize that the immune system of vertebrates has evolved to

Table 16.11   Infection Human MAbs to Infection Pathogens from Immune Libraries

Antigen	Library (Size)	K_d (nM)	Therapeutic Potential	References
HIV-1 gp120	Fab (1×10^7)	10	Virus neutralization	[275]
RSV	Fab (5×10^7)	nd	Virus neutralization	[294]
Measles virus	Fab (10^7)	10	nd	[295]
HCV	Fab (3×10^7)	151	nd	[296]
Rotavirus	Fab (2.5×10^7)	nd	nd	[297]
Ebola virus	Fab (6×10^6)	nd	Virus neutralization	[286]
Measles virus	Fab ($>10^9$)	nd	Virus neutralization	[298]
Botulinum toxin	Fab (10^7)	7.5	Toxin neutralization	[292]

Note: HCV, hepatitis C virus; HIV-1 gp120, human immunodeficiency virus type-1 glyco-protein 120; RSV, respiratory syncytial virus.

include inherent reactivity toward these protective domains of highly lethal toxins under stringent selection pressure over evolutionary time.

The antibodies selected against different infections are summarized in Table 16.11. First, it was demonstrated that broadly neutralizing MAbs to HIV-1 could be selected from the gp120-hyperstimulated B-cell pool of humans. Second, human MAbs to highly pathogenic Pox viruses and filoviruses can be selected from immune repertoires of boosted B-cell donors and convalescent bone marrow, respectively. Third, nanomolar affinity neutralizing scFvs can be obtained from antibody libraries from immune human repertoires but not from naive repertoires. The direct selection of human MAb fragments will continue to depend upon availability of convalescent or vaccine immunized donors.

16.3.4 Autoimmune

Rheumatoid arthritis (RA) is an autoimmune disease and currently has the largest number of patients being treated with MAbs [299]. Indeed, the study of human autoantibody responses is the field where only immune libraries can be used, and there is no competition from synthetic and naive libraries. For instance, dsDNA-specific Fabs with moderate affinities can be isolated from libraries prepared from healthy and systemic lupus erythematosus (SLE) donors. However, high-affinity Fabs were isolated only from an SLE library [276]. This is even more significant for some autoantibodies for which the frequency of specific B-cell precursors is very low. The B-cell precursor frequency for SLE-specific anti-Smith antibodies (anti-Sm) has been shown to be less than 1:30,000 splenocytes in the autoimmune mouse model [300]. The human anti-Sm IgG autoantibodies were successfully generated and characterized from an SLE patient [300,301]. Taking into account the limitations that conventional hybridoma technology impose on the generation of human MAbs, the phage display approach seems to be the main technology capable of generating functional human autoantibodies from immune donors. In 1994, Hexham et al. [302] first applied the pComb3 system to produce three thyroid peroxidase autoantibodies from a patient with Hashimoto's thyroiditis, thus demonstrating that the phage display system is an effective way to produce recombinant autoantibodies.

Because combinatorial antibody libraries randomly recombine heavy and light chains, the issue whether the selected antibodies are disease-relevant autoantibodies should be addressed. In 1996, Roben et al. [303] reported that anti-dsDNA autoantibodies could only be recovered from the library of an SLE patient and not from a library from a healthy identical twin of the patient. That study suggested that, in combinatorial libraries, the de novo pairing of heavy and light chains unrelated to the *in vivo* autoimmune response did not create the high-affinity disease-associated autoantibodies. More recently, Jury et al. [304] demonstrated that natural I_g heavy- and light-chain pairings of autoantibodies could be isolated from two patients at the onset of type 1 diabetes. An $IgG1_{\kappa\lambda}$ library of 2×10^6 independent clones was constructed from PBLs of two diabetic patients. After five rounds of panning on glutamate decarboxylase (GAD65), one of the major autoantigens, eight GAD65-reactive clones were isolated. Three of them reflected all typical features of naturally occuring GAD65 autoantibodies. Sequence comparison to monoclonal islet cell antibodies (MICAs) demonstrated that the heavy chain of GAD65ab was identical and light chain was nearly identical to the corresponding chain of MICA6. Thus, the authors demonstrated for the first time that selected clones reflect well the natural autoantibody response in type 1 diabetes.

Although the generation of human monoclonal autoantibodies is critical for understanding humoral immune response in autoimmunity, we would like to discuss two diseases where the selected autoantibodies have therapeutic potential. In the first example, Zeidel et al. [300] first reported the generation of functional human autoantibodies from PBLs of a patient with myastenia gravis (MG). Using the pComb3 vector, an $IgG1_{\kappa}$ library was constructed and panned against purified acetyl choline receptor (AChR). After five rounds of panning, four positive clones were selected and all demonstrated the ability to stain the muscle cells. The therapeutic potential of anti-AChR antibodies was demonstrated 2 years later by Graus et al. [305]. The $IgG1_{\kappa}$ and $IgG1_{\kappa}$ libraries were constructed from thymic tissue obtained immediately after therapeutic thymectomy. Panning was performed using human AChR expressed by TE671 rhabdomyosarcoma cells. Four different clones were selected after five rounds of panning, and all four Fabs stained human AChR expressed by the TE671 cells. More importantly, Fab 637 inhibited the binding of serum antihuman AChR from the MG patient by more than 90%, whereas the combination of Fabs 637 and 587 was able to limit AChR loss induced by MG serum to 20%. Since the recombinant antihumanAChR Fabs do not interfere with the receptor function and are unable to activate complement, it is feasible that they might be used to protect the AChR against degradation by intact autoantibodies *in vivo* during myastenic crisis.

In 1995, Ishida et al. first successfully selected recombinant human Fab against integrin $\alpha_{II}\beta_3$ from a phage library generated from PBLs of a patient with Glanzmann thrombasthenia (GT). Integrin $\alpha_{II}\beta_3$ is highly immunogenic in humans and remains the most frequently identified target of human autoantibodies that have been detected in a majority of patients with autoimmune thrombocytopenic purpura (AITP) and in patients with GT [306,307]. Ishida et al. [308] demonstrated a restricted usage of the V_{H4} gene family in the selected Fabs. In this context, the knowledge of the genetic status of anti-$\alpha_{II}\beta_3$ MAbs could help advance our understanding of the pathogenesis of autoantibody development. Jacobin et al. [309] adapted antibody library technology to determine the nature of the humoral immune response in patients with AITP and GT. Two scFv IgG1$_{\kappa\lambda}$ libraries were constructed from PBLs of GT patients and from spleen tissue of AITP patients. Several positive scFv clones were selected after two rounds of selection on activated platelets. After confirming the specificity of the selected clones on $\alpha_{II}\beta_3$ by ELISA, the scFv-binding affinities of two clones, TEG4 and EBB3, were determined by C-ELISA. The TEG4 exhibited a K_d value of 2.6×10^{-6} M, and EBB3 exhibited a K_d value of 1.8×10^{-7} M. The nucleotide sequence of variable regions of the selected clones revealed a polyclonal response in both patients. A large repertoire of V_H and V_L genes was used. The selected fully human Fabs reported in this article might be more suitable for repeated therapy as antagonist of $\alpha_{II}\beta_3$ integrin than the most commonly used chimeric Fab2′ 7E3.

The antibodies selected from autoimmune libraries are summarized in Table 16.12. We have summarized this section with the following points. First, it was demonstrated that high-affinity dsDNA-specific antibodies could be isolated for patients with SLE. Second, antithyroid peroxidase MAbs can be recovered from immune libraries made from the B cells of patients with Hashimoto's thyroiditis. Third, naturally paired autoantibodies to GAD65 can be isolated from patients with early onset of type 1 diabetes. Fourth, anti-AChR MAbs can be selected from immune libraries of patients with MG. Fifth, scFv MAb fragments with low nanomolar affinities can be selected from immune libraries of AITP patients. The ability to generate relevant human autoimmune MAbs by phage display allows investigators to define the antigenic epitopes targeted by autoimmune responses as well as to understand the genetic and structural bases of pathogenic autoantibody responses. Phage display has been used successfully to produce human autoantibodies from patients with diverse spectrums of autoimmune diseases, such as Hashimoto's thyroiditis [261], SLE [303], ulcerative colitis [310], idiopathic dilated cardiomyopathy [311], and autoimmune gastritis [312]. Some of the selected autoantibodies were in

Table 16.12   Autoimmune

Antigen	Library (Size)	K_d (nM)	Therapeutic Potential	References
TPO	Fab (10^5)	1	nd	[302]
dsDNA	Fab (8×10^6)	7.6	nd	[276]
Tg	Fab (5.3×10^7)	2.8	nd	[261]
AChR	Fab (1.1×10^6)	nd	Receptor protection	[305]
Sm	Fab (2×10^7)	10	nd	[301]
GAD65	Fab (2×10^6)	nd	nd	[304]
$\alpha_{II}\beta_3$	scFv (1.5×10^7)	180	Integrin antagonist	[309]

Note: $\alpha_{II}\beta_3$, integrin $\alpha_{II}\beta_3$; AChR, acetyl choline receptor; dsDNA, double-strand DNA; GAD65, decarboxylase; Sm, anti-Smith; Tg, thyroglobuline; TPO, thyroid peroxidase.

agreement with previous studies that suggested that the V_{H4} family is a major source of autoantibodies [313]. These and many other publications have greatly advanced our understanding of the pathogenesis of antibody-mediated autoimmune diseases.

16.3.4.1 Tumors

There are today five MAbs approved by the FDA for the treatment of cancer (Table 16.10), and numerous others are in advanced clinical development [314,315]. It is thought that the humoral antibody response in cancer patients may be selectively directed toward the "nonself" antigens expressed by the autologous tumor cells. Indeed, a humoral immune response directed to known tumor antigens has been demonstrated by the presence of serum antibodies in patients with cancer [260]. Thus, B cells from sensitized cancer patients who have mounted a response to altered or overexpressed antigens are a valuable source of mRNA to construct phage-displayed antibody libraries [68]. These cancer patients may provide an enriched source of disease-related antibodies that can be recovered from antibody libraries sorted by panning against specific tumor antigens. For example, a humoral immune response to tumor-related antigens such as p53 and erbB-2 has been demonstrated in some patients with cancer [316,317]. Additionally, it is of fundamental importance to develop new immunotherapeutic protocols designed to determine which antigens are the target of an immune response in the various types of cancer. This will lead to new therapies and treatments in particular for those cancers with poor prognoses.

Tumor-specific human-MAb fragments have been isolated against melanoma. In the first report, tumor-specific MAb fragments were isolated from individuals immunized with interferon-gamma (IFNγ)-transduced autologous melanoma cells [143]. The PBLs were isolated from two melanoma patients immunized with in vitro cultured, autologous-tumor cells infected with a retroviral vector carrying the human INFγ gene. Three scFv libraries were constructed using the fUSE5 phage vector for multivalent scFv display; the smallest of these two libraries contained 4×10^7 clones. These scFv fragments were synthesized from both the IgM and IgG, and λ and κ light-chain classes of mRNA. The selection involved panning against the autologous melanoma cell line, followed by extensive absorption against melanocytes to increase the chance of isolating antibodies with specificity for tumor. After three panning cycles, the majority of the selected clones reacted with all melanoma cell lines in ELISA and therefore recognize common melanoma antigens. However, clone V86 showed the tightest association with melanoma lines and demonstrated an intense staining of melanoma tissue but did not react with normal tissues present in the melanoma sections. This MAb clone is a potentially important tool for diagnostic screening for melanoma. The selection technique itself can be used to screen the antibody repertoire of any cancer patient, which may provide access to many more human antitumor MAbs.

Others have used a similar positive/negative selection strategy for antibody phage against melanoma. The selective removal of clones that react with normal tissue obviates the need for pure antigen [318]. For example, a human scFv IgGk library (9.3×10^7 clones) was constructed in pCANTAB5 vector from PBLs of 10 donors with a high titer of autoantibodies. After three rounds of using positive/negative selection strategy, they isolated two scFv clones, B3 and B4, that were positive on all melanoma sections obtained from several different patients. Neither B3 nor B4 cross-reacted with normal tissues, except for the weak reactivity of B3 with normal liver. The scFv clones reacted with antigens that are also expressed on tumors other than melanoma. Therefore, this approach should be applicable to the isolation of human antibodies against tumor markers or novel cell surface markers in general. The human antitumor scFvs B3 and B4 have a diagnostic value for the detection of metastatic disease by radioimaging of patients. Another potential application would be the use of these recombinant antibodies in targeted drug delivery to specifically kill the tumor cells *in vivo*.

Human MAb fragments against c-erbB-2 have been isolated from immune libraries of patients with colorectal cancer. Clark et al. [319] described the isolation of Fabs to c-erbB-2 from an immune library constructed from the isolated pericolic lymph node of a colorectal cancer patient.

An $IgG1_\kappa$ library contained 2×10^7 clones and was biopanned with purified recombinant c-erbB-2. After three rounds of selection, 16 clones showed a unique restriction pattern and reproducible reactivity against c-erbB-2 protein. Five Fabs with good expression levels were futher assessed for immunoreactivity against tumor cell lines. Remarkably, these MAbs bound strongly to the c-erbB-2 positive cell line, SKBR3, but displayed no significant staining of the negative MDA-MB-231 cells. All of the anti-c-erbB-2 Fabs isolated in this study used commonly represented V_H genes (V_{H4}, V_{H3}, and V_{H1}) with apparent V_{H4} overrepresentation, which suggests clonal selection. These results indicate that a naturally occuring immune response to tumor-related antigens can be exploited with phage display libraries for understanding of immune response to tumor cells and for the isolation of Fabs to predefined target antigens.

Human MAb fragments against p53 have also been isolated from immune libraries of patients with colorectal cancer. The same group selected anti-p53 Fabs from $IgG1_\kappa$ libraries also contructed from a similar library made from the pericolic lymph nodes taken from six colorectal cancer patients [260]. After five rounds of panning against recombinant p53, 14 unique clones were isolated from one library of 4.5×10^7 clones. The selected Fabs were encoded by germline (unmutated) V genes, predominantly from the V_{H1} family. Four of these Fabs were purified and further analyzed in detail. All four recognized p53 in ELISA and showed no reactivity against other antigens including ErbB2, MUC-1, CEA, TT, insulin, KLH, and BSA. Inhibition ELISA using serum from donor patients demonstrated various degrees of inhibition for each of the Fabs. The Fab 163.1 had high affinity for p53 and exhibited a K_d of 11.9 nM by surface plasmon resonance analysis. As there are no other human anti-p53 MAbs available from conventional cell immortalization, the information gained on human anti-p53 antibody V gene usage using immune antibody libraries, specifically regarding the degree of somatic mutation, is critical to any understanding of the nature and significance of the humoral immune response to p53. The Fabs selected in this study also have advantage over murine MAbs when used as anti-idiotypic vaccines. In 2001, the same group isolated a human Fab against the central DNA-binding domain of p53 from a large $IgG1-4_{\kappa\lambda}$ immune antibody library [320]. This Fab, 1159.8, provides additional insight into the nature of the tumor-specific immune responses and can be used as a reagent for functional studies of p53. Of note, this study also revealed that Fab 1159.8 binds to an epitope in close proximity to murine MAb Fab 240 that was able to prevent tumor growth and metastasis upon immunization [321]. Therefore, Fab 1159.8 may prove useful as an antitumor idiotypic vaccine while avoiding the undesirable features of the murine MAb.

Other studies have been performed using antibody libraries from colorectal cancer patients to investigate the nature and specificity of the humoral immune response. The lymphocytes infiltrating the primary colorectal tumor and lymph nodes draining the tumor were used for the construction of a Fab $IgG1_{\kappa\lambda}$ antibody libraries containing greater than 10^8 clones [322]. The antibody repertoires constructed from these two tissues were screened for the presence of antibodies directed to colorectal cancer cells by cell-based selection upon the cancer cell line $CaCo_2$. For comparison, the same selections were performed with a phage antibody repertoire made from B cells of healthy donors, which would in this case represent the relative "naïve" library [77]. Striking differences were observed in the panel of specificities selected from these different repertoires. Although a large panel of antibodies reactive with patient-derived primary tumors was obtained from the immune repertoires after two rounds of panning, antibodies selected from the local immune sources were directed to intracellular (cytoplasmic and nuclear) targets only. However, selections using the nonimmune library did result in numerous antibodies that recognized cell surface markers on $CaCo_2$. Although these data do not rule out the existence of humoral responses to certain cell surface antigens, they suggest a bias in the local humoral immune response in this colorectal cancer patient, directed primarily toward intracellular target antigens.

High-affinity human MAbs against the Lewis^X antigen have been isolated from immune repertoires. In 1999, an antibody library derived from the PBLs of 20 patients with various cancer diseases was successfully exploited to isolate scFv antibodies specific for the carbohydrate antigens sialyl Lewis^X (sLe^X) and Lewis^X (Le^X) [323]. An scFv antibody library was prepared from the assembly of V_H, V_κ, and V_λ PCR-amplified gene pools to yield approximately 2×10^8 clones. After four rounds of panning, four unique scFvs were then selected by using synthetic sLe^X and Le^X BSA conjugates. The selected scFv fragments were specific for sLe^X and Le^X, as demonstrated by ELISA, BIAcore, and flow cytometry binding to the cell surface of pancreatic adenocarcinoma cells. The K_d value of the best sLe^X binder, S6, was equal to 110 nM, which is comparable to the affinities of MAbs normally derived from the secondary immune response. These selected scFvs could be valuable reagents for probing the structure and function of carbohydrate antigens and in the treatment of human tumor diseases.

High-affinity human MAbs against the ganglioside G_{M3} antigen have been isolated from immune repertoires. The same multietiology library was used again to select scFvs against ganglioside G_{M3} overexpression, which is associated with a number of different cancers, including skin, colon, breast, and lung [324]. Several scFvs were affinity selected. One scFv, GM3A8, was purified and found to exhibit a K_d of 1200 nM by surface plasmon resonance analysis. The *in vitro* affinity of the scFv was estimated to be comparable to a previously reported G_{M3}-binding murine IgG and anti-G_{M3} polyconal antibodies [325,326], but with the significant advantages of

Table 16.13   Human MAbs to Tumor Targets from Immune Libraries

Antigen	Library (Size)	K_d (nM)	Therapeutic Potential	References
Melanoma	scFv (2×10^8)	nd	Diagnostic drug/delivery	[143]
ErbB2	Fab (2×10^7)	nd	Immunotherapy	[319]
Melanoma	scFv (9.3×10^7)	nd	Diagnostic drug/delivery	[318]
sLe^x	scFv (2×10^8)	110	Drug delivery	[323]
p53	Fab (4.5×10^7)	11.9	Anti-Id vaccine	[260]
p53	Fab (1.6×10^7)	nd	Anti-Id vaccine	[320]
CaCo$_2$	Fab ($>10^8$)	nd	nd	[322]
G$_{M3}$	scFv (2×10^8)	1200	Drug delivery	[324]

possessing a human antibody sequence and having the ability to specifically recognize highly metastalic cancer cells vs. normal cells. The GM3A8, with its favorable binding properties for melanoma and breast tumor cells, provides a solid foundation for further development.

The antibodies selected from cancer patient libraries are summarized in Table 16.13. We have summarized this section with the following points. First, it was demonstrated that autologous cells can be used in antibody selection strategies, thus circumventing a major weakness of the phage display technique such as the requirement of pure antigen for phage selection [143,318]. Second, anti-c-erbB-2 Fabs can be selected from immune libraries made from B cells of colorectal cancer patients [319]. Third, nanomolar affinity antip53 Fabs can be selected from immune libraries made from B cells of colorectal cancer patients [260]. Fourth, a large bias toward intracellular antigens was demonstrated in the local humoral immune response in a colorectal cancer patient [322]. Finally, high-affinity human antibodies against several tumor-associated carbohydrate antigens could be selected from a phage library constructed from the PBLs of various cancer patients [323,324]. That approach did not depend on the necessity to repeatedly construct phage antibody libraries.

16.3.5 Nonhuman Primates

The primate immune system and B-cell repertoire are highly similar to those of humans. Therefore, everything discussed in Section 16.3.3 about the immune repertoire and primer design applies for primates as well. The close genetic relationship between nonhuman primates and humans makes primates the paramount species for development of therapeutic antibodies for use in humans. However, research into this area has been severely limited due to the enormous expense associated with breeding colonies, and the price of a single primate, for example, a chimpanzee, is quite high. Moreover, there remains a lot of public scrutiny as well as criticism from private groups, which is invariably associated with research involving primates. Several publications have tried to explore the substitution of primate serological products in human prophylaxis and therapy by introducing human immune molecules into primates [327–329]. Collectively, these data demonstrate that human immune molecules are essentially nonimmunogenic when introduced to chimpanzees relative to other primates, and that human antibodies are recognized as being "self" by the chimpanzee immune system. Indeed, the half-life of a human MAb in a chimpanzee was found to be equivalent to the estimated half-life of IgG in humans [329]. Conversely, it is likely that chimpanzee antibodies would be like "self" in humans or at least nonimmunogenic and useful without further modifications (for example, to glycosylation patterns) in human therapy [82].

In the first example, Glamann et al. [330] produced an antibody phage display library constructed from RNA extracted from lymph node cells of a SIV-infected long-term nonprogressor macaque (rhesus). The Fab library was fully "macaque-like" in that even the C_κ and the C_{H1} domains were cloned from macaques using human primers. From this library with a primary diversity of 3×10^7 clones, seven gp120-reactive Fabs were obtained by selection of the library against SIV monomeric gp120. Although each of the Fabs was unique in molecular sequence, they all had highest homology to the human V_{H4} gene family. Furthermore, they formed two distinct groups based on epitope recognition, neutralizing activity *in vitro*, and molecular analysis. The first group of Fabs did not neutralize SIV and bound to a linear epitope in the V3 loop of the SIV envelope. In contrast, two of the group 2 Fabs neutralized homologous, neutralization-sensitive SIVsm isolates with high efficiency but failed to neutralize heterologous SIVmac isolates. Based on C-ELISAs with mouse MAbs of known specificity, these Fabs reacted with a conformational epitope that includes domains V3 and V4 of the SIV envelope. These macaque Fabs not only provide valuable standardized and renewable reagents for studying the role of antibody in SIV-associated disease, but they also set the stage for future use of macaques to develop MAbs for use in human therapy and prophylaxis.

In the second example, Schofield et al. [82] produced a Fab phage display library constructed from the RNA extracted from bone marrow lymphocytes of a chimpanzee, which had been previously experimentally infected with all five hepatitis-causing viruses, hepatitis A, B, C, D, and E. The Fab library IgG1$_\kappa$ was made with a primary diversity of 1.9×10^7 clones using again human C_κ and C_{H1} domain primers. Two hepatitis E virus (HEV) ORF2 capsid protein-reactive Fabs were obtained by selection of the library against SAR-55 ORF2. Competition experiments revealed that both Fabs reacted to the same epitope with affinities in the single-digit nanomolar range. These Fabs had the highest homology to the human V_{H3} family, and both neutralized the SAR-55 strain of HEV *in vitro* as shown by the complete prevention of infection of chimps inoculated with live HEV following preincubation with either Fab. Despite the high cost of using chimpanzees as a donor for immune repertoires, there are several advantages [82]. The first is that chimpanzees can be infected with many important human viral pathogens, and the chimp is the most closely related primate to humans; thus, chimpanzee antibodies may be able to be used directly in immune prophylactic treatment of infectious disease.

Antibody libraries from immune repertoires can be used to select MAbs against multiple targets if the lymphocytes were sensitized to these targets prior to tissue collection. Alternatively, the library can be used to select for binders to antigens conserved between strains of variant organisms. The same antibody library, which was used in the hepatitis E example [82], was used to select MAbs against hepatitis A virus (HAV). Two years later, the authors reported using whole inactivated HAV particles as the panning antigen to isolate four Fabs to the HAV capsid [331]. Following three rounds of panning on HAV particles, four unique HAV-specific clones were identified. Isolated Fabs competed with one of the murine MAbs that were used to define the HAV antigen site. These results seem to be in accordance with the earlier studies that suggested that there is a single immunodominant antigenic site on the HAV capsid. Overall, the Fab library used to isolate MAbs to both HEV and HAV is a potential repository for antibodies to all five recognized human hepatitis viruses, since the donor chimpanzee had been infected with each of the hepatitis viruses. Such chimpanzee-derived I_g sequences differ from human-derived sequences no more than genetically distinct human sequences differ from each other, and thus, they have direct therapeutic potential.

Melanoma-specific scFvs have been derived from nonhuman primates. Two Cynomolgus monkeys were immunized with a crude suspension of metastatic melanoma [332]. An scFv antibody phage library was generated from the lymph node mRNA with approximately 3×10^7 primary clones. Several clones producing scFvs that reacted with melanoma antigens were identified after three rounds of panning using melanoma cells and tissue sections. One of these scFvs, K305, demonstrated high-affinity binding and selectivity, supporting its use for tumor therapy in conjunction with T cell-activating superantigens. Comparison with the human germline sequence of the V_λ and V_H genes demonstrated 89% and 92% sequence identity on the nucleic acid level. Clone K305 was fused as a primate Fab to staphylococcal enterotoxin A. The affinity for melanoma tissue was in the low to subnanomolar range. T cell-mediated lysis of melanoma cells and *in vivo* tumor reduction mediated by this antibody in SCID mice were demonstrated, suggesting applicability for immunotherapy of malignant melanoma.

The antibodies selected from nonhuman primate libraries are summarized in Table 16.14. We conclude this section with the following points. First, it was demonstrated that specific MAbs could be selected against enveloped viruses from immunized macaques. Second, MAbs from chimpanzees were selected against multiple hepatitis virus strains by simply including these in the initial immunization strategy. Third, cancer cell-specific MAb fragments can be derived from immune antibody libraries from nonhuman primates. These successful examples of the use of antibody libraries from nonhuman primates for the development of

Table 16.14 Nonhuman Primates MAbs from Immune Libraries

Antigen	Library (Size)	K_d (nM)	Therapeutic Potential	References
SIV gp120	Fab (3×10^7)	nd	Virus neutralization	[330]
HEV ORF2	Fab (1.9×10^7)	1.7	Virus neutralization	[82]
Melanoma	scFv (3×10^7)	1.6	Drug delivery	[332]
HAV	Fab (1.9×10^7)	nd	Virus neutralization	[331]

Note: HAV, hepatitis A virus; HEV ORF2, hepatitis E virus open reading frame 2 protein; SIV gp120, simian immunodeficiency virus glycoprotein 120.

potential immunotherapeutic interventions for infectious diseases and tumors are rare, but they are expected to continue into the future as primate antibodies are extremely similar to human antibodies and can rapidly translate into clinical treatments.

16.4 THE FUTURE

Antibody libraries from immune repertoires have developed into a new and exciting field with broad applications in biotechnology and the pharmaceutical industry. Use of this technology in centralized laboratories will be crucial to the development of modern and rapid diagnostics for confirmatory diagnostic tests. These tests will, in the next decade, be transformed into small high-throughput antigen detection chips, where a single test can screen for thousands of pathogens. However, the key to developing these protein chips and other nanodetection technology is to first develop quality-controlled reagents capable of detecting pathogenic proteins. Clearly, antibody libraries from immune sources will be key to the production of many of the toxin, pathogen, autoantigen, and tumor-specific antibodies required for this technology. B-cell immortalization techniques and phage display offer complementary approaches to the development of antigen-specific MAbs. It has become increasingly important for scientists in many fields including neurology, cancer biology, autoimmunity, placentology, infectious disease, and pure biotechnology to be capable of alternating between hybridoma growth in tissue culture and antibody cloning in order to completely capitalize on new lead molecules from immunized sources.

The development of MAbs by traditional methods continues to evolve. New high-throughput screening systems are on the horizon for hybridoma and *E. coli*-produced MAbs. The generation of antibody libraries from immune repertoires provides a unique tool for fundamental research on the immunology of human and animal diseases. Moreover, immune libraries and the power of phage selection allow for the selection of antibodies to complex antigens.

Antibody libraries will continue to be developed anew from more diverse species such as fish, as genomic information about I_g repertoires is unveiled. For example, a naturally occurring V_H-like domain, with characteristics similar to camel V_HH antibodies [333], has recently been described in nurse sharks as the new antigen receptor (NAR) [334]. The NARs from the wobbegong shark have recently been used as scaffolds for the construction of phage-displayed libraries [96]. Similarly, new inroads have been made in the genetics of Atlantic salmon.

The recent cloning and sequencing of the cDNA of around 50 V_H (VDJ) and 15 V_L genes have quickly led to the selection of anti-TNP and anti-FITC-specific scFvs [335]. This work opens the possibility of using the immune repertoires of fish for MAb generation.

We predict a rapid growth in the area of infectious diseases given the huge problem related to antibiotic resistance and the need for new therapies. There is actually a remarkable death of human MAbs in clinical trials to infectious agents. Indeed, only

7 of 128 MAbs in clinical trial are against infectious agents [78]. The vast majority of these clinical antibodies are against cancer targets. There are also very few MAbs from animals for validating diagnoses of animal diseases. This seems counterlogical given the fact that antibodies evolved in order to target foreign antigens. However, this is more likely a reflection of first world health concerns.

The use of immune libraries from XenoMice will be an ideal source of cDNA for *in vitro* modifications including affinity maturation or humanization of existing potent murine MAbs, once again highlighting the complementary nature of these two techniques. The exquisite ability of the immune system to produce antibodies to a specific immunogen *in vivo*, whether on a pathogen, a cryptic self-antigen, or an overexpressed or altered tumor antigen, will ensure that antibody libraries from immune repertoires will continue to be exploited in the future.

ACKNOWLEDGMENTS

The authors thank Dev Sidhu for his time and patience. JDB is supported by CBRN research and technology initiative.

REFERENCES

1. Groves DJ, Morris BA. Veterinary sources of nonrodent monoclonal antibodies: Interspecific and intraspecific hybridomas. *Hybridoma* 2000; 19:201–214.
2. Flynn JN, Harkiss GD, Hopkins J. Generation of a sheep x mouse heterohybridoma cell line (1C6.3a6T.1D7) and evaluation of its use in the production of ovine monoclonal antibodies. *J Immunol Methods* 1989; 121:237–246.
3. Hoogenboom HR, Chames P. Natural and designer binding sites made by phage display technology. *Immunol Today* 2000; 21:371–378.
4. Rader C. Antibody libraries in drug and target discovery. *Drug Discov Today* 2001; 6:36–43.
5. Soderlind E, Strandberg L, Jirholt P, Kobayashi N, Alexeiva V, Aberg AM, Nilsson A et al. Recombining germline-derived CDR sequences for creating diverse single-framework antibody libraries. *Nat Biotechnol* 2000; 18:852–856.
6. Knappick A, Ge L, Honegger A, Pack P, Fischer M, Wellnhofer G, Hoess A et al. Fully synthetic human combinatorial antibody libraries (HuCAL) based on modular consensus frameworks and CDRs randomized with trinucleotides. *J Mol Biol* 2000; 296:57–86.
7. Smith G. Filamentous fusion phage: Novel expression vectors that display cloned antigens on the virion surface. *Science* 1985; 228:1315–1317.
8. Parmley SF, Smith GP. Antibody-selectable filamentous fd phage vectors: Affinity purification of target genes. *Gene* 1988; 73:305–318.
9. Scott JK, Smith GP. Searching for peptide ligands with an epitope library. *Science* 1990; 249:386–390.
10. Devlin JJ, Panganiban LC, Devlin PE. Random peptide libraries: A source of specific protein binding molecules. *Science* 1990; 249:404–406.
11. Cwirla SE, Peters EA, Barrett RW, Dower WJ. Peptides on phage: A vast library of peptides for identifying ligands. *Proc Natl Acad Sci U S A* 1990; 87:6378–6382.

12. Skerra A, Pluckthun A. Assembly of a functional immunoglobulin Fv fragment in *Escherichia coli*. *Science* 1988; 240:1038–1041.

13. Better M, Chang CP, Robinson RR, Horwitz AH. *Escherichia coli* secretion of an active chimeric antibody fragment. *Science* 1988; 240:1041–1043.

14. Chiang YL, Sheng-Dong R, Brow MA, Larrick JW. Direct cDNA cloning of the rearranged immunoglobulin variable region. *Biotechniques* 1989; 7:360–366.

15. Gavilondo-Cowley JV, Coloma M, Vazquez J, Ayala M, Macias A, Fry K, Larrick J. Specific amplification of rearranged immunoglobulin variable region genes from mouse hybridoma cells. *Hybridoma* 1990; 9:407–417.

16. Coloma MJ, Larrick JW, Ayala M, Gavilondo-Cowley JV. Primer design for the cloning of immunoglobulin heavy-chain leader-variable regions from mouse hybridoma cells using the PCR. *Biotechniques* 1991; 11:152–156.

17. Huse WD, Sastry L, Iverson S, Kang A, Alting-Mees M, Burton DR, Benkovic SJ et al. Generation of a large combinatorial library of the immunoglobulin repertoire in phage lambda. *Science* 1989; 246:1275–1281.

18. Caton A, Koprowski H. Influenza virus hemagglutinin specific antibodies isolated from a combinatorial expression library are closely related to the immune response of the donor. *Proc Natl Acad Sci U S A* 1990; 87:6450–6454.

19. Persson MA, Caothien RH, Burton DR. Generation of diverse high-affinity human monoclonal antibodies by repertoire cloning. *Proc Natl Acad Sci U S A* 1991; 88:2432–2436.

20. McCafferty J, Griffiths AD, Winter G, Chiswell DJ. Phage antibodies: Filamentous phage displaying antibody variable domains. *Nature* 1990; 348:552–554.

21. Clackson T, Hoogenboom HR, Griffiths AD, Winter G. Making antibody fragments using phage display libraries. *Nature* 1991; 352:624–628.

22. Kang AS, Barbas CF, Janda KD, Benkovic SJ, Lerner RA. Linkage of recognition and replication functions by assembling combinatorial antibody Fab libraries along phage surfaces. *Proc Natl Acad Sci U S A* 1991; 88:4363–4366.

23. Hoogenboom H, Griffiths A, Johnson K, Chiswell D, Hudson P, Winter G. Multisubunit proteins on the surface of filamentous phage: Methodologies for displaying antibody Fab heavy and light chains. *Nucleic Acids Res* 1991; 19:4133–4137.

24. Barbas CF III, Kang A, Lerner RA, Benkovic S. Assembly of combinatorial antibody libraries on phage surfaces: The gene III site. *Proc Natl Acad Sci U S A* 1991; 88:7978–7982.

25. Marks JD, Hoogenboom HR, Bonnert TP, McCafferty J, Griffiths AD, Winter G. By-passing immunization. Human antibodies from *V* gene libraries displayed on phage. *J Mol Biol* 1991; 222:581–597.

26. Barbas CF III, Crowe JE Jr, Cababa D, Jones TM, Zebedee SL, Murphy BR, Chanock RM et al. Human monoclonal Fab fragments derived from a combinatorial library bind to respiratory syncytial virus F glycoprotein and neutralize infectivity. *Proc Natl Acad Sci U S A* 1992; 89:10164–10168.

27. Burton D, Barbas CF III. Human antibodies from combinational libraries. *Adv Immunol* 1994; 57:191–280.

28. Winter G, Griffiths A, Hawkins R, Hoogenboom H. Making antibodies by phage display technology. *Annu Rev Immunol* 1994; 12:433–455.

29. Barbas CF, Burton D, Scott J, Silverman G, eds. *Phage Display: A Laboratory Manual*. Cold Spring Harbour, NY: Cold Spring Harbour Laboratory Press, 2001.

30. Boder ET, Wittrup KD. Yeast surface display for screening combinatorial polypeptide libraries. *Nat Biotechnol* 1997; 15:553–557.

31. Schaffitzel C, Hanes J, Jermutus L, Pluckthun A. Ribosome display: An *in vitro* method for selection and evolution of antibodies from libraries. *J Immunol Methods* 1999; 231:119–135.

32. Daugherty PS, Olsen MJ, Iverson BL, Georgiou G. Development of an optimized expression system for the screening of antibody libraries displayed on the *Escherichia coli* surface. *Protein Eng* 1999; 12:613–621.

33. Azzazy M, Highsmith W Jr. Phage display technology: Clinical applications and recent innovations. *Clin Biochem* 2002; 35:425–445.

34. Golub R, Fellah JS, Charlemagne J. Structure and diversity of the heavy chain VDJ junctions in the developing Mexican axolotl. *Immunogenetics* 1997; 46:402–409.

35. Dammers PM, Kroese FG. Evolutionary relationship between rat and mouse immunoglobulin IGHV5 subgroup genes (PC7183) and human IGHV3 subgroup genes. *Immunogenetics* 2001; 53:511–517.

36. Wagner B, Greiser-Wilke I, Wege AK, Radbruch A, Leibold W. Evolution of the six horse IGHG genes and corresponding immunoglobulin gamma heavy chains. *Immunogenetics* 2002; 54:353–364.

37. Momoi Y, Nagase M, Okamoto Y, Okuda M, Sasaki N, Watari T, Goitsuka R et al. Rearrangements of immunoglobulin and T-cell receptor genes in canine lymphoma/leukemia cells. *J Vet Med Sci* 1993; 55:775–780.

38. Patel M, Selinger D, Mark GE, Hickey G, Hollis G. Sequence of the dog immunoglobulin alpha and epsilon constant region genes. *Immunogenetics* 1995; 41:282–286.

39. Cho KW, Satoh H, Youn HY, Watari T, Tsujimoto H, O'Brien SJ, Hasegawa A. Assignment of the cat immunoglobulin heavy chain genes IGHM and IGHG to chromosome B3q26 and T cell receptor chain gene TCRG to A2q12→q13 by fluorescence in situ hybridization. *Cytogenet Cell Genet* 1997; 79:118–120.

40. Najakshin AM, Belousov ES, Alabyev BY, Christensen J, Storgaard T, Aasted B, Taranin AV. Structure of mink immunoglobulin gamma chain cDNA. *Dev Comp Immunol* 1996; 20:231–240.

41. Sinkora M, Sinkorova J, Butler JE. B cell development and VDJ rearrangement in the fetal pig. *Vet Immunol Immunopathol* 2002; 87:341–346.

42. Duchosal MA, Eming SA, Fischer P, Leturcq D, Barbas CF III, McConahey PJ, Caothien RH et al. Immunization of hu-PBL-SCID mice and the rescue of human monoclonal Fab fragments through combinatorial libraries. *Nature* 1992; 355:258–262.

43. Nguyen H, Hay J, Mazzulli T, Gallinger S, Sandhu J, Teng Y, Hozumi N. Efficient generation of respiratory syncytial virus (RSV)-neutralizing human MoAbs via human peripheral blood lymphocyte (hu-PBL)-SCID mice and scFv phage display libraries. *Clin Exp Immunol* 2000; 122:85–93.

44. Jakobovits A. Production of fully human antibodies by transgenic mice. *Curr Opin Biotechnol* 1995; 6:561–566.

45. Rajewsky K. Clonal selection and learning in the antibody system. *Nature* 1996; 381:751–758.

46. Klinman NR, Linton PJ. The clonotype repertoire of B cell subpopulations. *Adv Immunol* 1988; 42:1–93.

47. Kohler G, Milstein C. Continuous cultures of fused cells secreting antibody of predefined specificity. *Nature* 1975; 256:495–497.

48. Duggan JM, Coates DM, Ulaeto DO. Isolation of single-chain antibody fragments against Venezuelan equine encephalomyelitis virus from two different immune sources. *Viral Immunol* 2001; 14:263–273.

49. Gherardi E, Milstein C. Original and artificial antibodies. *Nature* 1992; 357:201.

50. Kettleborough CA, Ansell KH, Allen RW, Rosell-Vives E, Gussow DH, Bendig MM. Isolation of tumor cell-specific single-chain Fv from immunized mice using phage–antibody libraries and the re-construction of whole antibodies from these antibody fragments. *Eur J Immunol* 1994; 24:952–958.

51. Ames RS, Tornetta M, McMillan L, Kaiser K, Holmes S, Applebaum E, Cusimano D et al. Neutralizing murine monoclonal antibodies to human IL-5 isolated from hybridomas and a filamentous phage Fab display library. *J Immunol* 1995; 154:6355–6364.

52. Krebber A, Bornhauser S, Burmester J, Honegger A, Willuda J, Bosshard HR, Pluckthun A. Reliable cloning of functional antibody variable domains from hybridomas spleen cell repertoires employing a reengineered phage display system. *J Immunol Methods* 1997; 201(1):35–55.

53. Larrick J, Danielsson L, Brenner C, Abrahamson M, Fry K, Borrebaeck C. Rapid cloning of rearranged immunoglobulin genes from human hybridoma cells using mixed primers and the polymerase chain reaction. *Biochem Biophys Res Commun* 1989; 160:1250–1256.

54. Orlandi R, Gussow DH, Jones PT, Winter G. Cloning immunoglobulin variable domains for expression by the polymerase chain reaction. *Biotechnology* 1989; 1992:527–531.

55. Dattamajumdar AK, Jacobson DP, Hood LE, Osman GE. Rapid cloning of any rearranged mouse immunoglobulin variable genes. *Immunogenetics* 1996; 43(3):141–151.

56. Gilliland LK, Norris N, Marquardt H, Tsu T, Hayden M, Neubauer M, Yelton D et al. Rapid and reliable cloning of antibody variable regions and generation of recombinant single chain antibody fragments. *Tissue Antigens* 1996; 47:1–20.

57. Orum H, Andersen PS, Oster A, Johansen LK, Riise E, Bjornvad M, Svendsen I et al. Efficient method for constructing comprehensive murine Fab antibody libraries displayed on phage. *Nucleic Acids Res* 1993; 21:4491–4498.

58. Andris-Widhopf J, Rader C, Steinberger P, Fuller R, Barbas CF III. Methods for the generation of chicken monoclonal antibody fragments by phage display. *J Immunol Methods* 2000; 242:159–181.

59. Janda KD, Lo L-C, Lo C-HL, Sim M-M, Wang R, Wong C-H, Lerner RA. Chemical selection for catalysis in combinatorial antibody libraries. *Science* 1997; 275:945–948.

60. Moreno de Alboran I, Martinez-Alonso C, Barbas CF III, Burton DR, Ditzel HJ. Human monoclonal Fab fragments specific for viral antigens from combinational IgA libraries. *Immunotechnology* 1995; 1:21–28.

61. Steinberger P, Kraft D, Valenta R. Construction of a combinatorial IgE library from an allergic patient. Isolation and characterization of human IgE Fabs with specificity for the major timothy grass pollen allergen, Phl p 5. *J Biol Chem* 1996; 271:10967–10972.

62. Boel E, Verlaan S, Poppelier MJ, Westerdaal NA, Van Strijp JA, Logtenberg T. Functional human monoclonal antibodies of all isotypes constructed from phage display library-derived single-chain Fv antibody fragments. *J Immunol Methods* 2000; 239:153–166.

63. de Wildt R, Mundy C, Gorick B, Tomlinson I. Antibody arrays for high-throughput screening of antibody antigen interactions. *Nat Biotechnol* 2000; 18:989–994.

64. Rader C, Cheresh DA, Barbas CF III. A phage display approach for rapid antibody humanization: Designed combinatorial V gene libraries. *Proc Natl Acad Sci U S A* 1998; 95:8910–8915.

65. Hudson PJ, Souriau C. Engineered antibodies. *Nat Med* 2003; 9:129–134.

66. Adair F. Monoclonal antibodies—Magic bullet or a shot in the dark? *Drug Discov World* 2002; 3:53–59.

67. De Boer M, Ten Voorde G, Ossendorp F, Van Duijn G, Tager JM. Requirements for the generation of memory B cells *in vivo* and their subsequent activation *in vitro* for the production of antigen specific hybridomas. *J Immunol Methods* 1988; 113:143–149.

68. Williamson RA, Burioni R, Sanna PP, Partridge LJ, Barbas CF 3rd, Burton DR. Human monoclonal antibodies against a plethora of viral pathogens from single combinatorial libraries. *Proc Natl Acad Sci U S A* 1993; 90:4141–4145.

69. Berry JD, Licea A, Popkov M, Cortez X, Fuller R, Elia M, Kerwin L et al. Rapid monoclonal antibody generation via dendritic cell targeting in vivo. *Hybrid Hybridomics* 2003; 22:23–31.

70. Wang H, Griffiths MN, Burton DR, Ghazal P. Rapid antibody responses by low-dose, single-step, dendritic cell-targeted immunization. *Proc Natl Acad Sci U S A* 2000; 97:847–852.

71. Roost H, Bachmann M, Haag A, Kalinke U, Pliska V, Hengartner H, Zinkernagel R. Early high-affinity neutralizing anti-viral IgG responses without further overall improvements of affinity. *Proc Natl Acad Sci U S A* 1995; 92:1257–1261.

72. Barber BH. The immunotargeting approach to adjuvant-independent subunit vaccine design. *Semin Immunol* 1997; 9:293–301.

73. Kalinke U, Bucher E, Ernst B, Oxenius A, Roost H, Geley S, Kofler R et al. The role of somatic mutation in the generation of the protective humoral immune response against vesicular stomatitis virus. *Immunity* 1996; 5:639–652.

74. Kalinke U, Bucher E, Ernst B, Oxenius A, Roost H, Geley S, Kofler R et al. Virus neutralization by germ-line vs. hypermutated antibodies. *Proc Natl Acad Sci U S A* 2000; 97:10126–10131.

75. Kim SH, Park SY. Selection and characterization of human antibodies against hepatitis B virus surface antigen by phage display. *Hybrid Hybridomics* 2002; 21:385–392.

76. Marks JD, Hoogenboom HR, Griffiths AD, Winter G. Molecular evolution of proteins on filamentous phage. Mimicking the strategy of the immune system. *J Biol Chem* 1992; 267:16007–16010.

77. Vaughan TJ, Williams AJ, Pritchard K, Osbourn JK, Pope AR, Earnshaw JC, McCafferty J et al. Human antibodies with sub-nanomolar affinities isolated from a large non-immunized phage display library. *Nat Biotechnol* 1996; 14:309–314.

78. Gavilondo JV, Larrick JW. Antibody engineering at the millennium. *Biotechniques* 2000; 29:128–145.

79. Holt L, Enever C, de Wildt R, Tomlinson I. The use of recombinant antibodies in proteomics. *Curr Opin Biotechnol* 2000; 11:445–449.

80. O'Connell D, Becerril B, Roy-Burman A, Daws M, Marks JD. Phage versus phagemid libraries for generation of human monoclonal antibodies. *J Mol Biol* 2002; 321:49–56.

81. Muyldermans S. Single domain camel antibodies: Current status. *Rev Mol Biotechnol* 2001; 74:277–302.

82. Schofield DJ, Glamann J, Emerson SU, Purcell RH. Identification by phage display and characterization of two neutralizing chimpanzee monoclonal antibodies to the hepatitis E virus capsid protein. *J Virol* 2000; 74:5548–5555.

83. Glamann J, Hirsch VM. Characterization of a macaque recombinant monoclonal antibody that binds to a CD4-induced epitope and neutralizes simian immunodeficiency virus. *J Virol* 2000; 74:7158–7163.

84. Yuan D, Tucker PW. Regulation of IgM and IgD synthesis in B lymphocytes. I. Changes in biosynthesis of mRNA for mu and gamma chains. *J Immunol* 1984; 132:1561–1565.

85. Lefkovits I. ... and such are little lymphocytes made of. *Res Immunol* 1995; 146:5–10.

86. Bachmann MA, Kundig TM, Kalberer CP, Hengartner H, Zinkernagel RM. How many B cells are needed to protect against a virus? *J Immunol* 1994; 152:4235–4241.

87. Dilosa R, Maeda K, Masuda A, Szakal A, Tew J. Germinal center B cells and antibody production in the bone marrow. *J Immunol* 1991; 146:4071–4077.

88. Slifka MK, Matloubian M, Ahmed R. Bone marrow is a major site of long-term antibody production after acute viral infection. *J Virol* 1995; 69:1895–1902.

89. Yip Y, Hawkins N, Clark M, Ward R. Evaluation of different lymphoid tissue sources for the construction of human immunoglobulin gene libraries. *Immunotechnology* 1997; 3:195–203.

90. Thompson MA, Cancro MP. Dynamics of B cell repertoire formation: Normal patterns of clonal turnover are altered by ligand interaction. *J Immunol* 1982; 129:2372–2376.

91. O'Brien R, Aitken PM, O'Neil BW, Campo MS. Generation of native bovine mAbs by phage display. *Proc Natl Acad Sci U S A* 1999; 96:640–645.

92. Lee W, Kohler H. Decline and spontaneous recovery of the monoclonal response to phosphorylcholine during repeated immunization. *J Immunol* 1974; 113:1644–1654.

93. Neron S, Lemieux R. Negative effect of multiple antigen injections on the yield of murine monoclonal antibodies obtained by hybridoma technology. *Hybridoma* 1992; 11:639–644.

94. Burton DR, Barbas CF III. Antibodies from libraries. *Nature* 1992; 359:782–783.

95. O'Brien P, Aitken R. Broadening the impact of antibody phage display technology. Amplification of immunoglobulin sequences from species other than humans or mice. *Methods Mol Biol* 2002; 178:73–86.

96. Nuttall SD, Krishnan UV, Hattarki M, de Gori R, Irving RA, Hudson PJ. Isolation of the new antigen receptor from wobbegong sharks, and use as a scaffold for the display of protein loop libraries. *Mol Immunol* 2001; 38:313–326.

97. Max EE. Immunoglobulin molecular genetics. In: Paul WE, ed. *Fundamental Immunology*, 3rd ed. New York: Raven Press, 1993:315–382.

98. Tonegawa S. Somatic generation of antibody diversity. *Nature* 1983; 302:575–581.

99. McCormack WT, Tjoelker LW, Carlson LM, Petryniak B, Barth CF, Humphries EH, Thompson CB. Chicken IgL gene rearrangement involves deletion of a circular episome and addition of single nonrandom nucleotides to both coding segments. *Cell* 1989; 56:785–791.

100. Sanz I. Multiple mechanisms participate in the generation of diversity of human H chain CDR3 regions. *J Immunol* 1991; 147:1720–1729.

101. Ohlin M, Borrebaeck C. Characteristics of human antibody repertoires following active immune responses *in vivo*. *Mol Immunol* 1996; 33:583–592.

102. Casellas R, Shih TA, Kleinewietfeld M, Rakonjac J, Nemazee D, Rajewsky K, Nussenzweig MC. Contribution of receptor editing to the antibody repertoire. *Science* 2001; 291:1541–1544.

103. Du Pasquier L. Phylogeny of B-cell development. *Curr Opin Immunol* 1993; 5:185–193.

104. Thompson CB. New insights into V(D)J recombination and its role in the evolution of the immune system. *Immunity* 1995; 3:531–539.

105. Berek C, Brandl B, Steinhauser G. Human lambda light chain germline genes: Polymorphism in the IGVL2 gene family. *Immunogenetics* 1997; 46:533–534.

106. Solin ML, Kaartinen M. Immunoglobulin constant kappa gene alleles in twelve strains of mice. *Immunogenetics* 1993; 37:401–407.

107. Ulrich HD, Moore FL, Schultz PG. Germline diversity within the mouse Igk-V9 gene family. *Immunogenetics* 1997; 47:91–95.

108. Solin ML, Kaartinen M. Allelic polymorphism of mouse Igh-J locus, which encodes immunoglobulin heavy chain joining (JH) segments. *Immunogenetics* 1992; 36:306–313.

109. Mattila PS, Schugk J, Wu H, Makela O. Extensive allelic sequence variation in the J region of the human immunoglobulin heavy chain gene locus. *Eur J Immunol* 1995; 25:2578–2582.

110. Blankenstein T, Bonhomme F, Krawinkel U. Evolution of pseudogenes in the immuno-globulin VH-gene family of the mouse. *Immunogenetics* 1987; 26:237–248.

111. Adderson EE, Azmi FH, Wilson PM, Shackelford PG, Carroll WL. The human VH3b gene subfamily is highly polymorphic. *J Immunol* 1993; 151:800–809.

112. Milner EC, Hufnagle WO, Glas AM, Suzuki I, Alexander C. Polymorphism and utiliza-tion of human V_H genes. *Ann N Y Acad Sci* 1995; 764:50–61.

113. Juul L, Hougs L, Barington T. A new apparently functional IGVK gene (VkLa) present in some individuals only. *Immunogenetics* 1998; 48:40–46.

114. Flajnik MF. Comparative analyses of immunoglobulin genes: Surprises and portents. *Nat Rev Immunol* 2002; 2:688–698.

115. Sayegh CE, Drury G, Ratcliffe MJ. Efficient antibody diversification by gene conver-sion in vivo in the absence of selection for V(D)J-encoded determinants. *EMBO J* 1999; 18:6319–6328.

116. Reynaud CA, Anquez V, Weill JC. The chicken D locus and its contribution to the immu-noglobulin heavy chain repertoire. *Eur J Immunol* 1991; 21:2661.

117. Reynaud CA, Mackay CR, Muller RG, Weill JC. Somatic generation of diversity in a mam-malian primary lymphoid organ: The sheep ileal Peyer's patches. *Cell* 1991; 64:995–1005.

118. Hein WR, Dudler L. Diversification of sheep immunoglobulins. *Vet Immunol Immunopathol* 1999; 72:17–20.

119. Lanning D, Sethupathi P, Rhee KJ, Zhai SK, Knight KL. Intestinal microflora and diver-sification of the rabbit antibody repertoire. *J Immunol* 2000; 165:2012–2019.

120. French DL, Laskov R, Scharff MD. The role of somatic hypermutation in the generation of antibody diversity. *Science* 1989; 244:1152–1157.

121. Storb U. Progress in understanding the mechanism and consequences of somatic hyper-mutation. *Immunol Rev* 1998; 162:5–11.

122. Mage RG. Diversification of rabbit V_H genes by gene-conversion-like and hypermuta-tion mechanisms. *Immunol Rev* 1998; 162:49–54.

123. Muramatsu M, Kinoshita K, Fagarasan S, Yamada S, Shinkai Y, Honjo T. Class switch recombination and hypermutation require activation-induced cytidine deaminase (AID), a potential RNA editing enzyme. *Cell* 2000; 102:553–563.

124. Revy P, Muto T, Levy Y, Geissmann F, Plebani A, Sanal O, Catalan N et al. Activation-induced cytidine deaminase (AID) deficiency causes the autosomal recessive form of the Hyper-IgM syndrome (HIGM2). *Cell* 2000; 102:565–575.

125. Arakawa H, Hauschild J, Buerstedde JM. Requirement of the activation-induced deami-nase (AID) gene for immunoglobulin gene conversion. *Science* 2002; 295:1301–1306.

126. Kutemeier G, Harloff C, Mocikat R. Rapid isolation of immunoglobulin variable region genes from cell lysates of rat hybridomas by polymerase chain reaction. *Hybridoma* 1992; 11:23–31.

127. Johnson G, Wu TT. Kabat database and its applications: 30 years after the first variabil-ity plot. *Nucleic Acids Res* 2000; 28:214–218.

128. Kofler R, Geley S, Kofler H, Helmberg A. Mouse variable-region gene families: Complexity, polymorphism and use in non-autoimmune responses. *Immunol Rev* 1992; 128:5–21.

129. Burton DR. Antibody libraries. In: Barbas CF, Burton D, Scott J, Silverman G, eds. *Phage Display: A Laboratory Manual.* Cold Spring Harbour, NY: Cold Spring Harbour Laboratory Press, 2001.

130. Andris-Widhopf J, Steinberger P, Fuller R, Rader C, Barbas CF III. Generation of antibody libraries: PCR amplification and assembly of light- and heavy-chain coding sequences. In: Barbas CF, Burton D, Scott J, Silverman G, eds. *Phage Display: A Laboratory Manual.* Cold Spring Harbour, NY: Cold Spring Harbour Laboratory Press, 2001.

131. Huston JS, Levinson D, Mudgett-Hunter M, Tai MS, Novotny J, Margolies MN, Ridge RJ et al. Protein engineering of antibody binding sites: Recovery of specific activity in an anti-digoxin single-chain Fv analogue produced in *Escherichia coli*. *Proc Natl Acad Sci U S A* 1988; 85(16):5879–5883.

132. Whitlow M, Bell BA, Feng SL, Filpula D, Hardman KD, Hubert SL, Rollence ML et al. An improved linker for single-chain Fv with reduced aggregation and enhanced proteolytic stability. *Protein Eng* 1993; 6:989–995.

133. Thiel MA, Coster DJ, Standfield SD, Brereton HM, Mavrangelos C, Zola H, Taylor S et al. Penetration of engineered antibody fragments into the eye. *Clin Exp Immunol* 2002; 128:67–74.

134. Yokota T, Milenic DE, Whitlow M, Schlom J. Rapid tumor penetration of a single-chain Fv and comparison with other immunoglobulin forms. *Cancer Res* 1992; 52(12):3402–3408.

135. Russel M, Linderoth NA, Sali A. Filamentous phage assembly: Variation on a protein export theme. *Gene* 1997; 192(1):23–32.

136. Iannolo G, Minenkova O, Petruzzelli R, Cesareni G. Modifying filamentous phage capsid: Limits in the size of the major capsid protein. *J Mol Biol* 1995; 484:335–344.

137. Enshell-Seijffers D, Smelyanski L, Gershoni JM. The rational design of a "type 88" genetically stable peptide display vector in the filamentous bacteriophage fd. *Nucleic Acids Res* 2001; 29:E50, 1–13.

138. Rakonjac J, Model P. Roles of pIII in filamentous phage assembly. *J Mol Biol* 1998; 282(1):25–41.

139. Crissman JW, Smith GP. Gene-III protein of filamentous phages: Evidence for a carboxyl-terminal domain with a role in morphogenesis. *Virology* 1984; 132(2):445–455.

140. Burton DR, Pyati J, Koduri R, Sharp SJ, Thornton GB, Parren PW, Sawyer LS et al. Efficient neutralization of primary isolates of HIV-1 by a recombinant human monoclonal antibody. *Science* 1994; 266:1024–1027.

141. Ghahroudi MA, Desmyter A, Wyns L, Hamers R, Muyldermans S. Selection and identification of single domain antibody fragments from camel heavy-chain antibodies. *FEBS Lett* 1997; 414:521–526.

142. Frenken LGJ, van der Linden RHJ, Hermans PWJJ, Bos JW, Ruuls RC, de Geus B, Verrips CT. Isolation of antigen specific llama V_{HH} antibody fragments and their high level secretion by *Saccharomyces cerevisiae*. *J Biotechnol* 2000; 78:11–21.

143. Cai X, Garen A. Anti-melanoma antibodies from melanoma patient immunized with genetically modified autologous tumor cells: Selection of specific antibodies from single-chain Fv fusion phage libraries. *Proc Natl Acad Sci U S A* 1995; 92: 6537–6541.

144. Ward RL, Clark MA, Lees J, Hawkins NJ. Retrieval of human antibodies from phage-display libraries using enzymatic cleavage. *J Immunol Methods* 1996; 189:73–82.

145. Guidry AJ, Srikumaran S, Goldsby RA. Production and characterization of bovine immunoglobulins from bovine multi murine hybridomas. *Methods Enzymol* 1986; 121:244–265.

146. Li Y, Kilpatrick J, Whitlam GC. Sheep monoclonal antibody fragments generated using a phage display system. *J Immunol Methods* 2000; 236:133–146.

147. Charlton KA, Moyle S, Porter AJR, Harris WJ. Analysis of the diversity of a sheep antibody repertoire as revealed from a bacteriophage display library. *J Immunol* 2000; 164:6221–6229.

148. O'Brien P, Maxwell G, Campo MS. Bacterial expression and purification of recombinant bovine Fab fragments. *Protein Expr Purif* 2002; 24:43–50.

149. Reynolds C-A. The genesis, tutelage and exodus of B cells in the ileal Peyer's patch of sheep. *Int Rev Immunol* 1997; 15:265–299.

150. Masteller EL, Pharr GT, Funk PE, Thompson CB. Avian B cell development. *Int Rev Immunol* 1997; 15:185–206.

151. Knight KL, Winstead CR. Generation of antibody diversity in rabbits. *Curr Opin Immunol* 1997; 9:228–232.

152. Duenas M, Vazquez J, Ayala M, Soderlind E, Ohlin M, Perez L, Borrebaeck CA et al. Intra- and extracellular expression of an scFv antibody fragment in *E. coli*: Effect of bacterial strains and pathway engineering using GroES/L chaperonins. *Biotechniques* 1994; 16:476–477, 480–483.

153. Rader C, Popkov M, Neves JA, Barbas CF III. Integrin avb3 targeted therapy for Kaposi's sarcoma with an *in vitro* evolved antibody. *FASEB J* 2002; 16:2000–2002.

154. Brodeur PH, Riblet R. The immunoglobulin heavy chain variable region (Igh-V) locus in the mouse. I. One hundred Igh-V genes comprise seven families of homologous genes. *Eur J Immunol* 1984; 14:922–930.

155. Kurosawa Y, Tonegawa S. Organization, structure, and assembly of immunoglobulin heavy chain diversity DNA segments. *J Exp Med* 1982; 155:201–218.

156. Gough NM, Bernard O. Sequences of the joining region genes for immunoglobulin heavy chains and their role in generation of antibody diversity. *Proc Natl Acad Sci U S A* 1981; 78:509–513.

157. Kirschbaum T, Jaenichen R, Zachau HG. The mouse immunoglobulin kappa locus contains about 140 variable gene segments. *Eur J Immunol* 1996; 26:1613–1620.

158. Kabat EA, Wu TT, Reid-Miller M, Perry HM, Gottesman KS, Foeller F. *Sequences of Proteins of Immunological Interest*, 4th ed. Washington, DC: United States Department of Health and Human Services, 1987.

159. Zhou H, Fisher RJ, Papas TS. Optimization of primer sequences for mouse scFv repertoire display library construction. *Nucleic Acids Res* 1994; 22:888–889.

160. Dihn Q, Weng N-P, Kiso M, Ishida H, Hasegawa A, Marcus DM. High affinity antibodies against Lex and Sialyl Lex from a phage display library. *J Immunol* 1996; 157:732–738.

161. Kimura H, Suzuki H, Matsuzawa S. Quantitation of hemagglutinins by the planimetry of precipitated erythrocyte patterns. 4. Application to Lewis blood group determination from saliva and urine specimens and their stains. *Nippon Hoigaku Zasshi* 1989; 43:469–477.

162. Ball ED, Mills L, Hurd D, McMillan R, Gringrich R. Autologous bone marrow transplantation for acute myeloid leukemia using monoclonal antibody-purged bone marrow. *Prog Clin Biol Res* 1992; 377:97.

163. Williamson RA, Peretz D, Smorodinsky N, Bastidas R, Serban H, Mehlhorn I, DeArmond S et al. Circumventing tolerance to generate autologous monoclonal antibodies to the prion protein. *Proc Natl Acad Sci U S A* 1996; 93:7279–7282.

164. Williamson RA, Peretz D, Pinilla C, Ball H, Bastidas R, Rozenshteyn R, Houghten R et al. Mapping the prion protein using recombinant antibodies. *J Virol* 1998; 72:9413–9418.

165. Brocks B, Garin-Chesa P, Behrle E, Park JE, Rettig WJ, Pfizenmaier K, Moosmayer D. Species-crossreactive scFv against the tumor stroma marker "fibroblast activation protein" selected by phage display from an immunized FAP$^{-/-}$ knock-out mouse. *Mol Med* 2001; 7:461–469.

166. Chowdhury PS, Chang K, Pastan I. Isolation of anti-mesothelin antibodies from a phage display library. *Mol Immunol* 1997; 34:9–20.

167. Reiter Y, Pastan I. Antibody engineering of recombinant Fv immunotoxins for improved targeting of cancer: Disulfide-stabilized Fv immunotoxins. *Clin Cancer Res* 1996; 2:245–252.

168. Lorimer IAJ, Keppler-Hafkemeyer A, Beers RA, Pegram CN, Bigner DD, Pastan I. Recombinant immunotoxins specific for a mutant epidermal growth factor receptor: Targeting with a single chain antibody variable domain isolated by phage display. *Proc Natl Acad Sci U S A* 1996; 93:14815–14820.

169. Rojas G, Almagro JC, Acevedo B, Gavilondo JV. Phage antibody fragments library combining a single human light chain variable region with immune mouse heavy chain variable regions. *J Biotechnol* 2002; 23:287–298.

170. Winthrop MD, DeNardo SJ, DeNardo GL. Development of a hyperimmune anti-MUC-1 single chain antibody fragments phage display library for targeting breast cancer. *Clin Cancer Res* 1999; 5(suppl 10):3088s–3094s.

171. Nguyen VK, Hamers R, Wins L, Muyldermans S. Camel heavy-chain antibodies: Diverse germline VHH and specific mechanism enlarge the antigen-binding repertoire. *EMBO J* 2000; 19:921–931.

172. Peipp M, Simon N, Loichinger A, Baum W, Mahr K, Zunino SJ, Fey GH. An improved procedure for the generation of recombinant single-chain Fv antibody fragments reacting with human CD13 on intact cells. *J Immunol Methods* 2001; 251:161–176.

173. Burmester J, Spinelli S, Pugliese L, Krebber A, Honegger A, Jung S, Schimmele B et al. Selection, characterization and x-ray structure of anti-ampicillin single-chain Fv fragments from phage-displayed murine antibody libraries. *J Mol Biol* 2001; 309:671–385.

174. Rozemuller H, Chowdhury PS, Pastan I, Kreitman RJ. Isola- tion of new anti-CD30 scFvs from DNA-immunized mice by phage display and biologic activity of recom- binant immunotoxins produced by fusion with truncated pseudomonas exotoxin. *Int J Cancer* 2001; 92:861–870.

175. Spieker-Polet H, Sethupathi P, Yam PC, Knight KL. Rabbit monoclonal antibodies: Generating a fusion partner to produce rabbit–rabbit hybridomas. *Proc Natl Acad Sci U S A* 1995; 92:9348–9352.

176. Ridder R, Schmitz R, Legay F, Gramm H. Generation of rabbit monoclonal antibody fragments from a combinatorial phage display library and their production in the yeast *Pichia pastoris*. *Biotechnology* 1995; 13:255–260.

177. Lang IM, Barbas CF III, Schleef R. Recombinant rabbit Fab with binding activity to type-1 plasminogen activator inhibitor derived from a phage-display library against human alpha-granules. *Gene* 1996; 172:295–298.

178. Knight KL, Crane MA. Generating the antibody repertoire in rabbit. *Adv Immunol* 1994; 56:179–218.

179. Foti M, Granucci F, Ricciardi-Castagnoli P, Spreafico A, Ackermann M, Suter M. Rabbit monoclonal Fab derived from a phage display library. *J Immunol Methods* 1998; 213:201–212.

180. Rader C, Ritter G, Nathan S, Elia M, Gout I, Jungbluth AA, Cohen LS et al. The rabbit antibody repertoire as a novel source for the generation therapeutic human antibodies. *J Biol Chem* 2000; 275(18):13668–13676.

181. Popkov M, Mage R, Alexander C, Thundivalappil S, Barbas CF, Rader C. Rabbit immune repertoires as sources for therapeutic monoclonal antibodies: The impact of kappa allotype-correlated variation in cysteine content on antibody libraries selected by phage display. *J Mol Biol* 2003; 325:325–335.

182. Knight KL, Crane MA. Development of the antibody repertoire in rabbits. *Ann N Y Acad Sci* 1995; 764:198–206.

183. Becker R, Knight K. Somatic diversification of immunoglobulin heavy chain VDJ genes: Evidence for somatic gene conversion in rabbits. *Cell* 1990; 63:987–997.

184. Knight KL, Becker RS. Molecular basis of the allelic inheritance of rabbit immunoglobulin V_H allotypes: Implications for the generation of antibody diversity. *Cell* 1990; 60:963–970.

185. Lai E, Wilson RK, Hood LE. Physical maps of the mouse and human immunoglobulin-like loci. *Adv Immunol* 1989; 46:1–59.

186. Pascual V, Verkruyse L, Casey ML, Capra JD. Analysis of I_g H chain gene segment utilization in human fetal liver. Revisiting the "proximal utilization hypothesis." *J Immunol* 1993; 151:4164–4172.

187. Sehgal D, Schiaffella E, Anderson AO, Mage RG. Generation of heterogeneous rabbit anti-DNP antibodies by gene conversion and hypermutation of rearranged VL and VH genes during clonal expansion of B cells in splenic germinal centers. *Eur J Immunol* 2000; 30:3634–3644.

188. Benammar A, Cazenave PA. A second rabbit kappa isotype. *J Exp Med* 1982; 156:585–595.

189. Emorine L, Max EE. Structural analysis of a rabbit immunoglobulin kappa 2 J-C locus reveals multiple deletions. *Nucleic Acids Res* 1983; 11:8877–8890.

190. Hole NJ, Harindranath N, Young-Cooper GO, Garcia R, Mage RG. Identification of enhancer sequences 3′ of the rabbit I_g kappa L chain loci. *J Immunol* 1991; 146:4377–4384.

191. Good PW, Notenboom R, Dubiski S, Cinader B. Basilea rabbit immunoglobulins: Detection and characterization by specific alloantiserum. *J Immunol* 1980; 125:1293–1297.

192. Litwin SD. Immunoglobulin allotypes. *Immunol Ser* 1989; 43:203–236.

193. Winstead CR, Zhai S, Sethupathi P, Knight KL. Antigen-induced somatic diversification of rabbit IgH genes: Gene conversion and point mutation. *J Immunol* 1999; 162:6602–6612.

194. Kruithof EK, Nicolosa G, Bachmann F. Plasminogen activator inhibitor 1: Development of a radioimmunoassay and observations on its plasma concentration during venous occlusion and after platelet aggregation. *Blood* 1987; 70:1645–1653.

195. Chi XS, Landt Y, Crimmins DL, Dieckgraefe BK, Ladenson J. Isolation and characterization of rabbit single chain antibodies to human Reg I protein. *J Immunol Methods* 2002; 266:197–207.

196. Hawlish H, Meyer zu Vilsendorf A, Bautsh W, Klos A, Kohl J. Guinea pig C3 specific rabbit single chain Fv antibodies from bone marrow, spleen and blood derived phage libraries. *J Immunol Methods* 2000; 236:117–131.

197. Friedman ML, Tunyaplin C, Zhai SK, Knight KL. Neonatal VH, D, and JH gene usage in rabbit B lineage cells. *J Immunol* 1994; 152:632–641.

198. Hoogenboom HR. Designing and optimizing library selection strategies for generating high-affinity antibodies. *TibTech* 1997; 15:62–70.

199. Rader C, Barbas CF III. Phage display of combinatorial antibody libraries. *Curr Opin Biotechnol* 1997; 8:503–508.

200. Hudson PJ. Recombinant antibody constructs in cancer therapy. *Curr Opin Immunol* 1999; 11:548–557.

201. Steinberger P, Sutton JK, Rader C, Elia M, Barbas CF III. Generation and characterization of a recombinant human antibody. A phage display approach for rabbit antibody humanization. *J Biol Chem* 2000; 275:36073–36078.

202. Goncalves J, Silva F, Freitas-Vieira A, Santa-Marta M, Malho R, Yang X, Gabuzda D et al. Functional neutralization of HIV-1 Vif protein by intracellular immunization inhibits reverse transcription and viral replication. *J Biol Chem* 2002; 277:32036–32045.

203. Jendreyko N, Popkov M, Beerli RR, Chung J, McGavern DB, Rader C, Barbas CF III. Intradiabodies, bispecific, tetravalent antibodies for the simultaneous functional knockout of two cell surface receptors. *J Biol Chem* 2003; 278:47812–47819.

204. Larsson A, Sjoquist J. Chicken IgY: Utilizing the evolutionary difference. *Comp Immunol Microbiol Infect Dis* 1990; 13:199.
205. Asaoka H, Nishinaka S, Wakamiya N, Matsuda H, Murata M. Two chicken monoclonal antibodies specific for heterophil Hanganutziu–Deicher antigens. *Immunol Lett* 1992; 32:91–96.
206. Sasai K, Lillehoj HS, Matsuda H, Wergin WP. Characterization of a chicken monoclonal antibody that recognizes the apical complex of *Eimeria acervulina* sporozoites and partially inhibits sporozoite invasion of CD8+ T lymphocytes in vitro. *J Parasitol* 1996; 82:82–87.
207. Matsushita K, Horiuchi H, Furusawa S, Horiuchi M, Shinagawa M, Matsuda H. Chicken monoclonal antibodies against synthetic bovine prion protein peptide. *J Vet Med Sci* 1998; 60:777–779.
208. Nishinaka S, Suzuki T, Matsuda H, Murata M. A new cell line for the production of chicken monoclonal antibody by hybridoma technology. *J Immunol Methods* 1991; 139:217–222.
209. Nishinaka S, Akiba H, Nakamura M, Suzuki K, Suzuki T, Tsubokura K, Horiuchi H et al. Two chicken B cell lines resistant to ouabain for the production of chicken monoclonal antibodies. *J Vet Med Sci* 1996; 58:1053–1056.
210. Thompson CB, Neiman PE. Somatic diversification of the chicken immunoglobulin light chain gene is limited to the rearranged variable gene segment. *Cell* 1987; 48:369.
211. Ratcliffe MJ, Jacobsen KA. Rearrangement of immunoglobulin genes in chicken B cell development. *Semin Immunol* 1994; 6:175–184.
212. Reynaud CA, Dufour V, Weill JC. Generation of diversity in mammalian gut associated lymphoid tissues: Restricted *V* gene usage does not preclude complex *V* gene organization. *J Immunol* 1997; 159:3093–3095.
213. Reynaud CA, Garcia C, Hein WR, Weill JC. Hypermutation generating the sheep immunoglobulin repertoire is an antigen-independent process. *Cell* 1995; 80:115–125.
214. Reynaud CA, Dahan A, Anquez V, Weill JC. Somatic hyperconversion diversivies the single V_H gene of the chicken with a high incidence in the D region. *Cell* 1989; 59:171.
215. Arakawa H, Furusawa S, Ekino S, Yamagishi H. Immunoglobulin gene hyperconversion ongoing in chicken splenic germinal centers. *EMBO J* 1996; 15:2540–2546.
216. Berek C, Milstein C. Mutation drift and repertoire shift in the maturation of the immune response. *Immunol Rev* 1987; 96:23.
217. Parvari R, Ziv E, Lantner F, Heller D, Schecher I. Somatic diversification of chicken immunoglobulin light chains by point mutations. *Proc Natl Acad Sci U S A* 1990; 87:3072.
218. McCormack WT, Thompson CB. Chicken IgL variable region gene conversion display pseudogene donor preference and 5' to 3' polarity. *Genes Dev* 1990; 4:548.
219. Davies EL, Smith J, Birkett C, Manser J, Andersen-Dear D, Young J. Selection of specific phage-display antibodies using libraries derived from chicken immunoglobulin genes. *J Immunol Methods* 1995; 186:125–135.
220. Michael N, Accavitti M, Masteller E, Thompson C. The antigen binding characteristics of MAbs derived from *in vivo* priming of avian B cells. *Proc Natl Acad Sci U S A* 1995; 95:1166–1171.
221. Yamanaka HI, Inoue T, Ikeda-Tanaka O. Chicken monoclonal antibody isolated by a phage display system. *J Immunol* 1996; 157:1156–1162.
222. Cary S, Lee J, Wagenknecht R, Silverman GJ. Characterization of superantigen-induced clonal deletion with a novel clan III-restricted avian monoclonal antibody: Exploiting evolutionary distance to create antibodies specific for a conserved V_H region surface. *J Immunol* 2000; 164:4730–4741.

223. Davies J, Riechmann L. Single antibody domains as small recognition units: Design and in vitro antigen selection of camelized, human VH domains with improved protein stability. *Protein Eng* 1996; 9:531–537.

224. Vu KB, Ghahroudi MA, Wyns L, Muyldermans S. Comparison of llama VH sequences from conventional and heavy chain antibodies. *Mol Immunol* 1997; 34:1121–1131.

225. Tucker EM, Dain AR, Wright LJ, Clark SW. Culture of sheep X mouse hybridoma cells in vitro. *Hybridoma* 1981; 1:77.

226. Dufour V, Nau F, Malings S. The sheep I_g variable region repertoire consists of a single V_H family. *J Immunol* 1996; 156:2163–2170.

227. Motyka B, Reynolds JD. Apoptosis is associated with the extensive B cell death in the sheep ileal Peyer's patch and the chicken bursa of Fabricius: A possible role in B cell selection. *Eur J Immunol* 1991; 21:1951–1958.

228. Weill JC, Reynaud CA. Rearrangement/hypermutation/gene conversion: When, where and why? *Immunol Today* 1996; 17:92–97.

229. Hein WR, Dudler L. Diversity of I_g light chain variable region gene expression in fetal lambs. *Int Immunol* 1998; 10:1251–1259.

230. Reynaud CA, Weill JC. Postrearrangement diversification processes in gut-associated lymphoid tissues. *Curr Top Microbiol Immunol* 1996; 212:7–15.

231. Jenne CN, Kennedy LJ, McCullagh P, Reynolds JD. A new model of sheep Ig diversification: Shifting the emphasis toward combinatorial mechanisms and away from hypermutation. *J Immunol* 2003; 70:3739–3750.

232. Li Y, Cockburn W, Kilpatrick J, Whitlam GC. Selection of rabbit single-chain Fv fragments against the herbicide atrazine using a new phage display system. *Food Agric Immunol* 1999; 11:5.

233. Charlton KA, Harris WJ, Porter AJR. The isolation of super-sensitive anti-hapten antibodies from combinatorial antibody libraries derived from sheep. *Biosens Bioelectron* 2001; 16:639–646.

234. Lauwereys M, Ghahroudi MA, Desmyter A, Kinne J, Holzer W, De Genst E, Wyns L et al. Potent enzyme inhibitors derived from dromedary heavy-chain antibodies. *EMBO J* 1998; 17:3512–3520.

235. Lucier M, Thompson R, Waire J, Lin A, Osborne B, Goldsby R. Multiple sites of V lambda diversification in cattle. *J Immunol* 1998; 161:5438–5444.

236. Sinclair MC, Gilchrist J, Aitken R. Bovine IgG repertoire is dominated by a single diversified V_H gene family. *J Immunol* 1997; 159:3883–3889.

237. Armour KL, Tempest PR, Fawcett PH, Fernie ML, King SI, White P, Taylor G et al. Sequences of heavy and light chain variable regions from four bovine immunoglobulins. *Mol Immunol* 1994; 31:1369–1372.

238. Berens SJ, Wylie DE, Lopez OJ. Use of a single V_H family and long CDR3s in the variable region of cattle I_g heavy chains. *Int Immunol* 1997; 9:189–199.

239. Parng C, Hansal S, Goldsby R, Osborne B. Gene conversion contributes to I_g light chain diversity in cattle. *J Immunol* 1996; 157:5478–5486.

240. Meyer A, Parng CL, Hansal SA, Osborne BA, Goldsby RA. Immunoglobulin gene diversification in cattle. *Int Rev Immunol* 1997; 15:165–183.

241. Hamers-Casterman C, Atarhouch T, Muyldermans S, Robinson G, Hamers C, Songa EB, Bendahman N et al. Naturally occurring antibodies devoid of light chains. *Nature* 1993; 363:446–448.

242. van der Linden R, de Gaus B, Stok W, Bos W, van Wassenaar D, Verrips T, Frenken L. Induction of immune responses and molecular cloning of the heavy chain antibody repertoire of *Lama glama*. *J Immunol Methods* 2000; 240:185–195.

243. Muyldermans S, Atarhouch T, Saldanha J, Barbosa JA, Hamers R. Sequence and structure of V_H domain from naturally occurring camel heavy chain immunoglobulins lacking light chains. *Protein Eng* 1994; 7:1129–1135.

244. Transue TR, de Genst E, Ghahroudi MA, Wins L, Muyldermans S. Camel single-domain antibody inhibits enzyme by mimicking carbohydrate substrate. *Proteins* 1998; 32:515–522.

245. van Djik MA, van de Winkel JGJ. Human antibodies as next generation therapeutics. *Curr Opin Chem Biol* 2001; 5:368–374.

246. Gorman JG. Rh immunoglobulin in prevention of hemolytic disease of newborn child. *N Y State J Med* 1968; 68:1270–1277.

247. Beasley RP, Hwang LY, Lee GC, Lan CC, Roan CH, Huang FY, Chen CL. Prevention of perinatally transmitted hepatitis B virus infections with hepatitis B virus infections with hepatitis B immune globulin and hepatitis B vaccine. *Lancet* 1983; 2:1099–1102.

248. Enria DA, Briggiler AM, Fernandez NJ, Levis SC, Maiztegui JI. Importance of dose of neutralising antibodies in treatment of Argentine haemorrhagic fever with immune plasma. *Lancet* 1984; 2:255–256.

249. Buglio AF, Wheeler R, Trang J, Haynes A, Rogers K, Harvey E, Sun L et al. Mouse/human chimeric monoclonal antibody in man: Kinetics and immune response. *Proc Natl Acad Sci U S A* 1989; 86:4220–4224.

250. Endo T, Wright A, Morrison SL, Kobata A. Glycosylation of the variable region of immunoglobulin G-site specific maturation of the sugar chains. *Mol Immunol* 1995; 32:931–940.

251. Brekke O, Sandlie I. Therapeutic antibodies for human diseases at the dawn of the twenty first century. *Nat Rev* 2003; 2:52062.

252. Kretzschmar T, von Ruden T. Antibody discovery: Phage display. *Curr Opin Biotechnol* 2002; 13:598–602.

253. Abrams PG, Rossio JL, Stevenson HC, Foon KA. Optimal strategies for developing human–human monoclonal antibodies. *Methods Enzymol* 1986; 121:107.

254. Matsuda F, Ishii K, Bourvagnet P, Kuma K, Hayashida H, Miyata T, Honjo T. The complete nucleotide sequence of the human immunoglobulin heavy chain variable region locus. *J Exp Med* 1998; 188:2151–2162.

255. Meffre E, Catalan N, Seltz F, Fischer A, Nussenzweig MC, Durandy A. Somatic hypermutation shapes the antibody repertoire of memory B cells in humans. *J Exp Med* 2001; 194:375–378.

256. Sasso E, Van Dijk K, Milner E. Prevalence and polymorphism of human VH3 genes. *J Immunol* 1990; 145:2751–2757.

257. van Dijk K, Sasso E, Milner EC. Polymorphism of the human I_g VH4 gene family. *J Immunol* 1991; 146:3646–3651.

258. Sasso EH, Buckner JH, Suzuki LA. Ethnic differences in V_H gene polymorphism. *Ann N Y Acad Sci* 1995; 764:72–73.

259. Rao SP, Huang SC, Milner EC. Analysis of the VH3 repertoire among genetically disparate individuals. *Exp Clin Immunogenet* 1996; 13:131–138.

260. Coomber DWJ, Hawkins NJ, Clark MA, Ward RL. Generation of anti-p-53 Fab from individuals with colorectal cancer using phage display. *J Immunol* 1999; 163:2276–2283.

261. McIntosh RS, Asghar MS, Watson PF, Kemp EH, Weetman AP. Cloning and analysis of IgGk and IgGl anti-thyroglobulin autoantibodies from a patient with Hashimoto's thyroiditis. *J Immunol* 1996; 157:927–935.

262. Clark MA. Standard protocols for the construction of Fab libraries. In: O'Brien PM, Aitken R, eds. *Antibody Phage Display. Methods and Protocols.* Totowa, NJ: Humana Press, 2002.

263. Winter G, Milstein C. Man-made antibodies. *Nature* 1991; 349:293–299.

264. Green LL. Antibody engineering via genetic engineering of the mouse: XenoMouse strains are a vehicle for the facile generation of therapeutic human monoclonal antibodies. *J Immunol Methods* 1999; 231:11–23.

265. Ishida I, Tomizuka K, Yoshida H, Tahara T, Takahashi N, Ohguma A, Tanaka S et al. Production of human monoclonal and polyclonal antibodies in TransChromo animals. *Cloning Stem Cells* 2002; 4:91–102.

266. Mosier DE, Gulizia RJ, Baird SM, Wilson DB. Transfer of a functional human immune system to mice with severe combined immunodeficiency. *Nature* 1988; 335:256–259.

267. Tsui P, Tornetta MA, Ames RS, Bankosky BC, Griego S, Silverman C, Porter T et al. Isolation of a neutralizing human RSV antibody from a dominant, non-neutralizing immune repertoire by epitope-blocked panning. *J Immunol* 1996; 157:772–781.

268. Chang Q, Zhong Z, Lees A, Pekna M, Pirofski L. Structure–function relationships for human antibodies to pneumococcal capsular polysaccharide from transgenic mice with human immunoglobulin loci. *Infect Immun* 2002; 70:4977–4986.

269. Mukherjee J, Chios K, Fishwild D, Hudson D, O'Donnell S, Rich SM, Donohue-Rolfe A et al. Production and characterization of protective human antibodies against Shiga toxin 1. *Infect Immun* 2002; 70:5896–5899.

270. He Y, Honnen WJ, Krachmarov CP, Burkhart M, Kayman SC, Corvalan J, Pinter A. Efficient isolation of novel human monoclonal antibodies with neutralizing activity against HIV-1 from transgenic mice expressing human I_g loci. *J Immunol* 2002; 169:595–605.

271. Yang X, Jia X, Corvalan J, Wang P, Davis C, Jakobovits A. Eradication of established tumors by a fully human monoclonal antibody to the epidermal growth factor receptor without concomitant chemotherapy. *Cancer Res* 1999; 59:1236–1243.

272. Davis C, Gallo M, Corvalan J. Transgenic mice as a source of fully human antibodies for the treatment of cancer. *Cancer Metastasis Rev* 1999; 18:421–425.

273. Meyers K, Allen J, Gehret J, Jacobovits A, Gallo M, Neilson E, Hopfer H et al. Human antiglomerular basement membrane autoantibody disease in XenoMouse II. *Kidney Int* 2002; 61:1666–1673.

274. Yang X, Corvalan J, Wang P, Roy CM, Davis C. Fully human anti-interleukin-8 monoclonal antibodies: Potential therapeutics for the treatment of inflammatory disease states. *J Leukoc Biol* 1999; 66:401–410.

275. Burton DR, Barbas CF III, Persson M, Koenig S, Chanock R, Lerner R. A large array of human monoclonal antibodies to type 1 human immunodeficiency virus from combinatorial libraries of asymptomatic seropositive individuals. *Proc Natl Acad Sci U S A* 1991; 88:10134–10137.

276. Barbas SM, Ditzel HJ, Salonen AM, Yang WP, Silverman GJ, Burton DR. Human autoantibody recognition of DNA. *Proc Natl Acad Sci U S A* 1995; 92:2529–2533.

277. Burton DR. Antibodies viruses vaccines. *Nat Immunol Rev* 2002; 2:706.

278. Chanock RM, Crowe JE Jr, Murphy BR, Burton DR. Human monoclonal antibody Fab fragments cloned from combinatorial libraries: Potential usefulness in prevention and/or treatment of major human viral diseases. *Infect Agents Dis* 1993; 2:118–131.

279. Barbas CF III, Hu D, Dunlop N, Sawyer L, Cababa D, Hendry RM, Nara PL et al. In vitro evolution of a neutralizing human antibody to human immunodeficiency virus type 1 to enhance affinity and broaden strain cross-reactivity. *Proc Natl Acad Sci U S A* 1994; 91:3809–3813.

280. Parren PW, Marx PA, Hessell AJ, Luckay A, Harouse J, Cheng-Mayer C, Moore JP et al. Antibody protects macaques against vaginal challenge with a pathogenic R5 simian/human immunodeficiency virus at serum levels giving complete neutralization in vitro. *J Virol* 2001; 75:8340–8347.

281. Mascola JR, Stiegler G, VanCott TC, Katinger H, Carpenter CB, Hanson CE, Beary H et al. Protection of macaques against vaginal transmission of a pathogenic HIV-1/SIV chimeric virus by passive infusion of neutralizing antibodies. *Nat Med* 2000; 6:207–210.

282. Baba W, Liska V, Hofmann-Lehmann R, Vlasak J, Xu W, Ayehunie S, Cavacini LA et al. Human neutralizing monoclonal antibodies of the IgG1 subtype protect against mucosal simian–human immunodeficiency virus infection. *Nat Med* 2000; 6:200–206.

283. Schmaljohn C, Cui Y, Kerby S, Pennock D, Spik K. Production and characterization of human monoclonal antibody fragments to vaccinia virus from a phage display combinatorial library. *Virology* 1999; 258:189–200.

284. Moss B. Poxviridae and their replication. In: Fields B, Knip D, Chanock R, Hirsch M, Melnick J, Monath T, Roizman B, eds. *Virology*, Vol. 2, 2nd ed. New York: Raven Press, 1990:2079–2111.

285. Peters CJ, Sanchez A, Rollin PE, Ksiazek T, Murphy F. Filoviridae: Marburg and Ebola viruses. In: Fields D, Knipe BN, Howley P, eds. *Virology*, 3rd ed. New York: Raven Press Inc., 1996:1161–1176.

286. Maruyama T, Rodriguez LL, Jahrling PB, Sanchez A, Khan AS, Nichol ST, Peters CJ et al. Ebola virus can be effectively neutralized by antibody produced in natural human infection. *J Virol* 1999; 73:6024–6030.

287. Arnon SS. Clinical trial of human botulism immune globulin. In: DasGupta BR, ed. *Botulinum and Tetanus Neurotoxins, Neurotransmission and Biomedical Aspects*. New York: Plenum Press, 1993:477–482.

288. Peters C, DeLuc J. An introduction. Ebola: The virus and the disease. *J Infect Dis* 1999; 179(suppl 1):ix–xvi.

289. Jahrling P, Geisbert J, Swearengen J, Jaax P, Lewis T, Huggins J, Schmidt J et al. Passive immunization of Ebola virus infected cynomologus monkeys with immunoglobulin from hyperimmune horses. *Arch Virol* 1999; 11(suppl):135–140.

290. Herreros J, Lalli G, Schiavo M, Schiavo G. Pathophysiological properties of clostridial neurotoxins. In: Alouf J, Freer J, eds. *The Comprehensive Sourcebook of Bacterial Protein Toxins*, 2nd ed. San Diego, CA: Academic Press, 1999:202–229.

291. Dobrindt U, Hacker J. Plasmids, phages and pathogenicity islands: Lessons on the evolution of bacterial toxins. In: Alouf J, Freer J, eds. *The Comprehensive Sourcebook of Bacterial Protein Toxins*, 2nd ed. San Diego, CA: Academic Press, 1999:3–26.

292. Amersdorfer P, Wong C, Smith T, Chen S, Deshpande S, Sheridan R, Marks JD. Genetics and immunological comparison of anti-botulinum type A antibodies from immune and non-immune human phage libraries. *Vaccine* 2002; 20:1640–1648.

293. Foote J, Milstein C. Kinetic maturation of an immune response. *Nature* 1991; 352:530–532.

294. Crowe JE Jr, Murphy BR, Chanock RM, Williamson RA, Barbas CF III, Burton DR. Recombinant human respiratory syncytial virus (RSV) monoclonal antibody Fab is effective therapeutically when introduced directly into the lungs of RSV infected mice. *Proc Natl Acad Sci USA* 1994; 91:1386–1390.

295. Bender E, Pilkington GR, Burton DR. Human monoclonal Fab fragments from a combinatorial library prepared from an individual with a low serum titer to a virus. *Hum Antibodies Hybrid* 1994; 5:3–8.

296. Zhai W, Davies J, Shang D, Chan S, Allan J. Human recombinant single-chain antibody fragments, specific for the hypervariable region 1 of hepatitis C virus, from immune phage-display libraries. *J Viral Hept* 1999; 6:115–124.

297. Itoh K, Nakagomi O, Inoue K, Tada H, Suzuki T. Recombinant human monoclonal Fab fragments against rotavirus from phage display combinatorial libraries. *J Biochem* 1999; 125:123–129.

298. De Carvalho Nicacio C, Williamson RA, Parren PW, Lundkvist A, Burton DR, Bjorling E. Neutralizing human Fab fragments against measles virus recovered by phage display. *J Virol* 2002; 76:251–258.

299. Andreakos E, Taylor PC, Feldman M. Monoclonal antibodies in immune and inflammatory diseases. *Curr Opin Biotechnol* 2002; 13:615–620.

300. Zeidel M, Rey E, Tami J, Fischbach M, Sanz I. Genetic and functional characterization of human autoantibodies using combinatorial phage display libraries. *Ann N Y Acad Sci* 1995; 764:559–563.

301. del Rincon I, Zeidel M, Rey E, Harley JB, James JA, Fischbach M, Sanz I. Delineation of the human systemic lupus erythematosus anti-Smith antibody response using phage-display combinatorial libraries. *J Immunol* 2000; 165:7011–7016.

302. Hexham JM, Partridge LJ, Furmaniak J, Petersen VB, Colls JC, Pegg C, Rees Smith B et al. Cloning and characterization of TPO autoantibodies using combinatorial phage display libraries. *Autoimmunity* 1994; 17:167–179.

303. Roben P, Barbas SM, Sandoval L, Lecerf J-M, Stollar BD, Solomon A, Silverman G. Repertoire cloning of lupus anti-DNA autoantibodies. *J Clin Invest* 1996; 98:2827–2837.

304. Jury K, Sohnlein P, Vogel M, Richter W. Isolation and functional characterization of recombinant GAD65 autoantibodies derived by IgG repertoire cloning from patients with type 1 diabetes. *Diabetes* 2001; 50:1976–1982.

305. Graus YF, deBaets MH, Parren PW, Berrih-Aknin S, Wokke J, van Breda Vriesman PJ, Burton DR. Human anti-nicotinic acetylcholine receptor recombinant Fab fragments isolated from thymus derived phage display libraries from Myasthenia Gravis patients reflect predominant specificities in serum and block the action of pathogenic serum antibodies. *J Immunol* 1997; 15:1919–1929.

306. Kunicki TJ, Newman PJ. The molecular immunology of human platelet proteins. *Blood* 1992; 80:1386–1404.

307. Clofent-Sanchez G, Lucas S, Laroche-Traineau J, Rispal P, Pellegrin JL, Nurden P, Nurden A. Autoantibodies and anti-mouse antibodies in thrombocytopenic patients as assessed by different MAIPA assays. *Br J Haemotol* 1996; 95:53.

308. Ishida F, Gruel Y, Brojer E, Nugent DJ, Kunicki TJ. Repertoire cloning of a human IgG inhibitor of aIIbb3 function. The OG idiotype. *Mol Immunol* 1995; 32:613–622.

309. Jacobin MJ, Laroche-Traineau J, Little M, Keller A, Peter K, Welschof M, Nurden A et al. Human IgG monoclonal anti-alpha(IIb)beta(3)-binding fragments derived from immunized donors using phage display. *J Immunol* 2002; 168:2035–2045.

310. Eggena M, Targan SR, Iwanczyk L, Vidrich A, Gordon LK, Braun J. Phage display cloning and characterization of an immunogenetic marker (perinuclear anti-neutrophil cytoplasmic antibody in ulcerative colitis. *J Immunol* 1996; 156:4005–4011.

311. Lee S, Ansari AA, Cha S. Isolation of human anti-branched chain alpha-oxo acid dehydrogenase-E2 recombinant antibodies by I_g repertoire cloning in idiopathic dilated cardiomyopathy. *Mol Cells* 1999; 9:25–30.

312. Reiche N, Jung A, Brabletz T, Vater T, Kirchner T, Faller G. Generation and characterization of human monoclonal scFv antibodies against *Helicobacter pylori* antigens. *Infect Immun* 2002; 70:4158–4164.

313. Pascual V, Capra JD. VH4-21, a human VH gene segment overrepresented in the autoimmune repertoire. *Arthritis Rheum* 1992; 35:11.

314. Carter P. Improving the efficacy of antibody-based cancer therapies. *Nat Rev Cancer* 2001; 1:118–129.

315. Trikha M, Yan L, Nakada MT. Monoclonal antibodies as therapeutics in oncology. *Curr Opin Biotechnol* 2002; 13:609–614.

316. Labrecque S, Naor N, Thomson D, Matlashewski G. Analysis of the anti-p53 antibody response in cancer patients. *Cancer Res* 1993; 53:3468–3471.

317. Pupa SM, Menard S, Andreola S, Colnaghi MI. Antibody response against the cerbB-2 oncoprotein in breast carcinoma patients. *Cancer Res* 1993; 53:5864–5866.

318. Kupsch J-M, Tidman NH, Kang NV, Truman H, Hamilton S, Patel N, Bishop JAN et al. Isolation of human tumor-specific antibodies by selection of an antibody phage library on melanoma cells. *Clin Cancer Res* 1999; 5:925–931.

319. Clark MA, Hawkins NJ, Papaioannou A, Fiddes RJ, Ward RL. Isolation of human anti-c-erbB-2 Fabs from a lymph node derived phage display library. *Clin Exp Immunol* 1997; 109:166–174.

320. Coomber DWJ, Ward RL. Isolation of human antibodies against the central DNA binding domain of p53 from an individual with colorectal cancer using antibody phage display. *Clin Cancer Res* 2001; 7:2802–2808.

321. Courtenay-Luck NS, Epenetos AA, Moore R, Larche M, Pectasides D, Dhokia B, Ritter MA. Development of primary and secondary immune responses to mouse monoclonal antibodies used in the diagnosis and therapy of malignant neoplasm. *Cancer Res* 1986; 46:6489–6493.

322. Roovers RC, van der Linden E, Zijlema H, de Bruine A, Arends J-W, Hoogenboom HR. Evidence for a bias toward intracellular antigens in the local humoral anti-tumor immune response of a colorectal cancer patient revealed by phage display. *Int J Cancer* 2001; 93:832–840.

323. Mao S, Gao C, Lo C-HL, Wirsching P, Wong C-H, Janda KD. Phage-display library selection of high-affinity human single-chain antibodies to tumor-associated carbohydrate antigens sialyl Lewis^x and Lewis^x. *Proc Natl Acad Sci U S A* 1999; 96:6953–6958.

324. Lee KJ, Mao S, Sun C, Gao C, Blixt O, Arrues S, Hom LG et al. Phage-display selection of a human single-chain fv antibody highly specific for melanoma and breast cancer cells using a chemoenzymatically synthesized G_{m3}—Carbohydrate antigen. *J Am Chem Soc* 2002; 124:12439–12446.

325. Dohi T, Nores G, Hakomori S. An IgG3 monoclomal antibody established after immunization with GM3 lactobe: Immunochemical specificity and inhibition of melanoma cell growth in vitro and in vivo. *Cancer Res* 1988; 48:5680–5685.

326. Zou W, Abraham M, Gilbert M, Wakarchuk WW, Jennings HJ. *Glycoconjugate J* 1999; 16:507–515.

327. Ehrlich PH, Moustafa ZA, Harfeldt KE, Isaacson C, Ostberg L. Potential of primate monoclonal antibodies to substitute for human antibodies: Nucleotide sequence of chimpanzee Fab fragments. *Hum Antibodies Hybridomas* 1990; 1:23–26.

328. Logdberg L, Kaplan E, Drelich M, Harfeldt E, Gunn H, Ehrlich P, Dottavio D et al. Primate antibodies to components of the human immune system. *J Med Primatol* 1994; 23:285–297.

329. Ogata N, Ostberg L, Ehrlich PH, Wong DC, Miller RH, Purcell RH. Markedly prolonged incubation period of hepatitis B in a chimpanzee passively immunized with a human monoclonal antibody to the a determinant of hepatitis B surface antigen. *Proc Natl Acad Sci U S A* 1993; 90:3014–3018.

330. Glamann J, Burton DR, Parren PW, Ditzel HJ, Kent KA, Arnold C, Montefiori D et al. Simian immunodeficiency virus (SIV) envelope-specific Fabs with high-level homologous neutralizing activity: Recovery from a long-term-nonprogressor SIV-infected macaque. *J Virol* 1998; 72:585–592.

331. Schofield DJ, Satterfield W, Emerson SU, Purcell RH. Four chimpanzee monoclonal antibodies isolated by phage display neutralize hepatitis A virus. *Virology* 2002; 292:127–136.

332. Tordsson JM, Ohlsson LG, Abrahmsen LB, Karlstrom PJ, Lando PA, Brodin TN. Phage-selected primate antibodies fused to superantigens for immunotherapy of malignant melanoma. *Cancer Immunol Immunother* 2000; 48:691–702.

333. Roux KH, Greenberg AS, Green L, Strelets L, Avilqa D, McKinney EC, Flajnik MF. Structural analysis of the nurse shark (new) antigen receptor (NAR): Molecular convergence of NAR and unusual mammalian immunoglobulins. *Proc Natl Acad Sci U S A* 1998; 95:11804–11809.

334. Greenberg AS, Avila D, Hughes M, Hughes A, McKinney EC, Flajnik MF. A new antigen receptor gene family that undergoes rearrangement and extensive somatic deversification in sharks. *Nature* 1995; 374:168–173.

335. Jorgenson TO, Solem ST, Espelid S, Warr GW, Brandsdal BO, Smalas A. The antibody site in Atlantic salmon; phage display and modeling of scFv with anti-hapten binding ability. *Dev Comp Immunol* 2002; 26:201–206.

CHAPTER 17

Naive Antibody Libraries from Natural Repertoires

Claire L. Dobson, Ralph R. Minter, and Celia P. Hart-Shorrock

CONTENTS

17.1 INTRODUCTION

Large nonimmune antibody repertoires ($>10^{11}$ antibodies) are now used routinely for the isolation of therapeutic antibodies since they contain high-affinity antibodies and are very diverse. Indeed, subnanomolar antibodies have been isolated directly from large nonimmune antibody libraries. These affinities are comparable with those of antibodies produced in a secondary or tertiary immune response (10^{-8}–10^{-10} M) [1,2]. The size of the library is important not only for affinity but also to determine the success rate of selections against a large set of antigens [1,3]. This is well illustrated by the panel of over 1000 antibodies, all different in amino acid sequence, that was generated to the B-lymphocyte stimulator (BLyS) protein [4]. Furthermore, in some cases, antibodies with the desired characteristics have been isolated from naive antibody libraries without the need for affinity maturation. Phage display is a powerful tool for generating therapeutic drugs, as illustrated by the significant number of human antibodies currently in clinical trials that are derived from naive antibody phage display, with many more antibodies in preclinical development.

This chapter gives an overview of the process of constructing naive phage display libraries and discusses a broad range of applications, focusing on the development of antibody therapeutics. Finally, it addresses the potential advances in the field and future applications.

17.2 CONSTRUCTION OF NAIVE LIBRARIES

17.2.1 Phage Display

Antibody phage display is now established as a robust alternative to hybridoma technology for the isolation of monoclonal antibodies. Since the first description of the display of antibody variable domains on phage [5], there have been many antibody phage display libraries created. The common aim of all these libraries has been to capture a large and diverse panel of antibodies, thus enabling the rapid isolation of specific antibodies to any antigen. Such libraries can be divided into three categories based on the source of antibody diversity: naive, immune, and synthetic libraries.

When constructing naive antibody libraries, the aim is to capture the maximum number of different antibody sequences from the *in vivo* repertoire. It is crucial that a large number of different antibody sequences are captured if the library is to be used as a "single-pot" resource from which antibodies to any antigen can be isolated. Naive antibody libraries have been cloned in both the single chain variable fragment (scFv) format, in which the variable heavy (V_H) and variable light (V_L) chains are linked together to form a single protein, and the larger Fab format, in which the heavy ($V_H C_H 1$) and light ($V_L C_L$) chains associate posttranslationally (Figure 17.1). The scFv format has been favored by most as it is smaller in size and therefore more amenable to large library construction [1]. For the same reason, scFv libraries are thought to be more genetically stable than Fab libraries [6] and to have expression advantages, which help maintain the diversity of the library following expression on the phage surface [2].

To date, large, single-pot, naive libraries have only been cloned from human repertoires, although there is no reason why they cannot be cloned from other animal

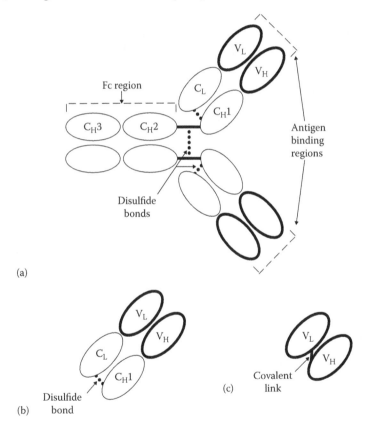

Figure 17.1 Schematic diagram to illustrate the basic structure of an immunoglobulin G (IgG) antibody molecule and the derivation of Fab and scFv antibody fragments. (a) Whole antibody molecule, with disulfide bonds linking the different chains, showing constant and variable regions on the heavy and light chains. The regions that carry out antibody functions (Fc and antigen binding regions) are indicated. (b) Fab, with heavy and light constant domains linked by a disulfide bond. (c) ScFv, with the linker between V_H and V_L.

sources. However, the human libraries have the distinct advantage that antibodies isolated from them can be used directly as therapeutic agents with minimal risk of rejection by the patient. This contrasts with mouse antibodies and human/mouse chimeric antibodies, which can elicit a human antimouse antibody (HAMA) response when administered to patients, reducing both the safety and efficacy of these drugs [7,8].

17.2.2 Overview of B-Cell Biology and Antibody Diversity

Considering that naive antibody libraries aim to clone the full spectrum of antibody sequences from the human repertoire, it is important to understand the biology of the cells that produce antibodies (B cells) in order to efficiently capture the maximum diversity. An overview of the important features of B-cell development and antibody production *in vivo* is shown in Figure 17.2.

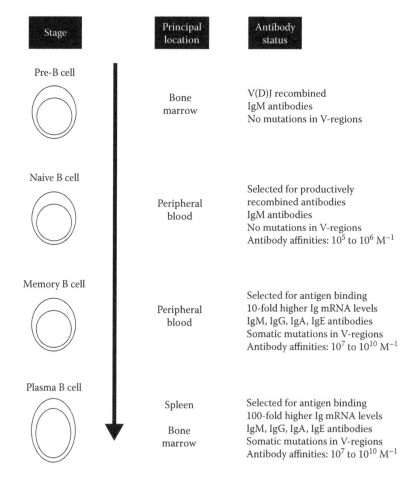

Figure 17.2   Illustration of the stages of B-cell development with reference to B-cell location and antibody status.

17.2.2.1 B Cells and Antibody Function

Antibody production by B cells is tailored to neutralizing extracellular pathogens by specifically targeting them for destruction. The specificity of antibodies is integral to their ability to only target foreign particles, such as bacteria and viruses, for destruction without causing damage to host cells.

This is possible because although there is a large antibody pool containing over 10^{10} different antibodies, each B cell only produces antibodies of a single specificity, and there are mechanisms in place to select for or eliminate B cells on the basis of the antibodies they produce [9].

17.2.2.2 Generation of the Primary Antibody Repertoire

The naive repertoire of variable heavy DNA sequences is generated by the recombination of three gene segments, which are termed V (variable), D (diversity), and J (joining). Analysis of the human IgH locus on chromosome 14 has revealed 51 functional V segments, 23 D segments, and 6 J segments, which can recombine to give 7038 different V_H sequences [10]. Further variation, known as junctional diversity, is introduced at the recombination sites where the V, D, and J segments are brought together. As both recombination sites fall within V_H complementarity determining region 3 (CDR), this concentrates the majority of diversity to this region and gives a primary V_H repertoire of approximately 106 different V_H sequences. A similar process occurs in the variable light chain where lambda or kappa V and J segments recombine to give 360 different V_L sequences [10]. With additional junctional diversity in the V_LCDR3, the V_L repertoire consists of approximately 10^4 different sequences, which can combine with the 10^6 V_H sequences to give a primary repertoire in excess of 10^{10} different antibody sequences (Figure 17.3) [11].

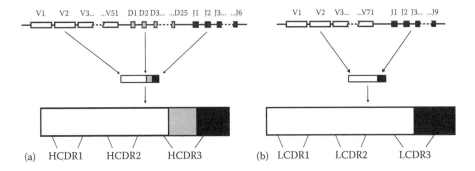

Figure 17.3 Schematic diagram to illustrate the generation of human antibody diversity by V(D)J recombination. (a) At the heavy-chain gene locus, one V, one D, and one J segment combine to give a heavy-chain variable domain, with additional junctional diversity at the recombining sites in V_HCDR3. (b) At the light-chain gene locus, one V and one J segment combine to give a light-chain variable domain with additional junctional diversity at the recombining site in V_LCDR3.

17.2.2.3 Naive B-Cell Population

Following V(D)J recombination in the bone marrow, the naive B cells enter the peripheral B-cell pool. These naive cells make up the majority of B cells circulating in peripheral blood at an estimated 45% of the B-cell total [12]. It is at this stage that naive B cells first encounter antigen; the antibodies they produce are of the IgM isotype and have low affinities for antigen (10^{-5}–10^{-6} M^{-1}), and their sequences are not mutated from the V, D, and J gene segments encoded in the germ line. As well as being present in peripheral blood, naive B cells are also found in lymphoid tissues, particularly the spleen, lymph nodes, tonsils, and Peyer's patches. However, the majority circulate in the periphery, where they are most likely to encounter foreign antigens [13].

17.2.2.4 Germinal Center and Affinity Maturation

Naive B cells, which encounter antigen and become activated, with the help of T cells, undergo affinity maturation in germinal centers prior to differentiation into memory cells or plasma cells. Germinal centers are found in various lymphoid tissues, particularly the spleen, lymph nodes, tonsils, and Peyer's patches. Each germinal center is seeded by one or very few B cells, which proliferate rapidly and, following stimulation with antigen, undergo a process known as somatic hypermutation. This process targets mutations to the DNA encoding the variable but not the constant regions of the antibody sequence. B cells containing antibody sequences with improved affinity for antigen as a result of the V-region mutations are selected, whereas low-affinity or nonfunctional counterparts undergo apoptosis. Mutations that cause affinity improvements are found more frequently within CDR regions than in framework regions.

Following somatic hypermutation, B cells that leave the germinal center, as memory cells or plasma cells, contain antibody sequences that have a number of mutations away from the germ-line-encoded V-gene DNA. The average mutation frequency is between 2% and 4%, which is equivalent to 15–30 mutations per variable region. An additional change occurs in the germinal center, and this is the isotype switch that replaces the IgM constant regions with IgG, IgA, or IgE constant regions. A large proportion of antibodies switch from IgM to IgG, IgA, or IgE at this stage and it was once thought that all somatically hypermutated antibodies underwent class switching. However, evidence has recently surfaced to suggest that a large proportion of cells leaving the germinal center have not undergone class switching and encode somatically hypermutated IgM molecules [14].

17.2.2.5 Memory and Plasma Cell Populations

Memory B cells leave the germinal centers and form a population that recirculates through the lymph and blood as long-lived cells. Of the total B-cell population in peripheral blood, between 25% and 40% are memory B cells [12,14]. It was originally thought that the majority of memory cells were either IgG or IgA producers, but recent evidence suggests that the proportion of IgM-producing memory cells could be as large as 40% of the circulating memory B-cell pool [14]. Another key feature of peripheral blood memory cells is that they have fivetold- to tenfold-increased Ig mRNA levels compared with naive cells [14].

Upon activation with antigen, memory cells proliferate and can then differentiate into antibody-secreting plasma cells. Plasma cells occur as a low percentage (0.1%) of peripheral blood lymphocytes (PBLs) but occur at relatively high levels in secondary lymphoid tissues, especially spleen and bone marrow [15]. Plasma cells are known to upregulate their Ig mRNA levels 100- to 180-fold over the levels in resting B cells [16,17].

17.2.3 Naive Antibody Library Construction

17.2.3.1 Overview of the Library Construction Process

Although several different strategies have been described for the construction of large, naive antibody libraries, there are six stages common to all approaches, as illustrated in Figure 17.4.

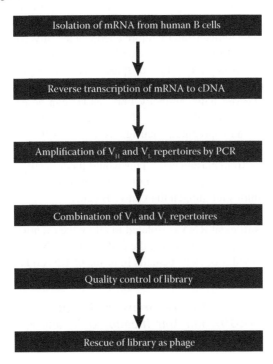

Figure 17.4 The six stages involved in the construction of a naive human antibody phage display library.

17.2.3.2 *Stage 1: Isolation of mRNA from Human B Cells*

The choice of lymphoid organ used in the isolation of human B cells can have significant effects on the array of antibodies that are cloned into the library. To date, all published naive libraries have used PBL as one of the sources of mRNA [1–3,18,19]. The reason for this is that the B-cell component of PBL contains a high proportion (approximately 45%) of naive B cells [12]. This naive B-cell population offers the advantage of being very diverse as it represents the raw output from V(D)J recombination in the bone marrow, that is, B cells that have yet to encounter antigen in the blood. The diversity of the naive population is beneficial when trying to clone as many antibody sequences as possible into the phage display library. The disadvantage of using PBL as a source is that the sequences have not undergone somatic hypermutation and are therefore of relatively low affinity. In contrast, approximately 25%–40% of PBL B cells have encountered antigen and undergone clonal expansion and somatic hypermutation. As a result, the memory compartment encodes a less diverse panel of antibody sequences, but they do generally have higher affinities due to the somatic mutations introduced.

Secondary lymphoid organs such as the spleen and tonsils, which are both centers for somatic hypermutation, have also been used as sources of mRNA for the production of naive antibody phage display libraries [1–3]. In contrast to PBL, the B cells in the spleen contain a high proportion of memory cells, which have already undergone selection on antigen and therefore encode a less diverse array of antibodies. However, as some of these sequences will have also undergone somatic hypermutation, the overall affinity for antigen of spleen-derived antibodies will be higher than that of peripheral blood B cells.

Bone marrow-derived naive antibody libraries [1] can be expected to contain antibodies derived from naive B cells and plasma cells. As active, antibody-secreting plasma cells are known to upregulate their Ig mRNA levels 100- to 180-fold over the levels in resting B cells [16,17], a large proportion of antibody sequences cloned from bone marrow will be from plasma cells. This would theoretically lead to a relatively low diversity in the cloned repertoire, but those antibodies present should have, on average, higher affinities for antigen than naive B cells.

From a practical point of view, it is crucial at this stage of the construction process to ensure that sufficient RNA is isolated to allow the eventual construction of a large library. Although time consuming, the collection of RNA from a large number of human donors provides obvious benefits in terms of RNA quantity and diversity. As many as 43 human donors have been used to construct a single naive library [1].

17.2.3.3 *Stages 2 and 3: Reverse Transcription of mRNA to cDNA and Amplification of V_H and V_L Repertoires by Polymerase Chain Reaction*

Following the isolation of RNA from B cells, it is necessary to reverse-transcribe the mRNA to cDNA and then amplify the V_H and V_L regions. The choice of oligonucleotide sequences used to prime these reactions has implications for the eventual library diversity. At the initial stage of reverse transcription to cDNA, it is possible to use either random hexamers or antibody-specific oligonucleotides to prime the cDNA synthesis. The use of IgM constant region primers for this initial reaction, in order to select for diverse, antibody sequences from naive IgM-expressing B cells, has been documented [2]. However, it is now apparent that as many as 40% of circulating memory cells also express IgM [14], and these cells will be a source of antigen-selected, somatically hypermutated antibody sequences. Conversely, the use of random hexamers to prime cDNA synthesis allows all five antibody classes to be represented and increases the potential diversity of the final library. As such, random hexamers have been used more frequently in naive library construction [1,3,19].

The choice of oligonucleotides for the amplification of V_H and V_L regions from cDNA has not changed considerably since the first attempts at cloning large numbers of human antibody variable regions [20]. The 31 oligonucleotide sequences cited in that paper were designed to amplify variable regions from all known heavy- and light-chain gene families and are summarized in Figure 17.5. Additional variable heavy-chain and variable lambda light-chain genes have since been discovered [21], so the set of required oligonucleotides to amplify all human heavy- and light-chain families has expanded slightly [1]. More recently, a thorough analysis of all functional germ-line V genes on the V region database (VBASE) [21] has enabled the design of a definitive set of primers that have been optimized to amplify all known V genes [22].

In all cases cited above, short oligonucleotides, with no flanking regions containing restriction sites, have been used for the initial amplification of all V genes, so that even the poorly represented templates are amplified. In order to capture maximum diversity, it is also important to perform separate polymerase chain reactions (PCRs) for each gene family, rather than performing PCRs with mixes of several primers [20]. When equimolar mixtures of several primers are used, it is thought that this can bias the V-gene representation in the eventual library [2].

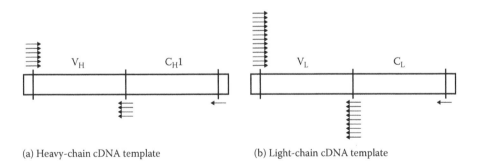

(a) Heavy-chain cDNA template   (b) Light-chain cDNA template

Figure 17.5 Illustration of the primer sites for amplification of human V_H (a) and V_L (b) repertoires. The V_H or V_L region is amplified from the cDNA template using a panel of primers that anneal to the 5′ end of all known V_H and V_L regions and primers that can anneal at the 3′ end to all known V_H and V_L J-regions. The J-region primers can be replaced with primers that anneal to regions within the constant domain.

17.2.3.4 Stage 4: Combination of V_H and V_L Repertoires

Up until this stage, the V_H and V_L repertoires have been handled separately. At some point in the process, it becomes necessary to combine the heavy- and light-chain DNA into the same vector. In the case of scFv libraries, the V_H and V_L need to be cloned either side of a short linker sequence, which, when translated, allows the V_H and V_L domains to assemble into a functional conformation. For Fab libraries, although the heavy and light chains do not need to be expressed as a single protein, it is usually desirable to clone them into a single phagemid to retain linkage between genotype and phenotype during selections.

A frequently used strategy for joining the V_H and V_L DNA together is a two-fragment PCR assembly [23]. By incorporating complementary regions at the ends of the V_H and V_L repertoires, the fragments can be spliced together and a DNA polymerase used to create a double-stranded joined product. A further PCR then amplifies the newly recombined V_H and V_L fragments prior to the final cloning step. This final PCR is important as it generates sufficient DNA to perform the multiple ligations and electroporations that are necessary to generate large library sizes. This method for joining V_H and V_L fragments has been used to create libraries with over 10^{10} transformants [1].

Alternatively, the heavy- and light-chain repertoires can be cloned separately into two vectors. Both vectors can be prepared as plasmid DNA and digested with two restriction enzymes to excise the V_H fragments from one repertoire and prepare the other plasmid, containing the V_L repertoire, as an acceptor vector. The V_H fragments can then be cloned into the V_L-containing acceptor vector to recombine the V_H and V_L repertoires. This method has also been used to generate libraries containing greater than 10^{10} clones [3].

All these methods maximize library diversity by creating as many new heavy- and light-chain combinations as possible *in vitro*. As an alternative strategy, the ability of bacteria to recombine fragments of DNA *in vivo* has been used to create new heavy- and light-chain combinations [19,24]. In the more recent of these papers, a relatively small primary repertoire of 7×10^7 scFv was cloned into a vector containing two different loxP sites, one between the V_H and V_L fragments and the second further downstream. Phagemids containing the scFv genes were used to infect bacteria expressing Cre recombinase, using a high multiplicity of infection of 200:1. This enabled multiple phagemids to infect a single cell, and the Cre recombinase activity catalyzed the exchange of DNA between different phagemids to allow new V_H/V_L combinations to be formed. It is estimated that the library size following this *in vivo* recombination step was 3×10^{11} [19].

17.2.3.5 Stage 5: Quality Control of Library

Once the library construction process is complete, there are some important quality control checks to carry out to ensure that the library will be capable of being used as a diverse single-pot resource. One of the obvious initial assessments of the usefulness of the library is that of library size. The first naive library constructed contained around 10^7 different antibody sequences, and with this level of diversity, the antibodies isolated from the library typically had affinities for antigen in the micromolar range [18]. A breakthrough was made with the construction of the first very large (10^{10}) libraries, from which clones in the nanomolar range can be isolated directly [1]. Since that point, all naive libraries that have been made have attempted to reach the same large library size to ensure that high-affinity antibodies can be isolated without the need for further engineering to improve affinity.

In order to obtain a large library size, it is often necessary to include extra PCR steps, such as those used to append restriction sites and to amplify the repertoire following the joining of V_H and V_L fragments. The result of these methods is that the scFv DNA may have undergone up to four PCR amplifications during library construction, and this introduces the possibility of PCR errors within the antibody sequences. In some cases, clones selected from naive antibody libraries contain up to 22 mutations away from the original germ-line gene [1]. As such, some libraries have been constructed using proofreading polymerases to reduce the number of PCR errors [2]. However, as it is virtually impossible to clone a large repertoire of fully germ-line naive antibody sequences, given the difficulty in isolating only naive mRNA from mixed B-cell populations, it is difficult to distinguish which variations from the germ line are due to somatic hypermutation *in vivo* and which are due to PCR errors *in vitro*.

PCR can be used to confirm that clones in the library contain full-length V_H and V_L domains. In most cases, greater than 90% of transformants contain inserts of the correct size, which indicates that a high efficiency was achieved at the ligation stage. Some libraries have also been tested for the expression of full-length recombinant antibody protein. Protein is expressed from individual library clones and assayed by Western or dot blot using a secondary antibody that binds to a detection handle at the C-terminus of the protein. Despite concerns that as few as 1%–10% of transformants can express full-length antibody fragments [25], it seems, in practice, that over 50% of transformants, in both scFv and Fab libraries, express full-length proteins [1,3].

Sequence analysis of libraries gives an initial indication of the level of antibody diversity that has been cloned. Unfortunately, relatively little data are available from the libraries constructed to date as most have estimated diversity by either *Bst*NI fingerprinting [18] or sequenced clones only after they have been selected on antigen. However, in one case, 36 library clones were sequenced, and all 36 were found to be unique [2]. In this relatively small sample of the library, 15 germ-line V_H genes and 8 germ-line V_L genes were represented. An observation that certain gene families, such as V_H3, $V_\kappa1$, and $V_\lambda3$, were overrepresented was thought to be a reflection of the natural bias found in human antibody repertoires *in vivo* [21,26] and also to be due to the use of pooled primers in the initial amplification [2].

Length variation in V_HCDR3 regions, a source of diversity arising from V(D)J recombination in B cells, offers further supportive evidence of the successful cloning of a diverse repertoire. A range of V_HCDR3 lengths, from 5 to 18 amino acids, was observed in the 36 library sequences analyzed [2].

Of course, the ultimate test of a naive antibody library is whether it can be used to isolate a large panel of specific, high-affinity antibodies to any given antigen. Some compelling evidence for the abilities of naive antibody libraries in this regard is given in Section 17.3.2.1.

17.2.3.6 Stage 6: Rescuing the Library as Phage

The final stage in the preparation of the antibody library is to express the antibody fragments on the surface of phage. All large libraries have been cloned into phagemid vectors to take advantage of the benefits of high transformation efficiency of small phagemids, as opposed to large phage vectors. ScFv or Fab libraries are typically cloned immediately upstream of the M13 gene III protein so that they can be expressed as a fusion protein and incorporated into the M13 phage surface. This process, which involves the growth of the phagemid library in bacteria followed by infection with helper phage, introduces the possibility of overrepresentation of clones that have a growth advantage. These could be "empty" phagemids that contain no antibody fragment, clones with truncated inserts, or antibody fragments that translate rapidly as they have no rare codons or fold rapidly due to their primary amino acid sequence. To avoid these clones becoming overrepresented, the expression of the fusion protein is under the control of a repressible promoter. When the library is grown in the presence of the repressor, the expression of antibody fragments is inhibited. Removal of the repressor at the same time as infection with helper phage then allows expression of the antibody fragment and incorporation into the M13 phage surface.

17.3 APPLICATIONS OF NAIVE LIBRARIES

17.3.1 Antibody Therapeutics

Antibodies selected from large, nonimmunized repertoires of scFv fragments have a multitude of applications by virtue of their specificity and affinity and the relative ease with which they can be derived. One such application is in the development of therapeutic antibodies, which have a more rapid discovery process and cost less to develop than traditional small molecule drugs. While traditional drugs generally require 3–5 years to refine and test in the laboratory, antibodies can take as little as 12–18 months to progress into full clinical development.

Therapeutic antibodies can mediate their action through a variety of different mechanisms [27]. First, antibodies can act by blocking the function of a target antigen. This is typically achieved by preventing access of a growth factor, a cytokine, or another soluble mediator by binding directly to the soluble factor itself or to its receptor. An example of this is the phage display-derived antibody, HUMIRA for the treatment of rheumatoid arthritis and other diseases [28]. This antibody acts by binding and blocking the action of the proinflammatory cytokine tumor necrosis factor alpha (TNFα).

Another mechanism of action is Fc-mediated targeting and is especially used for the treatment of cancer. Typically, IgG1 antibodies that bind proteins expressed in a particular disease state are used to harness the body's own immune system. Antibody-dependent cellular cytotoxicity (ADCC) occurs if antibody Fc regions are recognized by receptors present on cytotoxic cells, such as natural killer cells, macrophages, granulocytes, and monocytes [29]. Alternatively, instead of using the patient's immune system to destroy tumor cells, antibodies to the target antigen can be used to deliver radioactive isotopes or cytotoxic drugs [30–32].

Finally, antibodies can act by modulating the function of a cell by binding to an antigen capable of transducing intracellular signals. For example, phage display-derived agonistic antibodies to the muscle-specific kinase (MuSK) tyrosine kinase have been isolated [33]. This initial success has been extended to the isolation of agonistic antibodies to TNF-related apoptosis-inducing ligand receptor 1 (TRAIL-R1) and TRAIL-R2 [34], which are currently in clinical trials for the potential treatment of cancer. More recently, further agonistic antibodies have been described for the thyrotropin receptor [35] and Fas receptor [36], which have originated from naive phage display libraries.

17.3.1.1 Generation of Antibody Therapeutics

The generation of therapeutic antibodies usually occurs in two phases. In the first phase, naive antibody libraries are sampled by a series of selections and screens to identify antibodies with the desired characteristics. Generally, it is possible to isolate antibodies with the required function and specificity, but often, the affinity needs to be improved for use in therapy. This can be achieved by constructing, selecting, and screening secondary libraries, a process also known as affinity maturation.

The following sections give an overview of the key steps involved in the generation of therapeutic antibodies. These steps are highlighted in the schematic in Figure 17.6.

17.3.1.2 Selecting Therapeutic Antibodies from Naive Libraries

The selection of antibodies from phage libraries involves the sequential enrichment of specific binding phage from a large excess of nonbinding clones. This can be achieved by incubating the phage library with the target antigen. Unbound phage are removed by washing, and phage displaying scFvs that specifically bind the antigen are eluted by disrupting the phage–antigen interaction (e.g., by applying pH gradients, competitive elution conditions, or proteolytic reactions). Recovered phage are subsequently amplified by infecting *Escherichia coli* and further rounds of selection performed as illustrated in Figure 17.7.

There are many different types of possible selection methodologies, and these are described in detail in Chapters 4 and 5. Some of the main considerations for the selection of therapeutic antibodies are the quality of the antigen, its purity, and methodologies that enable the selection of high-affinity antibodies.

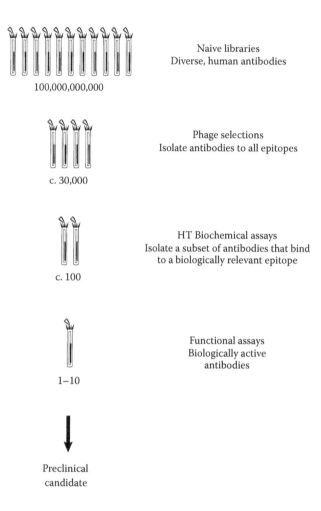

100,000,000,000

Naive libraries
Diverse, human antibodies

c. 30,000

Phage selections
Isolate antibodies to all epitopes

c. 100

HT Biochemical assays
Isolate a subset of antibodies that bind
to a biologically relevant epitope

1–10

Functional assays
Biologically active
antibodies

Preclinical
candidate

Figure 17.6   The stages involved in the generation of therapeutic antibodies from phage anti-body libraries.

When a purified target antigen is available, phage antibody selections are often carried out with the protein directly adsorbed onto a plastic surface, such as immunotubes (Maxisorb tubes, Nalge Nunc Intl., Naperville, IL), where it is noncovalently associated via electrostatic and Van der Waals interactions [18]. The main disadvantage of this method is that the adsorbed protein may be unfolded, and antibodies isolated often fail to bind the native protein. An improvement over the direct adsorption of proteins on plastic surfaces is their covalent coupling to plates or beads. This method of presentation enables the protein to maintain its native fold and thus increases the chances of selecting for antibodies binding to the native protein. The random orientation of the protein also increases the likelihood that all epitopes will be available for binding during selection. This is of particular importance for the selection of

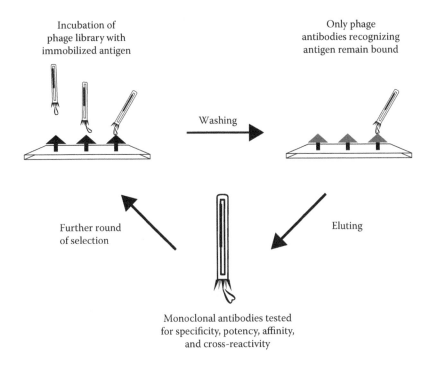

Incubation of
phage library with
immobilized antigen

Only phage
antibodies recognizing
antigen remain bound

Washing

Further round
of selection

Eluting

Monoclonal antibodies tested
for specificity, potency, affinity,
and cross-reactivity

Figure 17.7 Illustration of the different stages in a standard phage display selection cycle.

therapeutic antibodies since specific epitopes must be recognized for the antibody to alter the function of the target protein.

For therapeutic applications, it is often desirable to generate antibodies that bind the target antigen with high affinity. However, it is difficult to select for high-affinity antibodies when immobilizing the target protein on a solid surface due to rebinding and avidity effects. By performing the selection in solution, conditions can be chosen to favor affinity or kinetic parameters such as off rates [37,38]. For this type of selection, the target protein is usually biotinylated and subsequently captured using streptavidin-coated paramagnetic beads. Although affinity or off rate-driven soluble selections can be used to isolate antibodies from primary libraries, these methods are more often used to isolate high-affinity antibodies from secondary libraries during affinity maturation (see Section 17.3.1.5). The preferential selection of mutant antibodies of higher affinities is enabled by reducing the concentration of the target protein below the K_d of the parent clone [37,39].

Often, it is not possible to obtain the target protein in a purified form. This is the case for many proteins of therapeutic interest (receptors, ion channels) that only retain their functionality in lipid bilayers. Whole cells or plasma membrane preparations where the target protein is expressed can be used instead. However, the isolation of specific antibodies to target proteins on cell surfaces is usually challenging due to background binding of phage specific for nontarget proteins. Furthermore, many proteins are present on cells at very low densities, making the selection difficult as the antigen concentration is usually much lower than the K_d of any antibodies in the library. Depletion and/or subtraction methods can help with the first problem, as antibodies present in the libraries that bind these nontarget antigens are depleted [40,41]. Another technique for the selection of antibodies binding to cell surface antigens and potentially overcoming both problems of low antigen density and specificity is ProxiMol selection [42]. Antibodies binding to C-C chemokine receptor type 5 (CCR5) and blocking macrophage inflammatory protein 1 alpha (MIP-1a) binding were generated using this approach [43].

In summary, selection procedures are extremely flexible and continue to evolve to ensure that antibodies with the desired characteristics are isolated from phage display libraries. An example is described in Section 17.3.2.1 to support the theory that the choice of selection methods influences output numbers and diversity. There are multiple other selection methodologies not discussed here that could be useful for the generation of therapeutic antibodies. These include *in vivo* selections [44,45], selection for antibodies with intracellular activity [46,47], as well as selections for internalization [48] and in living animals [49]. Furthermore, methods have been developed to purify and stabilize membrane-associated proteins. Such methods could provide useful solutions to selections on "difficult" unpurified proteins and, in particular, to cell surface molecules. For example, tagged CCR5 was purified and reconstituted on proteoliposomes before selections [50].

17.3.1.3 Converting Nonhuman Antibodies to Human

The ability of phage display selections to isolate specific antigen binders from a large library has also been used to create a human equivalent of murine monoclonal antibodies. In these cases, a murine monoclonal antibody is available that has the desired properties of affinity and specificity but, being murine, would be immunogenic if used as a therapeutic. The murine antibody can be used as a template for the selection of human heavy- and light-chain variable regions by phage display in a process known as guided selection [51]. First, the murine heavy chain is paired with a library of naive human light chains, and selection on antigen is performed to isolate a human light chain that is capable of replacing the murine light chain. The next stage is to pair the selected human light chain with a repertoire of human heavy chains and select on antigen to obtain a human heavy- and light-chain combination that retain the original characteristics of the murine antibody (Figure 17.8a). In using the murine heavy chain to guide the selection, it is possible to isolate human antibodies that attain, or even improve, the affinity of the original murine antibody and also bind the same epitope [51].

A variation on this method, known as the parallel shuffle procedure, has also been used successfully [52]. Here, the murine light chain is paired with a human heavy-chain library, the murine heavy chain is paired with a human light-chain library, and both are selected on antigen in parallel (Figure 17.8b). Selected human heavy and light chains can then be paired together and tested for specificity and affinity.

Finally, a third method of guided selection has been devised that takes advantage of the important role that the V_HCDR3 region plays in determining antigen binding specificity. In this method, the C-terminal end of the murine heavy chain, including

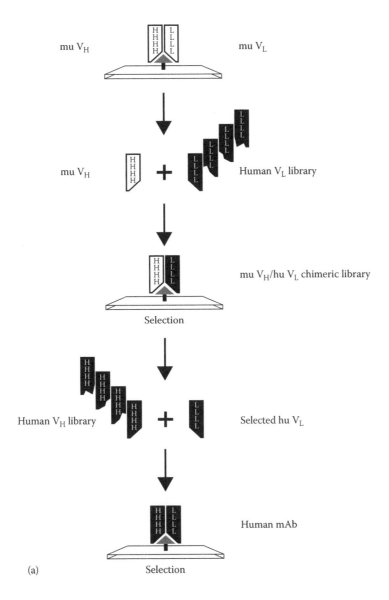

mu V_H

mu V_L

mu V_H +   Human V_L library

mu V_H/hu V_L chimeric library

Selection

Human V_H library   +   Selected hu V_L

Human mAb

(a)   Selection

Figure 17.8   An overview of methods for performing humanization of murine antibodies using phage display-guided selections. (a) The guided selection method in which human light and heavy chains are selected in two successive steps.

(Continued)

V_HCDR3 and framework 4, as well as the murine light chain are combined with a library of human heavy-chain regions from framework 1 to framework 3. Once a human–murine hybrid heavy chain has been selected, it is paired with a library of human light chains and selected on antigen (Figure 17.8c). The final heavy- and light-chain combinations are fully human, apart from the murine V_HCDR3 and

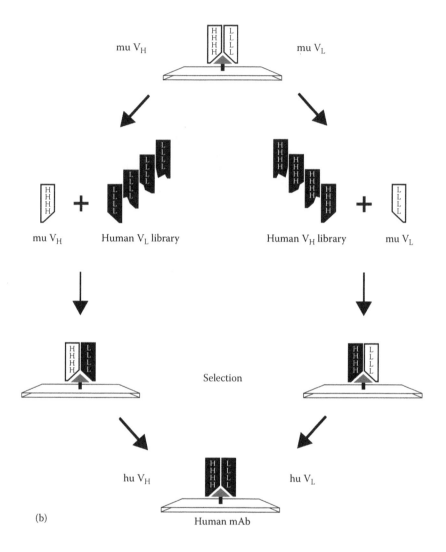

(b)

Figure 17.8 (Continued) An overview of methods for performing humanization of murine antibodies using phage display-guided selections. (b) The parallel shuffle procedure in which human light and heavy chains are selected in parallel and then paired together.

(Continued)

framework 4 region that was used to guide the selection. Using this method, a human scFv was obtained that retained the epitope specificity of the murine equivalent [53].

The phage display-derived anti-TNFα therapeutic antibody, HUMIRA, was generated by guided selections using a mouse monoclonal antibody [28]. Other guided selection-derived antibodies are currently in preclinical development, demonstrating that, in some cases, it is a valuable alternative to the de novo generation of antibody from antibody libraries (MedImmune, unpublished).

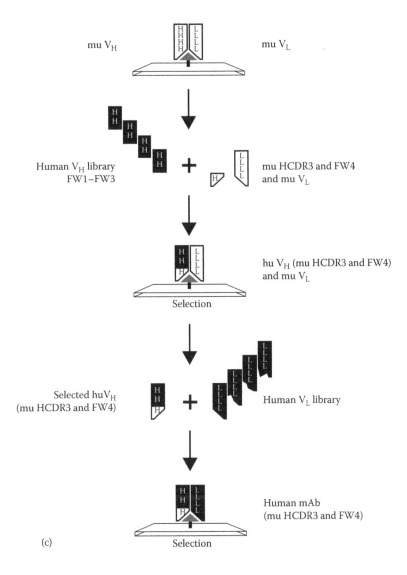

mu V$_H$ mu V$_L$

Human V$_H$ library   +   mu HCDR3 and FW4
FW1–FW3   and mu V$_L$

hu V$_H$ (mu HCDR3 and FW4)
and mu V$_L$

Selection

Selected huV$_H$ +   Human V$_L$ library
(mu HCDR3 and FW4)

Human mAb
(mu HCDR3 and FW4)

(c)   Selection

Figure 17.8 (Continued) An overview of methods for performing humanization of murine antibodies using phage display-guided selections. (c) A guided selection strategy that results in an antibody containing a fully human light chain and a human/murine hybrid heavy chain.

17.3.1.4 Screening for Therapeutic Antibodies

Selections are typically designed to yield antibodies that bind to a particular target protein. Since it is not always known which region of the target antigen needs to be recognized by an antibody to elicit a biological response, it is desirable to select multiple different antibodies to the same target that bind to different epitopes.

To do so, it is important to analyze selected clones after as small a number of selection rounds as possible, when the selection output is at its most diverse. The aim of a screening strategy is to identify, within a population of binders, those antibodies that have desirable therapeutic characteristics in terms of specificity, affinity, and function.

Initially, clones are tested for target antigen specificity in an enzyme-linked immunosorbent assay (ELISA) using unpurified phage antibodies. This is performed together with sequence analyses to identify a population of unique antigen binders. Maximal numbers (up to 10^5) of sequence diverse clones are then screened to identify a subset of antibodies that bind to a biologically relevant epitope. Such screens are usually biochemical assays that are fast, robust and automatable (e.g., 96-well or 384-well format) and use soluble antibody fragments (scFvs or Fabs) from the bacterial supernatant or periplasm. Biochemical assays can generally be divided into separation-based assays (in which the reaction product is detected after its separation from the starting material) and homogeneous assays (in which the detection of the product does not require a separation step). Separation-based assays include ELISA-based screens, which are commonly used to identify antibodies that can compete with the native ligand for binding to a target antigen. These assays have the advantage that the compound has usually been separated from the reaction product at the time of detection, which minimizes compound interference effects. Furthermore, these assay types tend to have larger signal windows than homogeneous formats as the reaction product is the only source of signal in the assay. Homogeneous assays can take a variety of formats depending on the target of interest. Fluorescence resonance energy transfer (FRET) is the basic principle behind a number of biochemical and cell-based assays that are widely used in high-throughput screening [54]. In FRET-based assays, a fluorescent molecule is excited by energy at a certain wavelength, and an acceptor molecule then captures the emission energy from the donor fluorophore. Homogeneous assays have the advantage that they require fewer addition or reagent transfer steps, making them easier to automate and miniaturize.

Once a panel of unique scFvs with the desired specificity has been isolated, functional assays are used to identify a subset of this population with the required biological activity. Such assays are generally cell based and can take a variety of formats depending on the biological activity under investigation. Functional assays can be used to study cellular responses (proliferation, chemotaxis, adhesion, apoptosis, receptor upregulation); cellular biochemistry (calcium signaling, kinase activation); and nuclear events (reporter genes, cell cycle status). In parallel, affinity studies are often carried out by surface plasmon resonance to characterize the antibodies further. At this point, it is desirable to test the antibodies in their final therapeutic format since differences in potency can be observed between the scFv and IgG proteins. In most cases, biological activity is retained, and it is often possible to obtain an improvement in potency on conversion to IgG format due to bivalent binding.

17.3.1.5 Secondary Libraries for Antibody Affinity Maturation

Following the selection and screening phase of the drug discovery process, a panel of antibodies will have been isolated with the desired specificity and biological characteristics. It is possible to isolate therapeutic-grade antibodies directly from a primary library without the requirement for any improvement in affinity. For example, antibodies to TRAIL-R1 and TRAIL-R2 for the potential treatment of cancer [34] are currently in early-phase clinical trials. In most cases, however, it is necessary to improve antibody affinity by a process analogous to affinity maturation. This has advantages for therapeutic applications; by increasing the affinity of an antibody to its target antigen, the therapeutic dose of a monoclonal antibody (mAb) is reduced [55], whereas its therapeutic duration is increased. In some cases, the benefit of very-high-affinity antibodies is still a matter of debate. For example, in the treatment of solid tumors, mAbs of too high affinity (low pM) may be counterproductive for tumor penetration [56].

The affinity maturation of antibodies consists of introducing diversity within the antibody V genes to create a secondary library, which can be selected and screened for antibody variants with higher affinities. Methods for antibody affinity maturation have been described in detail in Chapters 3, 15, and 17 and will be summarized briefly.

Secondary libraries are generated by introducing mutations in the V genes of the lead antibody, thereby creating a library of variants. Although antibody–antigen crystal structures can indicate which residues should be mutated to improve binding, atomic resolution structural data are not available for most antibodies.

Nondirected approaches whereby the V genes are mutated randomly have been successfully used to optimize antibodies with low starting affinities [57–60]. Methods to introduce diversity in the antibody V genes include error-prone PCR [58], mutator strains of bacteria [60,61], DNA shuffling [62], and chain shuffling [59,63].

For therapeutic antibodies, CDR-directed approaches are favored over random approaches. Antibodies isolated from large antibody libraries often have low nanomolar affinities, and CDR-directed approaches have been more successful for optimizing antibodies with such affinities. Furthermore, limiting the mutagenesis to the CDRs is less likely to generate immunogenic antibodies than changes in the more conserved framework regions. Ideally, residues that modulate affinity are targeted. These residues can be determined experimentally by alanine scanning of the CDRs [57], parsimonious mutagenesis [64,65], or modeling [57]. Structural information of antibody–antigen complexes and studies of the natural diversity of human antibodies created during the *in vivo* primary and secondary immune response also suggest key residues for targeting [66]. Typically, six CDR residues are targeted at a time, and residues involved in maintaining the CDR conformations are not altered. Approaches targeting CDRs sequentially (sequential CDR walking) are usually preferred to parallel targeting (parallel CDR walking) since additive effects of mutations are often unpredictable [67]. So far, the most significant gains in affinity have been obtained by optimizing CDR3 regions in the light and heavy chain [57,67].

Antibodies for clinical development have been optimized by using a range of strategies, including V_L shuffle [68], V_H and V_L CDR3 randomization [69], as well as targeting CDR residues known to participate in antibody–antigen interactions. Such antibodies often have affinities below 100 pM.

17.3.1.6 Therapeutic Antibody Format

ScFvs or Fabs with desired function, specificity, and affinity can be reformatted to whole IgG1, IgG2, or IgG4 isotypes depending on the therapeutic application. Eukaryotic expression vectors for rapid one-step cloning of antibody genes for either transient or stable expression in mammalian cells have been described [70].

The IgG1 isotype is typically used when cell killing is required, as in cancer. Human IgG1 in particular can trigger the classical complement cascade after binding to cell surfaces and promote ADCC, which is an important mediator of cell lysis by the bound mAb. The IgG4 isotype does not mediate activation of the immune system, and therapeutic antibodies in this format act mainly by blocking biological interactions. Although the majority of antibodies in clinical development are IgGs, when effector functions are not required for the therapeutic application, scFvs or Fabs may provide an alternative. These fragments are particularly suitable for tumor targeting, as they penetrate tumors more effectively than do whole IgGs [71]. Although the body clearance rates of unmodified scFvs and Fabs are much higher than those of full-length immunoglobulins, it has been shown that half-lives can be dramatically increased by coupling polyethylene glycol to Fabs [72].

In addition to the use of the conventional IgG1 isotype, there have been numerous recent advances to improve the biological properties of antibodies via engineering the Fc domain [73]. These include improvements to circulating half-life of the antibody drug in patients [74], as well as changes that enhance or knockout ADCC activity [75–77] and complement-dependent cytotoxicity (CDC). All these potential enhancements should be considered when deciding on the optimal antibody format for the given therapeutic application.

17.3.1.7 Preclinical and Clinical Development

Once a therapeutic lead has been identified from the drug discovery process, the antibody is further characterized using a series of *in vivo* models to assess the biological activity and pharmacokinetics. Preclinical activities also include the creation of antibody-expressing mammalian cell lines. The highest-producing cell lines are used to perform large-scale bioreactions to maximize antibody production. These provide antibody material for toxicology studies, and clinical trials and will ultimately be used to provide material for the treatment of patients.

17.3.2 Examples of Phage-Derived Antibodies in the Clinic

Naive phage display libraries have yielded many antibodies, which are currently in preclinical or clinical development as therapeutics (Table 17.1). HUMIRA (adalimumab), an anti-TNFα antibody for the treatment of rheumatoid arthritis, was the first phage display-derived antibody to receive approval for marketing (Abbott Laboratories). This antibody has also been launched as a treatment for a number of

Table 17.1   Summary of Antibodies Currently in Clinical or Preclinical Development That Have Been Isolated from Naive Phage Display Libraries

Product	Target	Disease	Originator/Developer	Status
HUMIRA (adalimumab)	TNFα	Rheumatoid arthritis	MedImmune/Abbott	Launched
HUMIRA (adalimumab)	TNFα	Juvenile idiopathic arthritis	MedImmune/Abbott	Launched
HUMIRA (adalimumab)	TNFα	Crohn's disease	MedImmune/Abbott	Launched
HUMIRA (adalimumab)	TNFα	Psoriatic arthritis	MedImmune/Abbott	Launched
HUMIRA (adalimumab)	TNFα	Ankylosing spondylitis	MedImmune/Abbott	Launched
HUMIRA (adalimumab)	TNFα	Ulcerative colitis	MedImmune/Abbott	Launched
HUMIRA (adalimumab)	TNFα	Chronic plaque psoriasis	MedImmune/Abbott	Launched
BENLYSTA (belimumab)	BLyS	Systemic lupus erythematosus	MedImmune/GSK	Launched
ABthrax (raxibacumab)	Anthrax protective antigen	Anthrax	MedImmune/GSK	Launched
ABT-874 (briakinumab)	IL-12	Autoimmune diseases	MedImmune/Abbott	Phase III
iCo-008 (bertilimumab)	Eotaxin1	Allergic disorders	MedImmune/iCo	Phase II
CAM-3001 (mavrilimumab)	GM-CSFRα	Rheumatoid arthritis	MedImmune	Phase II
CAT-354 (tralokinumab)	IL-13	Asthma	MedImmune	Phase II
CAT-354 (tralokinumab)	IL-13	Ulcerative colitis	MedImmune	Phase II
HGS-ETR-1 (mapatumumab)	TRAIL-R1	Oncology	MedImmune/GSK	Phase II
HGS-ETR-2 (lexatumumab)	TRAIL-R2	Oncology	MedImmune/GSK	Phase II
NI-0801	CXCL10	Inflammation	MedImmune/NovImmune	Phase II
MEDI-4212 (omalizumab)	IgE	Allergic disease	MedImmune	Phase I
MEDI-5117	IL-6	Rheumatoid arthritis	MedImmune	Phase I
MEDI-4736	PD-L1	Oncology	MedImmune	Phase I
NI-0501	Interferon-γ	Inflammation	MedImmune/NovImmune	Phase I
NI-0701	RANTES	Asthma	MedImmune/NovImmune	Phase I
MEDI-4893	*Staphylococcus aureus* alpha toxin	Infectious disease	MedImmune	Preclinical

(Continued)

Table 17.1 (Continued) Summary of Antibodies Currently in Clinical or Preclinical Development That Have Been Isolated from Naive Phage Display Libraries

Product	Target	Disease	Originator/ Developer	Status
GC-1008	TGF-β 1,2,3	Idiopathic pulmonary fibrosis	MedImmune/ Sanofi	Preclinical
Anti-CD148	CD148	Inflammation	MedImmune/ Amgen	Preclinical
Anti-CD30L	CD30L	Inflammation	MedImmune/ Amgen	Preclinical
Anti-IL-18R	IL-18R	Inflammation	MedImmune/ Amgen	Preclinical
Anti-Gastrin	Gastrin	Oncology	MedImmune/ Xoma	Preclinical
Kallikrein inhibitor, DX-2930	Kallikrein	Angiodema	Dyax	Preclinical
PC-mAb, M99-B05	Oxidized LDL	Restenosis; cardiovascular disease	Dyax	Preclinical
DX-2500	Neonatal Fc receptor	Autoimmune diseases	Dyax	Preclinical

Source: Thomson Reuters Pharma.

Note: BLyS, B-lymphocyte stimulating factor; CD, cluster of differentiation; CXCL10, C-X-C motif chemokine 10; GM-CSFRα, granulocyte macrophage colony stimulating factor receptor alpha; IL, interleukin; LDL, low density lipoprotein; PD-L1, programmed death ligand 1; RANTES, regulated on activation, normal T cell expressed and secreted; TGF-β, transforming growth factor-beta; TNF, tumour necrosis factor.

other diseases, including juvenile rheumatoid arthritis, Crohn's disease, and psoriatic arthritis. Another recent success story is the approval of the anti-BLyS receptor antibody belimumab (Benlysta) for the treatment of systemic lupus erythematosus (SLE). Belimumab is the first drug approved by the United States Food and Drug Administration (FDA) for the treatment of SLE in over 50 years [78].

To illustrate a typical drug discovery process using phage display, a project to isolate antibodies to the BLyS protein is described. This formed part of a collaboration between Cambridge Antibody Technology (now MedImmune) and Human Genome Sciences (now GlaxoSmithKline).

17.3.2.1 Therapeutic Antibodies to BLyS

17.3.2.1.1 Therapeutic Hypothesis

BLyS (also known as BAFF, zTNF4, TALL-1, TNFSF13B, and THANK) [79–84] is a TNF-related cytokine that plays a critical role in the regulation of B-cell maturation and development [80–82], through binding to specific receptors expressed predominantly on B cells [79,85]. Elevated levels of BLyS have been found in patients with SLE, rheumatoid arthritis [86,87], and Sjögren's syndrome [88,89]. These findings suggest that blocking the biological effects of BLyS with neutralizing antibodies may be an effective approach in the amelioration or long-term remission of B-cell-associated autoimmune disease. With the aim of developing a therapeutic agent for autoimmune disease, a human, neutralizing monoclonal antibody against human BlyS was generated [4,90].

17.3.2.1.2 Isolation of Antibodies by Phage Display

The isolation strategy was designed with the aim of generating a diverse panel of antibodies to the BLyS protein. Large, nonimmunized, human scFv phage display libraries were used for all selections. Three alternative selection strategies were adopted using purified recombinant BLyS (51 kDa, homotrimer) [80]. The antigen was (1) immobilized on immunotubes; (2) biotinylated and coupled to streptavidin-coated plates; or (3) biotinylated and used in soluble selection [37]. Three rounds of selection were carried out, and phage antibodies from the second and third round were screened by ELISA [19] for specific binding to BLyS and not to an unrelated bovine serum albumin (BSA) or related (TNFα) protein. Over 7500 antibodies were screened in this way, and 2730 (36%) specifically recognized the BLyS antigen. DNA sequencing subsequently identified 1287 sequence-unique scFvs, utilizing a wide range of antibody germ-line sequences. Therefore, by using diverse selection strategies, the authors were able to maximize antibody diversity, increasing the likelihood of identifying an antibody to a biologically relevant epitope. To illustrate the diversity present in the panel, the closest human germ line for each anti-BLyS antibody was identified by aligning its nucleotide sequence to germ-line V_H, V_L, D (diversity), and J (joining) databases using Basic Local Alignment Search Tool for Nucleic Acid Sequences (BLASTN) (version 2.0.9). The heavy-chain germ-line usage for a panel of antibodies to a single protein antigen was very extensive, as 6 of 7 V_H, 22 of 27 D, and 6 of 6 J_H germ lines were represented. Only the V_H2 subfamily, which is rarely used during an immune response, was not represented in the selected panel (Figure 17.9a and b). The light-chain usage, although less diverse than that observed for the heavy chain, was still considerable with 8 of 10 V_λ, 3 of 6 V_κ, 3 of 7 J_λ, and 4 of 6 J_κ germ lines exemplified (Figure 17.9c and d). Since any bias in germ-line usage is likely to be antigen dependent, this illustrates the importance of using antibody libraries with the highest levels of diversity.

Considerable diversity was also observed within the V_HCDR3 (complementarity determining region 3) sequences. These regions are nongerm-line encoded, and there is strong evidence to suggest that they are the key determinants of specificity in antigen recognition [91]. Five hundred and sixty-eight distinct V_HCDR3 sequences varying in length from 5 to 25 amino acid residues were identified within the panel of 1287 anti-BLyS antibodies, suggesting that antibodies may have been generated to many epitopes. This level of diversity is desirable when generating therapeutic antibodies since, at the start of a project, the biologically relevant epitope is not always known.

To assess the panel of antibodies for neutralizing activity, each scFv was tested for the ability to inhibit BLyS binding to receptors expressed on immunoglobulin-9 (IM9) cells, a myeloma cell line expressing significant levels of BLyS receptors. Approximately 40% of the 1287 antibodies inhibited BLyS binding to its receptor on B cells, and the most potent scFvs exhibited half maximal inhibitory concentration (IC_{50}) values in the

V$_H$ segments

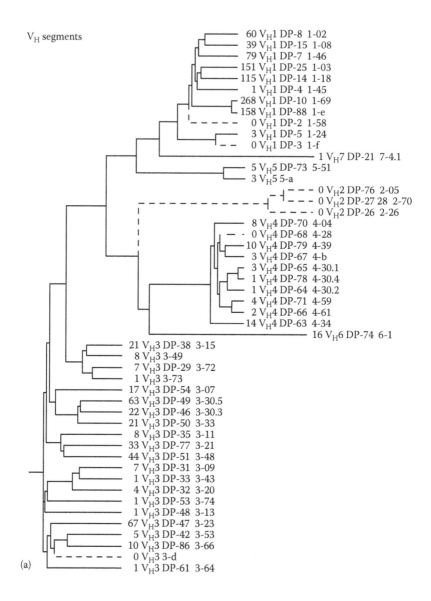

Figure 17.9 Anti-BLyS antibody germ-line usage. (a) The human immunoglobulin V$_H$. The total number of anti-BLyS antibodies (out of 1287) utilizing the different germ-line gene segments is listed together with the germ line locus name. Results are shown as dendrograms illustrating familial relationships. Solid lines indicate that the given germ line was observed. Note that for 571 antibodies in the panel, the V$_H$CDR3 domains were either too short or too novel to confidently assign a D gene segment (no D).

(Continued)

D segments

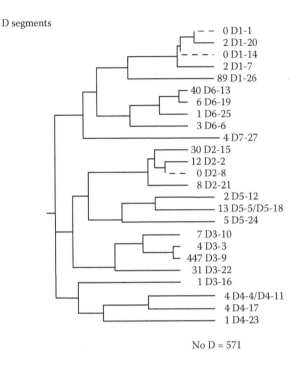

0 D1-1
2 D1-20
0 D1-14
2 D1-7
89 D1-26
40 D6-13
6 D6-19
1 D6-25
3 D6-6
4 D7-27
30 D2-15
12 D2-2
0 D2-8
8 D2-21
2 D5-12
13 D5-5/D5-18
5 D5-24
7 D3-10
4 D3-3
447 D3-9
31 D3-22
1 D3-16
4 D4-4/D4-11
4 D4-17
1 D4-23

No D = 571

J$_H$ segments

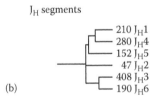

210 J$_H$1
280 J$_H$4
152 J$_H$5
47 J$_H$2
408 J$_H$3
190 J$_H$6

(b)

Figure 17.9 (Continued) Anti-BLyS antibody germ-line usage. (b) D and J$_H$. The total number of anti-BLyS antibodies (out of 1287) utilizing the different germ-line gene segments is listed together with the germ line locus name. Results are shown as dendrograms illustrating familial relationships. Solid lines indicate that the given germ line was observed. Note that for 571 antibodies in the panel, the V$_H$CDR3 domains were either too short or too novel to confidently assign a D gene segment (no D).

(Continued)

low nanomolar range. To assess the panel of antibodies for biological activity, a subset of scFvs was reformatted to IgG1 and tested for the ability to neutralize the activity of human BLyS protein in a murine splenocyte *in vitro* proliferation assay, as measured by ^{3}H-thymidine incorporation. Antibodies demonstrating the best inhibitory profile as full IgG molecules were hBLySmAb-1 (IC$_{50}$ = 0.05 nM) and hBLySmAb-2 (IC$_{50}$ = 0.08 nM). These data support findings that potent, high-affinity antibodies can be isolated directly from large nonimmune phage antibody libraries [1–3,19].

V_L segments

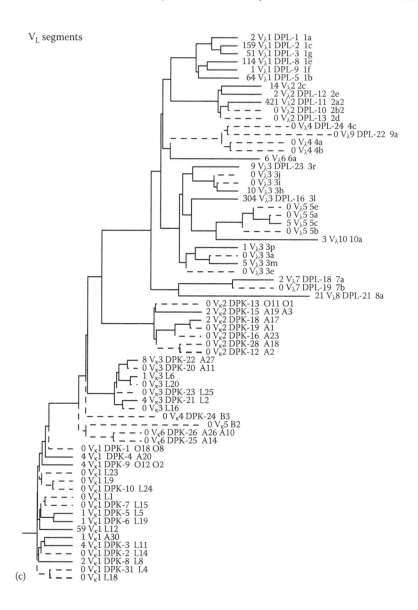

(c)

Figure 17.9 (Continued) Anti-BLyS antibody germ-line usage. (c) V_L (combined V_κ and V_λ). The total number of anti-BLyS antibodies (out of 1287) utilizing the different germ-line gene segments is listed together with the germ line locus name. Results are shown as dendrograms illustrating familial relationships. Solid lines indicate that the given germ line was observed. Note that for 571 antibodies in the panel, the V_HCDR3 domains were either too short or too novel to confidently assign a D gene segment (no D).

(Continued)

J$_L$ segments

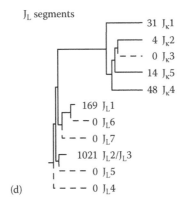

(d)

Figure 17.9 (Continued)   Anti-BLyS antibody germ-line usage. (d) J$_L$ (combined J$_\kappa$ and J$_\lambda$) germ-line family gene usage. The total number of anti-BLyS antibodies (out of 1287) utilizing the different germ-line gene segments is listed together with the germ line locus name. Results are shown as dendrograms illustrating familial relationships. Solid lines indicate that the given germ line was observed. Note that for 571 antibodies in the panel, the V$_H$CDR3 domains were either too short or too novel to confidently assign a D gene segment (no D). (Reprinted from *J. Mol. Biol.*, 334, Edwards, B.M. et al., The remarkable flexibility of the human antibody repertoire; isolation of over one thousand different antibodies to a single protein, BLyS. 103–118, Copyright 2003, with permission from Elsevier.)

17.3.2.1.3 Optimization of Lead Candidates

Optimization of the scFv corresponding to anti-BLyS antibodies hBLySmAb-1 and hBLySmAb-2 was performed to identify antibodies with enhanced inhibitory activity. This was achieved by randomizing blocks of six amino acid residues of the V$_H$CDR3 domain. Large secondary libraries were generated for hBLySmAb-1 and hBLySmAb-2. The randomized libraries were then subjected to stringency selections to isolate antibodies with improved affinities. Phage antibodies with higher affinity were enriched using successive rounds of selection by decreasing the concentration of BLyS, from 50 nM down to 100 pM. When selecting from a secondary phage library, the antigen concentration is typically reduced below the K_d of the parent clone to allow preferential selection of higher-affinity mutants [37].

From these selections, more than 30 scFvs were identified that were able to inhibit binding of biotinylated BLyS to its receptors on the surface of IM9 cells with improved inhibitory profiles. Figure 17.10 illustrates the improvements that were observed for the two most potent scFvs from each lineage, hBLySsc-1.1 and hBLySsc-2.1, compared with their respective parents.

These antibodies were reformatted as IgG1 and analyzed to confirm their specificity for the target antigen. ELISA analysis demonstrated that hBLySmAb-1 and hBLySmAb-2 were highly specific for BLyS, and no cross-reactivity was observed to a panel of other TNF ligand family members, including a proliferation-inducing ligand

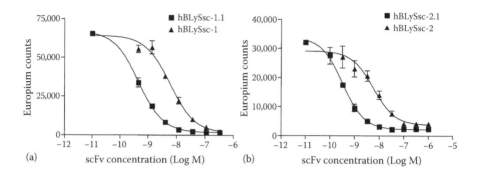

Figure 17.10   Improved potency of optimized BLyS single-chain antibodies in the receptor inhibition assay. Purified scFvs were evaluated for their ability to inhibit biotinylated BLyS binding to its receptors on IM9 cells, as measured by europium-labeled streptavidin. (a) Comparison of hBLySsc-1 and hBLySsc-1.1 (LymphoStat-B), showing that hBLySsc-1.1 results in a tenfold improvement in potency compared with the parental scFv hBLySsc-1. (b) Comparison of hBLySsc-2 and hBLySsc-2.1, showing that hBLySsc-2.1 results in a 20-fold improvement in potency compared with the parental scFv hBLySsc-2. The IC_{50} values are as follows: for hBLySsc-1, 6.3 nM; for hBLySsc-1.1, 0.5 nM; for hBLySsc-2, 5.86 nM; and for hBLySsc-2.1, 0.33 nM. A four-parameter logistic model was used for curve fitting and calculation of binding parameters. Values are the mean ± standard error of the mean (SEM) of triplicate samples. (Reprinted from Baker, K.P. et al., *Arthritis Rheum.*, 48, 2003.)

(APRIL), homologous to lymphotoxins, shows inducible expression, competes with HSV glycoprotein D for HVEM, a receptor expressed by T lymphocytes (LIGHT), TNFα, Fas ligand, TRAIL, and TNFβ, or to interleukin-4 (IL-4) or IL-18.

In addition to fine specificity for target antigen, lead antibodies must also demonstrate appropriate biological activity in key *in vitro* assays. The top candidates were assessed in the murine splenocyte proliferation assay, and hBLySmAb1.1 demonstrated the greatest inhibitory activity as IgG1. This was selected as the therapeutic candidate and taken forward into preclinical development as lead molecule LymphoStat-B, which was later renamed Benlysta.

17.3.2.1.4 Further Characterization of Anti-BLyS Antibody

Having demonstrated the activity of the antibody *in vitro*, further tests were performed *in vivo* to investigate its ability to neutralize the observable effects of exogenously administered human BLyS (0.3 mg/kg for 4 consecutive days). These effects include increases in murine spleen weight, increases in serum IgA levels, and increases in the population of mature B cells in the spleen. hBLySmAb1.1 (LymphoStat-B) at dosages of 1–5 mg/kg completely prevented these BLyS-induced activities, and no inhibition was observed by administration of an Ig isotype control monoclonal antibody (human IgG1) (Figure 17.11). These results demonstrate that hBLySmAb1.1 (LymphoStat-B) is an effective *in vivo* antagonist of human BLyS bioactivity.

The *in vivo* consequences of BLyS inhibition were evaluated in cynomolgus monkeys. This was considered a suitable animal model since cynomolgus BLyS

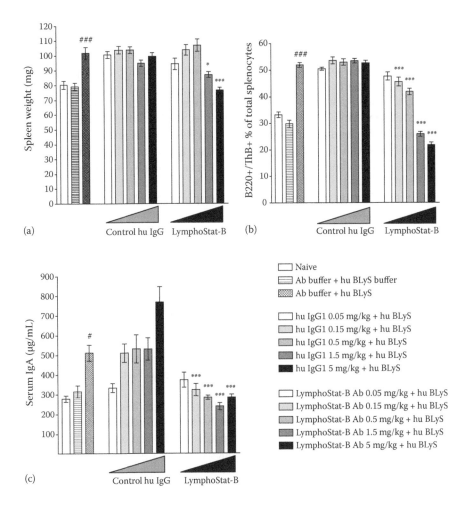

Figure 17.11 Inhibitory effects of LymphoStat-B on the responses of BALB/c mice stimulated with human B-lymphocyte stimulator (BLyS). BLyS was administered to mice over a 5-day period, with or without LymphoStat-B or control IgGl on days 1 and 3. The effects of BLyS on spleen weight, splenic B-cell representation, and serum IgA levels were determined on day 5. (a) Effect of LymphoStat-B on BLyS-induced increases in spleen weight. (b) Effect of LymphoStat-B on BLyS-induced increases in the number of mature splenic B cells (ThB+/B220+). These markers are the murine equivalents of Ly6D and CD45R, respectively. Data are reported as the mean of the B220+/ThB+ cell population, as determined by flow cytometry analysis. (c) Effect of LymphoStat-B on BLyS-induced elevations of serum IgA levels. Values are the mean ± SEM (n = 10 mice per treatment group). ### P < .0005 for recombinant human BLyS versus buffer; * P < .05 for LymphoStat-B versus the corresponding dose of human IgGl; *** P < .0005 for LymphoStat-B versus the corresponding dose of human IgGl; # P < .05 for recombinant human BLyS versus buffer. Ab, antibody; hu, human. (Reprinted from Baker, K.P. et al., *Arthritis Rheum.*, 48, 2003.)

is 96% identical to human BLyS, and hBLySmAb1.1 inhibits the *in vitro* activity of cynomolgus BLyS with similar potency as that of human BLyS. The results of the cynomolgus study demonstrated that hBLySmAb1.1 (LymphoStat-B) is able to inhibit BLyS *in vivo* and that inhibition results in depletion of B-cell populations after a relatively short course of treatment.

LymphoStat-B possesses many properties that make it ideally suited for use as a therapeutic agent. First, it binds with exquisite specificity to BLyS and does not recognize other TNF family members, including its closest homologue, human APRIL. Second, it binds with high affinity to its target antigen (human BLyS) and potently inhibits its activities both *in vitro* and *in vivo*. As an antibody, an additional advantage of LymphoStat-B is its long terminal half-life. In mice and monkeys, the half-life of LymphoStat-B is 2.5 and 11–14 days, respectively. This has advantages for therapeutic applications since the therapeutic dose of the mAb is reduced, whereas its therapeutic duration is increased.

LymphoStat-B, now known as belimumab (Benlysta), was developed by Human Genome Sciences (now GlaxoSmithKline) as a novel treatment for patients with SLE and achieved FDA approval in 2011. A remarkable observation about this drug is that the identification of the target, BLyS, was published as recently as 1998. Therefore, the timeline from target identification through antibody discovery and clinical development to drug approval was only 13 years [78] and is a compelling example of the phage display platform in drug discovery.

17.3.3 Further Applications of Naive Libraries and Future Directions

Antibodies isolated from large human scFv repertoires have a broad range of applications. Recent expansion of naive antibody repertoires has achieved a diversity of 10^{11} in some cases [92], which probably approaches the maximal size that can be achieved by current cloning methods. Besides offering a source of antibodies for therapy, phage display technology is very well suited for high-throughput generation of antibodies for research purposes, such as in the field of proteomics.

The combination of protein chips [93] and recombinant antibodies selected by phage display could enable the determination of the protein profile for any tissue or even whole organism. This could provide a useful tool for the identification of disease-specific antigens. Although the utilization of antibody libraries for target discovery has not yet been explored to its full potential, studies based on synthetic [41], naive [94], and immune repertoires have proven the concept. In a recent study, a naive phage antibody library was used to identify a novel target, the exopolysaccharide Psl, for the potential treatment of *Pseudomonas aeruginosa* infections [95]. The use of naive phage libraries in such phenotypic screening campaigns has the potential to simultaneously identify both novel targets and therapeutic lead antibodies and could be a growth area in the future. Although hybridoma technology has also been used for target discovery, phage antibody libraries can be positively and negatively selected against human cell lines, allowing specificity to become part of the selection [41]. Another promising strategy, in particular, for the discovery of targets that are expressed by the tumor vasculature, is the selection of antibody libraries *in vivo*. Johns et al. [49] utilized an *in vivo* selection process, intravenously injecting an antibody phage library into mice. ScFvs were obtained that bound specifically to thymic vascular endothelium and perivascular epithelium.

Anti-idiotypic antibodies are potentially useful as substitutes of antigens [96]. If anti-idiotypic antibodies bind to the antigen-combining site (paratope), they are displaying an "internal image" of the idiotypic antibody (Ab1). Such antibodies can potentially serve as a vaccine to induce an immune response to a pathogenic antigen, thus avoiding immunization with the pathogen itself. Goletz et al. [97] have efficiently selected anti-idiotypic antibodies from naive phage libraries by specific elution (in combination with trypsin treatment) of the idiotype bound with the soluble antigen as demonstrated with two carbohydrates and one conformational peptide epitope. Anti-idiotype antibodies are also used as reagents for the development of assays to monitor serum pharmacokinetics and assess any immune response to antibodies administered clinically [98].

Phage display is an established technology for creating human antibodies and has been successfully used to generate therapeutic antibodies. Alternative display technologies have also emerged, including ribosome display and covalent display. Libraries with a diversity of 10^{14}—significantly larger than phage display libraries—are possible using cell-free systems. By combining the technique of ribosome display with the use of purified *in vitro* translation components, scFvs have been efficiently selected from naive antibody repertoires [99]. However, ribosome display has more often been used as a tool for directed evolution studies due to the large repertoire sizes that can be accessed and the selection stringency possible in a purely monovalent system [100,101].

The widespread availability of next-generation sequencing technology, also known as massively parallel sequencing or deep sequencing, has enabled sequence data to be captured for more than a million gene templates in a single reaction. This has led to the application of this approach in conjunction with antibody repertoire mining. On the one hand, this technology can be used to assess the quality of antibody repertoires in naive libraries and, on the other, can enable the monitoring of enrichment within populations during phage display selection cycles [102]. It is anticipated that these developments will further improve both the quality of naive antibody repertoires and the efficiency with which they can be selected for a given function [103].

17.4 SUMMARY

This chapter gives an overview of the process of constructing naive phage display libraries and discusses a range of applications, focusing on the development of antibody therapeutics. The first and second phage display-derived antibodies to achieve FDA approval have been isolated from naive libraries, with adalimumab (HUMIRA) now approved for multiple indications and belimumab (Benlysta) the first new SLE drug in more than 50 years. Other applications for naive libraries are now emerging. For example, these libraries may be used for drug target identification, the development of anti-idiotypes, and the high-throughput generation of antibodies for research purposes. Phage display is clearly a powerful tool and will continue to be a major technique for the generation of antibodies in the future.

REFERENCES

1. Vaughan TJ, Williams AJ, Pritchard K, Osbourn JK, Pope AR, Earnshaw JC, McCafferty J et al. Human antibodies with sub-nanomolar affinities isolated from a large non-immunized phage display library. *Nat Biotechnol* 1996; 14:309–314.
2. Sheets MD, Amersdorfer P, Finnern R, Sargent P, Lindquist E, Schier R, Hemingsen G et al. Efficient construction of a large nonimmune phage antibody library: The production of high-affinity human single-chain antibodies to protein antigens. *Proc Natl Acad Sci U S A* 1998; 95:6157–6162.
3. de Haard HJ, van Neer N, Reurs A, Hufton SE, Roovers RC, Henderikx P, de Bruine AP et al. A large non-immunized human Fab fragment phage library that permits rapid isolation and kinetic analysis of high affinity antibodies. *J Biol Chem* 1999; 274:18218–18230.
4. Edwards BM, Barash SC, Main SH, Choi GH, Minter R, Ullrich S, Williams E et al. The remarkable flexibility of the human antibody repertoire; isolation of over one thousand different antibodies to a single protein, BLyS. *J Mol Biol* 2003; 334:103–118.
5. McCafferty J, Griffiths AD, Winter G, Chiswell DJ. Phage antibodies: Filamentous phage displaying antibody variable domains. *Nature* 1990; 348:552–554.
6. Griffiths AD, Duncan AR. Strategies for selection of antibodies by phage display. *Curr Opin Biotechnol* 1998; 9:102–108.
7. LoBuglio AF, Wheeler RH, Trang J, Haynes A, Rogers K, Harvey EB, Sun L et al. Mouse/human chimeric monoclonal antibody in man: Kinetics and immune response. *Proc Natl Acad Sci U S A* 1989; 86:4220–4224.
8. Kuus-Reichel K, Grauer LS, Karavodin LM, Knott C, Krusemeier M, Kay NE. Will immunogenicity limit the use, efficacy, and future development of therapeutic monoclonal antibodies? *Clin Diagn Lab Immunol* 1994; 1:365–372.
9. Rajewsky K. Clonal selection and learning in the antibody system. *Nature* 1996; 381:751–758.
10. Lefranc M. *The Immunoglobulin Factsbook*. San Diego, CA; London: Academic Press; 2001.
11. Winter G, Milstein C. Man-made antibodies. *Nature* 1991; 349:293–299.
12. Klein U, Rajewsky K, Kuppers R. Human immunoglobulin (Ig)M+IgD+ peripheral blood B cells expressing the CD27 cell surface antigen carry somatically mutated variable region genes: CD27 as a general marker for somatically mutated (memory) B cells. *J Exp Med* 1998; 188:1679–1689.
13. Roitt IM. *Immunology*, 4th ed. London: Mosby; 1996.
14. Klein U, Kuppers R, Rajewsky K. Evidence for a large compartment of IgM-expressing memory B cells in humans. *Blood* 1997; 89:1288–1298.
15. McHeyzer-Williams MG, Ahmed R. B cell memory and the long-lived plasma cell. *Curr Opin Immunol* 1999; 11:172–179.
16. Kelley DE, Perry RP. Transcriptional and posttranscriptional control of immunoglobulin mRNA production during B lymphocyte development. *Nucleic Acids Res* 1986; 14:5431–5447.
17. Matthes T, Kindler V, Zubler RH. Semiquantitative, nonradioactive RT-PCR detection of immunoglobulin mRNA in human B cells and plasma cells. *DNA Cell Biol* 1994; 13:429–436.
18. Marks JD, Hoogenboom HR, Bonnert TP, McCafferty J, Griffiths AD, Winter G. By-passing immunization. Human antibodies from V-gene libraries displayed on phage. *J Mol Biol* 1991; 222:581–597.

19. Sblattero D, Bradbury A. Exploiting recombination in single bacteria to make large phage antibody libraries. *Nat Biotechnol* 2000; 18:75–80.

20. Marks JD, Tristem M, Karpas A, Winter G. Oligonucleotide primers for polymerase chain reaction amplification of human immunoglobulin variable genes and design of family-specific oligonucleotide probes. *Eur J Immunol* 1991; 21:985–991.

21. Tomlinson IM, Walter G, Marks JD, Llewelyn MB, Winter G. The repertoire of human germline VH sequences reveals about fifty groups of VH segments with different hyper-variable loops. *J Mol Biol* 1992; 227:776–798.

22. Sblattero D, Bradbury A. A definitive set of oligonucleotide primers for amplifying human V regions. *Immunotechnology* 1998; 3:271–278.

23. Horton RM, Hunt HD, Ho SN, Pullen JK, Pease LR. Engineering hybrid genes without the use of restriction enzymes: Gene splicing by overlap extension. *Gene* 1989; 77:61–68.

24. Waterhouse P, Griffiths AD, Johnson KS, Winter G. Combinatorial infection and in vivo recombination: A strategy for making large phage antibody repertoires. *Nucleic Acids Res* 1993; 21:2265–2266.

25. McCafferty J, Johnson KS. *Phage Display of Peptides and Proteins: A Laboratory Manual.* San Diego, CA; London: Academic Press, 1996:79–111.

26. Griffiths AD, Williams SC, Hartley O, Tomlinson IM, Waterhouse P, Crosby WL, Kontermann RE et al. Isolation of high affinity human antibodies directly from large synthetic repertoires. *EMBO J* 1994; 13:3245–3260.

27. Glennie MJ, Johnson PW. Clinical trials of antibody therapy. *Immunol Today* 2000; 21:403–410.

28. Weinblatt ME, Keystone EC, Furst DE, Moreland LW, Weisman MH, Birbara CA, Teoh LA et al. Adalimumab, a fully human anti-tumor necrosis factor alpha monoclonal antibody, for the treatment of rheumatoid arthritis in patients taking concomitant methotrexate: The ARMADA trial. *Arthritis Rheum* 2003; 48:35–45.

29. Breedveld FC. Therapeutic monoclonal antibodies. *Lancet* 2000; 355:735–740.

30. Saleh MN, LoBuglio AF, Trail PA. *Monoclonal Antibody-Based Therapy of Cancer.* New York: Marcel Dekker; 1998:397–416.

31. Bodey B, Bodey B, Jr., Siegel SE, Kaiser HE. Genetically engineered monoclonal antibodies for direct anti-neoplastic treatment and cancer cell specific delivery of chemotherapeutic agents. *Curr Pharm Des* 2000; 6:261–276.

32. Grossbard ML, Press OW, Appelbaum FR, Bernstein ID, Nadler LM. Monoclonal antibody-based therapies of leukemia and lymphoma. *Blood* 1992; 80:863–878.

33. Xie MH, Yuan J, Adams C, Gurney A. Direct demonstration of MuSK involvement in acetylcholine receptor clustering through identification of agonist ScFv. *Nat Biotechnol* 1997; 15:768–771.

34. Dobson CL, Main S, Newton P, Chodorge M, Cadwallader K, Humphreys R, Albert V et al. Human monomeric antibody fragments to TRAIL-R1 and TRAIL-R2 that display potent in vitro agonism. *MAbs* 2009; 1:552–562.

35. Majumdar R, Railkar R, Dighe RR. Insights into differential modulation of receptor function by hinge region using novel agonistic lutropin receptor and inverse agonistic thyrotropin receptor antibodies. *FEBS Lett* 2012; 586:810–817.

36. Chodorge M, Zuger S, Stirnimann C, Briand C, Jermutus L, Grutter MG, Minter RR. A series of Fas receptor agonist antibodies that demonstrate an inverse correlation between affinity and potency. *Cell Death Differ* 2012; 19:1187–1195.

37. Hawkins RE, Russell SJ, Winter G. Selection of phage antibodies by binding affinity. Mimicking affinity maturation. *J Mol Biol* 1992; 226:889–896.

38. Duenas M, Malmborg AC, Casalvilla R, Ohlin M, Borrebaeck CA. Selection of phage displayed antibodies based on kinetic constants. *Mol Immunol* 1996; 33:279–285.

39. Schier R, Bye J, Apell G, McCall A, Adams GP, Malmqvist M, Weiner LM et al. Isolation of high-affinity monomeric human anti-c-erbB-2 single chain Fv using affinity-driven selection. *J Mol Biol* 1996; 255:28–43.

40. Siegel DL, Chang TY, Russell SL, Bunya VY. Isolation of cell surface-specific human monoclonal antibodies using phage display and magnetically-activated cell sorting: Applications in immunohematology. *J Immunol Methods* 1997; 206:73–85.

41. de Kruif J, Terstappen L, Boel E, Logtenberg T. Rapid selection of cell subpopulation-specific human monoclonal antibodies from a synthetic phage antibody library. *Proc Natl Acad Sci U S A* 1995; 92:3938–3942.

42. Osbourn JK, Derbyshire EJ, Vaughan TJ, Field AW, Johnson KS. Pathfinder selection: In situ isolation of novel antibodies. *Immunotechnology* 1998; 3:293–302.

43. Osbourn JK, Earnshaw JC, Johnson KS, Parmentier M, Timmermans V, McCafferty J. Directed selection of MIP-1 alpha neutralizing CCR5 antibodies from a phage display human antibody library. *Nat Biotechnol* 1998; 16:778–781.

44. Pasqualini R, Koivunen E, Ruoslahti E. Alpha v integrins as receptors for tumor targeting by circulating ligands. *Nat Biotechnol* 1997; 15:542–546.

45. Arap W, Kolonin MG, Trepel M, Lahdenranta J, Cardo-Vila M, Giordano RJ, Mintz PJ et al. Steps toward mapping the human vasculature by phage display. *Nat Med* 2002; 8:121–127.

46. Gargano N, Cattaneo A. Rescue of a neutralizing anti-viral antibody fragment from an intracellular polyclonal repertoire expressed in mammalian cells. *FEBS Lett* 1997; 414:537–540.

47. Visintin M, Settanni G, Maritan A, Graziosi S, Marks JD, Cattaneo A. The intracellular antibody capture technology (IACT): Towards a consensus sequence for intracellular antibodies. *J Mol Biol* 2002; 317:73–83.

48. Becerril B, Poul MA, Marks JD. Toward selection of internalizing antibodies from phage libraries. *Biochem Biophys Res Commun* 1999; 255:386–393.

49. Johns M, George AJ, Ritter MA. In vivo selection of sFv from phage display libraries. *J Immunol Methods* 2000; 239:137–151.

50. Mirzabekov T, Kontos H, Farzan M, Marasco W, Sodroski J. Paramagnetic proteoliposomes containing a pure, native, and oriented seven-transmembrane segment protein, CCR5. *Nat Biotechnol* 2000; 18:649–654.

51. Jespers LS, Roberts A, Mahler SM, Winter G, Hoogenboom HR. Guiding the selection of human antibodies from phage display repertoires to a single epitope of an antigen. *Biotechnology (N Y)* 1994; 12:899–903.

52. Wang Z, Wang Y, Li Z, Li J, Dong Z. Humanization of a mouse monoclonal antibody neutralizing TNF-alpha by guided selection. *J Immunol Methods* 2000; 241:171–184.

53. Klimka A, Matthey B, Roovers RC, Barth S, Arends JW, Engert A, Hoogenboom HR. Human anti-CD30 recombinant antibodies by guided phage antibody selection using cell panning. *Br J Cancer* 2000; 83:252–260.

54. Selvin PR. Principles and biophysical applications of lanthanide-based probes. *Annu Rev Biophys Biomol Struct* 2002; 31:275–302.

55. Barbas CF, 3rd, Burton DR. Selection and evolution of high-affinity human anti-viral antibodies. *Trends Biotechnol* 1996; 14:230–234.

56. Adams GP, Schier R. Generating improved single-chain Fv molecules for tumor targeting. *J Immunol Methods* 1999; 231:249–260.

57. Schier R, McCall A, Adams GP, Marshall KW, Merritt H, Yim M, Crawford RS et al. Isolation of picomolar affinity anti-c-erbB-2 single-chain Fv by molecular evolution of the complementarity determining regions in the center of the antibody binding site. *J Mol Biol* 1996; 263:551–567.

58. Deng SJ, MacKenzie CR, Sadowska J, Michniewicz J, Young NM, Bundle DR, Narang SA. Selection of antibody single-chain variable fragments with improved carbohydrate binding by phage display. *J Biol Chem* 1994; 269:9533–9538.

59. Marks JD, Griffiths AD, Malmqvist M, Clackson TP, Bye JM, Winter G. By-passing immunization: Building high affinity human antibodies by chain shuffling. *Biotechnology (N Y)* 1992; 10:779–783.

60. Low NM, Holliger PH, Winter G. Mimicking somatic hypermutation: Affinity maturation of antibodies displayed on bacteriophage using a bacterial mutator strain. *J Mol Biol* 1996; 260:359–368.

61. Irving RA, Kortt AA, Hudson PJ. Affinity maturation of recombinant antibodies using E. coli mutator cells. *Immunotechnology* 1996; 2:127–143.

62. Crameri A, Cwirla S, Stemmer WP. Construction and evolution of antibody-phage libraries by DNA shuffling. *Nat Med* 1996; 2:100–102.

63. Clackson T, Hoogenboom HR, Griffiths AD, Winter G. Making antibody fragments using phage display libraries. *Nature* 1991; 352:624–628.

64. Schier R, Balint RF, McCall A, Apell G, Larrick JW, Marks JD. Identification of functional and structural amino-acid residues by parsimonious mutagenesis. *Gene* 1996; 169:147–155.

65. Balint RF, Larrick JW. Antibody engineering by parsimonious mutagenesis. *Gene* 1993; 137:109–118.

66. Tomlinson IM, Walter G, Jones PT, Dear PH, Sonnhammer EL, Winter G. The imprint of somatic hypermutation on the repertoire of human germline V genes. *J Mol Biol* 1996; 256:813–817.

67. Yang WP, Green K, Pinz-Sweeney S, Briones AT, Burton DR, Barbas CF, 3rd. CDR walking mutagenesis for the affinity maturation of a potent human anti-HIV-1 antibody into the picomolar range. *J Mol Biol* 1995; 254:392–403.

68. Thompson JE, Vaughan TJ, Williams AJ, Wilton J, Johnson KS, Bacon L, Green JA et al. A fully human antibody neutralising biologically active human TGFbeta2 for use in therapy. *J Immunol Methods* 1999; 227:17–29.

69. Minter RR, Cohen ES, Wang B, Liang M, Vainshtein I, Rees G, Eghobamien L et al. Protein engineering and preclinical development of a GM-CSF receptor antibody for the treatment of rheumatoid arthritis. *Br J Pharmacol* 2013; 168:200–211.

70. Persic L, Roberts A, Wilton J, Cattaneo A, Bradbury A, Hoogenboom HR. An integrated vector system for the eukaryotic expression of antibodies or their fragments after selection from phage display libraries. *Gene* 1997; 187:9–18.

71. Chester KA, Hawkins RE. Clinical issues in antibody design. *Trends Biotechnol* 1995; 13:294–300.

72. Weir AN, Nesbitt A, Chapman AP, Popplewell AG, Antoniw P, Lawson AD. Formatting antibody fragments to mediate specific therapeutic functions. *Biochem Soc Trans* 2002; 30:512–516.

73. Vincent KJ, Zurini M. Current strategies in antibody engineering: Fc engineering and pH-dependent antigen binding, bispecific antibodies and antibody drug conjugates. *Biotechnol J* 2012; 7:1444–1450.

74. Dall'Acqua WF, Kiener PA, Wu H. Properties of human IgG1s engineered for enhanced binding to the neonatal Fc receptor (FcRn). *J Biol Chem* 2006; 281:23514–23524.

75. Horton HM, Bernett MJ, Pong E, Peipp M, Karki S, Chu SY, Richards JO et al. Potent in vitro and in vivo activity of an Fc-engineered anti-CD19 monoclonal antibody against lymphoma and leukemia. *Cancer Res* 2008; 68:8049–8057.

76. Kolbeck R, Kozhich A, Koike M, Peng L, Andersson CK, Damschroder MM, Reed JL et al. MEDI-563, a humanized anti-IL-5 receptor alpha mAb with enhanced antibody-dependent cell-mediated cytotoxicity function. *J Allergy Clin Immunol* 2010; 125:1344–1353.e2.

77. Schlaeth M, Berger S, Derer S, Klausz K, Lohse S, Dechant M, Lazar GA et al. Fc-engineered EGF-R antibodies mediate improved antibody-dependent cellular cyto-toxicity (ADCC) against KRAS-mutated tumor cells. *Cancer Sci* 2010; 101:1080–1088.

78. Stohl W, Hilbert DM. The discovery and development of belimumab: The anti-BLyS-lupus connection. *Nat Biotechnol* 2012; 30:69–77.

79. Gross JA, Johnston J, Mudri S, Enselman R, Dillon SR, Madden K, Xu W et al. TACI and BCMA are receptors for a TNF homologue implicated in B-cell autoimmune dis-ease. *Nature* 2000; 404:995–999.

80. Moore PA, Belvedere O, Orr A, Pieri K, LaFleur DW, Feng P, Soppet D et al. BLyS: Member of the tumor necrosis factor family and B lymphocyte stimulator. *Science* 1999; 285:260–263.

81. Schneider P, MacKay F, Steiner V, Hofmann K, Bodmer JL, Holler N, Ambrose C et al. BAFF, a novel ligand of the tumor necrosis factor family, stimulates B cell growth. *J Exp Med* 1999; 189:1747–1756.

82. Shu HB, Hu WH, Johnson H. TALL-1 is a novel member of the TNF family that is down-regulated by mitogens. *J Leukoc Biol* 1999; 65:680–683.

83. Mukhopadhyay A, Ni J, Zhai Y, Yu GL, Aggarwal BB. Identification and characteriza-tion of a novel cytokine, THANK, a TNF homologue that activates apoptosis, nuclear factor-kappaB, and c-Jun NH2-terminal kinase. *J Biol Chem* 1999; 274:15978–15981.

84. Tribouley C, Wallroth M, Chan V, Paliard X, Fang E, Lamson G, Pot D et al. Characterization of a new member of the TNF family expressed on antigen presenting cells. *Biol Chem* 1999; 380:1443–1447.

85. Thompson JS, Bixler SA, Qian F, Vora K, Scott ML, Cachero TG, Hession C et al. BAFF-R, a newly identified TNF receptor that specifically interacts with BAFF. *Science* 2001; 293:2108–2111.

86. Zhang J, Roschke V, Baker KP, Wang Z, Alarcon GS, Fessler BJ, Bastian H et al. Cutting edge: A role for B lymphocyte stimulator in systemic lupus erythematosus. *J Immunol* 2001; 166:6–10.

87. Cheema GS, Roschke V, Hilbert DM, Stohl W. Elevated serum B lymphocyte stimula-tor levels in patients with systemic immune-based rheumatic diseases. *Arthritis Rheum* 2001; 44:1313–1319.

88. Mariette X, Roux S, Zhang J, Bengoufa D, Lavie F, Zhou T, Kimberly R. The level of BLyS (BAFF) correlates with the titre of autoantibodies in human Sjogren's syndrome. *Ann Rheum Dis* 2003; 62:168–171.

89. Groom J, Kalled SL, Cutler AH, Olson C, Woodcock SA, Schneider P, Tschopp J et al. Association of BAFF/BLyS overexpression and altered B cell differentiation with Sjogren's syndrome. *J Clin Invest* 2002; 109:59–68.

90. Baker KP, Edwards BM, Main SH, Choi GH, Wager RE, Halpern WG, Lappin PB et al. Generation and characterization of LymphoStat-B, a human monoclonal antibody that antagonizes the bioactivities of B lymphocyte stimulator. *Arthritis Rheum* 2003; 48:3253–3265.

91. Xu JL, Davis MM. Diversity in the CDR3 region of V(H) is sufficient for most antibody specificities. *Immunity* 2000; 13:37–45.

92. Lloyd C, Lowe D, Edwards B, Welsh F, Dilks T, Hardman C, Vaughan T. Modelling the human immune response: Performance of a 10(11) human antibody repertoire against a broad panel of therapeutically relevant antigens. *Protein Eng Des Sel* 2009; 22:159–168.

93. Haab BB, Dunham MJ, Brown PO. Protein microarrays for highly parallel detection and quantitation of specific proteins and antibodies in complex solutions. *Genome Biol* 2001; 2:RESEARCH0004.

94. Ridgway JB, Ng E, Kern JA, Lee J, Brush J, Goddard A, Carter P. Identification of a human anti-CD55 single-chain Fv by subtractive panning of a phage library using tumor and nontumor cell lines. *Cancer Res* 1999; 59:2718–2723.

95. DiGiandomenico A, Warrener P, Hamilton M, Guillard S, Ravn P, Minter R, Camara MM et al. Identification of broadly protective human antibodies to Pseudomonas aeruginosa exopolysaccharide Psl by phenotypic screening. *J Exp Med* 2012; 209:1273–1287.

96. Jerne NK. Towards a network theory of the immune system. *Ann Immunol (Paris)* 1974; 125C:373–389.

97. Goletz S, Christensen PA, Kristensen P, Blohm D, Tomlinson I, Winter G, Karsten U. Selection of large diversities of antiidiotypic antibody fragments by phage display. *J Mol Biol* 2002; 315:1087–1097.

98. Tornetta M, Fisher D, O'Neil K, Geng D, Schantz A, Brigham-Burke M, Lombardo D et al. Isolation of human anti-idiotypic antibodies by phage display for clinical immune response assays. *J Immunol Methods* 2007; 328:34–44.

99. Villemagne D, Jackson R, Douthwaite JA. Highly efficient ribosome display selection by use of purified components for in vitro translation. *J Immunol Methods* 2006; 313:140–148.

100. Thom G, Cockroft A, Buchanan A, Canclotti C, Cohen E, Lowne D, Monk P et al. Probing a protein–protein interaction by in vitro evolution. *Proc Natl Acad Sci U S A* 2006; 103:7619–7624.

101. Groves M, Lane S, Douthwaite J, Lowne D, Rees DG, Edwards B, Jackson RH. Affinity maturation of phage display antibody populations using ribosome display. *J Immunol Methods* 2006; 313:129–139.

102. Ravn U, Gueneau F, Baerlocher L, Osteras M, Desmurs M, Malinge P, Magistrelli G et al. By-passing in vitro screening-next generation sequencing technologies applied to antibody display and in silico candidate selection. *Nucleic Acids Res* 2010; 38:e193.

103. 't Hoen PAC, Jirka SMG, ten Broeke BR, Schultes EA, Aguilera B, Pang KH, Heemskerk H et al. Phage display screening without repetitious selection rounds. *Anal Biochem* 2012; 421:622–631.

CHAPTER 18

Synthetic Antibody Libraries

Frederic A. Fellouse and Sachdev S. Sidhu

CONTENTS

18.1 INTRODUCTION

Phage display has proven to be an ideal technology for antibody engineering. The method has been used for affinity maturation of natural antibodies and also for antibody humanization (Chapter 15). In addition, large libraries from immunized donors (Chapter 16) or from naive natural repertoires (Chapter 17) have been used as valuable sources of antibodies with novel functions. However, while these applications use phage display for in vitro selection and optimization, they nonetheless rely on a natural source of antibody diversity. A particularly promising branch of antibody engineering research has focused on the design and application of synthetic antibody libraries, that is, libraries in which the diversity is derived from man-made sources, rather than from natural repertoires.

Libraries of this type would have numerous advantages from both theoretical and practical viewpoints. In practical terms, the use of synthetic repertoires should greatly expand the diversities accessible by phage display technology, since libraries would no longer be limited to the scope of the natural immune system. In particular, synthetic libraries would be completely naive and would not be subjected to the restrictions imposed by self-tolerance of natural repertoires. Thus, it should be relatively simple to obtain antibodies against even highly conserved proteins for which conventional hybridoma technologies are relatively ineffective. In addition, synthetic libraries permit the use of any framework of choice, which can be chosen for properties such as stability, high expression, or low immunogenicity. This is particularly valuable for the development of therapeutic antibodies, as it has become clear that the use of nonimmunogenic human frameworks is a key requisite for clinical success [1–3]. From a more fundamental point of view, the ability to precisely control the design of synthetic diversity offers unparalleled opportunities for the exploration of the fundamental principles governing the structure and function of antibodies. Not only can synthetic libraries be used to obtain antibodies with novel specificities and functions, but also, libraries can be designed to specifically address fundamental questions regarding the mechanisms of antigen recognition.

Despite the enormous potential of synthetic antibody libraries, the field developed more slowly than phage display applications that use natural repertoires. For the most part, this has been due to the fact that the development of useful synthetic repertoires is considerably more difficult than the simple exploitation of natural repertoires. In the latter case, libraries can be constructed from natural sources with only a minimal understanding of the fundamental principles of antibody structure and function, because the major challenge resides in transferring the in vivo repertoire to an in vitro display system with molecular biology techniques. In contrast, the development of synthetic repertoires amounts to building antibody diversity from scratch, and this requires not only molecular biology but also detailed and extensive knowledge of the antibody molecule.

Despite the difficulties inherent in the ab initio design of synthetic antibody repertoires, significant progress has been made in recent years. Most of the research in the field has involved four major groups: the Scripps Research Institute, the Medical Research Council (MRC), Morphosys AG, and Genentech Incorporated. In this chapter, we review the major efforts and findings from these centers and then give an update on recent progress.

18.2 SYNTHETIC ANTIBODY LIBRARIES BY RESEARCH CENTERS

18.2.1 Scripps Research Institute

Early work with synthetic antibody libraries at the Scripps Research Institute (La Jolla, CA) made use of a single antigen-binding fragment (Fab) as the framework. A simple library was constructed by replacing the DNA encoding for the third complementarity determining region of the heavy chain (CDR-H3) by a completely random stretch of 16 degenerate codons encoding for all 20 natural amino acids [4]. CDR-H3 was chosen for diversification because it lies at the center of the antigen-binding site, it is the most diverse CDR in natural repertoires, and it often makes extensive contact with antigen [5–8]. The resulting library was relatively modest in size (5×10^7 clones) and was limited in terms of both framework and CDR diversity. Nonetheless, the library was used successfully to obtain antibodies with novel binding specificities, thus demonstrating the utility of the concept.

The CDR-H3 library was used to select antibodies that bound to the small molecule fluoroscein. Many unique CDR-H3 sequences were obtained, but interestingly, many of these shared significant homology, suggesting a common mechanism of binding. Impressively, the affinities of the best antibodies (0.1 µM) approached the average K_d values for free fluoroscein observed during the secondary response of immunized mice. In a subsequent study, the concept was expanded to include different CDR-H3 lengths, and a collection of six libraries (10^8 clones each) was used to select for antibodies that coordinate metal ions [9]. Antibodies that selectively chelated specific metal ions were obtained; most of the functional antibodies were derived from libraries that encoded for long CDR-H3 loops, and the sequences were enriched for amino acids that chelate metals in natural proteins.

To further expand the diversity of the synthetic libraries, the next step involved the introduction of diversity into both CDR-H3 and the third CDR of the light chain (CDR-L3) [10]. A panel of libraries (10^8 clones each) was constructed with diversity designs of three basic types: randomization of CDR-H3 only with lengths of 5, 10, or 16 amino acids; randomization of CDR-L3 only with lengths of 8 or 10 amino acids; or simultaneous randomization of both CDR loops with all possible length combinations of the individual heavy- and light-chain libraries. This strategy allowed for the direct comparison of the different diversity designs against a set of three small molecule haptens. Selections against all three haptens were successful, and Fabs with affinities in the nanomolar range were obtained. It was found that neither the libraries with only light-chain diversity nor those with the shortest CDR-H3 sequences were successful in generating functional antibodies. Instead, all functional antibodies were derived from libraries with the longer CDR-H3 lengths, and most also contained diversity in CDR-L3. These results suggested that functional antibodies can be obtained more readily from relatively incomplete libraries that sample large regions of sequence space (i.e., simultaneous diversification of multiple, long CDR loops) rather than from more complete libraries that sample a small region of sequence space (e.g., diversification of a single, short CDR loop). The same libraries were also used to obtain antibodies that recognized DNA with high affinity [11], thus demonstrating the applicability of the synthetic antibody approach to antigens other than small molecule haptens.

Another interesting application of synthetic antibodies is the potential for generating targeted libraries that take into account known binding properties of a particular antigen. This approach was demonstrated at the Scripps Research Institute by developing antibodies against integrin adhesion receptors that recognize natural ligands predominantly through interactions with a core tripeptide

Table 18.1   Affinities of Anti-Integrin Fabs

Fab	CDR-H3[a]	Affinity (nM)[b]		
		$\alpha_v\beta_3$	$\alpha_{IIb}\beta_3$	$\alpha_v\beta_5$
4[c]	CTQGRGD**WRS**C	0.25	0.25	500
7[c]	CTYGRGD**TRN**C	0.20	0.50	NI
8[c]	CPIPRGD**WRE**C	0.20	0.35	NI
10[c]	CTWGRGD**ERN**C	0.25	0.25	NI
9[c]	CSFGRGD**IRN**C	1.6	5.0	NI
MTF-2[d]	CSFG**RTDQR**IC	130	17	
MTF-10[d]	CSFG**KGDNR**IC	500	7.0	
MTF-32[d]	CSFG**RRDER**NC	6.7	2.9	
MTF-40[d]	CSFG**RNDSR**NC	11	1.1	

[a]   Residues that were randomized in the library are shown in bold text.
[b]   Half inhibitory concentration (IC_{50}) of vitronectin binding (first four Fabs) or
      the surface-plasmon-resonance dissociation constant (K_d) are shown. NI
      stands for no inhibition detected.
[c]   From Barbas, C.F., III et al., *Proc. Natl. Acad. Sci. U S A*, 90, 1993.
[d]   Motif optimized (MTF) derived from Fab-9 by Smith, J.W. et al., *J. Biol.
      Chem.*, 269, 1994.

motif (Arg-Gly-Asp, RGD). A library was constructed in which the RGD motif was imbedded within CDR-H3 and was flanked on either side by three random codons and cysteine residues to promote the formation of a disulfide-constrained loop [12]. The library was selected for binding to integrin $\alpha_v\beta_3$, and several unique binding clones were obtained. All purified Fabs competed with the natural ligand vitronectin and exhibited subnanomolar binding affinities (Table 18.1). The Fabs also bound with high affinity to the integrin $\alpha_{IIb}\beta_3$ but did not recognize the integrin $\alpha_v\beta_5$, thus demonstrating some degree of specificity even amongst the closely related integrin family members. It was noted that small RGD-containing peptides recognize both $\alpha_v\beta_3$ and $\alpha_v\beta_5$, and thus, the specificity observed in the antibodies shows that appropriate display of the RGD motif within the CDR loop gives rise to receptor specificity. Subsequent work focused on engineering one of the selected Fabs (Fab9) to discriminate between $\alpha_v\beta_3$ and $\alpha_{IIb}\beta_3$ [13]. In this case, the RGD motif itself was subjected to randomization, and somewhat surprisingly, it was found that the motif is not absolutely required for integrin binding. Antibodies that showed moderate specificity for $\alpha_{IIb}\beta_3$ over $\alpha_v\beta_3$ were obtained, but antibodies selective for $\alpha_v\beta_3$ could not be obtained (Table 18.1). These studies showed that targeted synthetic libraries may be used to derive antibodies that exhibit affinities and specificities beyond those of natural ligands. Antibodies that selectively recognize particular integrins could have major applications as therapeutics, since various integrins have been implicated in osteoporosis, atherosclerosis, metastasis, and other pathological states [13].

18.2.2 Medical Research Council

Much pioneering work in antibody research has originated at the MRC at Cambridge in the United Kingdom. This work included the development of hybridoma technology [14] and key contributions to the development of methods for the humanization of murine antibodies [2]. Winter and colleagues [15,16] were amongst the first to display antibody fragments on phage, and this work led to the development of naive, phage-displayed libraries from natural repertoires (Chapter 17) and also to research into the combination of natural repertoires and synthetic diversity.

The first synthetic antibody library created at the MRC involved the introduction of diversity into only CDR-H3 with completely random degenerate codons that encoded for all 20 natural amino acids [17]. Two different libraries were created; one library contained a completely randomized five-residue CDR-H3, while the other contained an eight-residue CDR-H3, in which the first five residues were randomized and the last three were fixed as a sequence that commonly occurs in natural antibodies. For each library, the CDR-H3 diversity was combined with 49 germ-line V_H segments that encode most of the V_H repertoire and a single germ-line V_λ light chain, and repertoires of modest size (10^7 clones) were displayed on phage in a single-chain variable fragment (scFv) format. The libraries were used successfully to isolate antibodies that bound to two small molecule haptens with modest affinities (Table 18.2). However, the repertoires were less successful in experiments with protein antigens, as only a single antibody was obtained against human tumor necrosis factor and no antibodies were obtained against two other proteins. While these results demonstrated the feasibility of combining synthetic diversity with limited natural diversity, the libraries were not as robust as those derived from repertoires of V genes rearranged in vivo [18,19]. It was concluded that, to obtain antibodies against any antigen of interest, it was necessary to increase the diversity of antigen-binding site shapes by increasing the number of light chains and the diversity of CDR-H3 lengths [17].

The lessons learned from this initial attempt were incorporated into the design of an improved synthetic library [20]. The new library also utilized a single light chain and 50 human V_H segments, but diversity was significantly expanded by incorporating synthetic CDR-H3 sequences of all lengths between 4 to 12 residues and by increasing the overall library size by an order of magnitude (10^8 clones). The library

Table 18.2 Affinities of Antibody Fragments Isolated at the Medical Research Council

Antigen	K_d (nM)[a]
3-iodo-4-hydroxy-5-nitrophenylacetate (NIP)	700[b,d]
2-phenyl-5-oxazolone (phOx)	3000[b,d]
NIP	4.0[c,d]
Fluorescein	3.8[c,d]
F_v of monoclonal antibody NQ11	34[c,e]
F_c of monoclonal antibody NQ11	58[c,e]
Hepatocyte growth factor/scatter factor	7[c,e]

[a] The best affinity obtained for each antigen is shown.
[b] From Hoogenboom, H.R., and G. Winter, *J. Mol. Biol.*, 227, 1992.
[c] From Griffiths, A.D. et al., *EMBO J.*, 13, 1994.
[d] Determined by fluorescence quench titration.
[e] Determined by surface plasmon resonance.

was screened against a panel of 18 antigens (including haptens and both human and foreign proteins), and binding clones were obtained in all cases. Sequence analysis of antibodies against the range of antigens revealed that most of the V_H segments were used to some extent, as were all of the CDR-H3 loop lengths represented in the library. The affinities of the antibodies were not measured, but instead, both phage and purified scFvs were used as Western blotting reagents. In general, the scFv reagents were found to be as sensitive and specific as monoclonal antibodies. However, the effectiveness of phage-derived scFvs for immunological detection relied on self-aggregation, and it was surmised that multimeric avidity effects were required to compensate for the modest affinities of monomeric antibody fragments obtained from primary phage repertoires [20].

Further improvements were achieved by dramatically increasing the size of the naive repertoire using a process of combinatorial infection (described in Chapter 3) and displaying the repertoire in a Fab format [21]. Essentially, the heavy-chain repertoire of Nissim et al. [20] was combined with a diverse light-chain repertoire, which consisted of 47 light-chain segments with synthetic CDR-L3 sequences of variable length containing up to five randomized residues. A bacterial host harboring the "donor" heavy-chain repertoire was infected with the "acceptor" light-chain repertoire, and the two chains were combined by Cre-catalyzed recombination at *loxP* sites [22]. It was estimated that the recombined repertoire contained close to 6.5×10^{10} different antibodies, and importantly, the diversity was expanded both in terms of absolute numbers and by the addition of light-chain diversity.

The recombined library was highly successful in generating specific antibodies against a broad range of haptens and proteins. A total of 215 clones were sequenced, and of these, 137 were unique. A detailed analysis of the unique sequences revealed some interesting trends in the usage of V gene segments. While a range of V gene segments was used, there was a clear bias in favor of particular segments in both the heavy and light chains, as only 17 of 49 heavy-chain segments and 19 of 47 light-chain segments were used by the functional binding clones. In particular, almost half of the V_H genes used V_H segment DP-47, which is also most common in natural antibodies. A single V_κ segment (DPK-15) and a single V_λ segment (DPL-3) accounted for the majority of the light chains, but in contrast with the heavy chain, these segments are not common in natural antibodies. Overall, these results suggested that particular V genes (e.g., DP-47) are better able to generate functional antigen-binding sites that can recognize a diverse range of antigens. Thus, while it may be advantageous to include multiple V genes in synthetic libraries, it is likely that a small subset of the natural V genes can accommodate most, if not all, of the diversity required of a robust antibody library.

The binding affinities were determined for several Fab proteins raised against two small molecule haptens, fluorescein and 3-iodo-4-hydroxy-5-nitrophenyl-acetate (NIP). The best affinities for soluble hapten were found to be in the low nanomolar range (Table 18.2). Binding was also assayed for a limited number of Fabs isolated against two protein antigens (monoclonal antibody NQ11 and hepatocyte growth factor/scatter factor), and again, the best affinities were found to be in the low nanomolar range (Table 18.2). These results demonstrated that large synthetic repertoires may be used directly to produce high-affinity antibodies against both haptens and large proteins.

The concepts explored by Winter and colleagues have also been extended by other groups. de Kruif and colleagues [23] constructed a library that contained the same 49 germ-line V_H genes that Nissim et al. [20] used and seven light chains. Synthetic diversity was introduced into CDR-H3 by inserting randomized loops ranging from 6 to 15 residues in length. The central regions of these loops were completely randomized, but variability at the borders was restricted to only those sequences that are commonly observed in natural antibodies, as it was reasoned that the functional diversity of the library would be improved by favoring sequences that are abundant in natural repertoires. Specific binding clones were isolated for 13 different antigens, but, perhaps because the library was only of moderate size (3.6×10^8 clones), binding affinities were rather modest (2 μM to 100 nM).

Pini and colleagues [24] have used principles of protein design to further focus diversity at regions most likely to be involved in functional interactions with antigen. A library was constructed using a single V_H (DP-47) and a single V_κ (DPK-22) that dominate the natural repertoire [25]. Four positions each in CDR-H3 and CDR-L3 were chosen as sites for the introduction of diversity (Figure 18.1), as residues at these positions commonly make contacts with antigen [6]. The eight sites were completely randomized to produce a naive library of modest size (3×10^8 clones),

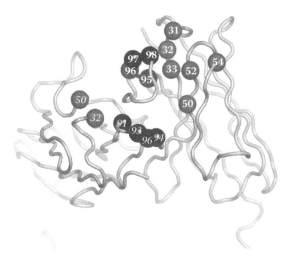

Figure 18.1 The library design of Pini et al. [24]. The main chains of an antibody variable domain are shown as gray ribbons. The Cα atoms of CDR residues included in the libraries are shown as spheres colored black (naive library) or gray (affinity maturation libraries). Numbering follows the Kabat nomenclature [49]. Heavy- or light-chain residues are numbered in bold text or italics, respectively. Structures were drawn with PyMOL (DeLano Scientific, San Carlos, CA). (From Pini, A. et al., *J. Biol. Chem.*, 273, 1998; Kabat, E. A. et al., *Sequences of Proteins of Immunological Interest*, 5th ed. National Institutes of Health, Bethesda, MD, 1991.)

and specific binding clones were isolated against a panel of six protein antigens. Antibodies isolated against the extra domain B (ED-B) domain of fibronectin were analyzed in detail, and affinities were found to be in the double-digit nanomolar range. To demonstrate the advantages of the simplified library design, one of these antibodies was subjected to affinity maturation by randomizing heavy-chain sites located on the periphery of the antigen-binding site (Figure 18.1). The new library yielded an antibody with 27-fold improved affinity (K_d = 1.5 nM). This improved antibody was used for a further round of affinity maturation in which two sites in the light chain were targeted for randomization, and ultimately, an ultra-high-affinity antibody was obtained (K_d = 54 pM). In an interesting extension of this approach, a similar library was constructed using an scFv framework that is stable in the reducing environment of the cytoplasm [26]. The library yielded specific antibodies against a variety of antigens, and importantly, the thermodynamic stabilities of these antibodies under reducing conditions were similar to that of the parent antibody. Taken together, these studies demonstrate the power of synthetic antibody libraries constructed on the basis of rational design principles both for the choice of scaffold and for sites of diversification.

A further extension of the utility of synthetic antibody libraries has been to engineer simultaneously both antigen-binding activity and the global structure of the antibody fragment. While the antigen-binding sites of most natural antibodies consist of a heterodimer of heavy and light chains, certain camelid antibodies are devoid of the light chain and, instead, contain V_H domains that fold and function autonomously [27–30]. It has been shown that the autonomous nature of camelid V_H domains is, at least in part, due to substitutions in the region of the V_H domain that formerly made contact with the light chain [31]. These substitutions are believed to increase the hydrophilicity of the former light-chain interface and thus reduce aggregation. Attempts to produce autonomous human V_H domains by incorporating camelid substitutions into the framework have been moderately successful, but the substitutions tend to cause thermodynamic destabilization and do not completely eliminate aggregation [32–35]. In addition, the introduction of camelid substitutions into human antibodies increases the risk of immunogenicity in vivo. In order to produce autonomous human V_H domains, Jespers and colleagues [36] have applied a synthetic library strategy. A library was generated by randomizing residues in all three heavy chains CDRs of a human V_H domain that was displayed on phage without a light chain. The large library (1.1 × 10^{10} clones) was screened for autonomous V_H domains that bound to hen egg-white lysozyme (HEL). Crystallographic analysis of one such autonomous V_H domain (HEL4) revealed that Trp47 at the former light-chain interface tucks into a hydrophobic cavity formed by Gly35, Val37, and Ala93 (Figure 18.2), and this significantly increases the hydrophilicity of the exposed interface. Gly35 resides at a position within CDR-H1 that was randomized in the library, and the lack of a side chain at this site was critical in forming a cavity for Trp47. Thus, a functional, autonomous human V_H domain was derived without resorting to camelid framework substitutions, and these results demonstrate how synthetic libraries can be used not only to engineer novel binding specificities but also to alter basic aspects of the immunoglobulin fold.

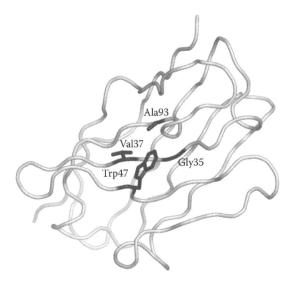

Figure 18.2   Crystal structure of the autonomous V$_H$ domain HEL4 (protein databank identification code [PDB ID] code 1OHQ). The hydrophobicity of the former light-
chain interface is reduced significantly by burial of the Trp47 side chain into a
cavity formed by Gly35, Val37, and Ala93. The main chain Cα trace is shown
as a ribbon, and selected side chains are shown as sticks. Residues are numbered according to the Kabat nomenclature [49], and labeled residues are colored black. (From Pini, A. et al., *J. Biol. Chem.*, 273, 1998; Kabat, E.A. et al.,
Sequences of Proteins of Immunological Interest, 5th ed. National Institutes of
Health, Bethesda, MD, 1991.)

18.2.3 Morphosys

Research at the biotechnology company Morphosys (Martinsried, Germany)
stems from the work of Knappik and colleagues [37] at the University of Zurich and
has focused on developing synthetic antibody libraries based on a structural analysis
of the human antibody repertoire. Antibody frameworks were chosen on the basis of
structural stability and high expression, and the genes were synthesized chemically.
High expression was ensured by the use of common *Escherichia coli* codons, and
subsequent affinity maturation was facilitated by the strategic placement of unique
restriction sites. Sequence and structural analysis of natural antibody databases was
used to identify a limited number of frameworks that would efficiently cover the full
range of CDR diversity in the human immune repertoire. In total, seven genes each
were chosen to represent heavy- and light-chain diversity, and these provided 49 combinations of light and heavy chains that were displayed on phage in an scFv format.

The first Morphosys human combinatorial antibody library (termed "HuCAL1")
was constructed by randomizing both CDR-H3 and CDR-L3 at the center of the
antigen-binding site [37]. To maximize the probability of obtaining functional
antibodies with CDR sequences resembling those of natural antibodies, sequences
from human antibodies were analyzed to obtain an amino acid distribution at each

position to be randomized. In the case of the light chain, V_κ and V_λ sequences were analyzed separately, while all V_H sequences were grouped together (Figure 18.3). Oligonucleotides were then synthesized to mimic the natural amino acid distribution, using a trinucleotide synthesis method that allows for precise tailoring of diversity at each position [38]. For the V_κ CDR-3, a single length that occurs most often in nature was used; three positions that point into the antigen-binding site were highly randomized, and four others were mildly randomized. For the V_λ CDR-3, three different lengths were used; one position was mildly randomized, and four, five, or six positions were highly randomized. As the V_H CDR-3 varies greatly in length and sequence, the library was designed to contain all possible loop lengths between 5 and 28 residues, and except for positions near the C-terminal base of the loop that were mildly randomized, all positions were highly randomized to contain the overall amino acid composition of natural CDR-H3 sequences. The HuCAL1 was a relatively large library (2.1×10^9 clones), and DNA sequencing of the naive repertoire

Amino acid	Frequency (%)												
	V_H sites			V_κ sites							V_λ sites		
	95–100s	100z	102	89	91	92	93	94	95	96	91	93–95b	96
A	3	3			2			7		3		3	2
C	1												
D	4	3		3	16	20		4		6		3	5
E	2	3			1			1		1		2	4
F	5	44	9	5	9			9		12		7	9
G	11	3		2		10	15	2		3		4	5
H	5		7	3	3			5		5		7	4
I	5	3	5		4			16		9		7	7
K	3				1			1		1		4	2
L	7	7	6	9	2			10		8		8	7
M	3	14		4	2			1		2		4	2
N	5		3	4	19	27		6		5		7	6
P	5	6	2					4	79	7		8	5
Q	3	2		80	1			5		1		3	4
R	4				1			6		3	5	5	7
S	6	7	4			27	37	4	21	9	21	8	8
T	5	2		3				4		8	73	6	6
V	4	3	26	3				4		5		6	7
W	2				1			1		1			2
Y	20	2	37	56	28			11		11		8	8

Figure 18.3 Composition of the CDR-3 sequences of HuCAL1. The frequency of each of the 20 natural amino acids at each site that was randomized in the heavy chains (V_H) and the light chains (V_κ and V_λ) was determined by sequencing 257 clones. Numbering follows the Kabat nomenclature [49]; for CDR-H3 the position preceding position 101 is numbered 100z, and the length variable region is numbered 95–100s. (From Pini, A. et al., *J. Biol. Chem.*, 273, 1998; Kabat, E.A. et al., *Sequences of Proteins of Immunological Interest*, 5th ed. National Institutes of Health, Bethesda, MD, 1991.)

revealed that diversity at each of the designed CDR sites closely mimicked that of the natural immune repertoire.

In the original report, the HuCAL1 was screened against a variety of antigens, but results were given for only a few examples [37]. Affinities for peptide antigens were found to be in the micromolar range, whereas affinities for protein antigens were in the nanomolar range (Table 18.3). Extremely high affinities in the picomolar range were achieved against bovine insulin, but the selections involved a ribosome display process during which point mutations accumulated and improved affinity up to 40-fold compared to the naive progenitor [37]. Subsequent reports have focused on exploiting the modular design of the library to facilitate the high-throughput production of antibodies [39,40]. Automated panning and screening systems that facilitated the analysis of multiple antigens were developed, and in addition, library screening was performed directly on mammalian cells expressing surface antigens. The modularity of the library was exploited to reformat the scFvs into different forms, including Fabs and full-length immunoglobulins. Overall, the library was found to be extremely robust, as the success rate was high, provided the antigen was of high quality. The isolated antibodies were used for a variety of applications, including flow cytometry, immunoprecipitation, Western blotting, and immunohistochemistry; affinities were determined for scFvs against three protein antigens and were found to be in the low to high nanomolar range (Table 18.3).

The major impetus for the HuCAL approach was to produce synthetic antibody libraries that closely resembled the human repertoire, so potential therapeutic

Table 18.3   Affinities of Antibody Fragments Isolated from the Morphosys HuCAL

Antigen	K_d (nM)[a]
Intracellular adhesion molecule-1	9.4[b]
Cluster of differentiation 11b (CD11b)	1.0[b]
Epidermal growth factor receptor	246[b]
Mac1 peptide	1130[b]
Hag peptide	610[b]
Nuclear factor kappa-light-chain-enhancer of activated B cells (NFκB) peptide	1600[b]
Bovine insulin	0.082[c]
Integrin Mac-1	1.7[d]
Fibroblast growth factor receptor-3	0.7[e]
Tissue inhibitor of metalloproteinase-1	13[f]
Citrate transporter CitS	4[g]

[a] Affinities were determined by surface plasmon resonance, and the best affinity for each antigen is shown.
[b] From Knappik, A. et al., *J. Mol. Biol.*, 296, 2000.
[c] From Hanes, J. et al., *Nature Biotechnol.*, 18, 2000, obtained by ribosome display and random mutagenesis.
[d] From Krebs, B. et al., *J. Immunol. Methods*, 254, 2001.
[e] From Rauchenberger, R. et al., *J. Biol. Chem.*, 278, 2003.
[f] From Parsons, C.J. et al., *Hepatology*, 40, 2004.
[g] From Rothlisberger, D. et al., *FEBS Lett.*, 564, 2004.

antibodies could be obtained rapidly [37]. Several reports have described efforts in this direction. The HuCAL1 was used to obtain scFvs that recognize human leukocyte antigen-DR (HLA-DR), as a first step in developing a therapeutic that induces programmed death of lymphoma/leukemia cells expressing the antigen [41]. Twelve unique clones were obtained, and one of these was subjected to an affinity maturation process that demonstrated the power of the modular HuCAL design, which allowed for the rapid generation of secondary libraries by using the strategically placed restriction sites to introduce additional CDR diversity. A clone scFv-B8 was converted to a Fab format, CDR-L3 was rerandomized, and the rather modest affinity of the parental clone (K_d = 350 nM) was improved by approximately sixfold. In a second round of optimization, additional diversity was introduced into CDR-L1, several clones with affinities in the low nanomolar range were isolated, and these were converted into IgG_4 antibodies with subnanomolar affinities (Table 18.4). The antibodies exhibited potent tumoricidal activity, both in vitro in assays with lymphoma and leukemia cell lines and in vivo in xenograft models of non-Hodgkin's lymphoma. A Fab version of the HuCAL1 library (HuCAL-Fab1) has also been constructed and was used to obtain potent neutralizing antibodies against human fibroblast growth factor receptor-3 (Table 18.3), a potential therapeutic antibody target that had previously proven recalcitrant to the generation of blocking antibodies [42]. The HuCAL-Fab1 has also been used to obtain neutralizing antibodies against tissue inhibitor of metalloproteinase-1 (Table 18.3), and these antibodies have been shown to be efficacious in attenuating liver fibrosis in an animal model, by relieving the inhibition of metalloproteinases that are involved in the degradation of extracellular matrix within injured liver tissue [43].

Aside from the development of therapeutic leads, Rothlisberger and colleagues [44] have also applied the HuCAL concept to novel applications. The determination of crystal structures for membrane proteins remains a daunting challenge for structural biology, and it is believed that the use of antibody fragments as crystallization chaperones may greatly aid these efforts. As a proof of concept, a Fab library in which all six CDRs were randomized according to the HuCAL strategy was used to derive Fabs against the detergent-solubilized citrate transporter CitS of *Klebsiella pneumoniae*. To ensure that the Fabs would be highly expressed and stable in detergent, the library

Table 18.4 Affinity Optimization of Anti-HLA-DR Antibodies

Antibody	Format	Optimization[a]	K_d (nM)[b]
B8	Fab	Parent	350
7BA	Fab	L3	60
305D3	Fab	L3 + L1	3.0
1C7277	Fab	L3 + L1	3.0
1D09C3	Fab	L3 + L1	3.0
305D3	IgG_4	L3 + L1	0.5
1C7277	IgG_4	L3 + L1	0.6
1D09C3	IgG_4	L3 + L1	0.3

[a] Indicates additional light-chain CDRs that were randomized.
[b] Determined by surface plasmon resonance.

was restricted to a subset of the HuCAL that contained only the most stable combination of light- and heavy-chain frameworks [45,46]. A Fab with low nanomolar affinity for CitS was obtained (Table 18.3), and although crystallization was not reported, it was shown that the Fab:CitS complex could be detected and purified by gel filtration chromatography. The HuCAL1 was also used to isolate catalytic antibodies with phosphatase activity, using a turnover-based in vitro selection that relied on a substrate trap for covalent capture of the catalytic antibody [47]. The naive repertoire yielded an scFv that exhibited a catalytic proficiency that was approximately 100-fold greater than that of the best phosphatase-like antibody obtained by hybridoma methods [48], and activity was improved a further tenfold by random mutagenesis and directed evolution of the parental clone. These results demonstrated that large synthetic repertoires and controlled in vitro selections can yield catalytic antibodies with activities far greater than those obtained from natural repertoires by immunization.

18.2.4 Genentech

Synthetic antibody libraries at Genentech (South San Francisco, CA) have been developed with a single human framework derived from the consensus sequences of the most abundant human subclasses, namely, V_H subgroup III and V_κ subgroup I [49]. This framework was originally developed for the humanization of murine antibodies, and in fact, it has been used in several antibodies that are successful therapeutics [50–53]. The anti-ErbB2 antibody humanized 4D5 [50] was chosen as the library scaffold, because Fab4D5 is well expressed in *E. coli* and had been displayed previously on phage [54], and in addition, the high-resolution crystal structure of the variable fragment has been determined [55].

In the first libraries, the antibody was displayed as a monomeric scFv, because the scFv format usually results in higher levels of display relative to the Fab format [56]. A bivalent scFv display format was also developed by inserting a dimerization domain between the scFv and the phage coat protein. Bivalent binding produces an avidity effect that increases apparent binding affinities for immobilized antigens [57,58], and it was reasoned that this would improve the recovery of rare or low-affinity clones from naive libraries.

To simplify library design and construction, only the heavy chain was subjected to randomization, and structural analysis was used to identify sites that would be suitable for supporting diversity (Figure 18.4). Within CDR-H1 and CDR-H2, solvent-exposed sites that were positioned for potential contact with antigen were chosen, and buried residues were excluded to minimize structural perturbations. Within the hypervariable CDR-H3, a continuous stretch of residues within the central region of the loop was chosen. For each site within CDR-H1 and CDR-H2, tailored degenerate codons were designed to bias the diversity in favor of amino acid types commonly found in the human repertoire (Figure 18.5), as it was reasoned that this would produce "humanlike" CDR sequences. Because there is little position-specific bias within natural CDR-H3 sequences, the entire loop was randomized with a degenerate codon that encoded for 12 amino acids but excluded most aliphatic side chains that may give rise to nonspecific hydrophobic interactions.

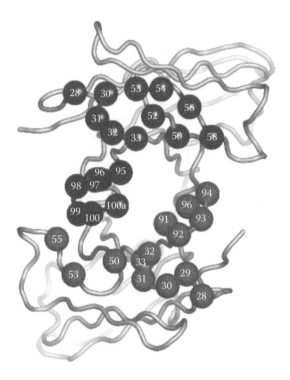

Figure 18.4 CDR residues chosen for diversification in the Genentech synthetic antibody libraries. The main chains of the humanized 4D5 V_H and V_L domains are shown as gray tubes. The Cα atoms of CDR residues included in the libraries are shown as spheres colored black (V_H) or gray (V_L). Numbering follows the Kabat nomenclature [49]. (From Pini, A. et al., *J. Biol. Chem.*, 273, 1998; Kabat, E.A. et al., *Sequences of Proteins of Immunological Interest*, 5th ed. National Institutes of Health, Bethesda, MD, 1991.)

Highly optimized library construction methods [59] were used to generate very large libraries (5×10^{10} clones) in both the monovalent and bivalent format. The libraries were highly successful against a panel of six protein antigens, in that numerous binding clones were obtained in each case. In particular, the bivalent format yielded dozens or even hundreds of unique clones against each antigen. Analysis of 86 unique antimurine vascular endothelial growth factor (mVEGF) clones revealed a broad range of affinities with the tightest binders in the 100 nM range. Interestingly, the bivalent format allowed for the selection of numerous clones, while the monovalent format selected only the few clones with the highest affinities. Thus, the two formats were shown to be complementary; the monovalent format is well suited for stringent affinity selections, while the bivalent format is useful for the recovery of rare clones against difficult antigens.

In a follow-up study, many different CDR designs were explored in a bivalent Fab format [60]. It was found that library performance was improved significantly by further reducing the diversity in CDR-H1 and CDR-H2 and concurrently increasing the

CDR-H1

Site	28	30	31	32	33
Codon	*AVT*	*RVM*	*RVM*	*WMY*	*KVK*
Diversity	Asn	Ala	Ala	Asn	Ala
	Ser	Arg	Arg	Ser	Asp
	Thr	Asn	Asn	Thr	Cys
		Asp	Asp	Tyr	Glu
		Glu	Glu		Gly
		Gly	Gly		Ser
		Lys	Lys		Trp
		Ser	Ser		Tyr
		Thr	Thr		

CDR-H2

Site	50	52	54	55	57	59
Codon	*KDK*	*DMT*	*NMY*	*DMK*	*DMK*	*DMT*
Diversity	Asp	Ala	Ala	Ala	Ala	Ala
	Cys	Asn	Asn	Asn	Asn	Asn
	Glu	Asp	Asp	Asp	Asp	Asp
	Gly	Ser	His	Glu	Glu	Ser
	Phe	Thr	Pro	Lys	Lys	Thr
	Leu	Tyr	Ser	Ser	Ser	Tyr
	Trp		Thr	Thr	Thr	
	Tyr		Tyr	Tyr	Tyr	
	Val					

Figure 18.5 Diversity of CDR-H1 and CDR-H2 in the Genentech synthetic libraries. For each site chosen for diversification, a tailored degenerate codon (italics) was designed to encode amino acid diversity similar to that found in natural antibodies. Numbering follows the Kabat nomenclature [49]. Equimolar DNA degeneracies are represented in accordance with the International Union of Biochemistry (IUB) nomenclature (D = A/G/T, K = G/T, M = A/C, N = A/C/G/T, R = A/G, S = G/C, V = A/C/G, W = A/T, Y = C/T). (From Pini, A. et al., *J. Biol. Chem.*, 273, 1998; Kabat, E.A. et al., *Sequences of Proteins of Immunological Interest*, 5th ed. National Institutes of Health, Bethesda, MD, 1991.)

diversity in CDR-H3. Ultimately, a highly robust library was constructed in which the CDR-H1 and CDR-H2 diversity was highly restricted to only those amino acids that are most common in natural antibodies, and the CDR-H3 diversity was greatly expanded to include combinations of all 20 natural amino acids in loop lengths of 7 to 19 residues. This library was screened against a panel of 13 antigens, and binding clones were found in all cases; for 9 of the antigens, Fabs with affinities in the low nanomolar range were obtained (Table 18.5). As only the heavy chain had been randomized in the naive library, it was shown that light-chain diversity could be exploited for subsequent affinity maturation. A high-affinity anti-mVEGF heavy chain (K_d = 0.9 nM) was combined with a light-chain library designed along principles similar to those used for the heavy-chain design (Figure 18.4). Many light-chain sequences were found to improve affinity, and the best Fab bound to mVEGF with an affinity in the low picomolar range (K_d = 20 pM). Taken together, these results demonstrated that a single-scaffold library with diversity restricted to the heavy chain

Table 18.5   Affinities of Antibody Fragments Isolated
              at Genentech

Antigen[a]	IC_{50} (nM)[b]
Vascular endothelial growth factor	0.6
Natural killer cell group 2D (NKG-2D) extracellular domain	3
April	1.4
Immunoglobulin E	9.8
Tissue factor	11

[a] All antigens were of murine origin.
[b] The best affinity for each antigen is shown and was deter-
     mined by competitive enzyme-linked immunosorbent
     assay (ELISA).

can yield high-affinity binders to most protein antigens, provided that the diversity is
tailored to favor functional antibody sequences. Furthermore, the light chain can be
recruited in a second step to obtain ultra-high-affinity synthetic antibodies.

The work described explores the capacity of synthetic antibody libraries to recapitulate natural antibody repertoires. The combined results of many groups demonstrate that synthetic naive libraries have been optimized to the point that high-affinity
antibodies can be generated against virtually any antigen of interest. Furthermore,
it has been shown that tailored design elements can be incorporated into synthetic
antibody libraries much more readily than into natural antibody libraries. As we
discuss in Section 18.3, this design versatility has been exploited in three major areas
that extend the impact of synthetic antibodies beyond the scope of natural antibodies.

18.3 BEYOND NATURAL ANTIBODY MIMICS

18.3.1 Tailored Libraries for Therapeutic Applications

The latest synthetic antibody repertoires have been optimized for downstream
therapeutic applications. Indeed, careful consideration of the framework sequences,
codon usage, and chemical diversity to be included in repertoires facilitates or
reduces the postselection protein engineering steps that are usually required for the
generation of therapeutic-grade antibodies.

For example, Ge et al. [61] reported the reduction of spurious mutations during
the library-making process with a method based on the ligation of long oligonucleotides and subsequent gap filling with a DNA polymerase. Also, the latest antibody
repertoires from Morphosys have been optimized for downstream therapeutic applications by using optimal frameworks, optimizing codons, and a novel trinucleotide-
based gene synthesis method [62,63].

Approaches have also been reported to introduce diversity in atypical regions
of the antibody, such as the constant heavy chain domain 2 (CH2) [64] or constant
heavy chain domain 3 (CH3) [65] IgG domains. Xiao et al. [64] diversified the BC
and FG loops of the CH2 IgG domains with a tetranomial combination of amino

acids [66]. From this library, they selected binders against the HIV-1 envelope gly-coprotein gp120, which showed HIV-1 infectivity neutralizing activity. Wozniak-Knopp et al. [65] generated bivalent antihuman epidermal growth factor receptor 2 (HER2) IgG by randomizing the AB and EF loops of the CH3 domain. Notably, despite the mutations in the CH3 domain, the antibody was still recognized by Fc receptors, and the vivo half life was not altered significantly.

For engineering bispecific antigen-binding sites, the diversification of nonoverlap-ping groups of residues in two distinct libraries enabled the Dutalys group to generate bispecific molecules by combining the sequences of the selectants obtained from each library [67]. Libraries based on a previously engineered functional antibody can also be used to generate antibody variable regions that can recognize two different antigens as shown by Genentech [68,69]. Bispecific antibodies generated by these and other methods hold considerable promise for the development of new therapeutics [70].

18.3.2 Exploring Molecular Recognition

Although natural antigen-binding sites are highly diverse, there are clear biases for particular amino acids [71–77]. Tyrosine and serine are particularly abundant, and tyrosine residues contribute a disproportionate number of antigen contacts [6,78,79]. In an effort to determine the minimal requirements for antigen recognition, a heavy-chain library was constructed by allowing random combinations of only four amino acids (tyrosine, aspartate, alanine, and serine) [66]. Despite the limited chemical diversity, specific antibodies with affinities in the micromolar range were obtained against vascular endothelial growth factor (VEGF), an angiogenic hormone that has been implicated in tumorigenesis [80,81]. The selected heavy chains were combined with a light-chain library, and Fabs with affinities in the low nanomolar range were obtained. Crystallographic analysis of two Fabs in complex with VEGF revealed that antigen recognition was mediated primarily by tyrosine residues, which comprised 71% of the structural epitope, and in contrast, aspartate was almost entirely excluded from the binding sites (Figure 18.6).

These results led to further restriction of the chemical diversity to a binary code containing only tyrosine and serine [82]. A degenerate codon that encodes for tyro-sine and serine was used to randomize solvent-accessible sites in CDR-H1 and CDR-H2, and random loops of variable lengths were inserted in place of CDR-H3. Two large libraries (10^{10} clones), which differed only in that one contained a random-ized CDR-L3, were constructed and screened against a panel of six protein antigens. Despite the extreme restrictions on chemical diversity, specific binding clones were obtained against each antigen. Detailed functional analysis of anti-VEGF antibod-ies revealed that the Fabs bound with surprisingly high affinity ($K_d = 60$ nM) and were as specific as natural antibodies in cell-based assays. Structural analysis of a binary Fab raised against human death receptor (hDR5), a cell-surface receptor that mediates apoptotic cell death [83], revealed the structural basis for the minimalist molecular recognition. While the Fab interacts with hDR5 in a normal manner, the unusually long CDR-H3 protrudes from the framework and makes extensive contact with antigen (Figure 18.7a). Interestingly, the CDR-H3 loops contains a "biphasic"

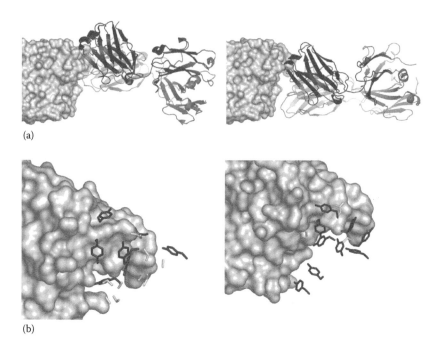

(a)

(b)

Figure 18.6   Anti-hVEGF antibodies obtained from a four-amino-acid code. (a) The complex
of hVEGF with Fab-YADS1 (left panel, PDB ID code 1TZH) and Fab-YADS2 (right
panel, PDB ID code 1TZI). The antigen is depicted as a white molecular surface. The main chains of the Fabs are shown as ribbons; the heavy and light
chains are colored black and gray, respectively. (b) The CDR side chains of
Fab-YADS1 (left panel) and Fab-YADS2 (right panel) that contact hVEGF. The
Fab side chains are shown as sticks colored black (tyrosine) or gray (all others).

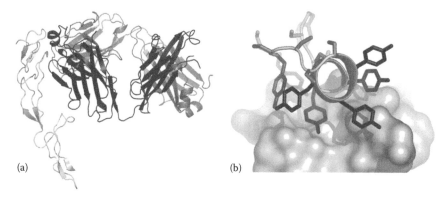

(a)   (b)

Figure 18.7   An anti-hDR5 antibody obtained from a binary code. (a) The complex of hDR5
with Fab-YSd1 (PDB 1ZA3). The antigen is depicted as a white molecular surface. The main chains of the Fab are shown as ribbons; the heavy and light
chains are colored black and gray, respectively. The CDR-H3 loop is circled.
(b) Contacts between CDR-H3 and hDR5. The CDR-H3 main chain is depicted
as a ribbon, and side chains are rendered as sticks.

helix with tyrosine and serine clustered on opposite faces (Figure 18.7b). The tyrosine face is buried against the surface of hDR5 and contributes all of the CDR-H3 accessible surface buried upon antigen binding, and serine residues likely play structural roles. Overall, these results demonstrate that functional antibodies can be created with even the simplest chemical diversity, and in fact, it is advantageous to restrict diversity to small subsets of functional groups that are particularly well suited for mediating molecular recognition.

Libraries with limited chemical diversity can be used as a benchmark to evaluate the individual propensity of different amino acid types to mediate binding affinity and specificity. Birtalan et al. [84] explored the effect of adding arginine or glycine into the tyrosine/serine binary libraries. They found that the addition of arginine was correlated with increased nonspecific binding, while the addition of glycine improved library performance. In another study, Birtalan et al. [85] used the tyrosine/serine binary diversity as a benchmark to explore the functionality of other binary amino acid combinations and to examine the functional effects of adding chemical diversity. The knowledge gained from these limited diversity libraries has been applied to the construction of optimized naive libraries for targeting a broad range of antigens [86,87] and also for tailored libraries optimized for targeting particular antigen types, as described in Section 18.3.3.

18.3.3 Tailored Antibody Libraries for Targeting Particular Antigen Types

Particular antibody features have been associated with the recognition of distinct types of antigens [88]. Some of these features have been incorporated into synthetic antibody library designs to target particular antigen types.

RNA-binding antibodies were found to have amino acid compositions and light-chain contributions that differed from those of protein-binding antibodies selected from the same synthetic libraries [89]. Based on these observations, Sherman et al. [90] designed a synthetic antibody library in which all six CDRs were diversified and arginine, which can interact favorably with the negatively charged RNA backbone, was added to the tyrosine/serine binary diversity [82]. Similarly, libraries were designed with CDR-H3 sequences biased in favor of short, charged loops that are typically found in carbohydrate-binding antibodies [91].

Peptide-binding antibodies are particularly challenging to obtain from in vitro selection methods, due to the entropy loss upon binding of a flexible peptide. Antibody libraries have been specifically designed to tackle this issue, as peptide-binding and especially phospho-specific antibodies are highly desirable reagents. Cobaugh et al. [92] reported peptide-specific antibodies derived from synthetic antibody libraries in which the heavy-chain CDRs were diversified in either a mouse antibody with CDR loop lengths and canonical structures often found in peptide-binding natural antibodies or a human antibody with a CDR-L1 sequence derived from a peptide-binding antibody. Koerber et al. [93] obtained several phospho-specific antibodies from synthetic antibody libraries with CDR-H2 loops designed to act as acceptors of anionic phosphorylated side chains.

Persson et al. [94] diversified cavity-lining residues of a hapten-binding antibody to generate a synthetic library that yielded specific antibodies targeting small molecules. The frequency of usage of the residues in natural hapten-binding antibodies [88,95] as well as structural integrity were taken into consideration in the design of the library. Also, the CDR-L3 and CDR-H2 loops included extensive chemical and length diversity, respectively.

18.4 CONCLUSIONS

While progress in synthetic antibody research was originally slow in comparison with that of natural phage-displayed repertoires, significant advances have been made. In fact, libraries developed over the last few years have achieved performance levels equivalent to those of their natural counterparts. As further insights are gained into the basic principles of library design and function, further progress can be expected. Thus, it seems likely that universal synthetic antibody libraries that can target any antigen of interest will be available in the near future. These libraries hold great promise for the development of highly optimized reagents and therapeutics and, in addition, will be extremely useful tools for elucidating the basic principles that govern antibody structure and function.

REFERENCES

1. Riechmann, L., M. Clark, H. Waldmann, and G. Winter. 1988. Reshaping human antibodies for therapy. *Nature* 332:323–327.
2. Jones, P. T., P. H. Dear, J. Foote, M. S. Neuberger, and G. Winter. 1986. Replacing the complementarity-determining regions in a human antibody with those from a mouse. *Nature* 321:522–525.
3. Jaffers, G. J., T. C. Fuller, A. B. Cosimi, P. S. Russell, H. J. Winn, and R. B. Colvin. 1986. Anti-idiotypic and non-anti-idiotypic antibodies to OKT3 arising despite intense immunosuppression. *Transplantation* 41:572–578.
4. Barbas, C. F., 3rd, J. D. Bain, D. M. Hoekstra, and R. A. Lerner. 1992. Semisynthetic combinatorial antibody libraries: A chemical solution to the diversity problem. *Proc. Natl. Acad. Sci. U.S.A.* 89 (10):4457–4461.
5. Xu, J. L., and M. M. Davis. 2000. Diversity in the CDR3 region of V(H) is sufficient for most antibody specificities. *Immunity* 13:37–45.
6. Padlan, E. A. 1994. Anatomy of the antibody molecule. *Mol. Immunol.* 31:169–217.
7. Wilson, I. A., and R. L. Stanfield. 1994. Antibody-antigen interactions: New structures and new conformational changes. *Curr. Opin. Struct. Biol.* 4:857–867.
8. Chothia, C., and A. M. Lesk. 1987. Canonical structures for the hypervariable regions of immunoglobulins. *J. Mol. Biol.* 196:901–917.
9. Barbas, C. F., 3rd, J. S. Rosenblum, and R. A. Lerner. 1993. Direct selection of antibodies that coordinate metals from semisynthetic combinatorial libraries. *Proc. Natl. Acad. Sci. U.S.A.* 90 (14):6385–6389.
10. Barbas, C. F., 3rd, W. Amberg, A. Simoncsits, T. M. Jones, and R. A. Lerner. 1993. Selection of human anti-hapten antibodies from semisynthetic libraries. *Gene* 137 (1):57–62.

11. Barbas, S. M., P. Ghazal, C. F. Barbas, 3rd, and D. R. Burton. 1994. Recognition of DNA by synthetic antibodies. *J. Am. Chem. Soc.* 116:2161–2162.

12. Barbas, C. F., 3rd, L. R. Languino, and J. W. Smith. 1993. High-affinity self-reactive human antibodies by design and selection: Targeting the integrin ligand binding site. *Proc. Natl. Acad. Sci. U.S.A.* 90 (21):10003–10007.

13. Smith, J. W., D. Hu, A. Satterthwait, S. Pinz-Sweeney, and C. F. Barbas, 3rd. 1994. Building synthetic antibodies as adhesive ligands for integrins. *J. Biol. Chem.* 269 (52):32788–32795.

14. Kohler, G., and C. Milstein. 1975. Continuous cultures of fused cells secreting antibody of predefined specificity. *Nature* 256:495–497.

15. Hoogenboom, H. R., A. D. Griffiths, K. S. Johnson, D. J. Chiswell, P. Hudson, and G. Winter. 1991. Multi-subunit proteins on the surface of filamentous phage: Methodologies for displaying antibody (Fab) heavy and light chains. *Nucleic Acids Res.* 19:4133–4137.

16. McCafferty, J., A. D. Griffiths, G. Winter, and D. J. Chiswell. 1990. Phage antibodies: Filamentous phage displaying antibody variable domains. *Nature* 348:552–554.

17. Hoogenboom, H. R., and G. Winter. 1992. By-passing immunization: Human antibodies from synthetic repertoires of germline V_H gene segments rearranged *in vitro*. *J. Mol. Biol.* 227:381–388.

18. Marks, J. D., H. R. Hoogenboom, T. P. Bonnert, J. McCafferty, A. D. Griffiths, and G. Winter. 1991. By-passing immunization: Human antibodies from V-gene libraries displayed on phage. *J. Mol. Biol.* 222:581–597.

19. Griffiths, A. D., M. Malmqvist, J. D. Marks, J. M. Bye, M. J. Embleton, J. McCafferty, M. Baier et al. 1993. Human anti-self antibodies with high specificity from phage display libaries. *EMBO J.* 12:725–734.

20. Nissim, A., H. R. Hoogenboom, I. M. Tomlinson, G. Flynn, C. Midgley, D. Lane, and G. Winter. 1994. Antibody fragments from a "single pot" phage display library as immunochemical reagents. *EMBO J.* 13:692–698.

21. Griffiths, A. D., S. C. Williams, O. Hartley, I. M. Tomlinson, P. Waterhouse, W. L. Crosby, R. E. Kontermann et al. 1994. Isolation of high affinity human antibodies directly from large synthetic repertoires. *EMBO J.* 13:3245–3260.

22. Waterhouse, P., A. D. Griffiths, K. S. Johnson, and G. Winter. 1993. Combinatorial infection and in vivo recombination: A strategy for making large phage antibody repertoires. *Nucleic Acids Res.* 21:2265–2266.

23. de Kruif, J., E. Boel, and T. Logtenberg. 1995. Selection and application of human single chain Fv antibody fragments from a semi-synthetic phage antibody display library with designed CDR3 regions. *J. Mol. Biol.* 248:97–105.

24. Pini, A., F. Viti, A. Santucci, B. Carnemolla, L. Zardi, P. Neri, and D. Neri. 1998. Design and use of a phage display library: Human antibodies with subnanomolar affinity against a marker of angiogenesis eluted from a two-dimensional gel. *J. Biol. Chem.* 273:21769–21776.

25. Kirkham, P. M., F. Mortari, J. A. Newton, and H. W. Schroeder, Jr. 1992. Immunoglobulin VH clan and family identity predicts variable domain structure and may influence antigen binding. *EMBO J.* 11:603–609.

26. Desiderio, A., R. Franconi, M. Lopez, M. E. Villani, F. Viti, R. Chiaraluce, V. Consalvi et al. 2001. A semi-synthetic repertoire of intrinsically stable antibody fragments derived from a single-framework scaffold. *J. Mol. Biol.* 310:603–615.

27. Hamers-Casterman, C., T. Atarhouch, S. Muyldermans, G. Robinson, C. Hamers, E. B. Songa, N. Bendahman et al. 1993. Naturally occurring antibodies devoid of light chains. *Nature* 363 (6428):446–448.

28. Ewert, S., C. Cambillau, K. Conrath, and A. Pluckthun. 2002. Biophysical properties of camelid V(HH) domains compared to those of human V(H)3 domains. *Biochemistry* 41 (11):3628–3636.

29. Dumoulin, M., K. Conrath, A. Van Meirhaeghe, F. Meersman, K. Heremans, L. G. Frenken, S. Muyldermans et al. 2002. Single-domain antibody fragments with high conformational stability. *Nat. Struct. Biol.* 11 (3):500–515.

30. Arbabi Ghahroudi, M., A. Desmyter, L. Wyns, R. Hamers, and S. Muyldermans. 1997. Selection and identification of single domain antibody fragments from camel heavy-chain antibodies. *FEBS Lett.* 414 (3):521–526.

31. Muyldermans, S., T. Atarhouch, J. Saldanha, J. A. R. G. Barbosa, and R. Hamers. 1994. Sequence and structure of VH domain from naturally occurring camel heavy chain immunoglobulins lacking light chains. *Protein Eng.* 7 (9):1129–1135.

32. Martin, F., C. Volpari, C. Steinkuhler, N. Dimasi, M. Brunetti, G. Biasiol, S. Altamura, R. Cortese et al. 1997. Affinity selection of a camelized V(H) domain antibody inhibitor of hepatitis C virus NS3 protease. *Protein Eng.* 10 (5):607–614.

33. Riechmann, L. 1996. Rearrangement of the former VL interface in the solution structure of a camelised, single antibody VH domain. *J. Mol. Biol.* 259 (5):957–969.

34. Davies, J., and L. Riechmann. 1994. "Camelising" human antibody fragments: NMR studies on VH domains. *FEBS Lett.* 339 (3):285–290.

35. Davies, J., and L. Riechmann. 1996. Single antibody domains as small recognition units: Design and in vitro antigen selection of camelized, human VH domains with improved protein stability. *Protein Eng.* 9 (6):531–537.

36. Jespers, L., O. Schon, L. C. James, D. Veprintsev, and G. Winter. 2004. Crystal structure of HEL4, a soluble refoldable human V_H single domain with a germ-line scaffold. *J. Mol. Biol.* 337:893–903.

37. Knappik, A., L. Ge, A. Honegger, P. Pack, M. Fischer, G. Wellnhofer, A. Hoess et al. 2000. Fully synthetic human combinatorial antibody libraries (HuCAL) based on modular consensus frameworks and CDRs randomized with trinucleotides. *J. Mol. Biol.* 296 (1):57–86.

38. Virnekas, B., L. Ge, A. Pluckthun, K. C. Schneider, G. Wellnhofer, and S. E. Moroney. 1994. Trinucleotide phosphoramidites: Ideal reagents for the synthesis of mixed oligonucleotides for random mutagenesis. *Nucleic Acids Res.* 22:5600–5607.

39. Krebs, B., R. Rauchenberger, S. Reifert, C. Rothe, M. Tesar, E. Thomassen, M. Cao et al. 2001. High-throughput generation and engineering of recombinant human antibodies. *J. Immunol. Methods* 254:67–84.

40. Frisch, C., B. Brocks, R. Ostendorp, A. Hoess, T. von Ruden, and T. Kretzschmar. 2003. From EST to IHC: Human antibody pipeline for target research. *J. Immunol. Methods* 75:203–212.

41. Nagy, Z. A., B. Hubner, C. Lohning, R. Rauchenberger, S. Reiffert, E. Thomassen-Wolf, S. Zahn et al. 2002. Fully human, HLA-DR-specific monoclonal antibodies efficiently induce programmed death of malignant lymphoid cells. *Nat. Med.* 8:801–807.

42. Rauchenberger, R., E. Borges, E. Thomassen-Wolf, E. Rom, R. Adar, Y. Yaniv, M. Malka et al. 2003. Human combinatorila Fab library yielding specific and functional antibodies against the human fibroblast growth factor receptor 3. *J. Biol. Chem.* 278: 38194–38205.

43. Parsons, C. J., B. U. Bradford, C. Q. Pan, E. Cheung, M. Schauer, A. Knorr, B. Krebs et al. 2004. Antifibrotic effects of a tissue inhibitor of metalloproteinase-1 antibody on established liver fibrosis in rats. *Hepatology* 40:1106–1115.

44. Rothlisberger, D., K. M. Pos, and A. Pluckthun. 2004. An antibody library for stabilizing and crystallizing membrane proteins—Selecting binders to the citrate carrier CitS. *FEBS Lett.* 564:340–348.

45. Ewert, S., T. Huber, A. Honegger, and A. Pluckthun. 2003. Biophysical properties of human antibody variable domains. *J. Mol. Biol.* 325:531–553.

46. Ewert, S., A. Honegger, and A. Pluckthun. 2004. Stability improvement of antibodies for extracellular and intracellular applications: CDR grafting to stable frameworks and structure-based framework engineering. *Methods* 34:148–199.

47. Hanes, J., C. Schaffitzel, A. Knappik, and A. Pluckthun. 2000. Picomolar affinity antibodies from a fully synthetic naive library selected and evolved by ribosome display. *Nat. Biotechnol.* 18:1287–1292.

48. Scanlan, T. S., J. R. Prudent, and P. G. Schultz. 1991. Antibody-catalyzed hydrolysis of phosphate monoesters. *J. Am. Chem. Soc.* 113:9397–9398.

49. Kabat, E. A., T. T. Wu, H. M. Perry, K. S. Gottesman, and C. Foeller. 1991. *Sequences of Proteins of Immunological Interest*, 5th ed. Bethesda, MD: National Institutes of Health.

50. Carter, P., L. Presta, C. M. Gorman, J. B. Ridgway, D. Henner, W. L. Wong, A. M. Rowland et al. 1992. Humanization of an anti-p185HER2 antibody for human cancer therapy. *Proc. Natl. Acad. Sci. U.S.A.* 89 (10):4285–4289.

51. Presta, L. G., S. J. Lahr, R. L. Shields, J. P. Porter, C. M. Gorman, B. M. Fendly, and P. M. Jardieu. 1993. Humanization of an antibody directed against IgE. *J. Immunol.* 151:2623–2632.

52. Presta, L. G., H. Chen, S. J. O'Connor, V. Chisholm, Y. G. Meng, L. Krummen, M. Winkler et al. 1997. Humanization of an anti-vascular endothelial growth factor monoclonal antibody for the therapy of solid tumors and other disorders. *Cancer Res.* 57:4593–4599.

53. Werther, W. A., T. N. Gonzalez, S. J. O'Connor, S. McCabe, B. Chan, T. Hotaling, M. Champe et al. 1996. Humanization of an anti-lymphocyte function-associated antigen (LFA)-1 monoclonal antibody and reengineering of the humanized antibody for binding to rhesus LFA-1. *J. Immunol.* 157:4986–4995.

54. Garrard, L. J., and D. J. Henner. 1993. Selection of an anti-IGF-1 Fab from a Fab phage library created by mutagenesis of multiple CDR loops. *Gene* 128:103–109.

55. Eigenbrot, C., M. Randal, L. Presta, P. Carter, and A. A. Kossiakoff. 1993. X-ray structures of the antigen-binding domains from three variants of humanized anti-p185HER2 antibody 4D5 and comparison with molecular modeling. *J. Mol. Biol.* 229: 969–995.

56. Sidhu, S. S., B. Li, Y. Chen, F. A. Fellouse, C. Eigenbrot, and G. Fuh. 2004. Phage-displayed antibody libraries of synthetic heavy chain complementarity determining regions. *J. Mol. Biol.* 338 (2):299–310.

57. Pack, P., and A. Pluckthun. 1992. Miniantibodies: Use of amphipathic helices to produce functional, flexibly linked dimeric Fv fragments with high avidity in *Escherichia coli. Biochemistry* 31:1579–1584.

58. Lee, C. V., S. S. Sidhu, and G. Fuh. 2004. Bivalent antibody phage display mimics natural immunoglobulins. *J. Immunol. Methods* 284:119–132.

59. Sidhu, S. S., H. B. Lowman, B. C. Cunningham, and J. A. Wells. 2000. Phage display for selection of novel binding peptides. *Methods Enzymol.* 328:333–363.

60. Lee, C. V., W.-C. Liang, M. S. Dennis, C. Eigenbrot, S. S. Sidhu, and G. Fuh. 2004. High-affinity human antibodies from phage-displayed synthetic Fab libraries with a single framework scaffold. *J. Mol. Biol.* 340:1073–1093.

61. Ge, X., Y. Mazor, S. P. Hunicke-Smith, A. D. Ellington, and G. Georgiou. 2010. Rapid construction and characterization of synthetic antibody libraries without DNA amplification. *Biotechnol. Bioeng.* 106 (3):347–357.

62. Rothe, C., S. Urlinger, C. Lohning, J. Prassler, Y. Stark, U. Jager, B. Hubner et al. 2008. The human combinatorial antibody library HuCAL GOLD combines diversification of all six CDRs according to the natural immune system with a novel display method for efficient selection of high-affinity antibodies. *J. Mol. Biol.* 376 (4):1182–1200.

63. Prassler, J., S. Thiel, C. Pracht, A. Polzer, S. Peters, M. Bauer, S. Norenberg et al. 2011. HuCAL PLATINUM, a synthetic Fab library optimized for sequence diversity and superior performance in mammalian expression systems. *J. Mol. Biol.* 413 (1): 261–278.

64. Xiao, X., Y. Feng, B. K. Vu, R. Ishima, and D. S. Dimitrov. 2009. A large library based on a novel (CH2) scaffold: Identification of HIV-1 inhibitors. *Biochem. Biophys. Res. Commun.* 387 (2):387–392.

65. Wozniak-Knopp, G., S. Bartl, A. Bauer, M. Mostageer, M. Woisetschlager, B. Antes, K. Ettl et al. 2010. Introducing antigen-binding sites in structural loops of immunoglobulin constant domains: Fc fragments with engineered HER2/neu-binding sites and antibody properties. *Protein Eng. Des. Sel.* 23 (4):289–297.

66. Fellouse, F. A., C. Wiesmann, and S. S. Sidhu. 2004. Synthetic antibodies from a four-amino-acid code: A dominant role for tyrosine in antigen recognition. *Proc. Natl. Acad. Sci. U.S.A.* 101 (34):12467–12472.

67. Jensen, K. 2013. DutaMabs: A novel bispecific monoclonal antibody platform. In *9th Annual Essential Protein Engineering Summit.* Boston, MA: Cambridge Healthtech Institute.

68. Schaefer, G., L. Haber, L. M. Crocker, S. Shia, L. Shao, D. Dowbenko, K. Totpal et al. 2011. A two-in-one antibody against HER3 and EGFR has superior inhibitory activity compared with monospecific antibodies. *Cancer Cell* 20 (4):472–486.

69. Bostrom, J., L. Haber, P. Koenig, R. F. Kelley, and G. Fuh. 2011. High affinity antigen recognition of the dual specific variants of herceptin is entropy-driven in spite of structural plasticity. *PLoS One* 6 (4):e17887.

70. Kontermann, R. E. 2012. Dual targeting strategies with bispecific antibodies. *MAbs* 4 (2):182–197.

71. Lea, S., and D. Stuart. 1995. Analysis of antigenic surfaces of proteins. *FASEB J.* 9:87–93.

72. Kabat, E. A., T. T. Wu, and H. Bilofsky. 1977. Unusual distributions of amino acids in complementarity-determining (hypervariable) segments of heavy and light chains of immunoglobulins and their possible roles in specificity of antibody-combining sites. *J. Biol. Chem.* 252 (19):6609–6616.

73. Padlan, E. A. 1990. On the nature of antibody combining sites: Unusual structural features that may confer on these sites an enhanced capacity for binding ligands. *Proteins: Struct. Funct. Genet.* 7:112–124.

74. Ivanov, I., J. Link, G. C. Ippolito, and H. W. Schroeder, Jr. 2002. Constraints on the hydropathicity and sequence composition of HCDR3 across evolution. In *The Antibodies,* edited by M. Zanetti, and J. Capra. New York: Taylor & Francis.

75. Zemlin, M., M. Klinger, J. Link, C. Zemlin, K. Bauer, J. A. Engler, H. W. Schroeder et al. 2003. Expressed murine and human CDR-H3 intervals of equal length exhibit distinct repertoires that differ in their amino acid composition and predicted range of structures. *J. Mol. Biol.* 334 (4):733–749.

76. Lo Conte, L., C. Chothia, and J. Janin. 1999. The atomic structure of protein–protein recognition sites. *J. Mol. Biol.* 285:2177–2198.

77. Collis, A. V., A. P. Brouwer, and A. C. Martin. 2003. Analysis of the antigen combining site: Correlations between length and sequence composition of the hypervariable loops and the nature of the antigen. *J. Mol. Biol.* 325 (2):337–354.

78. Davies, D. R., and G. H. Cohen. 1996. Interactions of protein antigens with antibodies. *Proc. Natl. Acad. Sci. U.S.A.* 93:7–12.

79. Mian, I. S., A. R. Bradwell, and A. J. Olson. 1991. Structure, function and properties of antibody binding sites. *J. Mol. Biol.* 217 (1):133–151.

80. Folkman, J. 1995. Angiogenesis in cancer, vascular, rheumatoid and other diesease. *Nat. Med.* 1:27–31.

81. Ferrara, N. 2001. Role of vascular endothelial growth factor in regulation of physiological angiogenesis. *Am. J. Physiol. Cell Physiol.* 280:C1358–C1366.

82. Fellouse, F. A., B. Li, D. M. Compaan, A. A. Peden, S. G. Hymowitz, and S. S. Sidhu. 2005. Molecular recognition by a binary code. *J. Mol. Biol.* 348 (5):1153–1162.

83. Ashkenazi, A., and V. M. Dixit. 1998. Death receptors: Signaling and modulation. *Science* 281:1305–1308.

84. Birtalan, S., Y. Zhang, F. A. Fellouse, L. Shao, G. Schaefer, and S. S. Sidhu. 2008. The intrinsic contributions of tyrosine, serine, glycine and arginine to the affinity and specificity of antibodies. *J. Mol. Biol.* 377 (5):1518–1528.

85. Birtalan, S., R. D. Fisher, and S. S. Sidhu. 2010. The functional capacity of the natural amino acids for molecular recognition. *Mol. Biosyst.* 6 (7):1186–1194.

86. Fellouse, F. A., K. Esaki, S. Birtalan, D. Raptis, V. J. Cancasci, A. Koide, P. Jhurani et al. 2007. High-throughput generation of synthetic antibodies from highly functional minimalist phage-displayed libraries. *J. Mol. Biol.* 373 (4):924–940.

87. Persson, H., W. Ye, A. Wernimont, J. J. Adams, A. Koide, S. Koide, R. Lam et al. 2012. CDR-H3 diversity is not required for antigen recognition by synthetic antibodies. *J. Mol. Biol.* 425 (4):803–811.

88. Almagro, J. C. 2004. Identification of differences in the specificity-determining residues of antibodies that recognize antigens of different size: Implications for the rational design of antibody repertoires. *J. Mol. Recognit.* 17 (2):132–143.

89. Ye, J. D., V. Tereshko, J. K. Frederiksen, A. Koide, F. A. Fellouse, S. S. Sidhu, S. Koide et al. 2008. Synthetic antibodies for specific recognition and crystallization of structured RNA. *Proc. Natl. Acad. Sci. U.S.A.* 105 (1):82–87.

90. Sherman, E. M., S. Holmes, and J. D. Ye. 2014. Specific RNA-binding antibodies with a four-amino-acid code. *J. Mol. Biol.* 426 (10):2145–2157.

91. Schoonbroodt, S., M. Steukers, M. Viswanathan, N. Frans, M. Timmermans, A. Wehnert, M. Nguyen et al. 2008. Engineering antibody heavy chain CDR3 to create a phage display Fab library rich in antibodies that bind charged carbohydrates. *J. Immunol.* 181 (9):6213–6221.

92. Cobaugh, C. W., J. C. Almagro, M. Pogson, B. Iverson, and G. Georgiou. 2008. Synthetic antibody libraries focused towards peptide ligands. *J. Mol. Biol.* 378 (3):622–633.

93. Koerber, J. T., N. D. Thomsen, B. T. Hannigan, W. F. Degrado, and J. A. Wells. 2013. Nature-inspired design of motif-specific antibody scaffolds. *Nat. Biotechnol.* 31 (10):916–921.

94. Persson, H., J. Lantto, and M. Ohlin. 2006. A focused antibody library for improved hapten recognition. *J. Mol. Biol.* 357 (2):607–620.

95. MacCallum, R. M., A. C. Martin, and J. M. Thornton. 1996. Antibody-antigen interactions: Contact analysis and binding site topography. *J. Mol. Biol.* 262 (5):732–745.

CHAPTER **19**

Engineering Antibody Fragments for Intracellular Applications

Jianghai Liu and Clarence Ronald Geyer

CONTENTS

19.1 INTRODUCTION

Antibodies possess many desired properties for validating protein function, and early evidence indicates that antibodies can be used to block the function of intracellular targets by injecting them into individual cells [1]. Antibodies, however, cannot fold correctly or form stable structures within the reducing environment of cells [2], and they are unable to penetrate the cell membrane, which hinders their ability to inhibit intracellular targets. Further, their large size and tetrameric structure limit their use as intracellular affinity reagents [2] and have driven the development of smaller antibody fragments (called "intrabodies") for intracellular applications (Figure 19.1). Intrabodies are designed to retain the antigen affinity of antibodies without the effector function normally associated with the antibody constant region, such as antibody-dependent cell-mediated cytotoxicity or complement-dependent cytotoxicity. The single-chain variable fragment (scFv) and the variable heavy (V_H) domain are the most commonly used intrabodies. The scFv consists of a V_H domain covalently linked to a variable light (V_L) domain via a short peptide linker, generally consisting of glycine-serine repeats. ScFvs have advantages over smaller variable domain fragments as the binding affinity/specificity of full-length immunoglobulins (IgGs) and antigen-binding fragments (Fabs) can be easily engineered into the scFv format. Many scFvs have been developed in recent years as reagents for studying and inhibiting cancer and other diseases, but only a relatively small number have been reported for intracellular applications [3,4].

Recombinant DNA technologies have been used to construct cytoplasm expression systems for intrabodies, which, in principle, allows them to be used for functional

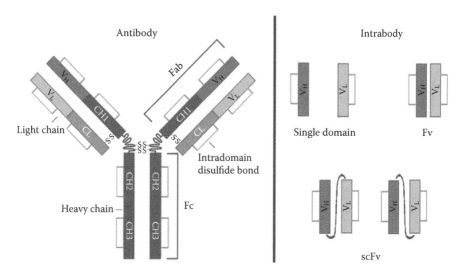

Figure 19.1 Schematic representation of an antibody and intrabodies. The full-length antibody or immunoglobulin (IgG) comprises two heavy and two light chains held together by both interchain disulfide bonds and noncovalent interactions. Each light chain has a variable domain (V_L) and a constant (CL) region, while each heavy chain has a variable domain (V_H) and three constant (CH1, CH2, and CH3) regions. The antigen-binding fragment (Fab) is a region on an antibody that binds to antigens and consists of one constant and one variable domain of each of the heavy and the light chain. The crystallizable fragment (Fc) is fragment of an antibody that interacts with immune effector functions. Intrabodies include single-variable domains (V_L and V_H) or two-variable domains, variable fragment (Fv), and the single-chain variable fragment (scFv). The Fv contains V_H and V_L domains held together by noncovalent interactions, and the scFv consist of V_H and V_L domains covalently joined via a flexible peptide linker.

genomics, proteomics, and gene therapy [5,6]. However, the intracellular environment posses many challenges for using intrabodies. The cell cytoplasm consists of a complex, crowded mixture of biomolecules that can promote aggregation [7]. In addition, the cytoplasm is a reducing environment that does not facilitate the formation of disulfide bonds, and thus, most intrabodies are destabilized in this environment.

The inability to reproducibly generate functional intracellular scFvs from synthetic libraries or hybridomas is a major limitation in expanding this field as the majority of scFvs do not survive the transition from in vitro isolation to intracellular expression. To be functional, an intracellular scFv needs to fold stably in the absence of disulfide bonds and avoid aggregation. Efforts to create intracellular scFvs can generally be classified into two strategies: (1) creation of artificial scFv frameworks and (2) isolation of naturally occurring variable domains that can be expressed as scFvs. Optimized artificial frameworks can be designed by systematically mutating framework residues, which can then be used to construct libraries or to graft complementarity-determining regions (CDRs). Alternatively, screening strategies can be used to isolate scFv framework regions from large antibody fragment libraries [8]. This method has the added advantage of revealing conserved structural elements from which a framework consensus sequence can be derived [9,10].

Previous studies have shown a strong correlation between intrabody stability and function inside cells [11–13]. Thus, considerable effort has gone into improving intrabody stability in terms of improving the intradomain stability of variable domains and the interdomain stability of scFvs. In this review, we discuss strategies to improve intrabody stability with the goal of developing better antibody fragments for intracellular applications.

19.2 INTRABODY STABILITY

ScFv and Fv stability primarily depend on intradomain stability of V_H and V_L domains. ScFv stability is also dependent on interdomain stability of the V_H/V_L interface [14]. Unfolding of variable domains or disruption of the V_H/V_L interface can cause scFvs to aggregate via multiple pathways [15,16] (Figure 19.2). The degree of aggregation depends on the competition between partially or completely unfolded scFvs to refold or aggregate. ScFvs can also aggregate when their variable domains are folded. This involves mechanisms where scFvs self-interact via attractive interactions between folded variable domains or by domain swapping between scFvs. Domain swapping occurs when the V_H domain from one scFv interacts with the V_L domain of another scFv.

19.2.1 Intradomain Stability of Variable Domains

Intradomain stability of most variable domains depends largely on the presence of an intradomain disulfide bond [14]. This disulfide bond connects β-sheets of the immunoglobulin fold and is highly conserved in all antibodies [17]. Disulfide bonds form between cysteine residues L23 and L88 in V_L and between H22 and H92 in

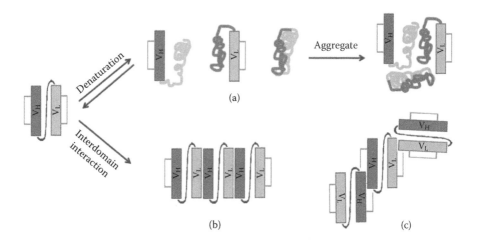

Figure 19.2   Intrabody aggregation. Intrabodies can aggregate in response to a variety of stresses including heat, pH, concentration, and so forth. (a) Unfolding of the V_H or V_L domains or both can result in aggregation or refolding. (b) Intrabodies can aggregate via domain swapping without unfolding. (c) Stresses such as high concentration and low temperature can lead to antibody aggregation without unfolding via attractive self-interactions.

V_H [18]. All known mouse and human germ-line variable domains contain conserved cysteine residues, and it is believed that the disulfide bond is maintained by selective pressure for correct folding and stability. Occasionally, variable domains lacking conserved cysteines are observed; however, these mutations are probably due to somatic mutation during B-cell development. For most cases, variable domains cannot tolerate removal of the disulfide bond. Since the disulfide bond contributes to the stability and folding of antibodies, it is not surprising that under reducing conditions, many antibody genes do not produce functional antibodies [19].

19.2.2 Framework Mutations to Enhance Intradomain Stability

To increase the stability of V_H and V_L domains, stabilizing mutations have been identified and introduced into framework positions of the variable domain. In most cases, efforts have focused on the V_H domain, since this domain is more thermostable than the V_L domain [20], and also, single-domain V_H antibodies are found in nature, such as in camelids [21] and cartilaginous fish [22]. In the absence of an interacting V_L domain, the V_H domain tends to unfold and aggregate as a consequence of exposing the hydrophobic interface between V_H and V_L domains. Despite this intrinsic instability shown by variable domains, several groups have shown that stable V_H domains can be engineered with enhanced stability and reduced tendency to aggregate [23–28].

Characterization of camelid antibodies, which consist only of heavy chains and bind antigen through their V_H domain [21], provided insights for designing stable human V_H domains. Camelid V_H domains (known as $V_H H$ domains) are thermal stable and can fold reversibly [29]. The stability of camelid V_HH domains is due in part to the reduced number of hydrophobic residues in the region analogous to the V_H/V_L interface in V_H domains [30–36]. The most significant differences between V_HH and V_H domains occurs at four positions, referred to as the "V_HH tetrad," which function to decrease the hydrophobicity of this region and to sequester hydrophobic side chains by facilitating their interactions with CDR3 and/or framework regions [30–36]. Camelid V_HH domains generally have larger CDR3 loops that pack against hydrophobic regions of the V_HH domain in the region analogous to the V_H/V_L interface of V_H domains [30,36,37].

Attempts have been made to construct soluble human V_H domains by introducing elements of camelid V_HH tetrads into the V_H domain. This "camelization" strategy, however, has only resulted in moderately stable V_H domains, which highlights the importance of additional factors, such as the requirement of the CDR3 loop for supplying additional stabilizing interactions [38]. The importance of the CDR3 loop in the stability of V_HH domains places constraints on the types of CDR3 loops that can be displayed on the camelid V_HH domains.

To comprehensively explore requirements for an autonomous V_H domain, Barthelemy and coworkers [27] used phage display and biophysical methods to simultaneously assess effects of many different V_H domain mutations. Stable V_H domains were selected from a large library of V_H domain mutants based on their ability to bind protein A. Previously, it has been shown that the binding of phage, which display antibody fragments, to protein A, is highly correlated with the intrinsic stability of the free antibody fragment [23,39,40]. Phage display V_H domain libraries were constructed by varying residues implicated in the stabilization of camelid V_HH domains as well additional sites in the V_H/V_L interface. Phage display selection against protein A was used to select autonomous V_H domain from these libraries and to analyze selected V_H domains using quantitative saturation scanning [41] and shotgun alanine-scanning methods [42]. Four mutations were identified that significantly increased the stability of the human V_H domain. Two of the mutations (Arg39 and Glu45) are in close proximity at the base of the V_H domain and likely enhance stability by creating a favorable electrostatic interaction. The second pair of mutations (Gly35 and Ser50) is located at the edges of CDR1 and CDR2 and impacts the orientation of adjacent hydrophobic side chains, which may explain the enhanced stability. Together, these four mutations enhanced the thermostability of the V_H domain, allowed it to fold reversibly, and provided stability that is independent of the CDR3 loop.

V_H domains with these mutations plus three additional mutations (W47L/S50R/W103S) identified to improve autonomous V_H domain behavior [27] were used to construct libraries of V_H domains [43]. These libraries were successfully screened to isolate V_H domains against vascular endothelial growth factor (VEGF) with low nanomolar affinity [43].

19.2.3 Influence of CDR Loops on Variable Domain Stability

Since variable domains with identical framework regions and different CDRs show dramatic differences in stability and aggregation, several groups have explored the properties of CDRs in mediating variable domain aggregation. Jespers and coworkers [44] used phage display to gain insight into the role of CDR sequences in V_H domain aggregation. They constructed libraries of V_H domains that contained random variations in CDR sequences and displayed V_H domains in a multivalent format on the pIII coat proteins of filamentous phage. Phage libraries were heated transiently at 80°C to unfold V_H domains and then cooled. V_H domains that were resistant to aggregation were selected based on their ability to bind protein A. Interestingly, aggregation-resistant V_H domains had similar folding stabilities relative to aggregation-prone V_H domains, despite being able to fold reversibly after heating [44,45]. Inspection of CDR sequences did not reveal any obvious similarities except for a slight increase in negatively charged amino acids. Similar phage display experiments using this strategy observed that CDR sequences of aggregation-resistant V_H domains had an increase in negatively charged amino acids [46,47] as well as an increase in disulfide-bonded CDRs [47]. Perchiacca and coworkers [48–50] have exploited the observed enrichment of negative charge in CDRs to engineer aggregation-resistant V_H domains. They investigated whether differences in the aggregation propensity of two homologous V_H domains that differ in CDR composition are a result of a specific CDR sequence, an amino acid within the CDR, or properties of all three CDRs [48]. By swapping CDRs between two V_H domains, they showed that CDR1 was responsible for aggregation resistance and that a specific "hot spot" within CDR1 was responsible for preventing aggregation of a V_H domain [48]. Specifically, a triad of three negatively charged amino acids (31-Asp-Glu-Asp-33) was sufficient to prevent aggregation [48]. To confirm whether these aggregation-resistant properties could be used to construct synthetic V_H domain libraries, a V_H domain with the aggregation-resistant CDR1 sequence was combined with an addition mutation proximal to CDR1 (I29D), which has also been shown to decrease aggregation of a V_H domain [48] and used to construct a V_H domain phage display library where CDR2 and CDR3 were randomized [51]. This V_H domain phage library was used to generate V_H domains that bound to lysozyme and to human brain vascular pericytes (HBVP) cells [51]. Isolated V_H domains had lower thermodynamic stability than I29D V_H domain mutant [48], and only four of six V_H domains characterized folded reversibly. These results suggest that the I29D mutation is not compatible with aggregation-resistant CDR1. In a similar study, an aggregation-resistant V_H domain with two negatively charged residues inserted into CDR1 (Y32D and A33D) was used as a template for constructing a V_H domain library where CDR2 and CDR3 were randomized [52]. The mean aggregation resistance of this phage V_H library was increased for aspartate mutants, and the aggregation resistance was mostly independent of the diversities in CDR2 and CDR3 [52]. These results show that stabilizing mutations in CDR1 is beneficial for constructing V_H domain libraries; however, mutation combinations that confer improved V_H domain stability still need to be tested empirically.

Perchiacca and coworkers [49,50] also examined whether negative-charged residues could be inserted into CDR3 to reduce the aggregation propensity of hydrophobic CDR3s. They found that insertion of two or more negative residues at edges of CDR3 prevented aggregation and surprisingly did not alter the affinity of the anti-Alzheimer's amyloid B peptide V_H domain [49]. Further, they found that negatively charged residues needed to be positioned at the edge of the CDR3 that was closest to the hydrophobic portion of the CDR [49]. In a second study, Perchiacca and coworkers [50] determined that the net charge on the V_H domain determines whether positive- or negative-charged residues inserted into CDR3 inhibit aggregation. They found that negatively charged residues are more effective in decreasing aggregation of V_H domains that are negative or near-neutrally charged, whereas positively charged residues are more effective at decreasing aggregation in positively charged V_H domains [50].

19.2.4 Engineering Disulfide Bonds to Enhance Variable Domain Stability

Several groups have explored the idea of increasing the stability of the V_H domain by inserting additional disulfide bonds to covalently link the two beta-sheets. Introducing cysteine mutations at positions Cys54 and Cys78 resulted in the formation of a disulfide bond in a human V_H domain [53,54] or at spatially equivalent positions in $V_H H$ [55] and V_L [56] domains. The addition of a second disulfide bond in the variable domain improved thermostability, reduced aggregation [53,54], and increased protease resistance [56].

19.2.5 Stabilization of V_L Domains

Fewer studies have focused on engineering V_L domains to improve stability. Several studies, however, have shown that V_L domains are more aggregation resistant than V_H domains [57–59]. To explore properties that account for aggregation resistance, Kim and coworkers [56] generated a naive human V_L phage display library and used a phage selection method that was previously used to select for nonaggregating V_H domains. This phage selection relies on the ability of folded V_L domains to bind protein L combined with the observation that phages displaying nonaggregating V_L domains form larger plaques on bacterial lawns than phages displaying V_L domains that aggregate [56,60]. The majority of V_L domains isolated were of the Vκ type, which is consistent with the better binding of Vκ to protein L and the higher stability of Vκ than Vλ [58]. Biophysical properties of isolated V_L domains were further enhanced by introducing a disulfide bond at positions 48 and 64 [56], similar to that seen with V_H domains [53,54]. In another study, Dudgeon and coworkers used phage display to identify aggregation-preventing mutations in V_L domains (Vκ1) [52]. Aggregation resistance properties clustered to the CDRL2 region of the V_L domain, where CDRL2s containing negatively charged mutations (Asp and Glu) showed the largest aggregation resistance. A combinatorial V_L domain library was constructed using a V_L framework with CDRL2 52D, 53D mutations, and randomizing CDRL1 and CDRL3. This V_L domain library showed increased mean aggregation resistance of the library relative to a library constructed without the CDRL2 mutations [52]. As a whole, this study highlights the potential for developing affinity reagents from V_L domains for basic science, therapeutics, and diagnostic applications. Future work in this area needs to be pursued to see if these stabilized V_H and V_L domains can be used in scFv formats.

19.2.6 V_H/V_L Interface Stability

V_H and V_L domains interact to form an Fv, the smallest antibody unit that retains the antigenic affinity of an IgG. Fv stability depends on an intradomain disulfide bond in each variable domain and on a hydrophobic interdomain interaction between V_H and V_L domains [14]. For most Fvs, interdomain stability is not sufficient to keep the V_H and V_L domains associated at low protein concentrations or under mild denaturing conditions [20]. Low stability of the V_H/V_L interface has been suggested to be the main cause of irreversible scFv inactivation [61]. V_H and V_L domains dissociate with K_ds ranging from 10^{-6} to 10^{-9} M [62], which allows transient dissociation and exposure of hydrophobic residues that cause aggregation.

The simplest approach to stabilize Fv domains is to link them together by a peptide linker, forming an scFv. Fv unfolding is dependent on the concentration of the V_H and V_L domains [14]. ScFvs are stabilized relative to Fvs since stability of the V_H/V_L interface is enhanced by the high effective concentration of variable domains provided by the peptide linker, which is estimated to be in the low millimolar range [63]. The linker, however, can reduce stability by allowing scFvs to adopt dimer and higher-order structures. Dimers occur when the V_H domain of one chain is paired with the V_L domain of another chain [64,65]. The fraction of dimer formed is dependent on linker length, with shorter linkers favoring dimer formation [66]. In general, monomeric scFvs are more stable than dimers, with dimers having stabilities similar to Fvs.

Another common method to stabilize the Fv is to engineer a disulfide bond at rationally defined locations within the V_H/V_L interface [67,68]. This strategy has been combined with the V_H–V_L linker to stabilize scFvs [69–71]. A variety of strategies have been used to identify locations at the V_H/V_L interface that tolerate insertion of a disulfide bond [67,72,73]. However, due to variations in variable domain orientation, it is difficult to design a "generic" disulfide bond for different variable domain combinations. This strategy of interdomain stabilization has been successful in increasing the stability of extracellular scFvs against thermal [69,70] and chemical denaturants [71,74]. Fvs and scFvs with interface disulfide bonds have proven more difficult to produce in *Escherichia coli* [69,71], and they are not effective in stabilizing intracellular Fvs or scFvs.

Another promising strategy to stabilize scFvs is cyclization. Proteins can be cyclized within cells using a split-intein approach [75,76], and there are several examples where protein cyclization enhances stability and solubility [77–80]. More recently, intein-mediated cyclization of two proteins stabilized the binary complex [81]. ScFvs can be cyclized by adding another linker between the free terminals of V_H and V_L domains. Cyclization constrains the protein in a conformation where the protein cannot unfold and aggregate, and thus, in theory, cyclized scFvs should have a higher stability than scFvs. Moreover, cyclized scFvs will lose the chance to form dimers and multimers. ScFvs can be constructed in the V_H–linker–V_L [82] or the V_L–linker–V_H [83] orientation, which suggests that there are no structural barriers to constructing cyclic scFvs that are constrained by two linker peptides. Using a split intein, we constructed cyclic scFv libraries recently. The screening results showed that in all cases, the cyclic scFv libraries had a higher binding capacity for various antigens tested than the linear scFv libraries [84].

19.2.7 Influence of Variable Domain Charge on Stability

Altering the net charge of antibody fragments has been shown to enhance their solubilities [44,46–49,85]. The optimal polarity of charged mutations depends on variety of factors, including net charge [86–90] and spatial distribution of charges [91–94]. Studies by Gribenko et al. [95] group showed that optimization of surface charge–charge interactions can stabilize a variety of proteins. Further, Lawrence et al. [86] showed that proteins are amenable to radical changes in their surface charge, which leads to enhanced solubility and reduced aggregation. In agreement, negatively charged mutations in or near CDRs have been shown to increase the solubility of antibody fragments [48–50]. Weidenhaupt et al. [96] showed that surface charge modifications on scFvs are tolerated with limited effect on kinetic interaction parameters. Onda et al. [97] have shown that the isoelectric point of scFvs can be significant lowered with no effect on their antitumor activity. Kvam et al. [85] showed that scFv solubility is improved by the net negative charge and reduced hydrophilicity. Recently, Ellington's group [89] showed that the scFv framework could be redesigned with increased surface charge to enhance solubility using a strategy referred to as "supercharging." This involves mutating solvent-exposed residues to charged residues of the same polarity. Several studies have shown that functional scFvs can be generated with lower isoelectric points by fusing scFvs to highly soluble proteins such as thioredoxin [98], small ubiquitin-like modifiers (SUMO) [99], maltose-binding protein [100,101], NusA [102], alkaline phosphatase [103], and others. Other relatively simple methods for increasing the net charge of scFvs have been developed that involve fusing charged peptides to the termini of antibodies and antibody fragments [85,104,105]. Together, these studies highlight the importance of optimal protein charge and charge distribution in preventing aggregation. Future studies incorporating charged mutations into antibody framework regions and CDRs should result in better antibody scaffolds for designing combinatorial antibody fragment libraries.

19.3 SELECTION METHODS FOR ISOLATING INTRABODIES

Antibody fragments derived from monoclonal antibodies or selected by phage display have been isolated for retaining activity under oxidizing conditions. These antibody fragments need to be tested on a case-by-case basis to identify those fragments that are stable and aggregation resistant in the reducing environment of a cell. This is a time-consuming, laborious process with a low probability success. Several strategies have been developed to increase the probability of obtaining intrabodies that retain activity within cells.

19.3.1 In Vitro Stability and Affinity Selection Using Phage Display

One strategy for obtaining intrabodies is to use phage display to select intrabody variants with enhanced stability. In this strategy, intrabodies are mutated and screened for binding under conditions that allow for selection of variants with enhanced stability [106,107]. Selection of stable variable framework regions is based on the irreversible denaturation of unstable variants by heating or by incubation with protein denaturants, such as urea or guanidine chloride [44,106,107]. Using phage display techniques, variants that remain folded and active can be isolated, propagated, and enriched within a few cycles of selection. Thereafter, stable variants are collected and verified for antigen-binding activity or phage infectivity. Moreover, the selective pressure can include the addition of reducing agents (e.g., dithiothreitol [DTT]) to mimic the conditions within cells where the disulfide bonds of an antibody are not formed [108,109]. Only the intrinsically very stable variants that tolerate the absence of intradomain disulfide bonds and fold correctly in the reducing conditions can be isolated by phage display.

The stability selection can start before (preselection) or after (postselection) antigen affinity selection (Figure 19.3). In preselection strategy, a library made by universal frameworks with randomized CDRs is first exposed to extreme physical and chemical conditions, such as high temperatures (e.g., 60°C, a temperature $>T_m$ of the variable domains) or low pH (e.g., pH 3) [107,108,110]. The selective pressures have to be optimized to remove phage particles that are prone to unfolding and aggregation, but not to affect their viability and infectivity [107]. After returning to native conditions (room temperature or neutral pH), phage particles that resist aggregation are collected and further selected against antigens of interest using a general selection procedure. In some cases, capture and purification of phage particles by a conformation-specific ligand (e.g., protein L [111] or protein A [112]) before antigen selection may be required to remove incorrectly folded intrabodies.

In the postselection strategy, the framework region of an intrabody against a known target is randomized to form a library of mutants that are then screened using phage display to isolate variants suitable for intracellular expression [106]. For example, V_L and V_H domains of an scFv have been diversified using a random mutagenesis approach (e.g., error-prone PCR [113] and/or DNA shuffling [114]). This library of scFv framework mutants is then selected for heat or acid resistance. ScFvs with high intrinsic stability are captured by antigens [106] or by protein L or protein A [110].

19.3.2 Selection of Functional Intrabodies within Cells

Despite progress in improving intrabody stability, most antibodies generated by hybridoma and phage display methods do not function effectively within cells. To reduce this problem, a combination of extracellular and/or intracellular screening strategies has been developed [8,115–117]. Perez-Martinez and coworkers [4] developed a two-step intrabody selection, referred to as "intracellular antibody capture" (IAC) (Figure 19.4). The first step in this selection involves screening the intrabody

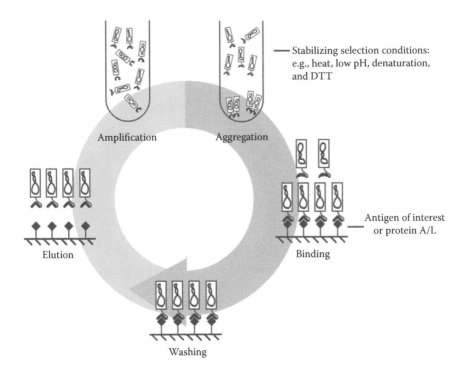

Stabilizing selection conditions:
e.g., heat, low pH, denaturation,
and DTT

Amplification   Aggregation

Antigen of interest
or protein A/L

Elution   Binding

Washing

Figure 19.3   Isolation of intrabodies with high stability and affinity using phage display. Intrabodies are displayed on the surface of phage particles as fusions to a coat protein. Each phage particle displays a unique intrabody and encapsulates the gene encoding the intrabody. An intrabody library with high diversity (>10^{10}) is selected for binding to immobilized or soluble antigens (e.g., proteins or peptides) in vitro following a general panning procedure that includes constitutive cycles of incubation with antigens of interest, elution, and amplification by infection of *E. coli* bacteria. High-affinity intrabodies are enriched after several selection cycles and further verified individually on their binding to targets. The majority of intrabodies isolated using the general procedure are unable to fold properly when expressed within cells, because the reducing environment of the cytoplasm prevents the formation of disulfide bonds. Thereby, a stability selection using phage display is employed, in which intrabodies are exposed to extremely physical and chemical conditions (e.g., high temperatures or low pH) before or after an affinity selection. During the stability selection, intrabodies that do not fold correctly aggregate and are subsequently removed by centrifugation, while intrabodies with increased stability are captured using an antigen of interest or protein A/L.

library for one round against a target using phage display. In the second step, this intrabody library, which is enriched for antigen-binding intrabodies, is subcloned into the yeast two-hybrid (Y2H) system, which screens for interactions against the target in the nucleus of a yeast cell. Y2H assays screen for intrabodies that bind their target and eliminate intrabodies that do not fold well or are not expressed in an intracellular environment [9,10,118,119].

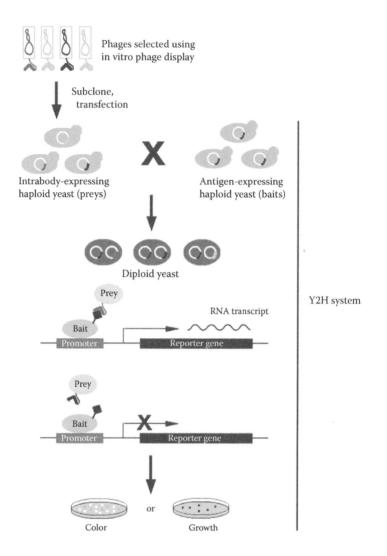

Figure 19.4   Intracellular intrabody isolation using phage display and the yeast two-hybrid (Y2H) assay. In the initial, step phage display is used to screen the intrabody library to enrich for intrabodies that interact with the target antigen. After one round of phage display, the intrabody library is subcloned into the Y2H prey expression plasmid as fusions to a transcription activation domain (e.g., Gal4 or VP16). The target antigen is cloned into the bait plasmid as a fusion to a DNA-binding domain (e.g., Gal4 or LexA). The bait and prey plasmids are transfected in to Y2H haploid yeast cells of opposite mating type. The library of intrabody preys is mated to the bait to generate diploid yeast, where an individual diploid yeast cell contains and intrabody prey and bait. If the intrabody prey interacts with the target antigen bait, then the prey containing the transcription activation is brought upstream of the reporter gene where the bait DNA-binding sequences are located, resulting in activation of reporter genes, which results in cell growth or other phenotypes such as production of color. After Y2H in vivo screen, posi-tive intrabody clones are isolated and used for intracellular applications.

In the Y2H system (Figure 19.4), libraries of intrabodies, referred to as preys, are fused with a transcription activation domain (e.g., Gal4 or herpes virus protein VP16), and each library member is expressed inside a different yeast cell. The target, referred to as a bait, is fused with a DNA-binding domain (e.g., Gal4 or LexA), which binds to the promoter region of a reporter gene. Prey libraries and the bait are expressed in two haploid strains of opposite mating types, that is, mating type a and mating type α [8,119]. Yeast cells containing bait or prey are mated to generate diploid yeast cells that contain both bait and prey. If an intrabody prey interacts with the bait, then the transcription activation domain is brought in the vicinity of the promoter and initiates expression of reporter genes, such as lacZ and his3 [4]. Thus, intrabodies that interact with the target bait can be isolated based on yeast cell growth or other phenotypes such as color.

A major limitation of the Y2H assay is that yeast transfection is inefficient relative to *E. coli*, which limits the maximum size of Y2H libraries to ~10^7 clones using standard lithium acetate transfection protocols [120]. This is much smaller than phage display libraries, which contain greater than 10^{10} clones. Thus, it is beneficial to screen a large intrabody library using phage display first, to reduce the diversity of the library prior to subcloning into Y2H prey plasmids. Typically, after one round of phage display, the library diversity is reduced to 10^5–10^6 clones, which is a manageable number for subcloning into yeast.

Rabbits have used the IAC assay to identify scFv framework consensus sequences that are stable and highly expressed in yeast and mammalian cells [10,121]. V_L and V_H domains containing consensus sequences have been used to graft known CDR loops to create intrabodies that retain high affinity for antigen inside cells [122,123]. Rabbitts and Cattaneo [121,124] created a second-generation IAC, in which a synthetic scFv library with consensus frameworks and randomized CDRs are directly screened in yeast and mammalian cells, without an initial phage display selection.

IAC has been adapted to isolate V_H domains using only the Y2H assay as V_H domains are more effective intracellular antibody fragments than scFvs [125]. In this assay, a predefined intrabody consensus V_H domain [9,121] with CDRs 1, 2, and 3 randomized is used that showed good intracellular expression and solubility without the requirement for the conserved intradomain disulfide bond [121].

Tanaka and Rabbitts [126] have also developed a two-step methodology to generate Fvs using the Y2H assay. The first step is identical to the IAC described in Figure 19.4, which produces a V_H domain intrabody. In the second step, a V_L domain library is generated, and the Y2H assay is used to isolate V_L domains from this library that interact with the target antigen when coexpressed with a V_H domain. This methodology, termed "CatcherAb," allows Fv generated with this assay to be easily converted to scFv, Fabs, and IgGs [126].

In the third-generation IAC, Tanaka and Rabbitts [127] developed a three-step Y2H assay to generate V_H domain intrabodies. In the first step, only CDR3 of the V_H domain is randomized. Positive clones from the Y2H assay are isolated. In the second step, CDR2 of the V_H domains isolated in the initial screen is randomized, and the Y2H assay is repeated. In the third step, CDR1 of the V_H domains isolated in the second screen is randomized, and the Y2H assay is repeated under more stringent conditions. This IAC method has the advantage of being an intracellular genetic screen, and thus, there is no need to purify antigens as they are expressed from plasmids within the yeast cell.

In order to develop a better method to use intrabodies to disrupt protein interactions, Visintin and Cattaneo developed a Y2H assay that directly selects for scFvs that block the interaction of a target protein with its binding partner [128]. The method, referred to as the "three-splint screen," involves a Y2H assay with a tetracycline repressor reporter gene. When a bait and prey interact, they induce the expression of the TetR, which binds to a TET operator upstream of a HIS3 gene, and suppress activation of the HIS3 gene. In the absence of histidine, yeast growth is prevented. If an intrabody is coexpressed in the Y2H assay and it blocks the interaction between the bait and prey, then the HIS3 gene is expressed, and yeast grow in the absence of histidine. This allows intrabodies that block a protein interaction to be directly selected.

19.4 APPLICATIONS

There are a growing number of publications describing the use of intrabodies in applications related to cancer and neurological disorders [4,129]. Intrabodies, unlike IgGs, are designed to target proteins within cells, which opens a whole new set of targets for developing antibody-based agents for basic science and therapy. It is estimated that only ~40% of potential drug targets are accessible by current drug technologies, which consists of catalytic proteins, cell surface receptors and transporters, and secreted proteins [130]. The remaining ~60% of potential targets are multiprotein complexes or in noncatalytic roles [130], which are ideally targeted by reagents like intrabodies. The ability of intrabodies to bind protein surfaces, inhibiting interactions and disrupting complexes, makes them useful as reagents in drug discovery. Intrabodies can have roles in target discovery and validation, and as potential therapeutics. In the discovery process, intrabodies can function as forward and reverse "genetic"-like agents. In the forward approach, libraries of intrabodies can be expressed in an organism to identify intrabodies that cause a desired phenotype. The target of the intrabody must then be identified using mass spectrometry in combination with techniques such as affinity purification and ion exchange or size exclusion chromatography [131–133] or by genetic screens such as the Y2H assay [134]. Alternatively, large sets of intrabodies can be used in a reverse analysis to identify known proteins that may be involved in a process. Intrabodies possess many desirable properties that make them useful as reagents for forward and reverse analyses of diploid organisms. They have a *trans* dominant mode of action as they inactivate gene products without altering the genetic material that encodes them. They are easily and rapidly generated against any given target. They can inhibit protein interactions and activities and block specific interactions with a protein while leaving other interactions unperturbed. For drug development, they have the advantage of inhibiting the target directly rather than blocking steps in transcription or translation. In addition, they can be used to inhibit specific protein isoforms or posttranslationally modified proteins.

Since intrabodies are generated using recombinant DNA methods, they can be easily modified to add additional functionalities. In analogy to IgGs, which have an Fc domain, which influences serum half-life and immune effector functions, such as complement-dependent cytotoxicity and antibody-dependent cellular cytotoxicity, intrabodies can be engineered as fusion proteins with a number of protein and peptide moieties. One promising area of research has been the fusion of antibody fragments to protein transduction domains (PTDs), which facilitate the transport of antibody fragments into cells [135]. Although some success has been reported for this strategy, there are problems with expressing proteins with PTD domains [136] and efficiently delivering them into the cytosol [137]. Future progress in this area may, one day, allow antibody fragments to be used as therapeutics, hugely expanding the range of druggable targets.

Intrabodies have been used to mislocalize their protein targets by fusing intrabodies to cellular localization sequences. For example, intrabodies fused to nuclear localization sequences have been used to translocate endogenous Bcr-Abl to the nucleus [138]. Oncogenic signaling by Bcr-Abl is localized to the cytoplasm, and when Bcr-Abl is forced to the nucleus, it causes apoptosis [139]. In another example, intrabodies have been anchored to the endoplasmatic reticulum by fusing them to a short amino acid sequence KDEL or SEKDEL derived from the endoplasmic reticulum (ER)-localized chaperone protein Bip [140]. These ER-anchored intrabodies have been used to target a number of proteins, including receptors and virus proteins [141]. This strategy may provide a general means to prevent transitory proteins from reaching their appropriate location of action. In a similar strategy, intrabodies have been fused to Pro/Glu/Ser/Thr-rich (PEST) domains, which target proteins for proteasomal degradation, reducing the half-life of the targeted protein [142]. This strategy has been used to target the hungingtin (htt) exon 1 protein fragments with 72 glutamine repeats (httex1-72Q) to the proteasome [143]. The mouse ornithine decarboxylase (mODC) mODC-PEST motif was fused to an scFv targeting httex1-72Q, which resulted in reduced steady-state levels [143]. Another strategy to cause intrabody-mediated proteolysis of a target protein, referred to as "suicide intrabody technology," involves tagging intrabodies with a ubiquitin proteasome targeting substrate [144]. This results in the recruitment of the intrabody and its target to the E2/E3 ubiquitination complex, where it is ubiquitinated and subsequently degraded by the proteasome.

Intrabodies have also been used to visualize endogenous proteins by fusing them to fluorescent proteins. This approach has been used to visualize the oncoprotein gankyrin in cells using pairs of intrabodies labeled with enhanced green fluorescent protein (eGFP) and red fluorescent protein (RFP). When the intrabodies simultaneously interact with gankyrin on different epitopes, gankyrin is detected using fluorescence lifetime imaging (FLIM)-fluorescence resonance energy transfer (FRET) microscopy [145].

Together, these examples provide strategies for using intrabodies that alone do not cause a desired effect, but a fusion can be used to alter protein function by mislocalization or proteasome destruction. Alternatively, they can be used to characterize their targets by allowing the tracking of endogenous proteins.

19.5 SUMMARY

Although the potential for using intrabodies for basic and translational science has not been fully realized, methods for enhancing their stability, improving their expression, and transporting them into cells highlight the promise of using them for basic science and therapeutics in the future. Phage display technology has played an integral part in improving properties of intrabodies and has helped improve the thermostabilty and reduce the aggregation properties of intrabodies. Given the significant progress in establishing strategies for producing stable intrabodies and importing them into cells, it is likely that in the near future, intrabodies will be used along with small molecules for the development of therapeutic and diagnostic reagents. This will greatly expand the number of targets available for antibody-based therapeutics, especially for targets that have traditionally been difficult for small molecule drugs such as protein interactions.

REFERENCES

1. Antman KH, Livingston DM. Intracellular neutralization of SV40 tumor antigens following microinjection of specific antibody. *Cell* 1980; 19:627–635.
2. Hoogenboom HR. Selecting and screening recombinant antibody libraries. *Nat Biotechnol* 2005; 23:1105–1116.
3. Williams BR, Zhu Z. Intrabody-based approaches to cancer therapy: Status and prospects. *Curr Med Chem* 2006; 13:1473–1480.
4. Perez-Martinez D, Tanaka T, Rabbitts TH. Intracellular antibodies and cancer: New technologies offer therapeutic opportunities. *Bioessays* 2010; 32:589–598.
5. Lo AS, Zhu Q, Marasco WA. Intracellular antibodies (intrabodies) and their therapeutic potential. *Handb Exp Pharmacol* 2008; 181:343–373.
6. Messer A, Lynch SM, Butler DC. Developing intrabodies for the therapeutic suppression of neurodegenerative pathology. *Expert Opin Biol Ther* 2009; 9:1189–1197.
7. Ellis RJ. Macromolecular crowding: An important but neglected aspect of the intracellular environment. *Curr Opin Struct Biol* 2001; 11:114–119.
8. Visintin M, Tse E, Axelson H, Rabbitts TH, Cattaneo A. Selection of antibodies for intracellular function using a two-hybrid in vivo system. *Proc Natl Acad Sci U S A* 1999; 96:11723–11728.
9. Tse E, Lobato MN, Forster A, Tanaka T, Chung GT, Rabbitts TH. Intracellular antibody capture technology: Application to selection of intracellular antibodies recognising the BCR-ABL oncogenic protein. *J Mol Biol* 2002; 317:85–94.
10. Visintin M, Settanni G, Maritan A, Graziosi S, Marks JD, Cattaneo A. The intracellular antibody capture technology (IACT): Towards a consensus sequence for intracellular antibodies. *J Mol Biol* 2002; 317:73–83.
11. Zhu Q, Zeng C, Huhalov A, Yao J, Turi TG, Danley D, Hynes T et al. Extended half-life and elevated steady-state level of a single-chain Fv intrabody are critical for specific intracellular retargeting of its antigen, caspase-7. *J Immunol Methods* 1999; 231:207–222.
12. Worn A, Auf der Maur A, Escher D, Honegger A, Barberis A, Pluckthun A. Correlation between in vitro stability and in vivo performance of anti-GCN4 intrabodies as cytoplasmic inhibitors. *J Biol Chem* 2000; 275:2795–2803.

13. Rajpal A, Turi TG. Intracellular stability of anti-caspase-3 intrabodies determines efficacy in retargeting the antigen. *J Biol Chem* 2001; 276:33139–33146.
14. Worn A, Pluckthun A. Stability engineering of antibody single-chain Fv fragments. *J Mol Biol* 2001; 305:989–1010.
15. Paborji M, Pochopin NL, Coppola WP, Bogardus JB. Chemical and physical stability of chimeric L6, a mouse-human monoclonal antibody. *Pharm Res* 1994; 11:764–771.
16. Breen ED, Curley JG, Overcashier DE, Hsu CC, Shire SJ. Effect of moisture on the stability of a lyophilized humanized monoclonal antibody formulation. *Pharm Res* 2001; 18:1345–1353.
17. Proba K, Honegger A, Pluckthun A. A natural antibody missing a cysteine in VH: Consequences for thermodynamic stability and folding. *J Mol Biol* 1997; 265:161–172.
18. Kabat EA, Wu TT, Perry HM, Gottesman KS, Foeller C. Variable region heavy chain sequences. In: *Sequences of Proteins of Immunological Interest*. U.S. Department of Health and Human Services, Public Health Service, National Institutes of Health. NIH Publication No. 91-3242: National Technical Information Service (NTIS), 1991; 310–539.
19. Glockshuber R, Schmidt T, Pluckthun A. The disulfide bonds in antibody variable domains: Effects on stability, folding in vitro, and functional expression in *Escherichia coli*. *Biochemistry* 1992; 31:1270–1279.
20. Rothlisberger D, Honegger A, Pluckthun A. Domain interactions in the Fab fragment: A comparative evaluation of the single-chain Fv and Fab format engineered with variable domains of different stability. *J Mol Biol* 2005; 347:773–789.
21. Hamers-Casterman C, Atarhouch T, Muyldermans S, Robinson G, Hamers C, Songa EB, Bendahman N et al. Naturally occurring antibodies devoid of light chains. *Nature* 1993; 363:446–448.
22. Hinds KR, Litman GW. Major reorganization of immunoglobulin VH segmental elements during vertebrate evolution. *Nature* 1986; 320:546–549.
23. Bond CJ, Marsters JC, Sidhu SS. Contributions of CDR3 to V H H domain stability and the design of monobody scaffolds for naive antibody libraries. *J Mol Biol* 2003; 332:643–655.
24. Dumoulin M, Conrath K, Van Meirhaeghe A, Meersman F, Heremans K, Frenken LG, Muyldermans S et al. Single-domain antibody fragments with high conformational stability. *Protein Sci* 2002; 11:500–515.
25. Saerens D, Pellis M, Loris R, Pardon E, Dumoulin M, Matagne A, Wyns L et al. Identification of a universal VHH framework to graft non-canonical antigen-binding loops of camel single-domain antibodies. *J Mol Biol* 2005; 352:597–607.
26. Vincke C, Loris R, Saerens D, Martinez-Rodriguez S, Muyldermans S, Conrath K. General strategy to humanize a camelid single-domain antibody and identification of a universal humanized nanobody scaffold. *J Biol Chem* 2009; 284:3273–3284.
27. Barthelemy PA, Raab H, Appleton BA, Bond CJ, Wu P, Wiesmann C, Sidhu SS. Comprehensive analysis of the factors contributing to the stability and solubility of autonomous human VH domains. *J Biol Chem* 2008; 283:3639–3654.
28. Wirtz P, Steipe B. Intrabody construction and expression III: Engineering hyperstable V(H) domains. *Protein Sci* 1999; 8:2245–2250.
29. Ewert S, Cambillau C, Conrath K, Pluckthun A. Biophysical properties of camelid V(HH) domains compared to those of human V(H)3 domains. *Biochemistry* 2002; 41: 3628–3636.
30. Desmyter A, Transue TR, Ghahroudi MA, Thi MH, Poortmans F, Hamers R, Muyldermans S et al. Crystal structure of a camel single-domain VH antibody fragment in complex with lysozyme. *Nat Struct Biol* 1996; 3:803–811.

31. Desmyter A, Decanniere K, Muyldermans S, Wyns L. Antigen specificity and high affinity binding provided by one single loop of a camel single-domain antibody. *J Biol Chem* 2001; 276:26285–26290.

32. Spinelli S, Frenken L, Bourgeois D, de Ron L, Bos W, Verrips T, Anguille C et al. The crystal structure of a llama heavy chain variable domain. *Nat Struct Biol* 1996; 3:752–757.

33. Decanniere K, Desmyter A, Lauwereys M, Ghahroudi MA, Muyldermans S, Wyns L. A single-domain antibody fragment in complex with RNase A: Non-canonical loop structures and nanomolar affinity using two CDR loops. *Structure* 1999; 7: 361–370.

34. Spinelli S, Frenken LG, Hermans P, Verrips T, Brown K, Tegoni M, Cambillau C. Camelid heavy-chain variable domains provide efficient combining sites to haptens. *Biochemistry* 2000; 39:1217–1222.

35. Spinelli S, Tegoni M, Frenken L, van Vliet C, Cambillau C. Lateral recognition of a dye hapten by a llama VHH domain. *J Mol Biol* 2001; 311:123–129.

36. Desmyter A, Spinelli S, Payan F, Lauwereys M, Wyns L, Muyldermans S, Cambillau C. Three camelid VHH domains in complex with porcine pancreatic alpha-amylase. Inhibition and versatility of binding topology. *J Biol Chem* 2002; 277:23645–23650.

37. Harmsen MM, Ruuls RC, Nijman IJ, Niewold TA, Frenken LG, de Geus B. Llama heavy-chain V regions consist of at least four distinct subfamilies revealing novel sequence features. *Mol Immunol* 2000; 37:579–590.

38. Riechmann L. Rearrangement of the former VL interface in the solution structure of a camelised, single antibody VH domain. *J Mol Biol* 1996; 259:957–969.

39. Bond CJ, Wiesmann C, Marsters JC, Jr., Sidhu SS. A structure-based database of antibody variable domain diversity. *J Mol Biol* 2005; 348:699–709.

40. de Wildt RM, Mundy CR, Gorick BD, Tomlinson IM. Antibody arrays for high-throughput screening of antibody-antigen interactions. *Nat Biotechnol* 2000; 18:989–994.

41. Pal G, Kouadio JL, Artis DR, Kossiakoff AA, Sidhu SS. Comprehensive and quantitative mapping of energy landscapes for protein–protein interactions by rapid combinatorial scanning. *J Biol Chem* 2006; 281:22378–22385.

42. Weiss GA, Watanabe CK, Zhong A, Goddard A, Sidhu SS. Rapid mapping of protein functional epitopes by combinatorial alanine scanning. *Proc Natl Acad Sci U S A* 2000; 97:8950–8954.

43. Ma X, Barthelemy PA, Rouge L, Wiesmann C, Sidhu SS. Design of synthetic autonomous VH domain libraries and structural analysis of a VH domain bound to vascular endothelial growth factor. *J Mol Biol* 2013; 425:2247–2259.

44. Jespers L, Schon O, Famm K, Winter G. Aggregation-resistant domain antibodies selected on phage by heat denaturation. *Nat Biotechnol* 2004; 22:1161–1165.

45. Jespers L, Schon O, James LC, Veprintsev D, Winter G. Crystal structure of HEL4, a soluble, refoldable human V(H) single domain with a germ-line scaffold. *J Mol Biol* 2004; 337:893–903.

46. Dudgeon K, Famm K, Christ D. Sequence determinants of protein aggregation in human VH domains. *Protein Eng Des Sel* 2009; 22:217–220.

47. Arbabi-Ghahroudi M, To R, Gaudette N, Hirama T, Ding W, MacKenzie R, Tanha J. Aggregation-resistant VHs selected by in vitro evolution tend to have disulfide-bonded loops and acidic isoelectric points. *Protein Eng Des Sel* 2009; 22:59–66.

48. Perchiacca JM, Bhattacharya M, Tessier PM. Mutational analysis of domain antibodies reveals aggregation hotspots within and near the complementarity determining regions. *Proteins* 2011; 79:2637–2647.

49. Perchiacca JM, Ladiwala AR, Bhattacharya M, Tessier PM. Aggregation-resistant domain antibodies engineered with charged mutations near the edges of the complementarity-determining regions. *Protein Eng Des Sel* 2012; 25:591–601.
50. Perchiacca JM, Lee CC, Tessier PM. Optimal charged mutations in the complementarity-determining regions that prevent domain antibody aggregation are dependent on the antibody scaffold. *Protein Eng Des Sel* 2014; 27:29–39.
51. Mandrup OA, Friis NA, Lykkemark S, Just J, Kristensen P. A novel heavy domain antibody library with functionally optimized complementarity determining regions. *PLoS One* 2013; 8:e76834.
52. Dudgeon K, Rouet R, Kokmeijer I, Schofield P, Stolp J, Langley D, Stock D et al. General strategy for the generation of human antibody variable domains with increased aggregation resistance. *Proc Natl Acad Sci U S A* 2012; 109:10879–10884.
53. Kim DY, Kandalaft H, Ding W, Ryan S, van Faassen H, Hirama T, Foote SJ et al. Disulfide linkage engineering for improving biophysical properties of human VH domains. *Protein Eng Des Sel* 2012; 25:581–589.
54. Kim DY, Ding W, Tanha J. Solubility and stability engineering of human VH domains. *Methods Mol Biol* 2012; 911:355–372.
55. Chan PH, Pardon E, Menzer L, De Genst E, Kumita JR, Christodoulou J, Saerens D et al. Engineering a camelid antibody fragment that binds to the active site of human lysozyme and inhibits its conversion into amyloid fibrils. *Biochemistry* 2008; 47:11041–11054.
56. Kim DY, To R, Kandalaft H, Ding W, van Faassen H, Luo Y, Schrag JD et al. Antibody light chain variable domains and their biophysically improved versions for human immunotherapy. *MAbs* 2014; 6:219–235.
57. Hussack G, Keklikian A, Alsughayyir J, Hanifi-Moghaddam P, Arbabi-Ghahroudi M, van Faassen H, Hou ST et al. A V(L) single-domain antibody library shows a high-propensity to yield non-aggregating binders. *Protein Eng Des Sel* 2012; 25:313–318.
58. Ewert S, Huber T, Honegger A, Pluckthun A. Biophysical properties of human antibody variable domains. *J Mol Biol* 2003; 325:531–553.
59. Dubnovitsky AP, Kravchuk ZI, Chumanevich AA, Cozzi A, Arosio P, Martsev SP. Expression, refolding, and ferritin-binding activity of the isolated VL-domain of mono-clonal antibody F11. *Biochemistry (Mosc)* 2000; 65:1011–1018.
60. To R, Hirama T, Arbabi-Ghahroudi M, MacKenzie R, Wang P, Xu P, Ni F et al. Isolation of monomeric human V(H)s by a phage selection. *J Biol Chem* 2005; 280:41395–41403.
61. Reiter Y, Brinkmann U, Kreitman RJ, Jung SH, Lee B, Pastan I. Stabilization of the Fv fragments in recombinant immunotoxins by disulfide bonds engineered into conserved framework regions. *Biochemistry* 1994; 33:5451–5459.
62. Pluckthun A. Mono- and bivalent antibody fragments produced in *Escherichia coli*: Engineering, folding and antigen binding. *Immunol Rev* 1992; 130:151–188.
63. Jager M, Pluckthun A. Domain interactions in antibody Fv and scFv fragments: Effects on unfolding kinetics and equilibria. *FEBS Lett* 1999; 462:307–312.
64. Desplancq D, King DJ, Lawson AD, Mountain A. Multimerization behaviour of single chain Fv variants for the tumour-binding antibody B72.3. *Protein Eng* 1994; 7:1027–1033.
65. Whitlow M, Filpula D, Rollence ML, Feng SL, Wood JF. Multivalent Fvs: Characterization of single-chain Fv oligomers and preparation of a bispecific Fv. *Protein Eng* 1994; 7:1017–1026.
66. Arndt KM, Muller KM, Pluckthun A. Factors influencing the dimer to monomer transition of an antibody single-chain Fv fragment. *Biochemistry* 1998; 37:12918–12926.

67. Glockshuber R, Malia M, Pfitzinger I, Pluckthun A. A comparison of strategies to stabilize immunoglobulin Fv-fragments. *Biochemistry* 1990; 29:1362–1367.
68. Brinkmann U, Reiter Y, Jung SH, Lee B, Pastan I. A recombinant immunotoxin containing a disulfide-stabilized Fv fragment. *Proc Natl Acad Sci U S A* 1993; 90:7538–7542.
69. Young NM, MacKenzie CR, Narang SA, Oomen RP, Baenziger JE. Thermal stabilization of a single-chain Fv antibody fragment by introduction of a disulphide bond. *FEBS Lett* 1995; 377:135–139.
70. Rajagopal V, Pastan I, Kreitman RJ. A form of anti-Tac(Fv) which is both single-chain and disulfide stabilized: Comparison with its single-chain and disulfide-stabilized homologs. *Protein Eng* 1997; 10:1453–1459.
71. Worn A, Pluckthun A. Different equilibrium stability behavior of ScFv fragments: Identification, classification, and improvement by protein engineering. *Biochemistry* 1999; 38:8739–8750.
72. Reiter Y, Brinkmann U, Jung SH, Pastan I, Lee B. Disulfide stabilization of antibody Fv: Computer predictions and experimental evaluation. *Protein Eng* 1995; 8:1323–1331.
73. Rodrigues ML, Presta LG, Kotts CE, Wirth C, Mordenti J, Osaka G, Wong WL et al. Development of a humanized disulfide-stabilized anti-p185HER2 Fv-beta-lactamase fusion protein for activation of a cephalosporin doxorubicin prodrug. *Cancer Res* 1995; 55:63–70.
74. Dooley H, Grant SD, Harris WJ, Porter AJ. Stabilization of antibody fragments in adverse environments. *Biotechnol Appl Biochem* 1998; 28 (Pt 1):77–83.
75. Scott CP, Abel-Santos E, Wall M, Wahnon DC, Benkovic SJ. Production of cyclic peptides and proteins in vivo. *Proc Natl Acad Sci U S A* 1999; 96:13638–13643.
76. Evans TC, Jr., Martin D, Kolly R, Panne D, Sun L, Ghosh I, Chen L et al. Protein trans-splicing and cyclization by a naturally split intein from the dnaE gene of Synechocystis species PCC6803. *J Biol Chem* 2000; 275:9091–9094.
77. Iwai H, Pluckthun A. Circular beta-lactamase: Stability enhancement by cyclizing the backbone. *FEBS Lett* 1999; 459:166–172.
78. Iwai H, Lingel A, Pluckthun A. Cyclic green fluorescent protein produced in vivo using an artificially split PI-PfuI intein from Pyrococcus furiosus. *J Biol Chem* 2001; 276:16548–16554.
79. Camarero JA, Fushman D, Sato S, Giriat I, Cowburn D, Raleigh DP, Muir TW. Rescuing a destabilized protein fold through backbone cyclization. *J Mol Biol* 2001; 308:1045–1062.
80. Williams NK, Liepinsh E, Watt SJ, Prosselkov P, Matthews JM, Attard P, Beck JL et al. Stabilization of native protein fold by intein-mediated covalent cyclization. *J Mol Biol* 2005; 346:1095–1108.
81. Jeffries CM, Graham SC, Stokes PH, Collyer CA, Guss JM, Matthews JM. Stabilization of a binary protein complex by intein-mediated cyclization. *Protein Sci* 2006; 15:2612–2618.
82. Huston JS, Levinson D, Mudgett-Hunter M, Tai MS, Novotny J, Margolies MN, Ridge RJ et al. Protein engineering of antibody binding sites: Recovery of specific activity in an anti-digoxin single-chain Fv analogue produced in *Escherichia coli*. *Proc Natl Acad Sci U S A* 1988; 85:5879–5883.
83. Bird RE, Hardman KD, Jacobson JW, Johnson S, Kaufman BM, Lee SM, Lee T et al. Single-chain antigen-binding proteins. *Science* 1988; 242:423–426.
84. Bernhard W. Engineering Intracellular Antibody Libraries, MSc Thesis, University of Saskatchewan, Saskatoon, SK, Canada, 2008.

85. Kvam E, Sierks MR, Shoemaker CB, Messer A. Physico-chemical determinants of soluble intrabody expression in mammalian cell cytoplasm. *Protein Eng Des Sel* 2010; 23:489–498.

86. Lawrence MS, Phillips KJ, Liu DR. Supercharging proteins can impart unusual resilience. *J Am Chem Soc* 2007; 129:10110–10112.

87. Trevino SR, Scholtz JM, Pace CN. Amino acid contribution to protein solubility: Asp, Glu, and Ser contribute more favorably than the other hydrophilic amino acids in RNase Sa. *J Mol Biol* 2007; 366:449–460.

88. Wu SJ, Luo J, O'Neil KT, Kang J, Lacy ER, Canziani G, Baker A et al. Structure-based engineering of a monoclonal antibody for improved solubility. *Protein Eng Des Sel* 2010; 23:643–651.

89. Miklos AE, Kluwe C, Der BS, Pai S, Sircar A, Hughes RA, Berrondo M et al. Structure-based design of supercharged, highly thermoresistant antibodies. *Chem Biol* 2012; 19:449–455.

90. Der BS, Kluwe C, Miklos AE, Jacak R, Lyskov S, Gray JJ, Georgiou G et al. Alternative computational protocols for supercharging protein surfaces for reversible unfolding and retention of stability. *PLoS One* 2013; 8:e64363.

91. Zbilut JP, Giuliani A, Colosimo A, Mitchell JC, Colafranceschi M, Marwan N, Webber CL, Jr. et al. Charge and hydrophobicity patterning along the sequence predicts the folding mechanism and aggregation of proteins: A computational approach. *J Proteome Res* 2004; 3:1243–1253.

92. Long WF, Labute P. Calibrative approaches to protein solubility modeling of a mutant series using physicochemical descriptors. *J Comput Aided Mol Des* 2010; 24:907–916.

93. Yadav S, Sreedhara A, Kanai S, Liu J, Lien S, Lowman H, Kalonia DS et al. Establishing a link between amino acid sequences and self-associating and viscoelastic behavior of two closely related monoclonal antibodies. *Pharm Res* 2011; 28:1750–1764.

94. Chari R, Singh SN, Yadav S, Brems DN, Kalonia DS. Determination of the dipole moments of RNAse SA wild type and a basic mutant. *Proteins* 2012; 80:1041–1052.

95. Gribenko AV, Patel MM, Liu J, McCallum SA, Wang C, Makhatadze GI. Rational stabilization of enzymes by computational redesign of surface charge–charge interactions. *Proc Natl Acad Sci U S A* 2009; 106:2601–2606.

96. Weidenhaupt M, Khalifa MB, Hugo N, Choulier L, Altschuh D, Vernet T. Functional mapping of conserved, surface-exposed charges of antibody variable domains. *J Mol Recognit* 2002; 15:94–103.

97. Onda M, Nagata S, Tsutsumi Y, Vincent JJ, Wang Q, Kreitman RJ, Lee B et al. Lowering the isoelectric point of the Fv portion of recombinant immunotoxins leads to decreased nonspecific animal toxicity without affecting antitumor activity. *Cancer Res* 2001; 61:5070–5077.

98. Jurado P, de Lorenzo V, Fernandez LA. Thioredoxin fusions increase folding of single chain Fv antibodies in the cytoplasm of *Escherichia coli*: Evidence that chaperone activity is the prime effect of thioredoxin. *J Mol Biol* 2006; 357:49–61.

99. Ye T, Lin Z, Lei H. High-level expression and characterization of an anti-VEGF165 single-chain variable fragment (scFv) by small ubiquitin-related modifier fusion in *Escherichia coli*. *Appl Microbiol Biotechnol* 2008; 81:311–317.

100. Hayhurst A. Improved expression characteristics of single-chain Fv fragments when fused downstream of the *Escherichia coli* maltose-binding protein or upstream of a single immunoglobulin-constant domain. *Protein Expr Purif* 2000; 18:1–10.

101. Bach H, Mazor Y, Shaky S, Shoham-Lev A, Berdichevsky Y, Gutnick DL, Benhar I. *Escherichia coli* maltose-binding protein as a molecular chaperone for recombinant intracellular cytoplasmic single-chain antibodies. *J Mol Biol* 2001; 312:79–93.

102. Zheng L, Baumann U, Reymond JL. Production of a functional catalytic antibody ScFv-NusA fusion protein in bacterial cytoplasm. *J Biochem* 2003; 133:577–581.

103. Martin CD, Rojas G, Mitchell JN, Vincent KJ, Wu J, McCafferty J, Schofield DJ. A simple vector system to improve performance and utilisation of recombinant antibodies. *BMC Biotechnol* 2006; 6:46.

104. Schaefer JV, Pluckthun A. Engineering aggregation resistance in IgG by two independent mechanisms: Lessons from comparison of Pichia pastoris and mammalian cell expression. *J Mol Biol* 2012; 417:309–335.

105. Tan PH, Chu V, Stray JE, Hamlin DK, Pettit D, Wilbur DS, Vessella RL et al. Engineering the isoelectric point of a renal cell carcinoma targeting antibody greatly enhances scFv solubility. *Immunotechnology* 1998; 4:107–114.

106. Proba K, Worn A, Honegger A, Pluckthun A. Antibody scFv fragments without disulfide bonds made by molecular evolution. *J Mol Biol* 1998; 275:245–253.

107. Jung S, Honegger A, Pluckthun A. Selection for improved protein stability by phage display. *J Mol Biol* 1999; 294:163–180.

108. Brockmann EC, Cooper M, Stromsten N, Vehniainen M, Saviranta P. Selecting for antibody scFv fragments with improved stability using phage display with denaturation under reducing conditions. *J Immunol Methods* 2005; 296:159–170.

109. Martineau P, Betton JM. In vitro folding and thermodynamic stability of an antibody fragment selected in vivo for high expression levels in *Escherichia coli* cytoplasm. *J Mol Biol* 1999; 292:921–929.

110. Dudgeon K, Rouet R, Famm K, Christ D. Selection of human VH single domains with improved biophysical properties by phage display. *Methods Mol Biol* 2012; 911:383–397.

111. Bjorck L. Protein L. A novel bacterial cell wall protein with affinity for Ig L chains. *J Immunol* 1988; 140:1194–1197.

112. Starovasnik MA, O'Connell MP, Fairbrother WJ, Kelley RF. Antibody variable region binding by Staphylococcal protein A: Thermodynamic analysis and location of the Fv binding site on E-domain. *Protein Sci* 1999; 8:1423–1431.

113. Fromant M, Blanquet S, Plateau P. Direct random mutagenesis of gene-sized DNA fragments using polymerase chain reaction. *Anal Biochem* 1995; 224:347–353.

114. Stemmer WP. Rapid evolution of a protein in vitro by DNA shuffling. *Nature* 1994; 370:389–391.

115. De Jaeger G, Fiers E, Eeckhout D, Depicker A. Analysis of the interaction between single-chain variable fragments and their antigen in a reducing intracellular environment using the two-hybrid system. *FEBS Lett* 2000; 467:316–320.

116. Portner-Taliana A, Russell M, Froning KJ, Budworth PR, Comiskey JD, Hoeffler JP. In vivo selection of single-chain antibodies using a yeast two-hybrid system. *J Immunol Methods* 2000; 238:161–172.

117. Gennari F, Mehta S, Wang Y, St Clair Tallarico A, Palu G, Marasco WA. Direct phage to intrabody screening (DPIS): Demonstration by isolation of cytosolic intrabodies against the TES1 site of Epstein Barr virus latent membrane protein 1 (LMP1) that block NF-kappaB transactivation. *J Mol Biol* 2004; 335:193–207.

118. Tanaka T, Rabbitts TH. Intracellular antibody capture (IAC) methods for single domain antibodies. *Methods Mol Biol* 2012; 911:151–173.

119. Visintin M, Quondam M, Cattaneo A. The intracellular antibody capture technology: Towards the high-throughput selection of functional intracellular antibodies for target validation. *Methods* 2004; 34:200–214.

120. Stocks MR. Intrabodies: Production and promise. *Drug Discov Today* 2004; 9:960–966.

121. Tanaka T, Rabbitts TH. Intrabodies based on intracellular capture frameworks that bind the RAS protein with high affinity and impair oncogenic transformation. *EMBO J* 2003; 22:1025–1035.

122. Edwardraja S, Neelamegam R, Ramadoss V, Venkatesan S, Lee SG. Redesigning of anti-c-Met single chain Fv antibody for the cytoplasmic folding and its structural analysis. *Biotechnol Bioeng* 2010; 106:367–375.

123. Honegger A, Malebranche AD, Rothlisberger D, Pluckthun A. The influence of the framework core residues on the biophysical properties of immunoglobulin heavy chain variable domains. *Protein Eng Des Sel* 2009; 22:121–134.

124. Visintin M, Meli GA, Cannistraci I, Cattaneo A. Intracellular antibodies for proteomics. *J Immunol Methods* 2004; 290:135–153.

125. Tanaka T, Lobato MN, Rabbitts TH. Single domain intracellular antibodies: A minimal fragment for direct in vivo selection of antigen-specific intrabodies. *J Mol Biol* 2003; 331:1109–1120.

126. Tanaka T, Rabbitts TH. Selection of complementary single-variable domains for building monoclonal antibodies to native proteins. *Nucleic Acids Res* 2009; 37:e41.

127. Tanaka T, Rabbitts TH. Protocol for the selection of single-domain antibody fragments by third generation intracellular antibody capture. *Nat Protoc* 2010; 5:67–92.

128. Visintin M, Melchionna T, Cannistraci I, Cattaneo A. In vivo selection of intrabodies specifically targeting protein–protein interactions: A general platform for an "undruggable" class of disease targets. *J Biotechnol* 2008; 135:1–15.

129. Messer A, Joshi SN. Intrabodies as neuroprotective therapeutics. *Neurotherapeutics* 2013; 10:447–458.

130. Stocks MR. Intrabodies: Turning the immune system inside out for new discovery tools and therapeutics. *Discov Med* 2005; 5:538–543.

131. Jensen KB, Jensen ON, Ravn P, Clark BF, Kristensen P. Identification of keratinocyte-specific markers using phage display and mass spectrometry. *Mol Cell Proteomics* 2003; 2:61–69.

132. Hansen MH, Nielsen H, Ditzel HJ. The tumor-infiltrating B cell response in medullary breast cancer is oligoclonal and directed against the autoantigen actin exposed on the surface of apoptotic cancer cells. *Proc Natl Acad Sci U S A* 2001; 98:12659–12664.

133. Robert R, Jacobin-Valat MJ, Daret D, Miraux S, Nurden AT, Franconi JM, Clofent-Sanchez G. Identification of human scFvs targeting atherosclerotic lesions: Selection by single round in vivo phage display. *J Biol Chem* 2006; 281:40135–40143.

134. Vielemeyer O, Nizak C, Jimenez AJ, Echard A, Goud B, Camonis J, Rain JC et al. Characterization of single chain antibody targets through yeast two hybrid. *BMC Biotechnol* 2010; 10:59.

135. Marschall AL, Frenzel A, Schirrmann T, Schungel M, Dubel S. Targeting antibodies to the cytoplasm. *MAbs* 2011; 3:3–16.

136. Nakajima O, Hachisuka A, Okunuki H, Takagi K, Teshima R, Sawada JI. Method for delivering radiolabeled single-chain Fv antibody to the brain. *J Health Sci* 2004; 50:159–163.

137. Fischer PM. Cellular uptake mechanisms and potential therapeutic utility of peptidic cell delivery vectors: Progress 2001–2006. *Med Res Rev* 2007; 27:755–795.

138. Dixon AS, Constance JE, Tanaka T, Rabbitts TH, Lim CS. Changing the subcellular location of the oncoprotein Bcr-Abl using rationally designed capture motifs. *Pharm Res* 2012; 29:1098–1109.

139. Vigneri P, Wang JY. Induction of apoptosis in chronic myelogenous leukemia cells through nuclear entrapment of BCR-ABL tyrosine kinase. *Nat Med* 2001; 7:228–234.

140. Munro S, Pelham HR. A C-terminal signal prevents secretion of luminal ER proteins. *Cell* 1987; 48:899–907.

141. Boldicke T, Somplatzki S, Sergeev G, Mueller PP. Functional inhibition of transitory proteins by intrabody-mediated retention in the endoplasmatic reticulum. *Methods* 2012; 56:338–350.

142. Li X, Zhao X, Fang Y, Jiang X, Duong T, Fan C, Huang CC et al. Generation of destabilized green fluorescent protein as a transcription reporter. *J Biol Chem* 1998; 273:34970–34975.

143. Butler DC, Messer A. Bifunctional anti-huntingtin proteasome-directed intrabodies mediate efficient degradation of mutant huntingtin exon 1 protein fragments. *PLoS One* 2011; 6:e29199.

144. Melchionna T, Cattaneo A. A protein silencing switch by ligand-induced proteasome-targeting intrabodies. *J Mol Biol* 2007; 374:641–654.

145. Rinaldi AS, Freund G, Desplancq D, Sibler AP, Baltzinger M, Rochel N, Mely Y et al. The use of fluorescent intrabodies to detect endogenous gankyrin in living cancer cells. *Exp Cell Res* 2013; 319:838–849.

Index

Page numbers followed by f and t indicates figures and tables respectively.